Advances in Scalable and Intelligent Geospatial Analytics

Geospatial data acquisition and analysis techniques have experienced tremendous growth in the last few years, providing an opportunity to solve previously unsolved environmental- and natural resource-related problems. However, a variety of challenges are encountered in processing the highly voluminous geospatial data in a scalable and efficient manner. Technological advancements in high-performance computing, computer vision, and big data analytics are enabling the processing of big geospatial data in an efficient and timely manner. Many geospatial communities have already adopted these techniques in multidisciplinary geospatial applications around the world. This book is a single source that offers a comprehensive overview of the state of the art and future developments in this domain.

Features

- Demonstrates the recent advances in geospatial analytics tools, technologies, and algorithms
- Provides insight and direction to the geospatial community regarding the future trends in scalable and intelligent geospatial analytics
- Exhibits recent geospatial applications and demonstrates innovative ways to use big geospatial data to address various domain-specific, real-world problems
- Recognizes the analytical and computational challenges posed and opportunities provided by the increased volume, velocity, and veracity of geospatial data

This book is beneficial to graduate and postgraduate students, academicians, research scholars, working professionals, industry experts, and government research agencies working in the geospatial domain, where GIS and remote sensing are used for a variety of purposes. Readers will gain insights into the emerging trends on scalable geospatial data analytics.

Advances in Scalable and Intelligent Geospatial Analytics

Challenges and Applications

Edited by
Surya S Durbha
Jibonananda Sanyal
Lexie Yang
Sangita S Chaudhari
Ujwala Bhangale
Ujwala Bharambe
Kuldeep Kurte

CRC Press is an imprint of the
Taylor & Francis Group, an **informa** business

Designed cover image: © Shutterstock

First edition published 2023
by CRC Press
6000 Broken Sound Parkway NW, Suite 300, Boca Raton, FL 33487-2742

and by CRC Press
4 Park Square, Milton Park, Abingdon, Oxon, OX14 4RN

CRC Press is an imprint of Taylor & Francis Group, LLC

ISBN: 978-1-032-20031-6 (hbk)
ISBN: 978-1-032-22032-1 (pbk)
ISBN: 978-1-003-27092-8 (ebk)

DOI: 10.1201/9781003270928

Typeset in Times
by codeMantra

Contents

SECTION VI Case Studies from the Geospatial Domain

Preface

In the last decade, rapid evolution in geospatial data sources such as remote sensing technologies, crowdsourcing mechanisms, and social media platforms has enabled several opportunities to analyze the geospatial data that can serve the community needs such as sustainable agriculture, monitoring, and modeling environmental changes, urban planning, disaster detection, management, and many more. Moreover, great advancements in data storage and computing capabilities have improved the efficiency of spatial data use by enabling scalable and near-real-time processing that yields timely responses. The collective process of gathering, analyzing, and visualizing geospatial data is referred to as geospatial analytics.

This book mainly delivers the recent trends and technologies involved in geospatial analytics that is informative to academicians, researchers, and practitioners to understand the state-of-the-art methods and further, the possibility of exploration and application in the geospatial domain. The book covers various paradigms, including Geo-AI, which applies artificial intelligence to geospatial data, and scalable geospatial computation intelligence, which incorporates technological advances in several fields, including scalable storage on the cloud to handle huge amounts of geospatial data, and high-performance computing to speed up computations. By leveraging these advances, big geospatial data can be processed rapidly and help the end-users access the extracted information in a timely manner. The recent development of technologies such as WebGIS, IoT, and 3D visualization empowers many applications in the geospatial domain. Several successful applications of geospatial analytics in agricultural, healthcare, and disaster management are also covered in this book, to inspire readers for more related applications by leveraging the potential of geospatial data.

ORGANIZATION OF THE BOOK

This book is composed of 20 chapters divided into 6 sections. Below is a brief explanation of each section.

SECTION I: INTRODUCTION TO GEOSPATIAL ANALYTICS

This section covers an invited chapter on the fundamentals of geospatial analytics. The author of this chapter Prof. P. Venkatachalam is a retired faculty from IIT Bombay, India, and a well-renowned researcher in the field of geospatial science and technology. This chapter focuses on spatial data concepts, data models, the role of remote sensing and GPS, analytics, visualization, and the open research challenges in geospatial technology.

SECTION II: GEO-AI

Recent advancements in artificial intelligence are enabling us to solve previously unsolved problems. This section covers the chapters discussing the progress made in geospatial analytics that is fueled by the recent developments in AI. The spectrum of topics covered by the chapters in this section includes Geo-AI for multimodal spatiotemporal data, temporal dynamics of place and human mobility, knowledge-graph-based workflow for remote sensing scene understanding, geosemantics-based information retrieval for 3D LiDAR data, and geospatial analytics using natural language processing. The contributions to the chapters in this section come from the AI/ML research group at Apple, Cupertino, California, USA; Geospatial Sciences and Human Security Division and Computational Urban Sciences Group at Oak Ridge National Laboratory, USA; GeoSysIoT Group at the Indian Institute of Technology, Bombay, India; Shah and Anchor Kutchhi Engineering College, Mumbai, India; and Thadomal Shahani Engineering College, Mumbai, India.

SECTION III: SCALABLE GEOSPATIAL ANALYTICS

Along with the developments in Geo-AI, the advancements in storage and computing are providing scalable solutions to store multimodal geospatial data and processes using high-performance computing platforms allowing geospatial analytics to be done in a near-real-time fashion. The chapters in this section cover the topics of scalable geospatial data acquisition and pre-preprocessing pipeline, challenges in building an end-to-end pipeline for geospatial intelligence, distributed deep-learning for geospatial applications, and embedded high-performance computing for disaster assessments. The contributions to the chapters come from the Tata Consultancy Services Research and Innovation Lab, Mumbai, India; Geospatial Sciences and Human Security Division at Oak Ridge National Laboratory, USA; GeoSysIoT Group at the Indian Institute of Technology, Bombay, India; K. J. Somaiya College of Engineering, Mumbai, India; Veermata Jijabai Technological Institute, Mumbai, India; and the Machine Learning and Visual Computing Group at the Indian Institute of Technology, Bombay, India.

SECTION IV: GEOVISUALIZATION: INNOVATIVE APPROACHES FOR GEOVISUALIZATION AND GEOVISUAL ANALYTICS FOR BIG GEOSPATIAL DATA

It is known that proper visualization of the data helps us to better understand the data. However, geospatial and earth observation data need a little different treatment due to the associated location and time information, and inherent spatial autocorrelations. This section covers the chapters discussing topics of earth observation data visualization and visual exploration of semantic classification point-cloud data. The contributions to these chapters come from earth and data science researchers from NASA's Marshall Space Flight Center, Huntsville, AL, USA; Development Seed, a Washington DC-based company that provides geospatial solutions to social and environmental problems; the University of Alabama, Huntsville, Alabama, USA; and the Graphics-Visualization-Computing Lab, IIIT Bangalore, India.

SECTION V: OTHER ADVANCES IN GEOSPATIAL DOMAIN

This section covers a chapter that lays a vision on how the emerging metaverse technique can be used for an immersive visual experience of the big geospatial data, specifically in the context of the digital twin of a city. Also, the second chapter describes recent developments in unmanned aerial surveys technology that is enabling rapid data acquisition through unmanned aerial vehicles (UAVs) which is beneficial during disaster situations. The contributions to the chapters in this section come from Computational Sciences and Engineering Division, the Electrification and Energy Infrastructure Division, and Buildings and Transportation Science Division, and the Geospatial Science and Human Security Division at Oak Ridge National Laboratory.

SECTION VI: CASE STUDIES FROM THE GEOSPATIAL DOMAIN

This section presents five case studies of geospatial analytics ranging from the landslide, urban dynamics, land-use/cover prediction, and forest applications. These chapters describe how the recent advancements in geospatial analytics can be used in addressing various real-world applications. The contributions to these case studies come from the researchers from CGARD, Ministry of Agriculture, Livestock and Fisheries, Madagascar; CGARD, National Institute of Rural Development and Panchayati Raj, India; Academy of Scientific and Innovative Research, India; CSIR Central Building Research Institute, India; Geospatial World, Noida, India; Maulana Azad National Institute of Technology Bhopal, India; University of Peradeniya, Sri Lanka; Universiti Teknologi Malaysia; Eastern University, Sri Lanka; Central University of Rajasthan, India; North Eastern Space Application Centre, India; and the Indian Institute of Technology, Kanpur, India.

Editors

Surya S Durbha, PhD, is a Professor at the Centre of Studies in Resources Engineering (CSRE), Indian Institute of Technology Bombay (IITB). Before joining IITB, he worked as an Assistant Research Professor at the Center for Advanced Vehicular Systems (CAVS), Geosystems Research Institute (GRI), and also held an adjunct faculty position with the Electrical and Computer Engineering Department at Mississippi State University (MSU). Earlier, he worked as a scientist at the Indian Institute of Remote Sensing (IIRS), ISRO, Government of India. He earned an MTech in remote sensing at Andhra University, India in 1997 and a PhD in computer engineering at Mississippi State University (MSU), Starkville, Mississippi, USA, in 2006. At MSU, he researched and developed image information mining tools for content-based knowledge retrieval from remote sensing imagery and the retrieval of biophysical variables from multiangle satellite data. He has published over 80 peer-reviewed articles, served on program committees of several international conferences including SSKI, SSTDM, and IGARSS, cochaired sessions at various conferences, and an invited speaker for HPC, Geospatial technologies, and IOT training programs in various institutes. He is a manuscript reviewer for IEEE (TGRS, JSTARS, GRSL) and other high-impact journals. He organized the Distributed and Embedded High Performance Computing International symposium at IIT Bombay in 2016. He received the Excellence in Teaching award at IIT Bombay and the NVIDIA Innovation Award 2016 for his work in image information mining and high-performance computing (HPC). He also received the State Pride Faculty and Outstanding Research awards at MSU. He has recently written a book on the Internet of Things (IoT), which was published by Oxford University Press in 2021. His current research interests are geospatial knowledge-based systems, GPU-based high-performance computing (HPC), Big Data, the Internet of Things (IoT), image information mining, remote sensing, and sensor web.

Jibonananda Sanyal, PhD, serves as the Group Manager for the Hybrid Energy Systems Group at the U.S. Department of Energy's National Renewable Energy Laboratory. The group's focus is on pushing the frontier across a combination of renewable energy technologies to accelerate a global clean energy transition. In the past, he has served as a Group Leader for the Computational Urban Sciences research group at the Oak Ridge National Laboratory. His work falls at the intersection of extreme scale data and analytics, HPC, modeling and simulation, AI, visualization, and controls in various application areas. He has extensive experience applying these techniques toward developing insights and solutions for hybrid energy systems, renewables deployment, smart city applications, urban mobility, energy applications, situational awareness tools, as well as emergency response and resiliency across local, regional, and national scales. He is an IEEE Senior Member, an ACM Distinguished Speaker, and a 2017 Knoxville's 40 under 40 honoree.

Lexie Yang, PhD, is a lead research scientist in the GeoAI Group at Oak Ridge National Laboratory. Her research interests focus on advancing high performance computing and machine learning approaches for geospatial data analysis. She leads several AI-enabled geoscience data analytics projects with large-scale multimodality geospatial data. Her team's recent work has been widely used to support national-scale disaster assessment and management by federal and local agencies.

Sangita S Chaudhari, PhD, earned an ME (Computer Engineering) at Mumbai University, Maharashtra, India in 2008 and a PhD in GIS and remote sensing at the Indian Institute of Technology, Bombay, Mumbai, India in 2016. She is a Professor in the Department of Computer Engineering, Ramrao Adik Institute of Technology Nerul, Navi Mumbai, India. She has published several papers in international and national journals and conferences, book chapters, and two books.

She is an IEEE Senior Member and a Life Member of CSI, ISTE, and ISRS. She is the vice chair of IEEE GRSS Mumbai Chapter and is instrumental in conducting various activities in GIS and remote sensing domain under the umbrella of IEEE GRSS. Her research interests include image processing, information security, GIS, and remote sensing.

Ujwala Bhangale, PhD, earned a BE in computer science and engineering in 2000, an ME in computer engineering at Mumbai University in 2008, and a PhD at IIT, Bombay, Mumbai, India in 2018. She is an Associate Professor in the Department of Information Technology at K. J. Somaiya College of Engineering, Somaiya Vidyavihar University, Mumbai, India. Her current research interests include data exploration, analytics and big data analytics, high-performance computing (multicore systems), remote sensing data analytics (satellite image processing), and geographical information systems. She has published several papers in IEEE/ACM conferences and reputed journals. She has 2 years of industrial experience on data-warehousing projects and web application development projects. She is a life member of ISTE and an IEEE member of CSI and IEEE GRSS Society.

Ujwala Bharambe, PhD, earned a BTech in information technology, an ME in computer engineering at Mumbai University in 2009, and a PhD at IIT Bombay, Mumbai, India in 2019. She is an Assistant Professor in the Department of Computer Engineering at Thadomal Shahani Engineering College, Mumbai, India. Her current research interests include geospatial semantics, natural language processing, machine learning, multiobjective optimization, and geographical information systems. She has published several papers in IEEE/ACM conferences and reputed journals. She is a life member of ISTE and an IEEE member.

Kuldeep Kurte, PhD, is a research scientist in the Computational Urban Sciences Group (CUSG) at Oak Ridge National Laboratory. He earned a PhD in image information mining at the Indian Institute of Technology Bombay, India in 2017. While working on his PhD, he generated an interest in the field of machine learning and high-performance computing for various geoscience and remote sensing applications. He joined Oak Ridge National Laboratory as a postdoctoral researcher in scalable geocomputation in January 2018. In his first project at the lab, he worked on scalable end-to-end settlement detection workflow and its deployment on a Titan supercomputer, which was further used to detect swimming pools in Texas from satellite imagery. As a part of the Urban Exascale Computing Project (UrbanECP), he worked on the program's capability to facilitate running several instances of Transims simulations on the Titan supercomputer. He also worked on analyzing the regional-scale impact of inclement weather on traffic speed and coupled Transims' output with a building energy simulation through an efficient spatial indexing approach. Continuing his interest in data-driven urban computing, he is working on intelligent HVAC control using reinforcement learning for building energy optimization. Dr. Kuldeep has experience with various applications on different HPC platforms, from NVIDIA Jetson Tk1 to the Summit supercomputer.

Contributors

Rishabh Agrawal
TCS Research and Innovation
Thane, India

Ashish Babu
Centre on Geo-Informatic Applications in Rural Development
Ministry of Agriculture, Livestock and Fisheries
Antananarivo, Madagascar

Biplab Banerjee
Indian Institute of Technology
Bombay, India

Aimee Barciauskas
Development SEED
Garland, Texas

Andy Berres
Oak Ridge National Laboratory
Oak Ridge, Tennessee

Shubham Chaudhary
Academy of Scientific and Innovative Research
Chennai, India

Dibyajyoti Chutia
North Eastern Space Applications Centre
Meghalaya, India

David L. Cotten
Oak Ridge National Laboratory
Oak Ridge, Tennessee

Suvam Das
Academy of Scientific and Innovative Research
Chennai, India

Onkar Dikshit
Indian Institute of Technology Kanpur
Kanpur, India

Andrew Duncan
Oak Ridge National Laboratory
Oak Ridge, Tennessee

Brian Freitag
National Aeronautics and Space Administration
Washington, District of Columbia

Swetava Ganguli
Apple Inc.
Cupertino, California

Rajit Gupta
School of Earth Sciences
Central University of Rajasthan
Rajasthan, India

Iksha Gurung
University of Alabama
Huntsville, Alabama

Andrew Harter
Oak Ridge National Laboratory
Oak Ridge, Tennessee

Mohamed Haniffa Fathima Hasna
Postgraduate Institute of Science
University of Peradeniya
Peradeniya, Sri Lanka

David Hughes
Oak Ridge National Laboratory
Oak Ridge, Tennessee

C.V. Krishnakumar Iyer
Apple Inc.
Cupertino, California

Debi Prasanna Kanungo
Council of Scientific and Industrial Research
Central Building Research Institute
Roorkee, India

Aaron Kaulfus
National Aeronautics and Space Administration
Washington, District of Columbia

Penchala Vineeth Kurapati
Centre on Geo-Informatic Applications in Rural Development
Ministry of Agriculture, Livestock and Fisheries
Antananarivo, Madagascar

Matt Larson
Oak Ridge National Laboratory
Oak Ridge, Tennessee

Melanie Laverdiere
Oak Ridge National Laboratory
Oak Ridge, Tennessee

Dalton Lunga
Oak Ridge National Laboratory
Oak Ridge, Tennessee

Jacob McKee
Oak Ridge National Laboratory
Oak Ridge, Tennessee

Manimala Mahato
Shah and Anchor Kutchhi Engineering College
Mumbai, India

Venkata Ravi Babu Mandla
Centre for Geo-Informatic Applications in Rural Development
National Institute of Rural Development and Panchayati Raj
Hyderabad, India

Manil Maskey
National Aeronautics and Space Administration
Washington, District of Columbia

Jayantrao Mohite
TCS Research and Innovation
Thane, India

Nilkamal More
Somaiya Vidyavihar University
Mumbai, India

Don March
Oak Ridge National Laboratory
Oak Ridge, Tennessee

Anugrah Anilkumar Nagaich
Maulana Azad National Institute of Technology
Bhopal, India

Liz Neunsinger
Oak Ridge National Laboratory
Oak Ridge, Tennessee

Joshua R. New
Oak Ridge National Laboratory
Oak Ridge, Tennessee

V.B. Nikam
Veermata Jijabai Technological Institute
Mumbai, India

Prasad NSR
Centre for Geo-Informatic Applications in Rural Development
National Institute of Rural Development and Panchayati Raj
Hyderabad, India

Olufemi A. Omitaomu
Oak Ridge National Laboratory
Oak Ridge, Tennessee

Vipul Pandey
Apple Inc.
Cupertino, California

Ankur Pandit
TCS Research and Innovation
Thane, India

Srinivasu Pappula
TCS Research and Innovation
Thane, India

Vipul Parmar
School of Planning and Architecture
Bhopal, India

Jesse Piburn
Oak Ridge National Laboratory
Oak Ridge, Tennessee

Abhishek V. Potnis
Indian Institute of Technology
Bombay, India

Kesava Rao Pyla
Centre for Geo-Informatic Applications in Rural Development
National Institute of Rural Development and Panchayati Raj
Hyderabad, India

Naveen Ramachandran
Indian Institute of Technology Kanpur
Kanpur, India

Rahul Ramachandran
National Aeronautics and Space Administration
Washington, District of Columbia

Muthukumaran Ramasubramanian
University of Alabama, Huntsville
Huntsville, Alabama

Rekha Ramesh
Shah and Anchor Kutchhi Engineering College
Mumbai, India

Andrew Reith
Oak Ridge National Laboratory
Oak Ridge, Tennessee

Amy Rose
Oak Ridge National Laboratory
Oak Ridge, Tennessee

Shantanu Sarkar
Council of Scientific and Industrial Research
Central Building Research Institute
Roorkee, India

K.K. Sarma
North Eastern Space Applications Centre
Meghalaya, India

Suryakant Sawant
TCS Research and Innovation
Thane, India

Mathanraj Seevarethnam
Department of Geography
Faculty of Arts and Culture
Eastern University
Chenkalady, Sri Lanka

Vasanthakumary Selvanayagam
Department of Geography
Eastern University
Chenkalady, Sri Lanka

Jash Shah
Somaiya Vidyavihar University
Mumbai, India

Yunli Shao
Oak Ridge National Laboratory
Oak Ridge, Tennessee

Laxmi Kant Sharma
School of Earth Sciences
Central University of Rajasthan
Rajasthan, India

Rajat C. Shinde
Indian Institute of Technology
Bombay, India

Satendra Singh
International Institute of Information Technology
Bangalore, India

Kevin Sparks
Oak Ridge National Laboratory
Oak Ridge, Tennessee

Jaya Sreevalsan-Nair
International Institute of Information Technology
Bangalore, India

Brad Stinson
Oak Ridge National Laboratory
Oak Ridge, Tennessee

Benjamin Swan
Oak Ridge National Laboratory
Oak Ridge, Tennessee

Pratyush V. Talreja
Indian Institute of Technology
Bombay, India

Gautam Thakur
Oak Ridge National Laboratory
Oak Ridge, Tennessee

Leo Thomas
Development SEED
Garland, Texas

Marie Urban
Oak Ridge National Laboratory
Oak Ridge, Tennessee

Laurie Varma
Oak Ridge National Laboratory
Oak Ridge, Tennessee

Olaf Veerman
Development SEED
Garland, Texas

Parvatham Venkatachalam
Indian Institute of Technology
Bombay, India

Sophie Voisin
Oak Ridge National Laboratory
Oak Ridge, Tennessee

Chieh (Ross) Wang
Oak Ridge National Laboratory
Oak Ridge, Tennessee

Haowen Xu
Oak Ridge National Laboratory
Oak Ridge, Tennessee

Section I

Introduction to Geospatial Analytics

1 Geospatial Technology – Developments, Present Scenario and Research Challenges

Parvatham Venkatachalam
Centre of Studies in Resources Engineering,
Indian Institute of Technology Bombay

CONTENTS

1.1 INTRODUCTION

Data are the raw figures collected from varied sources for specific task. When data is structured, processed, analysed and presented to make them useful and relevant to a context it becomes information. Information technology helps in data collection, classification, storage, analysis and dissemination of information. Geospatial technology is a branch of information technology that deals with geographic/spatial data. Geospatial technology helps to integrate large volume of spatial data derived from a variety of sources, store and analyse them and

DOI: 10.1201/9781003270928-2

provide a range of possible scenarios to decision-makers. It is a multidisciplinary field with contribution from several disciplines.

Use of spatial data in planning had been in practice perhaps hundreds of years back. The early explorers investigated the spatial distribution of people, plants, animals and natural resources to form new settlements. Then general-purpose maps came focusing on topography, hand layout, water bodies, etc. In nineteenth century, thematic maps showing specific themes like soils, geology, and administrative units came to use. Maps were integrated quantitatively to arrive at a decision. With the advent of digital computers, quantitative analysis of maps started in 1960s and Geographic Information System (GIS) got evolved as a software tool to integrate and analyse spatial data.

From a humble beginning in the academic environment, today GIS has matured as a geospatial technology to become a part of Information Technology. With the technological innovations of Internet and WWW, geospatial technology has entered into the houses of millions and in the hands of billions.

The components of geospatial technology can be grouped under hardware, software, data and liveware. The hardware component can be from high-end workstation to handheld mobile devices. The software is the geoprocessing engine that handles data capture, storage, analysis and visualisation. It can be from a professional multifunctional tool to a web query tool. The most important component of geospatial technology is data. Geospatial data are expensive to collect, store and manipulate and large volumes of data are needed for a good study. The final component is the liveware, the people responsible for designing, implementing and using geospatial technology. They can be simple viewers, general users and professional specialists. Efforts are being taken by several institutes to generate skilled manpower in geospatial technology. Figure 1.1 depicts an overview of geospatial technology.

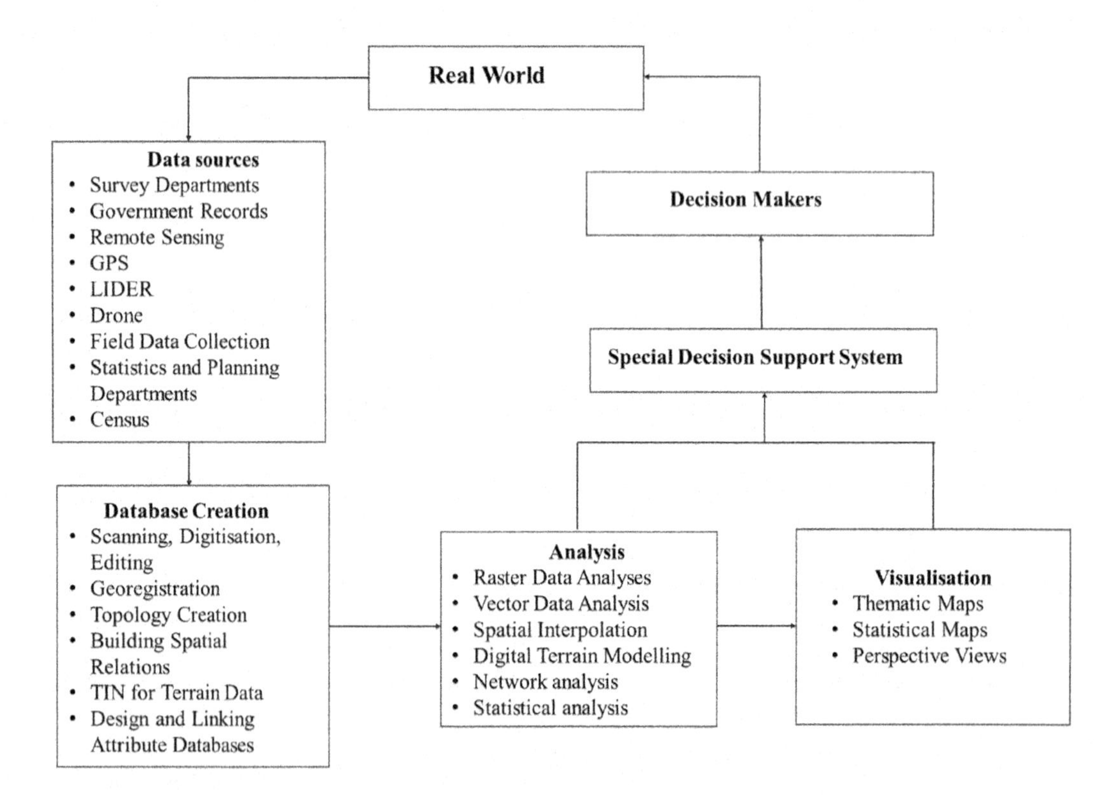

FIGURE 1.1 Geospatial technology—An overview.

1.1.1 Concept of Spatial Data

Geography of the real world is infinitely complex. It varies continuously without distinct boundaries [1]. To describe a spatial object, we simplify the inherent complexity and try to identify it with a key feature of that object with the help of conceptual model. The spatial objects are identified as building, road or hill and the spatial relations among them are described as in front of, to the right of, etc. These conceptual models are converted into digital form using spatial data models and stored in spatial database. The real world is seen through this digital spatial database. For example, a building is described by its location (where it is) and its characteristics (name, address, etc.). The spatial features are represented as point features (wells), line features (roads) and area features (boundary of a village). Some spatial variables that vary continuously like elevation or temperature are represented by dividing the study area into zones assuming the variable is constant within each zone and measuring at sample points within each zone or by drawing contours connecting points of equal values with lines.

A spatial object refers to a phenomenon that cannot be subdivided into like units and indivisibility depends on the properties used in the definition. There are three basic types of spatial objects used in geospatial database. A point is an object that has a position in space and no length and has zero dimension. A line is a one-dimensional object and has the property of length. Area is a two-dimensional object that has the property of length and width. An area object may be alone or shared boundaries with other areas and may have hole inside. An area is often referred as polygon. The map scale of the source document helps to decide the level of detail represented by a spatial feature. For example, a city may be point on a small-scale map (1:250,000) and can be polygon on a large-scale map (1:2,500). In a spatial database, spatial object's location, its characteristics (attributes) and its spatial relations with neighbouring objects are stored. Using georelational data model, the location information is stored in vector or raster data structure in a file while the attributes are stored in a relational database. Recent GIS software tools store both spatial objects and attributes in a single relational table.

1.1.2 Spatial Data Sources

One of the important parts in geospatial technology is the creation of geospatial database and it is an expensive task. Spatial data sources are topographic sheets, government survey maps, government records, remote sensing satellite images, orthophotos, drones and global positioning systems. Plenty of data may be available, but one may not know the quality and accuracy of the data. Available and accessible data may not be in the digital form. Collecting the data from primary and secondary data sources, several steps have to be taken to digitise, pre-process and build the spatial database. Globally national agencies and private industries are working to build spatial data warehouses including meta information. Open Geospatial consortium tries to bring spatial data standards and data exchange formats.

1.1.3 Geographic Coordinate System

Although the earth appears relatively flat at close range, we all know that the earth is relatively spherical. The geographical coordinate system (latitude and longitude) helps to specify the location of any spatial feature on the earth surface. Latitudes (also known as parallels) are measured in degrees towards north and south from the equatorial plane. Keeping the equator as 0° latitude, latitudes of spatial features are measured from 0° to 90° towards north or south. Longitudes (also known as meridians) are measured towards east or west from the Greenwich prime meridian. Keeping the prime meridian as a reference, longitudes are measured from 0° to 180° towards east or west. Any two parallels are the same distance apart. Meridians are the farthest apart at the equator and converge to a single point at the poles. The shape of the earth is approximated to an ellipsoid and each country and nearly region approximate the earth's shape to an ellipsoid that fits best to their region.

A datum is a mathematical model (ellipsoid) consists of an origin, the parameters of ellipsoid selected for computation and the separation of the ellipsoid from the centre of the earth. There are several datums across the world. Presently an earth-centric (geocentric) datum using earth centre of mass as the origin is the most widely used datum known as WGS 84.

The maps are the reduction of reality. The amount of reduction depends on the level of detail we need for a study. Scale is a common term used to show the amount of reduction found on maps. It is the ratio of the map distance to the ground distance. 1:5,000 scale map is called larger scale map and 1:500,000 scale map is called a smaller scale map as the former gives more details of a smaller area than 1:500,000 scale map. Enlarging a map during the analysis or in the output will not improve the initial accuracy of the map.

1.1.4 Map Projections

A set of mathematical techniques developed to depict with reasonable accuracy of the spherical earth (3-D surface) in two-dimensional plane surface are called map projections. A two-dimensional map on a planer surface helps us to work on plane or projected coordinates rather than latitudes and longitudes. But the map projection involves distortion in terms of shape, area and distance and no map projection is perfect. Choice of a map projection depends on the purpose of use, type of area covered, extent, location, the properties desired, needed accuracy and so on. Three types of map projections are conformal, equivalent and equidistant. A conformal projection preserves local angle and shape, and equivalent projection preserves areas and equidistant projection preserves distance in a direction or at a place.

The principle used is by keeping a light source inside a globe and projecting the spatial features onto a two-dimensional surface surrounding the globe. Mathematical principles of geometry and trigonometry are used and a cylinder or a cone or a plane paper is used as a projected surface. Map projections are grouped as cylindrical projections, conical projections and azimuthal projections. The orientation of the projection surface can be changed as needed. A cylinder is normally oriented tangent along the equator, a cone is normally oriented tangent along a parallel and a plane is normally oriented tangent at the pole. If the projection surface is changed to 90° from normal, the result is transverse projection. To get the coordinate on a projected map, the midpoint of the central parallel and central meridian becomes the origin of the coordinate system and coordinates are calculated as per four quadrants.

The most common projection used globally is Universal Transverse Marketer (UTM) projection. It uses varied cylinders with different central meridians for different zones to minimise the distortions. There are 60 zones, each zone covering 6° longitude and numbers sequentially starting with zone one at 180° west and moving towards east. Each zone is further divided into north and south hemispheres.

1.1.5 Spatial Data Modelling

Geographical variations in the real world are continuous in nature and infinitely complex. These continuous geographical variations have to be approximated and generalised for converting them into quantitative or digital form in finite-dimensional space. A spatial data model helps to convert these variations into discrete objects. It gives a set of guidelines for the representation of these discrete objects and the spatial relations between them in finite-dimensional space. Two major models adopted for handling spatial objects are raster and vector data models. The spatial data structure provides the way in which the spatial objects are coded and digitally analysed. A geographical space is visualised as a container with a series of a thematic maps such as administrative boundaries, land use, drainage, soils, topography, etc., placed one over the other. Spatial features on a map can be categorised into points, lines and areas. Points represent location features such as wells, electric poles, schools, banks, etc. Lines represent roads, drainage, topographic contours, etc., while areas represent features with close boundaries such as village boundaries, lakes, land use boundaries, etc.

In raster data model, grid or pixel is a unit and the value in the grid gives the characteristics of the spatial features in that grid. Remote sensing images and scanned maps are good examples of raster data. A raster layer can be considered as a two-dimensional array with rows and columns and grid values are stored in them. A grid value can be 2-bit, 8-bit or higher-bit integer or decimal value and the grid size gives the resolution of the spatial data. If the resolution is smaller, the data accuracy is higher. Digital Elevation Model (DEM) uses raster data structure to represent the high spatial variability in elevation data. When spatial variability is not high like village boundary or water body then data compression methods like run-length encoding, value point coding or quadtree are used to represent the data. One of the advantages of raster data model is that the spatial relation among objects is intrinsically getting stored in terms of rows and columns.

The vector data model helps in the accurate positioning of spatial features and the features are represented in terms of X, Y coordinate in Cartesian coordinate system. A point feature is represented as a single X, Y coordinates, a line as a sequence of X, Y coordinate and an area as a closed boundary of X, Y coordinates. In order to store the spatial relationship among the objects such as point in a polygon, polygon inside a polygon, polygon adjacent to a polygon, the concept of topology is used. Topology is the study of those properties of geometric objects that remain invariant under transformation such as skewing, bending, and stretching. Basic entity in topological model is an arc with a start and end node. Along with arc, the attributes indicating which polygon falls on each side of the arc are also stored in the case of polygon layer. In the case of network data, the connecting arcs on both ends of the arc are stored. In addition to points, lines and polygons, a region consisting of many polygons and a route consisting of many arcs are also stored. In vector data structures, along with spatial objects, the attributes are also stored using suitable data bases.

Triangulated Irregular Network (TIN) is a topological model to handle elevation data. The terrain surfaces are stored as a set of non-overlapping triangles with the concept that each triangle has a constant gradient. Delaunay triangulation method is generally used to create TIN model.

1.1.6 Spatial Database Creation

In any geospatial technology application, most part of the time and cost goes in data collection and database creation. Many countries have started building spatial data warehouses from where the data can be downloaded. When original maps are collected from different sources and scanned, all the maps should be brought to common reference system and projection in the preprocessing step. During digitisation, cleaning becomes an important task to remove errors such as overshoots, undershoots and duplicate segments, and sliver polygon. This helps to build spatial relations among the spatial features using topology. In addition to map data sets, attribute data are stored in tables using relational database models. Links are provided between the spatial objects and corresponding attributes inside the databases.

1.1.7 Spatial Relations

A spatial relation gives how an object is located in relation to another object in space in terms of geometric properties. Relations can be directional, distance and topological relations among the spatial objects such as points, lines and polygons. Topological relations are characterised by properties to be preserved under topological transformations such as translation, rotation and scaling [2]. Generally, topological relations are defined in terms of disjoint, meet, intersect, equal, contains, inside, covers and covered by. These relations help in integrating map players and running spatial queries.

1.1.8 Spatial Data Analysis

Spatial data analysis can be defined as a set of techniques derived for the manipulation of data whose outcomes are not invariant under relocation of the objects of interest in some space [3]. Using the spatial database, simple data retrieval to complex modelling can be carried out. In vector database, the simplest analysis is to run a query in the attribute database and see the result in the tabular form and corresponding spatial objects in map form. Several new thematic maps such as infrastructure distribution map, population map, etc., can be generated. As spatial objects are stored as points, lines and polygons, the vector data queries can be carried out in terms of union, intersect, identity, clip, update, erase and split among two layers. In a single layer, operation such as buffer, eliminate and dissolve can be done.

In raster data layers, operations are done using the grid values and they are categorised as local, focal and zonal operations. In local operations, the concept of map algebra is used to combine several input layers with algebraic, logical and relational operators. It is a very powerful tool to carry out applications on land use mapping, water resources, studies, soil erosion and resource management. In focal operations, single input layer is used to get buffering, proximity and distance maps. In zonal operations, a zone map like a village map is combined with a thematic map like an elevation map to compute parameters such as minimum, maximum, arithmetic mean, standard deviation, etc., for each zone.

In applications related to natural resources management, raster data model is used while in network analysis related to shortest path creation, location allocation problem, travelling salesman problem, etc., vector data model is used.

1.1.9 Spatial Data Interpolation

Whenever the data are available at few sample points, spatial data interpolation techniques are used to estimate the values at other points in the study area. A global interpolation technique uses every known data point to estimate the value at other points. Local interpretation technique uses a small number of known data points to estimate the values in nearly points. Generally, inverse distance weighted method is used where the estimated value of a point depends more on nearby points than far off points. It is a deterministic technique. Kriging is a geostatistical technique and uses the concept of variogram modelling to generate the interpolated surface.

1.1.10 Digital Terrain Modelling

The terrain is an undulating continuous surface and digital terrain modelling represents the terrain in a digital map form. The elevation data is often given by Z value above the mean sea level. In raster data modelling, using suitable interpolation technique, the elevation value for all grids is computed and shown as digital elevation model. In vector data modelling, the elevation data are shown in the form of TIN. A TIN shows the terrain data as a series of non-overlapping triangles. Delaunay triangulation is well-known method to generate TIN surface and it uses the concept that all the triangles are as equiangular as possible and the circumcircle of a triangle should not contain another sample point inside it. Slope and aspect are important parameters that are generated from the terrain data. Slope gives the rate of change of elevation at a location and aspect gives the direction of the slope. Digital terrain modelling is used in several applications like watershed mapping, site suitability mapping, canal/road alignment, intervisibility map generation, and three-dimensional view of the terrain.

1.1.11 Network Analysis

In geospatial information system, a network represents a set of spatial locations interconnected by many routes. Generally, the network data is represented in vector form as linear features with nodes and directions. Network analysis helps in shortest path problem, finding the closest facility, location-allocation problem, service area mapping, etc. It is a very useful tool in transportation planning and management.

1.1.12 Statistical Analysis

As the locations and their parameters are stored in spatial databases, many statistical measures can be derived to understand the data. Descriptive statistics is used to calculate the mean, median, mode, standard deviation, etc., using the attribute data. Spatial autocorrelation helps to measure the dependence on nearby values and generate the spatial patterns. Trend surface analysis brings the spatial trend locally or regionally for properties such as income distribution, infrastructure facility, etc. Gravity models are used to study the spatial interpolation between two locations such as population migration and commodity flows.

1.1.13 Visualisation of Spatial Data Analysis

Spatial data visualisation is very important to make the end user understand easily the results of the analysis. Maps are the most effective tool for communicating the results. In vector data visualisation, point features are shown with suitable symbols, lines with varied width of lines and areas by different fill patterns. Generally, elevation data is shown on a three-dimensional surface. In raster data visualisation, colours are used to present the features and properties of hue, saturation and intensity help to improve the choices of colours. A dot density map shows quantitatively the property such as population. Statistical maps use bar and pie diagrams to quantitatively show multiple properties. While preparing the map visualisation, the map body is placed in the centre and other map elements such as title, legend, scale, direction, projection, source, etc., are placed in such a way that the map looks balanced and communicates well to the end users.

1.1.14 Spatial Decision Support Systems

Spatial decision support systems (SDSS) are the tools that are designed and implemented to support decision-makers in solving complex spatial problems. A SDSS provides capabilities to import spatial data, represent complex spatial relationships, do spatial analysis and provide the results in the form of maps and tables. The process involved in SDSS is iterative, integrative and participative [4]. The system architecture must be in such a way that it can acquire knowledge from the decision-makers, support modelling and analysis and provide alternative solutions to the end users. Several SDSS are getting developed for environmental analysis, industry risk assessment, watershed management, regional planning, etc.

1.1.15 Spatial Data Accuracy

One of the important issues in geospatial technology is accuracy. It includes data quality, error, uncertainty, scale, resolution, method of data models, analysis methods, visualisation of results and the interpretation of final output by the end user. As the input spatial data are collected from various sources and many are downloaded from the websites, it becomes difficult to know the quality of the data set at the production level. One cannot make an assumption that all available data are perfect. Hence it becomes necessary to give some indications and a confidence level on data quality while presenting the final results.

There are several components that can be used to indicate the spatial data quality such as positional accuracy, attribute accuracy, geographical extent of data set, completeness and collection of data, scale, projection and coordinate information system, data source, condition, source documentations, classification, sampling methods, boundary definitions and lineage. On the original map, a line width of 0.5 mm is equivalent to 12 m on a 1:24,000 scale map or 125 m on a 1:250,000 scale map. Combining the errors in source document, geo-registration and digitisation process, a root mean square value can estimate the positional accuracy.

Attribute accuracy can be found based on the time of data collection and the method used for data collection. The extent of data set must cover the user's complete study area. The data completeness is affected by the rules of selection, generalisation and scale. The information about the projection and datum of the source data must be known.

Paper media tends to stretch and swell over time which can cause distortion during the map's georegistration. Often data may be for different time frames which may impact the results. In raster data, the variation within a pixel cannot be shown. So, the choice of pixel resolution is very important. In vector data, classification of data may be provided as low, medium and high which becomes user dependent. A lineage report can provide documentation on the source data, method of preparation, area of coverage, symbolisation used, etc. The data quality is directly proportional to the data cost. Presently metadata standards are being worked out globally to provide information on data set. In addition to input data quality, errors occur during scanning, digitisation and cleaning, error propagation in analysis, improper visualisation and misinterpretation of the results.

1.2 APPLICATIONS

The use of geospatial technology is to solve real-world problems and provide alternate scenarios to the decision-makers. The applications cover from resource planning and management to high-end modelling in defence sector.

Major areas of applications are natural resources management, infrastructure planning, transportation, health and human services, civil engineering, surveying, cartography, telecommunications, disaster management, defence, business, etc. Under natural resources, spatial technology is applied in agriculture, hydrology, land use planning, forestry, archaeology, environmental management, marine and coastal mapping, mining and earth sciences, petroleum, etc. In transportation sector, it is termed as GIS–T and is used for modelling travel demand, planning of road network, monitoring traffic, managing the transportation infrastructure, etc. In the government sector, the spatial technology helps in security, law enforcement, crime mapping, disaster mitigation and management, economic development, urban and regional planning and sustainable development. Applications are ever growing and with the availability of trained skilled manpower, the spatial technology will become one of the powerful tools to solve the problems faced by humanity and the solutions will provide benefits to the society.

1.3 RESEARCH CHALLENGES

An offshoot from academic research, geospatial technology has become a core technology for information resource management and decision-making in government and businesses. With the availability of Internet and Web technology, it has entered in the homes of millions and into hands of billions. There are several research challenges in terms of data, technology, applications and manpower that need our attention. Geospatial data is inherently complex and continuous in nature. Huge amount on spatial data is available and there is a rapid growth in the technology front. The scope in applications is ever-growing and there is a need for well-trained skilled manpower to handle the data, design the system and implement the applications [5].

Today vast amount of geospatial data is available with government, commercial suppliers and through public participation. All these data may be with different cartographic specifications, varied accuracy and incompatible formats. Integrating this multi-source data is a complex task. Some government organisations are attempting to build spatial data infrastructures following data standards. Data ownership and copyright are major concerns to use public data. Spatial data security is a major issue. There has to be a balance between what public's right to know and an individual's right to maintain privacy. Inherent uncertainties, geometric incompatibilities, mismatching of boundaries across the spatial features, unavailability of updated data and lack of meta information give a big challenge to the user while integrating the data from multiple sources.

The three Vs (volume, variety and velocity) are the properties of today's big data. Under geospatial technology, new data models and new computational algorithms are needed to use this structured and semi-structured data sets. Annotating semantically the big data may make both human and machine to understand this data. Machine learning and process models have to interact and for real-time analysis, there is a need for parallel/stream/cloud/grid computing engines. Modelling uncertainty and ambiguity in big data is a challenge. The technological advances in information and communication technology and Internet are changing the way the spatial data can be accessed, used, analysed and visualised. Mobile computing helps in vehicle tracking, remote data viewing, field data collection, etc. Spatial data warehouses are getting created and interoperability helps to tackle incompatibility between data formats. Open Geospatial Consortium (OGC) prepares specifications for data standards, data sharing and applications across GIS implementations.

1.4 OPEN AREAS OF RESEARCH

Study of geospatial information both in system and science perspectives has created new research areas in computer science, computational geometry, spatial statistics, geography, etc. With the availability of high-resolution and better accuracy data from LIDAR, drone and remote sensing platforms, integration of this multisource data set is a big challenge. Spatial data mining may help to extract useful information from these large data sets. Figure 1.2 summarises the present research challenges in geospatial technology.

Research is needed to study geographical processes independent of geometry based on human cognition. Handling multidimensional geospatial data, study of dynamic spatial phenomena, work on spatiotemporal analysis and representation of variable resolution data still remain as research challenges. Prediction on moving object is an open area of research. One of the important topics where attention is needed is the estimation of uncertainty and modelling of error propagation in spatial data analysis.

In space-time data modelling, attempts are being made to carry out physical modelling, geostatistical modelling, spatial processing of time series and hierarchical modelling. In order to minimise the error propagation, it is preferable to use spherical coordinates rather than projected coordinates in the analysis. Statistical methods may help to quantify positional uncertainty and fuzzy boundaries. New data structures and algorithms are needed to speed up computations in large data sets.

Geospatial data generation is a complex and expensive task. Variety of models and techniques are available to manage and analyse geospatial data. Very little attention is given to address the spatial data security issues such as data access control, security and privacy policies at storage and dissemination level, protection of data ownership, authorised users to access to the data, etc. Research challenges are protection of ownership of geospatial data at dissemination step, task based and feature-based control at storage level and algorithms for secure, efficient and computationally inexpensive data outsourcing.

One of the attempts being made is the use of watermarking. In vector data, the watermarks must be invisible and robust against geometric transformations, cropping, format change, etc. It should maintain the positional accuracy and preserves topological relationships [6]. In case of raster watermarking, the original data must be near lossless and maintain the classification. Watermarking can be combined with cryptography to protect the spatial data. At the storage level, there must be specifications and policies for secured access control to the data. While outsourcing the data, efficient transformation can be used in the data set and getting the key from

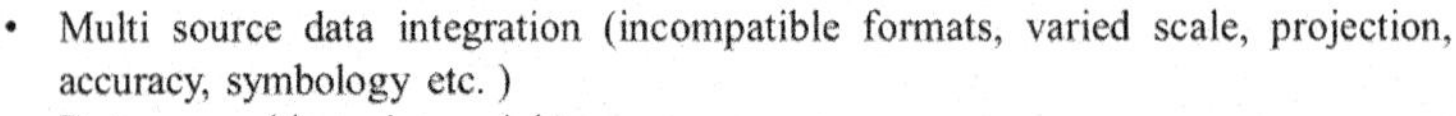

FIGURE 1.2 Geospatial technology—Research challenges.

the data owner, the authorised user can run the query and view the results with suitable inverse transformations [7].

Combining with web technology, geospatial technology has grown as web GIS and has reached to millions of users. It has resulted in huge amount of user-generated content and remixing of web services with cloud has helped to build new applications. Google maps, Google Earth, Yahoo map, etc., are well-used web mapping applications. Web GIS is a powerful tool in E-governance and can provide geospatial intelligence to decision-makers with real-time data. Web GIS demonstrates immense value to government, business, science and daily life [8]. With volunteered Geographic Information (VGI) getting generated in web by individual, usage of this data in terms of quality, integrity, accuracy and personal privacy becomes an issue for the end users. Geocollaboration can be useful in disaster management. Geotagging can enrich the spatial data and geoparsing can turn the documents into useful geospatial data. Sensor and sensor web helps in near real-time analysis.

Semantic Web provides meaning to the information in the web such that the web data can be processed intelligently by machines.

1.5 CONCLUSION

Geospatial technology can model and analyse the data and present the results and the people are the decision-makers. Users are from varied technical backgrounds. Applications vary from simple desktop mapping to high-end modelling. The decision-makers must understand the capability of geospatial technology. There is a need for skilled specialists who can design the system and develop geospatial applications. In academic front, the educational structure must include the theoretical aspects of mathematics, statistics, computer science, geography, surveying and application areas. Research areas are getting evolved on concept, techniques, applications and implications pertaining to humanities and social sciences. Academic and research institutions must participate in government initiatives and do research so that the use of geospatial technology can benefit the society. The real success of geospatial technology is the massive benefits that it can provide to the society.

REFERENCES

1. Venkatachalam, P. (2009). Geographic Information System (GIS) as a Tool for Development. Information Technology and Communication Resources for Sustainable Development. http://www.eolss.net/sample-chapters/c15/E1-25-01-12.pdf.
2. Egenhofer, M.J. (1989). A formal definition of binary topological relationships. *In International Conference on Foundations of Data Organization and Algorithms* (pp. 457–472). Springer, Berlin, Heidelberg.
3. Goodchild, M.F. (2018). Spatial data analysis. In: Liu, L. and Özsu, M.T. (eds) *Encyclopedia of Database Systems*. Springer, New York, doi: 10.1007/978-1-4614-8265-9_353.
4. Venkatachalam, P. (2021). *Geographic Information Science*, Encyclopedia of Earth Sciences Series (EESS), Encyclopedia of Mathematical Geosciences, Springer.
5. Kumar, P., Rani, M., Pandey, P.C., Sajjad, H., & Chaudhary, B.S. (Eds.). (2019). *Applications and Challenges of Geospatial Technology: Potential and Future Trends*. Springer International Publishing, Cham.
6. Zope-Chaudhari, S., & Venkatachalam, P. (2012). Evaluation of spatial relations in watermarked geospatial data. *In Proceedings of the 3rd ACM SIGSPATIAL International Workshop on GeoStreaming* (pp. 78–83), Redondo Beach, California, USA.
7. Chaudhari, S., Venkatachalam, P., & Buddhiraju, K.M. (2019). Secure outsourcing of geospatial vector data. *In IGARSS 2019-2019 IEEE International Geoscience and Remote Sensing Symposium* (pp. 871–874).
8. Fu, P. (2018). Getting to know Web GIS. *Photogrammetric Engineering & Remote Sensing*, 84(2), 59–60.

Section II

Geo-AI

2 Perspectives on Geospatial Artificial Intelligence Platforms for Multimodal Spatiotemporal Datasets

C. V. Krishnakumar Iyer, Swetava Ganguli, and Vipul Pandey
Apple Inc.

CONTENTS

DOI: 10.1201/9781003270928-4

2.1 INTRODUCTION

Geospatial datasets are inherently multimodal and spatiotemporal. In this context, data modality refers to the different data types encountered such as imagery, time series, sequences and trajectories, text, point features and associated metadata, line features and associated metadata, and features represented by vector geometry, to name a few. The spatiotemporal nature arises from the fact that each data point in these datasets is associated with a timestamp corresponding to when data was acquired and a geographic location where the data was acquired.

Geospatial data is voluminous and its size continues to grow. With the omnipresent deployment of novel sensing technologies, advanced earth monitoring systems, modeling and forecasting technologies, documentation of events around the world in news databases and social media, and large volumes of other unstructured geo-tagged data generated from mobile devices, social networks, edge devices, and sensor/IoT (Internet of Things) networks, there has been an explosion in the volume of geospatial (including data that is geo-tagged) data. For example, satellites like Landsat 7 (USGS, 2022) generate about 1 terabyte of data per day, weather models created by the European Centre for Medium-Range Weather Forecasts (ECMWF) amount to about 12 terabytes of data per day (Klein et al., 2015), over 500 million tweets (a good source of contextual information especially during disasters such as wildfires, hurricanes, and floods) and over 350 million photos uploaded to Facebook every day each being associated with a timestamp and rough geographic location. While jointly distilling information from each of these data modalities is highly desirable and can benefit many tasks in geospatial analysis and remote sensing, the feasibility and viability of multimodal modeling require tools that can (1) transform different data modalities so that they align with established geospatial data modalities like road network data, satellite imagery, and topographic data, (2) index acquired data for efficient storage and scalable retrieval, and (3) fuse different data modalities to enable joint reasoning.

Conventional database technologies need to be significantly modified to adapt to the complexities of spatiotemporal and multimodal geospatial datasets while ensuring efficient storage and scalable retrieval for the large data volumes. At the same time, machine learning modeling strategies developed for commonly encountered data modalities such as images, video, text, and speech need to be adapted to handle the nuances of geospatial datasets. The technical and modeling debt accrued by building task-specific solutions (Sculley et al., 2015) can be substantially reduced by building a data-centric geospatial platform that can process and fuse multimodal data and allow for end-to-end machine learning, model building, experimentation, evaluation, and model deployment at scale. In this chapter, we lay out the design considerations that guide the development of machine learning platforms for geospatial applications.

Section 2.2 discusses the unique challenges and key opportunities afforded by various geospatial data modalities. Given the promise of deploying data-driven, machine learning-based solutions to solve problems of various types in geospatial analysis, Section 2.3 delineates the motivation for building a data-centric machine learning platform capable of solving multiple types of geospatial tasks using multimodal geospatial datasets as opposed to building individual, problem-specific solutions. This section also introduces Trinity (Iyer et al., 2021) as an example of such a platform. Throughout this chapter, Trinity is used as a case study to demonstrate the impact of different design choices and provide a constructive example of an implementation of the presented ideas. Three critical challenges in multimodal machine learning are representation, alignment, and fusion of different data modalities (Baltrusaitis et al., 2018). For a wide selection of geospatial data modalities, Section 2.4 presents strategies found in the literature for modeling these modalities while also presenting the strategy adopted by Trinity to support multimodal geospatial modeling that addresses the aforementioned three challenges. Section 2.5 gives a brief overview at a high level of abstraction of the key components that comprise a geospatial machine learning platform. Each of these components is described in detail in the sections that follow. Specifically, Sections 2.6–2.9 describe the platform components that deal with raw and computed machine learning features (usable by supervised and unsupervised models; can be task-specific or task-agnostic), acquired and generated task-specific labels (for supervised models), and machine learning modeling, respectively.

Section 2.10 details the requirements and design principles guiding the development and deployment lifecycles of machine learning models. In each of these sections, generic practices that can be adopted for geospatial ML platforms have been discussed with Trinity exemplifying one approach for implementing the proposed design principles. Section 2.10.5 demonstrates solutions to a variety of real-world applications that have been solved using Trinity.

2.2 CHALLENGES AND OPPORTUNITIES OF DIFFERENT GEOSPATIAL DATA MODALITIES

Mobility data, typically obtained in the form of GPS trajectories, is sequential in nature with each GPS record having floating point representations of location in the form of latitude, longitude, altitude, speed, and direction of heading, among other sensor attributes. Mobility data can be acquired from subjects performing a broad range of activities and movements such as driving, walking, biking, running, and swimming, thereby encoding various motion modalities. With the growth of the number and types of wearable sensors and mobile devices, there has been growing interest in analyzing trajectory datasets for mobility pattern mining, user activity recognition, location-based social networks, and location recommendation. These tasks, either individually or in combination with others, are critical to many important downstream applications (Wang et al., 2020) like efficient traffic modeling and management (Zheng, 2015; Zheng et al., 2008), smart city and urban planning (Zheng et al., 2014), public safety (Zhang et al., 2017), epidemiological monitoring for public health policy and decision making (Smith & Mennis, 2020), building and maintaining mapping services (Xiao et al., 2020; Iyer et al., 2021; Ganguli et al., 2021), and environmental sustainability (de Souza et al., 2016). Mobility data is key to building better, more efficient, climate-friendly, sustainable, and robust transportation systems, but must be used responsibly such that privacy of the travelers is preserved. Recently, many cities, mobility service providers, technology companies, privacy advocates, and academics have created a guidebook for responsibly handling mobility data (NUMO et al., 2021). The sensitivity of location data and the resulting challenges around preserving location privacy remain a key challenge to the availability of large mobility datasets in the public domain that can be used for multimodal modeling of geospatial problems. Promising progress has been made in dealing with the privacy trade-off (Martelli et al., 2021) when dealing with large mobility datasets due to advances in cryptography, synthetic data generation (Berke et al., 2022), privacy-preserving machine learning techniques such as federated learning (Briggs et al., 2021) and homomorphic encryption (Chen et al., 2021), and privacy-preserving data release mechanisms based on differential privacy (Ho & Ruan, 2011; Acs & Castelluccia, 2014) or context-aware privacy metrics (Jiang et al., 2021; Zhang et al., 2022). Despite the progress, there is ample scope for novel research ideas to be proposed to address the challenge of the privacy-utility trade-off for mobility data. Machine learning models operating on mobility data can help city operators, transportation service providers, and real-time mapping applications identify and enable timely intervention in incidents on the transport networks and provide advice back to travelers on the real-time status of the network. At the same time, transportation management agencies can enforce transportation regulations in real time. In this chapter, it is assumed that available mobility data has been preprocessed to ensure anonymization and privacy.

Road networks are usually represented as a graph, $G_r(V_r, E_r)$, that encodes the topology, connectivity, and spatial structure of roads. Vertices, V_r, denote intersections, end of roads or passages, or the starting/ending of road segments. Edges, E_r, denote road segments between vertices of the graph. Nodes and edges can themselves be associated with different types of metadata. Many important applications like traffic forecasting (Yu et al., 2018), speed limit attribution (Jepsen et al., 2018, 2020), traffic direction attribution, and travel time estimation, to name a few, can be cast as canonical tasks on graphs such as node or edge classification, shortest path estimation, and edge prediction. Given this correspondence, there has been increasing interest in applying traditional node or edge embedding techniques (e.g., Perozzi et al., 2014; Grover & Leskovec, 2016; Dong et al., 2017) or developing novel techniques (Jepsen et al., 2018; Ganguli et al., 2021) to generate representations of components of the road network so that they can be used in coordination with neural architectures targeting graphs such as graph neural networks and graph convolutional networks. Despite the correspondence of road network graphs with graphs in other domains, there are a few key differences (Jepsen et al., 2019) that are manifestations of the underlying geography in which the road network is located. Road network graphs usually have low density, i.e., the number of edges actually present in the graph is much less than the maximum number of edges that can be present for the same number of nodes since most road segments in the graph have few adjacent road segments. In addition to intra-node, intra-edge, and node-edge attributes and relationships, there are significant relationships between neighboring edges in road network graphs that need to be modeled since they affect performance on downstream applications. For example, the angle between edges affects the estimated travel time (Jepsen et al., 2019), the elevation change between neighboring segments can affect traffic patterns, the change of directionality of traffic, or the difference in the temporal nature of traffic restrictions between two neighboring edges can affect traffic routing, etc. Road network graphs also distinguish themselves in the rate at which node or edge attributes vary, often called network homophily. While for many nodes and edges, neighboring nodes and edges have similar attributes, instances, where abrupt changes in attributes occur, are fairly common. For example, road segments belonging to ramps of a freeway have much lower speed limits than segments belonging to the freeway, frontage roads leading to a freeway are many times narrower than the width of the freeway it connects to, etc. These distinguishing characteristics of road network graphs require careful attention when solving a downstream task by modeling it as a canonical task on the road network graph.

Another important method of representing many types of geospatial data is as vector features with its shape represented as a geometry. Real-world geospatial features, both naturally occurring or man-made structures, such as houses, buildings, roads, trees, rivers, and lakes, can be represented as vector features. Vector features have attributes consisting of textual or numerical information that describe the features and a shape represented using geometry. Geometric features are made up of individual vertices that correspond to a location (specifically, a latitude-longitude pair) on the earth's surface. Features associated with points such as locations of trees, entries/exits of buildings, and locations of points-of-interest (POI) are called point features. Vertices can be connected to form lines and lines can be connected to form polygons that bound or enclose geospatial features with discrete boundaries such as water bodies, roads, and building footprints. The main advantage of vector data is that the representation of geospatial features and topologies is sharp, clean, scalable, and can be stored without any loss thereby preserving accurate geolocation information. However, complex features can require many vertices and even simple changes to the feature can require complex, time-intensive processing and algorithms. Continuous type data (e.g., surface elevations) is not efficiently presented and stored in vector format.

Imagery of geographic locations may be obtained from active or passive sources. Sensing systems that use naturally available energy sources (e.g., solar energy) for illumination of the earth's surface for imaging are called passive imaging sensors while systems that provide their own energy source for illumination (e.g., light waves of different frequencies) are called active imaging sensors. Examples of passive imaging include remote sensing systems that are sensitive to the visible, near-infrared, and thermal infrared bands of the electromagnetic spectrum (wavelengths at which the magnitude of solar radiation is greatest) such as the Ikonos, QuickBird, WorldView, and GeoEye commercial earth observation satellites from DigitalGlobe, the LandSat satellites from NASA, and the Sentinel satellites that are part of ESA's Copernicus earth observation program, to name a few. In addition to optical imagery in the visible range, these satellites collect high-resolution imagery across many different spectral bands. The obtained imagery is called multispectral when imagery is collected across 3 to 10 (in general, O(10)), fairly wide spectral bands (usually with descriptive names), while the imagery is referred to as hyperspectral when imagery is collected across hundreds or thousands of narrow (usually 10–20 nm) spectral bands. A challenge with multispectral or hyperspectral imagery is that the spatial resolution of pixels may differ across the bands. For example, LandSat eight produces images in 11 bands (corresponds to channels in image processing). All bands except bands 8, 10, and 11 have spatial resolution of 30 m. Band 8 has a spatial resolution of 15 m while bands 10 and 11 have a spatial resolution of 100 m. Passive imaging, given its requirement of solar energy, can only be done during daylight hours and can be occluded due to factors such as cloud cover. To overcome these challenges, active airborne, satellite, or ground-based sensors beam particular wavelengths of electromagnetic energy toward Earth's surface or geospatial features and record the pulse returns. The computed time interval of return and the intensity of the returning pulses can be processed to generate imagery of geographic locations or geospatial features. Compared to passive systems, active systems can sense in wavelengths that are not naturally available at sufficient intensities for imaging (e.g., active systems can use microwaves which can penetrate cloud cover), can direct radiation for targeted illumination, and can image at any time of the day. Since active systems also measure the time interval of pulse returns, they can measure the depth or elevation of natural geographic features or manmade structures. Examples of active imaging include imagery obtained from synthetic aperture radars (SAR), laser fluorosensors used to detect deliberate or accidental discharges of oil into water bodies, and point-cloud data for topographic mapping obtained from LiDAR (Light Detection and Ranging), to name a few. SAR imagery captured by ESA's Sentinel-1 mission is publicly available while LiDAR, in addition to capturing earth observation and mapping imagery, has become an indispensable navigation tool for autonomous systems on earth (e.g., self-driving cars) and in space (e.g., NASA's Ingenuity helicopter on Mars).

Aggregated social media posts can be a rich source of real-time information with a plethora of applications including traffic congestion mitigation, natural disaster monitoring and response, food security measurement and response (Dunnmon et al., 2019), and disease outbreak tracking, detection, and prevention. Social media posts are a combination image, video, audio, and text modalities. Due to their practical application in many other domains, there has been increasing interest and promising research in jointly modeling image, audio, and text modalities (Baltrusaitis et al., 2018; Zhang et al., 2020; Guo et al., 2019).

In addition to the established geospatial data modalities described above, there are many other sources of data that can be meaningful for geospatial analysis. A Digital Elevation Model (DEM) is a representation of the bare ground (bare earth) topographic surface of the Earth excluding natural or man-made geospatial features like trees, buildings, and any other surface objects. DEMs are created from a variety of sources such as photogrammetry, LiDAR, Interferometric SAR, and land surveys and therefore have varying resolutions. For example, the DEMs provided by the United States Geological Survey (USGS) have resolutions varying from 3, 10, 30, and 60 m. DEMs are useful for landscape modeling, city modeling and visualization applications, flood or drainage modeling, land-use studies, and geological applications, to name a few. In addition to DEMs, various other datasets with important topographic information such as historical seismic activity, historical natural hazard events (e.g., wildfire, floods, tsunamis, hurricanes, and cyclones), historical evolution of water feature depth and boundaries, can be pertinent to various geospatial applications. Demographic datasets comprise another rich source of information for many geospatial applications geared toward sociological modeling, public health, and public safety. Examples of such datasets are government census data that help allocate government resources based on population densities, mobility data joined with aggregated data from public health agencies used for epidemiological modeling of diseases, historical population dynamics, and economic and econometric datasets, to name a few. Common challenges when dealing with both topographic and demographic datasets include inconsistency of spatial resolution in different geographies, data corruption due to varying collection conditions, varying resolutions between underlying data collection methods, sparse geographic coverage, and temporal inconsistencies (i.e., data collected from different geographies at different times). In general, there is no common technique that addresses these challenges (sometimes, these challenges cannot be addressed satisfactorily) and application-specific solutions need to devised. However, as described in the later sections of this chapter, having a ML platform with well-defined Service Level Agreements (SLAs) for data ingestion into its data platform and feature store makes the limitations of a given dataset transparent to modelers and stakeholders. Demographic and topographic datasets may be image-based, textual, sparse numerical, sparse categorical, quantized into numerical buckets, quantized into qualitative buckets, and aggregated over non-uniform areas like states or countries. All of these characteristics individually present unique challenges for combining these datasets with other data modalities thereby explaining the dearth of multimodal models for geospatial applications.

One distinguishing feature of geospatial datasets compared to natural language (e.g., SQuAD, IMDB Reviews) or natural RGB imagery datasets (e.g., ImageNet, CIFAR-10) is the inherent differences between the data distributions. For example, natural images have similar normalized data distributions (Ruderman, 1994) irrespective of the dataset they are obtained from. This is far from true in geospatial imagery datasets. The pixel distribution of satellite imagery is vastly different from that of its representation as a map on an online mapping service which is vastly different from the image-like representations of other data modalities shown in Figure 2.2. Furthermore, the dynamic range and pixel distribution statistics vary as a function of the geographic region, season of the year, and many other hidden variables. This inherent diversity across geographic regions and data modalities pose an open challenge for the generalizability of machine learning models. Multimodal modeling may be a contributing factor to solving the generalizability challenge since the model can learn the correlations and relationships between different modalities while identifying the complementary and supplementary information present across modalities. In Sections 2.6.1 and 2.7.1.4, we discuss data-centric approaches such as building, maintaining, and evaluating training datasets using stratified sampling and active learning so that the training data reflects the rich diversity in the true data distribution and can be class-balanced during creation of the dataset. Such tools are crucial parts of a data-centric geospatial AI platform. On the other hand, sampling or filtering of data may not be a feasible option in low-resource machine learning tasks, e.g., natural disaster modeling in remote areas where remote sensing imagery may have poor quality and fewer satellite revisit rates imply that available imagery cannot be filtered out. A thorough discussion of related issues is provided in (Weir et al., 2019; Lunga et al., 2021).

2.3 MOTIVATION FOR A DATA-CENTRIC, MULTIMODAL GEOSPATIAL ARTIFICIAL INTELLIGENCE PLATFORM

As described in Section 2.2, with the explosion of smart devices, there has been a rapid rise in collection and harvesting of geo-referenced data that contains locations on or near the surface of the Earth (Hu et al., 2019b). Examples of geo-referenced datasets include GPS traces and trajectories, satellite imagery, weather data, and location-based data from social media. Such datasets are crucial for several tasks including geo-feature detection, urban planning, and remote sensing. We define a geo-feature as an entity or attribute that has a geo-reference. For instance, buildings, road networks, residential areas, and stop signs can be considered as geo-features. Map building usually involves encoding such geo-features in a map. Traditionally, maps were manually built by human editors which are a tedious and time-consuming processes. Map building fundamentally relies on extracting a representation of a location from the fusion of multiple modalities of data that are complementary or supplementary. In other words, it relies on multimodal modeling of the real world. Recently, there has been a rapid rise in the application of machine learning and deep learning techniques, both supervised and unsupervised, in the field of GIS to make the pain-staking process scalable. For instance, efficient traffic modeling and mobility data mining (Zheng et al., 2008; Zheng, 2015), high-resolution satellite image interpretation (Zhang et al., 2015), and hyperspectral image analysis (Chen et al., 2014), among others have been efforts toward this goal. With the advent of Transformer-based (Vaswani et al., 2017) models that can be used for different data modalities, common model architectures show high performance across a wide range of tasks across different domains. Scaling, productionizing, deploying, and maintaining trained machine learning models, sub-topics under the umbrella of Machine Learning Operations (MLOps), have their own set of challenges and are some of the critical bottlenecks (United Nations, 2020) limiting the large-scale deployment of machine learning solutions for geospatial applications. The onus of training good models then falls to the ability of the modeler to systematically engineer the data used to build an Artificial Intelligence (AI) system. This shift in paradigm has been referred to as data-centric AI (Ng, 2022; Strickland, 2022). The design principles outlined in this chapter for a geospatial AI platform, that among other things can enable building, maintaining, and evaluating datasets in an easier, cheaper, and more repeatable fashion, are motivated by this data-centric AI philosophy. Trinity (Iyer et al., 2021), which is presented as a case study of a data-centric geospatial AI platform, provides frameworks for data collection, data generation, data ingestion, data labeling, data pre-processing/augmentation, data quality evaluation, and data governance.

2.3.1 Current Challenges in ML-Based Geospatial Analysis

Adopting the recent progress in machine learning-based modeling techniques to geospatial datasets is challenging since (1) there is a high barrier of entry for domain experts, (2) real-world problems usually require non-standard, problem-specific solutions, and (3) the multiple resource bottlenecks and inefficiencies in building and maintaining machine learning solutions. Geospatial domain experts have a deep and intuitive understanding of the geospatial domain, the available datasets, and the pros and cons of using the different available signals for a given application. However, building solutions using machine learning techniques, scalably analyzing and processing large-scale, multimodal, and spatiotemporal datasets, and scalably deploying trained machine learning models needs specialized technical skills. Therefore non-technical domain experts encounter a steep learning curve before they can address geospatial problems independently. Consequently, they frequently work in collaboration with data scientists and engineers to translate their ideas into scalable solutions that can be implemented, introducing another level of dependency. Another challenge that is frequently encountered is the diversity of the problems that need to be solved. On the surface, many of these problems appear different and disconnected with each other. For instance, detecting residential and commercial areas appears as a significantly different task compared to detecting road attributions such as one-way or U-Turn restrictions. In these cases, typically data scientists and engineers select solutions from their technical repertoire to solve individual tasks which they then optimize on a case-by-case basis. Notably, for each new task, they need to start from scratch, and perform a series of tasks that includes data pre-processing, feature engineering, model building, and model validation, followed by prediction and visualization, and culminating in product deployment and maintenance. The case-by-case repetition of steps customized to each specific task also has the potential of giving rise to non-standard workflows that are costly to build, maintain and manage. Furthermore, the dependency introduced between domain experts and data scientists tends to slow down collaboration across the variety of problems to build scalable, shared workflows with the active and hands-on involvement of domain experts.

Geospatial AI platforms are best poised to solve the above challenges since they provide ways for solving problems by sharing information across products, by solving disparate looking but deeply connected problems in similar fashion with shared workflows, promoting knowledge and skills sharing between stakeholders, and most importantly, with active and potentially hands-on involvement of domain experts leading to close collaboration between ML experts, geospatial domain experts, software and data engineers, and management stakeholders. While the dominant paradigm in machine learning systems is to fix the dataset and iterate on the model (sometimes referred to as model-centric approach) and training objectives, geospatial applications can greatly benefit from using the modeling developments and rapid progress from other domains such as computer vision, self-supervised learning, reinforcement learning, natural language, and speech processing, among others in conjunction with novel ways for learning better representations (e.g., following the guidelines and "general-purpose" priors in Section 2.3 of (Bengio et al., 2013)) of multimodal geospatial datasets, i.e., the data-centric approach (Strickland, 2022). Geospatial platforms with modular architectures like the one discussed in this chapter are best posed to adopt this approach. For example, data-centric principles (SnorkelAI, 2022) like label consistency, consensus labeling to spot inconsistencies, repeatedly clarifying labeling instructions by tracking down ambiguous labels, and identification and removal of noisy/bad examples, among others, can be efficiently taken care of by the label platform component (see Section 2.7) of a geospatial AI platform. On the other hand, using error analysis to focus on a subset rather than the aggregate data can be undertaken by the feature platform, kernel, and workflow management components of the geospatial AI platform.

2.3.2 An Example of a Geospatial AI Platform: Trinity

Trinity (Iyer et al., 2021) is built to tackle the challenges mentioned in Section 2.3.1 from three dimensions. First, Trinity's upstream data ingestion and feature platform (alternatively called feature store) brings information in disparate spatiotemporal datasets comprising of diverse data modalities to a uniform format by applying various types of data transformations (Section 2.4 discusses some of these transformations). Second, by leveraging this uniform format, Trinity standardizes the workflow of solving disparate-looking problems by casting these different applications as canonical tasks in computer vision. For example, alluding back to the example in Section 2.3.1, identifying commercial zones and detecting locations with restricted U-Turns can be cast as a supervised semantic segmentation problem. Trinity's supervised learning platform allows users to cast a geospatial problem into one of four canonical computer vision tasks: classification (e.g., identifying forested areas), object detection (e.g., detecting buildings), semantic segmentation (e.g., identifying landcover types), and instance segmentation (e.g., detecting individual crosswalks). Since bucketed measures are almost always more important than exact measures in geospatial analysis, regression problems are converted to classification problems (e.g., speed limits). Finally, Trinity provides an intuitive, no-code environment for rapid experimentation and prototyping thereby lowering the technical entry barrier for non-technical domain experts. Trinity leverages the rapid advancements in the area of deep learning techniques for computer vision to solve problems using spatiotemporal datasets.

For most geo-feature detection problems (e.g., detecting traffic lights, stop signs, roundabouts, one-way roads, and road casements), semantic segmentation is the method of choice. Semantic segmentation (Long et al., 2015) is a canonical computer vision task where each pixel of an image is tagged with a class label. Computer vision tasks operate on image-like tensor representations of the data. Geospatial datasets such as satellite imagery (RGB, multispectral or hyperspectral), aerial imagery, panoramic drive imagery, SAR-based imagery are naturally represented as image-like tensors. However, many other geospatial datasets are not naturally represented as imagery, e.g., sequential GPS trajectories, points-of-interest, geo-tagged social media posts, and various sensor datasets. These datasets need to be transformed from their natural representations into image-like tensor representations to make them amenable to analysis using computer vision techniques. Trinity leverages all these complex, spatiotemporal datasets by transforming them into image-like tensors (described in detail in Section 2.4). Some of these transformations are simple pixel-wise aggregation. For instance, computation of densities of walking and driving GPS traces per pixel of a certain size (Panzarino, 2018). Going one step further, self-supervised representations called embeddings are learnt from the datasets and transformed to embedding fields having an image-like structure. These embedding fields help to retain relevant information in the data while also making the data representation amenable to analysis by computer vision techniques.

Pre-generation of data representations (also called channels when in image-like form and profiles when in the form required for storage in the feature store) alleviates the need for repeated processing of the data for different models. By formulating geospatial problems as canonical computer vision tasks, Trinity helps standardize the solution strategies and workflows used to solve the problems thereby enabling reuse of off-the-shelf deep learning algorithms and state-of-the-art neural architectures (such as convolutional or transformer-based architectures like VGG, UNet, RCNN, and its optimized variants, ResNet, Mask R-CNN, ViT) thereby benefiting from the advances made in this area in both industry and academia. To this end, Trinity packages selected neural network architectures in its deep learning kernel. Users can choose an architecture, run manual or automated (using AutoML-based neural architecture selection) ablation studies over different architectures, or contribute their own architecture experimentation. By building a robust software system around the deep learning kernel and profile store (also called feature store), Trinity can be used by anybody with a problem statement in mind to run their experiments at scale while choosing from a variety of input data combinations and neural architectures without writing a single line of code. Moreover, an intuitive user interface (UI) for experiment management, scalable data storage for features, containerization of the kernel for deploying on GPU clusters, and a suite of standard pipelines for data preparation and large-scale predictions on Apache Hadoop using Apache Spark facilitates sharing of data and models leading to low time-to-market for trained models, standardized MLOps workflows, and seamless collaboration between geospatial domain experts and scientists. In short, Trinity is designed for rapid and parallel prototyping and experimentation to obtain the best deployable solution for a given problem as quickly as possible with minimal technical overhead. Sections 2.6–2.9 detail the different components of a geospatial system using Trinity as a case study.

2.3.3 Key Advantages and Observed Benefits of Trinity

Below, we delineate the observed benefits provided by a platform like Trinity for different categories of users. We have identified three main categories of beneficiaries are geospatial domain experts, ML data scientists, and data and ML engineers. These advantages are summarized in Figure 2.1a while their observed usage patterns are summarized in Figure 2.1b.

1. **Benefits for the Domain Expert**: Domain experts usually have the best intuitive understanding of a problem and insights into the shortcomings of potential data-driven solutions. As a zero-coding platform where the complexities of ML workflow are abstracted out, Trinity offers a self-service model that enables a wide range of non-technical users to train, predict, and scale their own models. The lowered barrier of entry thus made possible by an AI platform such as Trinity enables domain experts to run experiments firsthand, something that would otherwise require a steep learning curve. Furthermore, Trinity makes it possible for the domain experts to make a quick transition from having a promising intuitive idea to validation of their hypotheses. It also facilitates quick iterations through an intuitive UI that leverages underlying infrastructure efficiently without requiring to deal

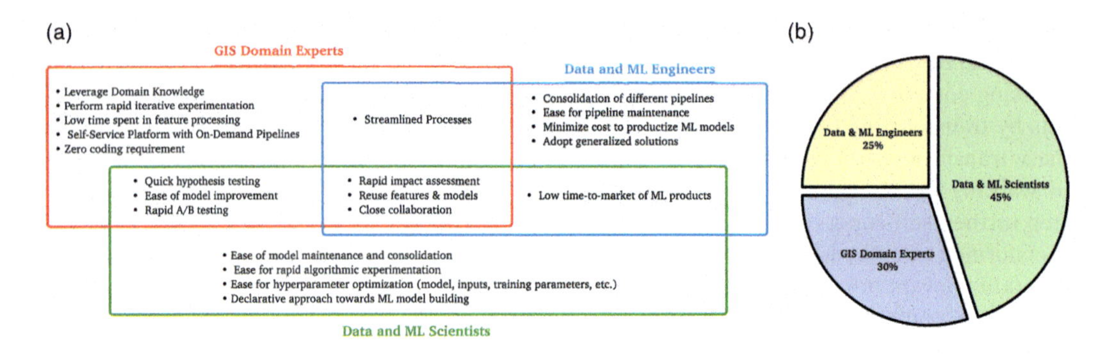

FIGURE 2.1 Observed (a) benefits provided by a geospatial AI platform like Trinity to the three categories of users defined in Section 2.3.3 and (b) distribution of users in the three categories.

with the underlying complexity of the models, feature engineering, and data processing. Trinity lets the domain experts ideate, share, and collaborate on experiments and predictions with scientists as equal partners.

2. **Benefits for the Data or ML Scientist**: Machine learning practitioners work actively in extracting information from the raw data in the form of channels that are later used by segmentation models. This decouples channel generation from downstream products. Scientists find it easy to work on channel generation and scale them independently if needed. Others may choose to use the channels already pre-populated in the channel store. Complex and repeated feature engineering constituted the biggest workload of scientists before. They now reuse and mix & match channels from other scientists and engineers while building their models. Scientists who are not trained in deep learning also get to leverage advancements in the field directly. The ability to perform rapid and parallel experimentations for a problem helps scientists move and develop faster. Moreover, the ability to compare experiments easily helps them shorten the path to an optimal solution. Due to quick and easy experimentation, and minimal feature processing, the observed latency from idea to initial prototype is low. Standardized problem formulation, workflows, and reusable channels lead to enhanced productivity and greater focus on the solutions and business problems. As discussed above, having a shared vocabulary in terms of channels, models and labels facilitate better collaboration between team members working on related projects in Trinity. Experiments are often shared between scientists and domain experts or their ownership is transferred allowing seamless communication. Trinity is designed to be horizontally scalable during both training and inference. This alleviates the need for the scientist to be dependent on the engineer to scale up the jobs before they can be tested at scale, provides a standardized productization workflow to the scientists, and lays out a clear path to the productionization of a trained model.
3. **Benefits for the Data or ML Engineer**: Trinity decouples engineers working on scalability issues from individual case-by-case problem statement and model building. Instead, engineers contribute pluggable platform components such as modules for scalable channel creation and generic data processors that can be reused across multiple applications and products. Therefore, such a platform enables engineers to avoid duplication of work and easier maintenance by promoting standardized workflows for diverse problems to avoid tedious repetitive tasks. This consolidation ensures that any effort spent in optimizing workflows now benefits all the users and use cases.

Another observed outcome is that platforms like Trinity allow other non-expert and management stakeholders to better understand and actively participate in the planning and development life-cycle of solutions to critical problems using a clear and well-defined path from problem statements to a deployed solution.

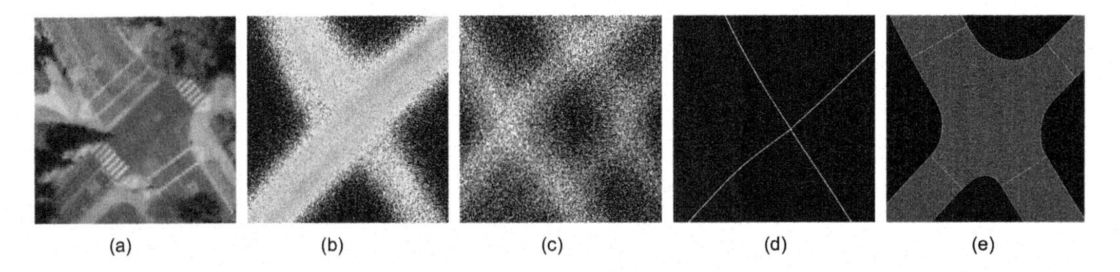

FIGURE 2.2 An example of the complementary and supplementary information provided by various data modalities at a given location. The walking mobility data corroborates the presence of pedestrian crosswalks in the satellite image. The road polygon width is corroborated from the satellite imagery and the width of the driving mobility data. The road network segments provide the centerline of the roads. (a) Satellite image, (b) drive mobility, (c) walk mobility, (d) road network, and (e) road polygon.

2.4 REPRESENTATION, ALIGNMENT, AND FUSION OF MULTIMODAL GEOSPATIAL DATASETS

Having seen the intricacies of some common geospatial data modalities in Section 2.2, it is clear that each data modality has unique characteristics. For a chosen location, each data modality encodes different aspects, not mutually exclusive, of the events, objects, geographic features, and phenomena associated with the chosen geographic location. In other words, despite their syntactic dissimilarities, the different data modalities are synchronized and describe different aspects of the same semantic reality. When the descriptions overlap, they serve to corroborate each other. When they are exclusive, they serve to overcome each others' information deficit. A visual example is provided in Figure 2.2. Multimodal machine learning models aim to exploit the synchronization of various data modalities to build a more holistic picture compared to unimodal models with the hope of improving performance of important downstream tasks.

As mentioned previously, the first step in multimodal modeling is to address the three key challenges of multimodal machine learning, i.e., representation, alignment, and fusion of the different data modalities (Baltrusaitis et al., 2018). Learning good representations (Bengio et al., 2013) from raw data of different modalities has been a critical and active research area. Deep networks are robust feature learners and have contributed to the recent gains in performance of machine learning models across different domains. Multimodal representation learning aims to reduce the heterogeneity gap between different data modalities that correspond to the same semantic reality by integrating the complementary and supplementary information encoded in the unimodal representations. The plethora of techniques proposed in the literature for multimodal representation learning can be broadly categorized into three types (Baltrusaitis et al., 2018; Guo et al., 2019): joint representations, coordinated representations, and encoder-decoder-based representations. Joint representations (e.g., (Liu et al., 2016; S. Wu et al., 2014; Poria et al., 2016)) project related or corresponding examples of different data modalities into a common semantic manifold while coordinated representations (e.g., Lazaridou et al., 2015; Frome et al., 2013; Karpathy & Fei-Fei, 2015) learn separate but constrained representations for each modality. In contrast, the encoder-decoder representations (e.g., Xu et al., 2015; Venugopalan et al., 2014; Reed et al., 2016) aim to learn an intermediate representation that contains information from multimodal inputs which is then used to map individual modalities into one another (Ngiam et al., 2011). Finding correspondences, correlations, and relationships between sub-components of the representations of two or more data modalities is defined as multimodal alignment (Baltrusaitis et al., 2018). When the relationship between two data modalities is known apriori, the representations can be explicitly aligned to reflect these known relationships. For example, temporal alignment of the audio and visual components in a video. In contrast, many techniques model the relationships implicitly. For example, Learning to align word embeddings and image regions from a dataset of images and corresponding image captions (Bayoudh et al., 2022). Multimodal fusion (Atrey et al., 2010) describes the strategy used for integrating information from different input data modalities that may themselves have different types of representations. When the correlations in the low-level features (e.g., tensor representation of an image) of different modalities are combined to generate the multimodal representation, it is broadly categorized as early fusion (Barnum et al., 2020). When more holistic features (e.g., feature vector of an image resulting from a deep neural network) of the unimodal inputs are combined (e.g., weighted averaging (Potamianos et al., 2003), concatenation, voting (Morvant et al., 2014), learning a feature combination model (Ngiam et al., 2011)), it is broadly categorized as late fusion (Gadzicki et al., 2020). Fusion techniques attempting to exploit advantages of both are broadly categorized as hybrid fusion (Amer et al., 2018). Explainability and interpretability of the learned multimodal representations is an important and active research area. It can be observed that representation, alignment, and fusion of multimodal data are not independent challenges but are coupled — the approach for representation and alignment depends or informs the approach for fusion and the vice versa. Additionally, the target problem under consideration informs the choice of the chosen approach. For example, early fusion might benefit anomaly detection tasks where models must be discriminative to minor changes and perturbations in the input data. On the other hand, a plethora of contrastive self-supervision strategies (discussed in Section 2.4.7) train models to learn holistic multimodal representations (usually using late fusion) that are invariant to small perturbations in the input data. The inherent and diverse multimodality of geospatial datasets presents an exciting and challenging landscape for applying and extending some of the existing techniques proposed in the literature while leaving a wide open field for proposing new ideas to solve impactful problems in domains spanned by these datasets.

Casting geospatial tasks as canonical tasks in computer vision is a promising approach and has enjoyed wide success. There has been rapid progress in deep learning techniques for solving computer vision tasks which have been coupled with the development of efficient hardware that natively support operations crucial to vision-based neural networks. A common theme in computer vision-based solutions that has emerged in the last few years is to develop reusable models (can be thought as components that produce compressed representations of the input, e.g., ResNet (He et al., 2016), EfficientNet (Tan & Le, 2019), Densenet (G. Huang et al., 2017)) and reusable features (e.g., which is the principal goal of self-supervised learning). While different problems can be solved using different methods, a successful strategy has been to remodel the problem so as to cast it as a canonical computer vision task like classification, object detection, semantic or instance segmentation, etc. A simple ablation study over architectures, optimizers, and loss functions to train supervised models is then undertaken to find an optimal solution. Most modern deep learning frameworks and libraries like TensorFlow and PyTorch provide commonly used neural architectures through simple APIs. The Trinity (Iyer et al., 2021) platform is built to take advantage of all these readily available components.

In this section, we demonstrate the approach that Trinity uses for addressing the three challenges in multimodal machine learning mentioned above across a diverse set of geospatial data modalities as a case study. We propose methods to generate image-like representations of all data modalities such that pixels of all modalities have fixed spatial resolution. The image-like tensors for all the modalities have the same size which depends on the downstream task. Having pixels of the same spatial resolution across modalities implies that pixel-wise alignment is equivalent to spatial alignment. Different modalities can then be concatenated along the channel dimension, also called early fusion, to yield a multimodal image-like tensor. Alternatively, neural networks can be trained for individual modalities (either pre-trained networks can be used or the networks for each modality can be trained jointly with the task-specific network) to produce compressed feature representations that are fused (e.g., concatenation) at a later stage to solve a task (late fusion). The remainder of this section is dedicated to different transformations that convert data of various modalities into image-like artifacts. In the Trinity platform, we have scalable and distributed Scala Spark (Zaharia et al., 2016) based pipelines that generate these transformed image-like tensors for each modality, align them pixel-wise, and fuse transformations of modalities chosen by the user on-demand for use with downstream tasks. Separating the feature representation, alignment, and fusion process from the training and prediction process allows a modular design of the geospatial platform. Details of the machine learning feature platform (also referred to as the ML feature store) including versioning, feature aggregation, drift detection, storage, and cataloging are discussed in Section 2.6.

The choice of restricting representations of modalities to image-like tensors for all downstream tasks may not always be the optimal choice. However, the immense benefits afforded by this data-centric design choice such as modular platform components, a common framework for scalable and distributed generation, ease of aggregation, clear semantic meaning, and flexibility for early or late fusion can mightily outweigh the downsides of this choice, especially in an industrial applications environment where rapid prototyping, reducing the time to first (feasible but possibly sub-optimal) solution, and rapid deployment are critical. The success of the Trinity (Iyer et al., 2021) platform in solving multiple business problems and its wide use by a spectrum of stakeholders (e.g., domain experts, machine learning scientists, data engineers) demonstrates this trade-off. In each of the sections that follow, the design principles for generating representations of the data modality follow the guidelines and "general-purpose" priors described in Section 2.3 of (Bengio et al., 2013).

2.4.1 Preliminary: Spherical Mercator Projection and Zoom-q Tiles

Before we delve into the spatial transformations of different modalities of data, we start by taking a quick look at the concept of Mercator projection and tiling that underpins the alignment and fusion of diverse modalities. The remaining subsections are built on this representation. The spherical Mercator projection coordinate system (EPSG:3857, 2022) is a mathematical transformation of the earth's surface into a flat, two-dimensional map. In the process of projecting the surface of a sphere onto a square, it stretches the map at the two poles. This stretching non-uniformly distorts the representation of the earth's surface so that locations farther from the equator appear larger. However, there are two important properties of this projection that make it a convenient choice for web-mapping and machine learning applications. First, the Mercator projection is a conformal projection meaning that the projection correctly preserves the shapes of small areas. This is crucial in showing aerial views of locations where the shapes of buildings, roads, parks, etc., are preserved. Second, the Mercator projection is a cylindrical projection so that the top and bottom of the projection correspond to north and south directions while right and left correspond to east and west. A zoom-q tile is the cell resulting from a $2^q \times 2^q$ raster representation of the spherical Mercator projection of the earth's surface.

2.4.2 Spatial Transformations of Mobility Data

Mobility data can be obtained either in a moving reference frame or in a static reference frame. GPS trajectories obtained from GPS-enabled devices embedded in travelers (e.g., pedestrians, cyclists, drivers, and their vehicles) are naturally represented as sequences. On the other hand, frequency of travelers passing traffic infrastructure (e.g., sensors on highways, sensors on road signage and traffic control devices, traffic cameras) is also a measure of the traffic passing through a given location as a function of time. A natural representation of data obtained from static sensors is a heatmap over a chosen time interval ($[t, t+\Delta t]$) of the data as a function of the starting time t.

Due to their natural representation as sequences, many approaches have been proposed in the literature to generate embeddings from trajectories using natural language processing-inspired sequence modeling techniques. The general theme consists of modeling trajectories as a chronological sequence of GPS records and analogizing a trajectory and its component GPS records to a sentence and its component words. With this analogy, embeddings for individual records are obtained with methods similar to word2vec and related methods (Mikolov et al., 2013a, b; Pennington et al., 2014), or the entire trajectory is modeled using sequence modeling and embedding methods such as (Sutskever et al., 2014). Example approaches include (Gao et al., 2017) which proposes learning and embedding of user trajectories for the task of trajectory-user linking in location-based social networks, identification of underlying mobility modes in a given trajectory (Jiang et al., 2017), prediction of destination based on observed partial trajectories (Ebel et al., 2020), trip semantic modeling and clustering based on geospatial context (Chen et al., 2019), generating semantic compressed representations of points-of-interest (PoIs) based on trajectories passing through them (Cao et al., 2019b), among many others. Most approaches use recurrent neural networks (with or without gated memory units like LSTM (Hochreiter & Schmidhuber, 1997) or GRU (Cho et al., 2014)) to construct embeddings of either individual records, the location corresponding to the record, or the entire trajectory itself. The feature vector for each GPS record either is the metadata associated with the record such as latitude, longitude, altitude, speed, and heading or is a projection of this data in a different learned space. In the supervised approach, the individual GPS records or GPS trajectories have labels associated with them which can be used to define loss functions, e.g., a cross-entropy loss per GPS record for the multi-class classification task of assigning the record to a particular motion modality. In the case where no labels are present, a common choice for the training objective is minimizing the negative log-likelihood computed for the task of next GPS record prediction having observed a partial trajectory similar to the methods proposed in (Graves, 2013). Other methods of representing trajectories that have been used in literature include a graph-based representation for generating the underlying road network from observed trajectories (Huang et al., 2018). With the recent developments in contextual embedding and sequence modeling techniques with transformer-based (Vaswani et al., 2017) neural architectures (Devlin et al., 2018; Brown et al., 2020; Smith et al., 2022), there is immense potential for the trajectory-sentence analogy to be exploited further to obtain better-compressed representations of GPS trajectories or its component GPS records.

Trinity adopts the approach of generating image-like tensor representations from mobility data that can be used by itself or in combination with other modalities (multimodal data fusion) for solving downstream tasks using deep learning techniques in computer vision. Irrespective of whether mobility data is collected using static or dynamic sensors, Trinity uses count-based raster maps so that mobility datasets resulting from either type of sensors can be represented in the same way. The first step is to decide the spatial resolution of the pixels in the image-like representations for each modality. Typical choices in Trinity are zoom-24 (~2 m at the equator), zoom-25 (~1.25 m at the equator), or zoom-26 (~0.6 m at the equator) tiles. For the rest of this section, we will assume that pixels correspond to zoom-24 tiles unless otherwise specified. Let each pixel be associated with c real numbers (represented as c-channels so that the image has dimensions $h \times w \times c$) that meaningfully correspond to the zoom-24 tile represented by the pixel. Figure 2.3b provides a schematic of the proposed method for constructing the image-like tensor. A count-based raster map (CRM) is a single-channel representation ($c = 1$) where the value of each pixel is the number of occurrences of a GPS record in the zoom-24 tile corresponding to the pixel during a time period $[t - \Delta t/2, t + \Delta t/2]$ that belong to a chosen motion modality. A schematic example is provided in the left column in Figure 2.3a where the top subfigure shows observed trajectories within the time period $[t - \Delta t/2, t + \Delta t/2]$ and the bottom figure shows the resulting computed CRM. Instead of aggregating all the occurrences of GPS records in a zoom-24 tile into a single number, if they are bucketed based on the direction in which the record is heading into 12 buckets of 30°, and each bucket is represented as an individual channel ($c = 12$), we obtain the heading count-based raster map (HCRM) representation. Similarly, we obtain the speed-count raster map (SCRM) representation by bucketizing the speed of the GPS records into 14 buckets of 5 miles per hour starting from 0, and each bucket is represented as an individual channel ($c = 14$),. The number of buckets for HCRM and SCRM is a user-specified number and 12 and 14 have been chosen here as examples. Each of these representations can either be computed for the entire mobility dataset or can be computed for a chosen motion modality, e.g., Figure 2.2b and c provide the CRMs obtained from driving and walking mobility data respectively. For a crosswalk detection task, the walking modality data can provide a strong signal of the

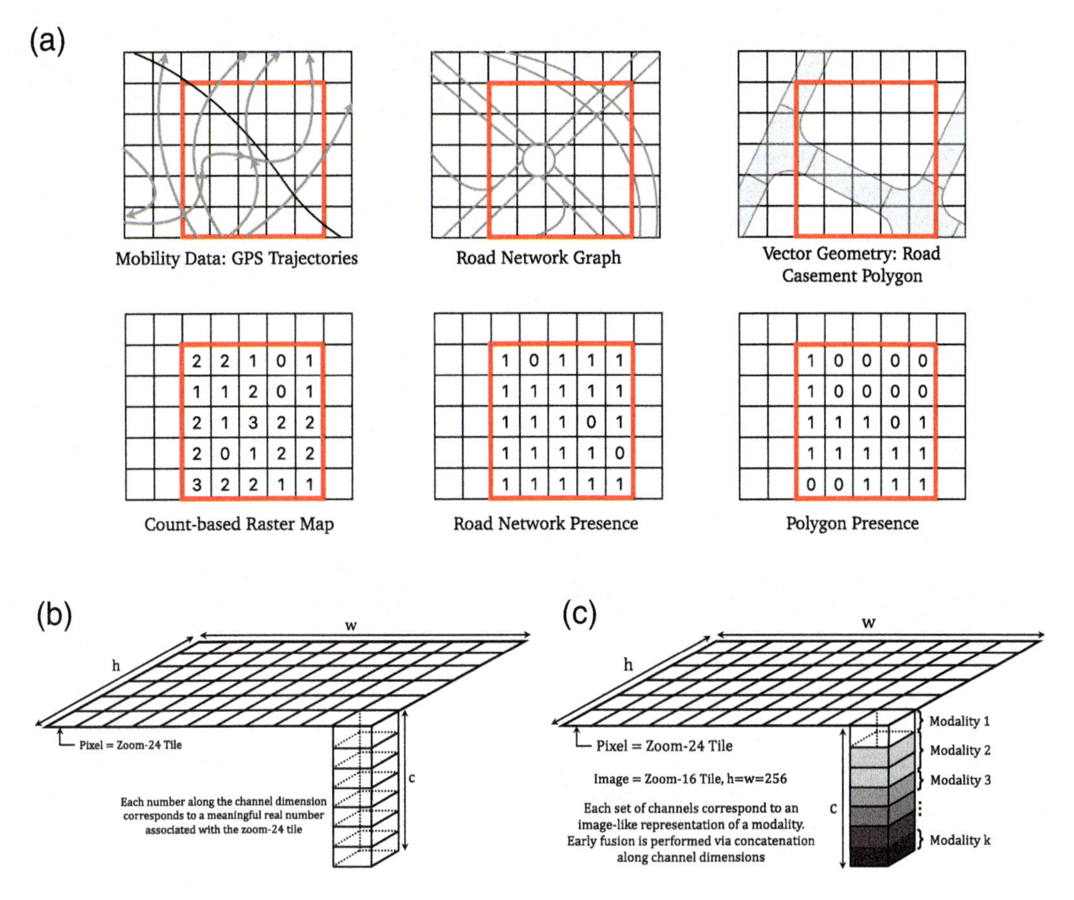

FIGURE 2.3 (a) An exaggerated schematic showing the transformation of mobility data, road network graph, and vector geometry into image-like representations discussed in Sections 2.4.2–2.4.4. For brevity, only single-channel derivatives are shown. (b) Any $h \times w$ neighboring pixels in the rasterized Mercator grid may be viewed as an $h \times w \times c$ image-like tensor representation of the data modality. (c) A multimodal image-like tensor is obtained by concatenating along the channel dimension multiple image-like unimodal artifacts.

presence of a crosswalk while the driving CRM can remove false positives by corroborating the presence and width of a road.

The proposed scheme for generating image-like representations from mobility data has many advantages: (1) each pixel (equivalent to the corresponding zoom-24 tile) can be processed independently (embarrassingly parallel) and hence can be computed in a scalable and distributed fashion; (2) similar raster map representations can be deduced for other attributes of mobility data such as buckets for altitude at a given pixel, bucketing based on hour of the day to produce a 24-channel representation. In Trinity, scalable and distributed feature ingestion pipelines (Section 2.6.2) are implemented in the Scala-based API for Apache Spark.

2.4.3 Spatial Transformations of Road Network Geometry

Road networks provide crucial information for multiple tasks related to geospatial modeling, transportation and route planning, network design, and management for supply chains, among many others. For each application, the interpretation of road networks can be different. If a network science-based graph modeling approach is used, road networks may be interpreted as trees or grids formed by graph edges. Nodes can either be major points-of-interest or could correspond to logical points where line segments need to be broken to represent the road geometry. Edges represent roads that connect two nodes. This is sometimes refers to in literature as the line graph transformation (Gharaee et al., 2021). Based on the connectivity between nodes, the edges can be associated with a hierarchical representation such as edges forming a path between main and subsidiary nodes, or edges forming a path between strategic and local nodes, or paths being through and side roads. The graph components can be associated with additional metadata based on network properties such as average degree, clustering coefficient, average shortest path, meshedness, betweenness centrality, webness or treeness, or general graph theoretic measures such as alpha index, and beta index (Marshall, 2016).

Given its natural graph representation, road network analysis in literature is dominated by methods that can be classified into one of two classes: (1) traditional graph analysis methods such as graph traversals, compressing matrix representations of the graph that exploit the structure of the graph (Masucci et al., 2009; Strano et al., 2012; Barthelemy & Flammini, 2008; Barthelemy, 2011; Barnes & Harary, 1983; Cardillo et al., 2006; Xie & Levinson, 2011), or (2) neural network-based methods for generating semantically meaningful compressed representations of graph components such as nodes, edges, subgraphs, or the entire graph. Recently, the traditional graph-based techniques have either been adapted to use neural networks or have been completely replaced with neural networks resulting in methods in class (2) given the end-to-end feature generation and task-based learning capabilities of deep neural networks. While a number of deep learning methods have been proposed for generating node embeddings of a known graph (Cai et al., 2018), they can be grouped into three distinct categories: (1) techniques that efficiently sample random walk paths from a given graph and generate node embeddings by processing the paths as sequences using NLP-inspired methods like skip-gram (Mikolov et al., 2013a, b) such as (Perozzi et al., 2014; Grover & Leskovec, 2016; Dong et al., 2017) or explicitly using recurrent neural networks (RNN) based on LSTMs or GRUs such as (Z. Liu et al., 2017); (2) techniques that generate node embeddings by compressing matrix representations derived from graphs (e.g., adjacency matrix, positive point-wise mutual information matrix) such as (Cao et al., 2016; Wang et al., 2016); (3) techniques that generate node embeddings directly from the graph using graph neural networks (GNNs) (Bruna et al., 2013; Defferrard et al., 2016; Kipf & Welling, 2016; Velickovic et al., 2017) and its variants (Zhou et al., 2018) (e.g., graph convolutional network (GCN)). Research in applying graph neural networks has focused on traffic prediction and forecasting (Yu et al., 2017; Zhao et al., 2019), identifying gaps in road network data (Hu et al., 2019a), forecasting demand of ride-hailing services (Geng et al., 2019), among other important geospatial modeling tasks.

For multimodal fusion, Trinity chooses to transform the road network graph into an image-like representation, called road network presence (RNP), via rasterization. The road network graph is projected onto the spherical Mercator projection of the earth's surface following which the projection is rasterized resulting in cells that correspond to zoom-24 tiles. Pixels of the rasterized image correspond to these zoom-24 tiles. Pixel values, initialized to 0, are incremented by 1 for every road network edge passing through the corresponding zoom-24 tile. Based on the application, pixel values may be transformed to either of 0 or 1 based on whether at least one road network edge passes through the corresponding zoom-24 tile. A schematic of this rasterization process is shown in Figure 2.3a.

2.4.4 Vector Geometry Data

Due to its convenience, geospatial data is often represented as vector geometries such as polygons, lines, and points. Vector geometries consist of individual latitude-longitude location coordinates. In most traditional machine learning applications that use geospatial polygon data, the geometry is not used directly. Instead, derived properties from the polygon are used. By far, the most popular strategy is rasterization of the polygons (Zhu et al., 2017b; Ball et al., 2017) so that polygons are projected to pixels with pixel values corresponding to the presence of polygons (e.g., value of 1 if the pixel belongs to the polygon, 0 otherwise. For the case with multiple polygons, either the absolute number of polygons (n_p) is chosen as the pixel value or, in cases where only polygon presence is semantically meaningful, the pixel value is set to min (1, n_p)). While rasterization is an obvious choice for applications that use computer vision-based techniques, there are some limitations that rasterization poses: (1) Geospatial vector data is highly versatile and may have complex shapes with varying levels of fine spatial details. Rasterization smoothens out some of the details, especially at the boundaries of polygons, and has a resolution of fixed and uniform size. (2) Since vector data is a set of coordinates, the size of the data does not scale with the area of the geometry and is therefore almost always more compact in comparison to the rasterized version of the polygon. Thus, the conversion of vector polygons into its rasterized version involves expansion of the data. (3) Operations such as intersection and spatial adjacency that are invariant to linear transformations on vector data may incur errors due to the smoothing resulting from the rasterization process. Some methods have been proposed to counter these problems such as preserving most of the polygon properties without rasterization (Veer et al., 2018) and methods derived from geometric algebra (Bhatti et al., 2020), among others. Despite its theoretical shortcomings, rasterization as a technique to represent vector data works well in practice, especially when the spatial resolution of pixels is high (e.g., an image with zoom-26 tiles as pixels will resolve vector polygons better than an image whose pixels are zoom-24 tiles).

Trinity uses the rasterization method for polygons since it conveniently transforms polygons into an image-like tensor. Pixels correspond to zoom-24 tiles (in typical Trinity applications, the input has size 256×256 with zoom-24 pixels corresponding to a spatial resolution of 2.4 m) and pixel values are obtained by projecting vector polygons onto the spherical Mercator projection of the earth's surface and rasterizing them based on the grid consisting of zoom-24 tiles. Pixels are assigned a value of 1 if the pixel belongs to the projected polygon and a value of 0 otherwise. Depending on the application, pixels may be assigned a value greater than 1 if it belongs to multiple polygons or the value is capped at 1. Figure 2.3a shows a schematic of the transformation of vector geometries into a rasterized image-like representation. As discussed in Section 2.4.1, the errors incurred by the Mercator projection increase near the poles since it distorts the relative size of landmasses thereby exaggerating the size of land near the poles more as compared to areas near the equator. These errors will propagate to the rasterized representation.

2.4.5 Temporal Transformations of Mobility Data

Mobility data is spatiotemporal. In Section 2.4.2, we explicitly modeled the spatial component of the data by aggregating the data over a given time interval $[t-\Delta t/2, t+\Delta t/2]$ thereby smoothing out the temporal variations. This section discusses techniques that explicitly model both the temporal and spatial variations. The design choice of whether both or either of the spatial and temporal variations need to be modeled depends on the application, cadence of change, and refresh latency of the geospatial feature under consideration. For example, for an application like monitoring mobility trends for public safety during pandemics, aggregating the temporal evolution over a suitable Δt (typically 1 day) in different geographic locations is sufficient. On the other hand, forecasting busyness of an area-of-interest (e.g., shopping mall) or point-of-interest (e.g., popular landmarks) (D'Zmura, 2020) requires modeling of granular temporal buckets (e.g., hourly activity forecasting) of mobility aggregated over the entire spatial extent under consideration.

The continued interest and research in techniques for temporal modeling of mobility data has been motivated by recent development of telecommunication networks, the growth in the number and type of sensors and Internet-of-Things (IoT) devices, and the increased fidelity of sensor datasets is producing an unprecedented wealth of mobility information. As a consequence, there has been an increasing interest in analyzing such data for public safety applications. Temporal trends in mobility play a crucial role in urban planning (Isaacman et al., 2011; Ahas et al., 2010), transportation network optimization (Song et al., 2010; Simini et al., 2012; Calabrese et al., 2010), demographic modeling (Soto et al., 2011), disaster management (Krisp, 2010), predicting stratification based on land-use (Soto & Frias-Martinez, 2011), crowd flow prediction (Cao et al., 2019a), environmental safety (Hatchett et al., 2021), and public health (Badr et al., 2020), among many other applications. A good review of the general techniques used in modeling the temporal trends in mobility data for computational transportation modeling is presented in (Campos-Cordobes et al., 2018). In most applications, sequences of counts or other features of mobility data bucketed over an application-specific time interval is modeled using classical time series methods such as ARIMA and GARCH (Shumway & Stoffer, 2000) to create predictive or forecasting models.

Given that Trinity aims to perform multimodal modeling, classical time series models provide no obvious way to create spatiotemporal features that can be used in the paradigm described in the previous sections (schematically described in Figure 2.3c). Trinity, therefore, adopts two different approaches to generate image-like representations for modeling temporal trends in mobility data. The goal of both these approaches is to learn the vector representation of each pixel as shown in Figure 2.3b. The first approach, temporal bucketing, is similar to the strategies described in Section 2.4.2. Constructively, assume that the observation time interval of mobility data is $[t, t+\Delta t]$. Bucket the interval into B buckets $[t, t+\Delta t/B]$, $[t+\Delta t/B, 2\Delta t/B]$, ..., $[t+(B-1)\Delta t/B, t+\Delta t]$. For each bucket, compute the motion-modality-specific CRM and concatenate in the bucket-wise temporal order along the channel dimension. This approach yields a $h \times w \times B$-dimensional image-like tensor where B is a user-specified parameter and may be treated as a hyperparameter for the specific downstream application. Advantages of CRM generation like embarrassingly parallel implementations on distributed, data-parallel frameworks like Apache Spark and ease of spatial alignment are inherited by this approach. Since CRMs have no bound on the dynamic range of the output, log-normalization of the output is carried out before fusing the artifact with other data modalities or as an input directly to a machine learning model. While this approach represents the temporal nature of the time series, the downstream model is tasked with inferring the spectral characteristics of the time series. The second approach, called spectral embedding, attempts to extract the information encoded in the frequency domain rather than the temporal domain. Reconsider the time series generated per pixel (zoom-24 tile) from the temporal bucketing approach. Using an integral transform method like Fourier Transform (implemented using the Fast Fourier Transform (FFT)) algorithm, a spectral vector representation is generated from a temporal vector. The spectral vector is used in one of three ways: (1) directly as the pixel representation with the size of the vector being the number of channels in the generated image-like tensor; (2) picking components in the vector that correspond to semantically meaningful frequencies (e.g., daily, weekly); (3) using highway networks (Srivastava et al., 2015) to learn a low-dimensional embedding from the spectral representation which is then used as the pixel representation.

2.4.6 Synthetic Generation of Geospatial Data Representations

Synthetic data is an increasingly popular tool for training deep learning models, especially in computer vision, but also in other areas (Nikolenko, 2021). The main trade-off with synthetic datasets is the accuracy of the generated data versus the size of the training data available to train downstream model. Historically, the biggest obstacle to adopting synthetically generated datasets for training models has been the reality gap—the minor differences between real and synthetic data (e.g., high-frequency artifacts, subtle systematic patterns in colors or pixel values in the generated data) that downstream models may learn to easily distinguish between real and synthetic data thereby harming the generalization capability of a trained model. On the other hand, synthetic data generation is usually much (sometimes order of magnitude) cheaper than getting labeled data for supervised model training especially for models trained to detect rare or infrequent events, especially in an industrial context (Nisselson, 2018). The most effective use case for synthetic datasets has been for pre-training models to perform transfer learning followed by fine-tuning for specific supervised tasks. In recent years, a surge in interest for better generation strategies of synthetic datasets has led to more sophisticated strategies such as GANs for domain adaption (e.g., CycleGAN (Zhu et al., 2017a) and others (Ganguli et al., 2019b) which translates satellite imagery to the hybrid layer of a map), domain randomization, adversarial losses on neural net activations to force synthetic/real feature invariance, among others. In other words, a promising strategy is to train models to generate synthetic samples whose feature representations are close to those of real samples thereby using synthetic data as complementary to real data. Synthetic data can also be used to generate non-identifiable datasets thereby helping train machine learning models with privacy guarantees and may be a tool in the privacy enhancing technologies (PET) toolkit. A thorough discussion of the applications of synthetic data to train deep learning models with privacy guarantees is provided in (Nikolenko, 2021).

The rest of the section describes one constructive approach of synthetic data generation in geospatial analysis that is adopted within the Trinity ecosystem to synthetically generate the representations of mobility data described in Section 2.4.2. Since mobility data can have sparse coverage in different geographic areas, models can be trained on geographies with good coverage and used to augment real data in geographies with sparse coverage. This is the main goal of the VAE-Info-cGAN model proposed in (Xiao et al., 2020). The VAE-Info-cGAN (combines a Variational Autoencoder(VAE) and an information maximizing generative adversarial network (InfoGAN) in a conditional setting) is a deep conditional generative model that simultaneously combines pixel-level conditions (PLC) and feature-level conditions (FLC), which are provided as inputs, to generate semantically rich synthetic images from a target distribution. The neural architecture of the VAE-Info-cGAN model is shown in Figure 2.4a. The training dataset for the deep conditional generative model comprises of pairs of the pixel-level condition (PLC) and the corresponding true sample from the target distribution obtained from diverse geographic locations. The diversity of locations is ensured by following an important sampling approach (discussed in Section 2.7.1.4). The model's training objective comprises of an adversarial component, an evidence lower bound (ELBo) component which is the sum of probabilistic reconstruction and divergence from a heuristic prior sub-components and a pixel-wise reconstruction component. The training objective is designed to incentivize the model to accurately generate synthetic samples from the target distribution conditioned on the pixel-level condition while also exposing control of the generation to the user to modulate the generated sample. The user-defined modulation can be either at the (microscopic) pixel level via the pixel-level condition or at the (macroscopic) feature level via the feature-level condition. Controlling the generation of synthetic samples is critical for industrial and real-world applications of generative models. Enhanced control over the generated samples affords the ability of targeted data augmentation of the training dataset. For example, synthetically adding seasonality variations, appending the dataset with junctions changed to roundabouts for a class-balanced dataset, targeted changes of known turn-restrictions to train a turn-restriction anomaly detection model, etc.

In Xiao et al. (2020), the proposed VAE-Info-cGAN is shown to outperform comparable deep conditional generative baselines including conditional Variational Autoencoders (cVAE) and conditional Generative Adversarial Networks (cGAN). The metric for comparison is the realism of the generated samples since the ground truth is available for both the training and testing datasets. The target distributions for which results are shown are the tasks of generating the count-based raster map (CRM) and the heading count-based raster map as the only conditional input. The multi-scale nature of images and the implicit mapping of the conditional inputs to the desired regions in the target distribution is exploited by the model by simultaneous conditioning on both the pixel-level and feature-level conditions. Since the latent space of the model is exposed to the user, controllable generation of diverse samples is possible. An example of this where the density of GPS traces is varied is shown in (Xiao et al., 2020). Since conditional inputs can be scalably modulated in a programmatic fashion and the forward pass of the trained models is fast (one forward pass of the inference computation graph for a batch of 32 examples takes (on average) 0.03s for CRM and 0.07s for HCRM on a NVIDIA Tesla V100 GPU), efficient sampling from the target distribution is possible.

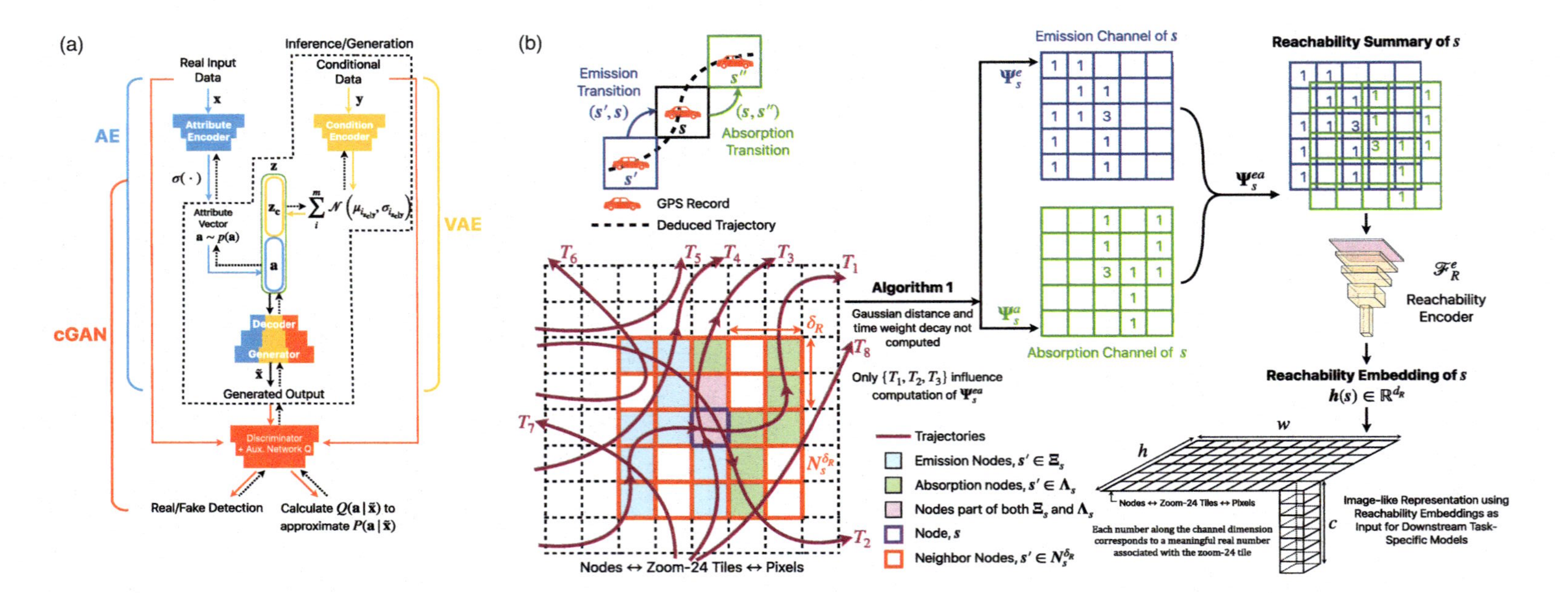

FIGURE 2.4 (a) A schematic of the neural architecture of the VAE-Info-cGAN model (Section 2.4.6) for generating synthetic versions of image-like representations of mobility data described in Section 2.4.2. (b) A schematic describing the process of generating reachability embeddings (Section 2.4.7). The self-supervision signal is provided by the task of contractive reconstruction of the reachability summaries.

Figures 2.2 and 2.3 in (Xiao et al., 2020) show visual comparisons of synthetic and real examples of CRM and HCRM.

The proposed VAE-Info-cGAN model allows for flexibility in the amount of information supplied as the conditional input. For example, in addition to the road network presence, additional information such as binary masks signifying turn-restrictions, direction-of-travel (e.g. one-way, two-way), form-of-way (e.g. pedestrian walkways, driveways, bike paths) can be concatenated to the input pixel-level condition to control the generated samples. Such controlled generation has multiple applications such as automated labeled training data generation, adversarial active learning for robust training of supervised models, data augmentation, and curated rare and extreme example generation, among others. In literature, many physics-aware or physics-informed deep generative models exploit the latent representation to supply critical physics-based constraints such as conservation laws and material properties. Analogously, applications in the geospatial domain can supply relevant real-world modeling information via the latent space such as traffic rules, traffic flow constraints, events and incident information, and seasonality parameters for more accurate and informative generation of samples.

2.4.7 Self-Supervised Representation Learning from Geospatial Data

Self-Supervised Learning (SSL) has been an area of active research and has achieved promising performance on both natural language processing (NLP) (Brown et al., 2020; Devlin et al., 2018; Mikolov et al., 2013a) and computer vision tasks (Chen et al., 2020a; Grill et al., 2020). Producing large labeled datasets with clean labels is expensive in time and money. Weakly-supervised techniques relax the requirement of clean labels but still require some labeled data. In contrast, self-supervised learning aims to utilize the easily available and large amount of unlabeled data to learn semantically meaningful, task-agnostic representations of the data. A common theme of self-supervised learning strategies is using predefined pretext tasks to derive supervision signals directly from unlabeled data by making neural networks predict with-held parts or properties of inputs. The learned representations can then be used by downstream (usually supervised) task-specific models to solve different applications. In other words, it decouples representation learning and task-specific modeling. SSL has successfully been used to learn context-independent (Mikolov et al., 2013b) and contextual (Brown et al., 2020; Devlin et al., 2018), task-agnostic word embeddings for NLP applications. Most popular SSL techniques for learning visual representations can be classified into two types: generative approaches that learn representations under the pretext of generating images by modeling the data distribution (Donahue & Simonyan, 2019; Kingma & Welling, 2013; Goodfellow et al., 2014), and discriminative approaches that use pretext tasks, designed to efficiently produce labels for inputs (e.g., based on heuristics (Doersch, Gupta, & Efros, 2015; Gidaris, Singh, & Komodakis, 2018; Noroozi & Favaro, 2016) or contrastive learning (Chen et al., 2020a; Grill et al., 2020)), coupled with a supervised objective.

There has been rapid progress on models and techniques for multimodal self-supervised learning (Bayoudh et al., 2022; Guo et al., 2019; Ngiam et al., 2011; Ramachandram & Taylor, 2017; Baltrusaitis et al., 2018) from combining subsets of natural language, visual, and audio signals given the relevance of jointly interpreting and reasoning from these signals in many practical applications such as voice assistants, robotics, and autonomous driving. These signals, although having immense intrinsic differences, are strongly correlated and unlabeled data for them is readily available, which makes cross-modal prediction a potentially rewarding SSL pretext task. Most multimodal training approaches use paired data (synchronized across different modalities) and one of two learning strategies: (1) separate feature extractors for different data modalities with different architectural choices for alignment and fusion of learned representations (Aytar et al., 2017; Alayrac et al., 2020; Akbari et al., 2021; Radford et al., 2021; Singh et al., 2021; Alwassel et al., 2020; Arandjelovic & Zisserman, 2017, 2018; Korbar et al., 2018; Owens & Efros, 2018; Owens et al., 2016), or (2) using a deep fusion encoder based on the Transformer architecture (Vaswani et al., 2017) with cross-modal attention across different data modalities (Tsai et al., 2019; Lu et al., 2019; Su et al., 2019b; Tan & Bansal, 2019; Chen et al., 2020b; Kim et al., 2021; Li et al., 2021; Wang et al., 2021). On a parallel front without performing multimodal training, recent works such as data2vec (Baevski et al., 2022) and Perceiver (Jaegle et al., 2021) aim to unify the learning objective and neural architectures used learning from different data modalities by systematically identifying and removing domain-specific inductive biases. Self-supervised multimodal learning has also been applied to other domains such as medical image analysis (Taleb et al., 2021), associating visual signals with touch (Lee et al., 2019), and improving perception in autonomous systems (Meyer et al., 2020), to name a few. Most of these methods can be adopted for multimodal geospatial datasets by transforming geospatial data into image-like tensors or sequential representations.

In this section, we briefly describe one example of a self-supervised strategy to generate representations of geographic locations from mobility data in the form of GPS trajectories that are adopted in Trinity. In Ganguli et al. (2021), a novel self-supervised method is presented for learning task-agnostic feature representations of geographic locations from GPS trajectories by modeling zoom-24 tiles as nodes of a graph (termed earth surface graph) and modeling the observed GPS trajectories as allowed Markovian paths on this graph. From the perspective of graph embeddings, the proposed algorithm learns node embeddings from observed paths in a graph; the nodes being geographic locations and the paths being observed GPS trajectories. There are two stages in the proposed algorithm as shown in Figure 2.4b. The first stage generates an image-like representation, termed reachability summary, for every node on the earth surface graph from the observed GPS trajectories. The second stage generates node embeddings (compressed, low dimensional representations) of these reachability summaries, called reachability embeddings, using a fully-convolutional, contractive autoencoder. For solving geospatial downstream tasks, the nodes of the earth surface graph (tiles in a raster representation) can be conveniently treated as pixels of an image and the generated node embeddings are treated as input features along the channel dimension of the image. Using reachability embeddings as pixel representations for five different downstream geospatial tasks cast as supervised semantic segmentation problems, (Ganguli et al., 2021) quantitatively demonstrate that reachability embeddings are semantically meaningful representations and result in 4%–23% gain in performance, as measured using area under the precision-recall curve metric, when compared to baseline models that use spatial transformation of mobility data described in Section 2.4.2. While generating reachability embeddings is more compute-intensive than the representations described in Section 2.4.2, reachability embeddings are global representations (i.e., they also take into account activity in the neighborhood of the tile) as opposed to the local representations described in Section 2.4.2 that only consider the activity in a given tile independently to generate its representation. Thus, for a given task at hand, there is a trade-off between the quality of the representation and the compute required to generate the representation.

For geospatial feature detection tasks like detecting the direction of travel on roads (e.g. one-way, two-way), detecting access points (e.g. locations where vehicles exit the road to enter a block), detecting turn restrictions on road intersections, detecting presence of road medians, detecting pedestrian crosswalks, understanding the flow (inflow and outflow) of traffic of various motion modalities (e.g. walking, driving, biking) at a given location is vitally important. Conditioned on a motion modality, traffic flow at a given location, ℓ, can be characterized by the frequency of transitions observed between ℓ and its neighboring locations, time taken and distance traveled during these transitions, and crucially the directionality of these transitions — neighbors reachable from ℓ and neighbors from which ℓ is reachable. Reachability embeddings are designed to capture how a given location is connected to locations in its neighborhood using observed GPS trajectories as evidence. Reachability embeddings are contextual embeddings by design, where the context comes from the observed GPS trajectories and may be different for different observation intervals.[1] Events like road construction and road closures will result in changes in observed trajectories and thus change the computed reachability summaries of the nodes in the vicinity of the event. This allows reachability embeddings to be good candidates as input features to models built for detecting changes in geospatial features.

Reachability embeddings are generated for every zoom-24 tile on the earth's surface where trajectories are present at a pre-defined cadence from GPS trajectory data and are stored in Trinity's feature store (refer Figure 2.5b). Zoom-24 tiles which do not have any GPS activity are assigned a zero-valued embedding. While reachability embeddings can have any dimension, Trinity uses 16-dimensional embeddings given the practical constraints of heterogeneous compute clusters having GPUs with varied memory. Reachability embeddings transform sequential, spatiotemporal mobility data into semantically meaningful image-like representations that can be combined with other sources of imagery and are designed to facilitate multimodal learning in geospatial computer vision. For instance, (Ganguli et al., 2021) demonstrate that combining reachability embedding representations of mobility data with road network presence (Section 2.4.3) and satellite imagery improves performance of three geospatial feature detection tasks cast as semantic segmentation problems irrespective of the chosen neural architecture.

2.4.8 Geospatial Imagery

The previous sections describe methods to represent non-imagery data as image-like artifacts via different transformations. In this subsection, we will take a quick tour of the different kinds of imagery datasets available for geospatial applications. One of the most common sources of imagery comes from the several imaging satellites orbiting the Earth. The satellites are typically operated by governments and private businesses. The second common source of imagery comes from images taken from different types of aircrafts—both manned and unmanned. These images are called as aerial imagery. Although both satellite and aerial images are taken from vantage points above the surface of the earth, the quality, resolution, frequency, and the post-processing can be different between the two sets of imagery. In general, since aerial imagery is taken much closer to the Earth's surface, the clarity and resolution can be higher. Also, since it is possible to take the aerial imagery using devices such as drones, they are much cheaper to get. On the other hand, since the satellites continuously orbit the Earth, it is easier to get multiple images at a pre-defined cadence. There have been several works in literature performing geospatial analysis on both satellite and aerial imagery and several applications have been developed for defense, remote sensing, and monitoring purposes. For instance (Mayer, 1999) focused on extracting buildings using aerial Imagery. More recently, (Ammour et al., 2017; Oghaz et al., 2019) have tried to use aerial imagery from Unmanned Aerial Vehicles (UAVs) for diverse use cases such as car detection and smart farming. Likewise, using satellite imagery, there have been diverse set of applications such as land cover and crop-type detection (Kussul et al., 2017), poverty mapping (Perez et al., 2019; Ganguli et al., 2019a), quantifying impact of policy decisions (Yeh et al., 2020), measuring road quality (Cadamuro et al., 2019), and building detection (Jin & Davis, 2005).

In addition to the three channels (red-green-blue RGB imagery), some systems also capture thermal, infrared, or multi-spectral imagery. Multi-spectral imagery consists of additional channels beyond the visual range and has also been used to extract features such as road networks (Doucette et al., 2004). In addition to imagery taken from different sources, recent advancements have made point cloud data extracted from LiDAR captured from both the aerial as well as surface vehicles an excellent source of highly accurate information (Morchhale et al., 2016). With the explosion in the computational capabilities, there has been an increase in the use of LiDAR point cloud data for geospatial applications such as building detection (Malpica et al., 2013).

2.4.9 Auxiliary Datasets and Data Provenance

In the prior sections, various techniques for representing different geospatial data modalities have been presented. These techniques can be extended to other auxiliary datasets such as density of social media posts bucketed by topic in each zoom-24 tile (similar to CRM), pixel values obtained from a digital elevation map for each zoom-24 tile, pixel values obtained as a vector from historical seismic activity for each zoom-24 tile, etc. While the possibilities are endless, the paradigm of transforming different data modalities into artifacts amenable by standard deep neural architectures can serve as a useful guiding principle.

For each of the generated data representations, data provenance refers to the lineage of how a specific dataset was produced (Di et al., 2013). In a machine learning platform, provenance enables reproducibility of experimental results and helps in debugging (Closa et al., 2019). In the context of dynamic geospatial platforms, data provenance is especially crucial due to the multitude of versions that exist for a specific type of dataset. For instance, satellites capture imagery at a specific cadence. Each version of the captured imagery can have different characteristics. For instance, imagery captured during stormy weather may have a higher cloud cover. This version may not be conducive to detect certain features on the surface of the earth. Tracking the version therefore becomes crucial to build a model that is effective. Although the example provided considered satellite imagery, similar scenarios exist for different versions of vector geometry data and mobility data as well. Provenance is typically handled by maintaining versions of code and the parameters used to generate the data. The initial source dataset is assigned a version and is considered immutable and accessible. From that point, the code and the configurations used for transforming the source dataset into features are captured by pipelines and are stored as provenance metadata in a database. This ensures reproducibility of data artifacts.

2.5 DESIGN OVERVIEW OF A GEOSPATIAL AI PLATFORM

2.5.1 Machine Learning Operations: MLOps

The rise in adoption of machine learning-based solutions in various industrial settings has been a reckoning for the field of software systems design leading to birth of the sub-field of machine learning systems design and MLOps (Huyen, 2022). Databricks (2021b) defines MLOps as "a core function of Machine Learning engineering, focused on streamlining the process of taking machine learning models to production, and then maintaining and monitoring them". While many practices from traditional software design seamlessly carry over to machine learning systems, the unique challenges and requirements of the development and deployment lifecycle of machine learning models compared to rule-based systems are addressed by MLOps (Treveil et al., 2020). Specifically, as elaborated in Gift and Deza (2021), MLOps involves building of a system that creates robust reproducible models that are easy to develop, deploy and validate. For geospatial machine learning platforms, certain additional considerations are required to ensure models are built, deployed, and maintained reliably and efficiently while dealing with the unique complexities associated with geospatial datasets. Such considerations are elaborated with the example of Trinity, which attempts to incorporate the concepts of MLOps into its different components. The rest of this section gives a brief overview of those components and they are described in greater detail in the upcoming sections.

2.5.2 Components of a Geospatial AI Platform

An AI platform typically consists of several crucial modules such as a feature management module, a feature store, a label management module, a data engineering workflow for experiment tracking and workflow management, a machine learning kernel, tools for evaluation and model selection, followed by a deployment management module for serving and continuous deployment of the models. Examples of such machine learning platforms that are general purpose and not specifically designed for geospatial datasets include TFX (Baylor et al., 2017), Azure (Chappell, 2015), and SageMaker (Joshi, 2020). On the other hand, platforms designed specifically for geospatial applications such as PAIRS Autogeo (Zhou et al., 2020) also consist of similar modules. We use Trinity, which is designed as a no-code geospatial AI platform, to describe to give an overview of the platform and interactions among the modules as described in Figure 2.5a. The following sections go into depth on the design of the different components. In Sections 2.6 and 2.7, we take a deeper look at the feature and label platforms respectively. Following that, in Section 2.9, we analyze the kernel to train as well as deploy the model. In Section 2.8, we cover the infrastructural components that underpin these components, and finally in Section 2.10, we perform a quick walkthrough through the different steps that a typical user follows in an experiment's lifecycle along with some real-world examples of applications from Trinity.

2.6 ML FEATURE MANAGEMENT AND FEATURE PLATFORM

A feature in machine learning is a part of the input to a model. For example, in the context of text analysis, bag of words can be considered as features. In a geospatial platform like Trinity, a feature is a measurable vector of values for a specific entity, usually a zoom-q tile as discussed in Section 2.4.1. As discussed in Section 2.4, features can represent transformation of a variety of input data. A feature platform is a data management layer for machine learning that allows users to create, ingest, store, retrieve, access, share and discover features and create effective machine learning pipelines (Hopsworks, 2018). In the following section, we will try to separate out the storage aspect of the platform (commonly called as a feature store) from the other aspects that are crucial to a machine learning system. There are a number of feature platforms prevalent in the industry. Some examples include Apple's Overton platform (Re et al., 2020), Hopsworks (Hopsworks, 2020), Uber's Michelangelo (Hermann & Balso, 2017), AirBnB's Zipline (AirBnB, 2020), Amazon's Sagemaker feature store (Amazon, 2017), Databricks's feature store (Databricks, 2021a), Bigabid (Zehori, 2020), Google's Feast (GO-JEK & Google, 2019), Salesforce's ML Lake (Levine, 2020), H2O AI feature store (H2O.ai, 2020), Iguazio feature store (Iguazio, 2021), and the Doordash feature store (Khan & Hassan, 2020), among others.

2.6.1 Why Do We Need a Feature Platform?

A feature platform facilitates reusability of channels among users. by enabling channels generated for one use case reused for other applications across multiple users and teams. Having access to multiple signals in the form of pre-generated channels paves the way for users to experiment with different combinations of channels. This in turn alleviates the need for the user to start generating features from scratch for every new experiment. Second, the feature platform decouples the generation from the usage of the features. As a result, users can work at a higher level of abstraction. So, the users developing an application can depend on scientists and engineers upstream to have scalable and efficient ways of generating features, feature storage, and provenance. Third, the feature platform enables rapid experimentation and prototyping that fuels faster development of models. Wherever possible, the platform enables the users to aggregate channels across a specified duration of time on the fly providing the flexibility to select the amount of data needed for their experiments. For example, a user can launch multiple experiments with different combinations of profiles for rapid experimentation to find an optimal solution faster.

2.6.2 Components of a ML Feature Platform

There are several key components that are required for a geospatial feature platform. A schematic of these components and their interactions at a high level of abstraction are shown in Figure 2.5b. While the specific implementation, business needs, and constraints of each component may vary across different platforms, the key aspects of each component are described below.

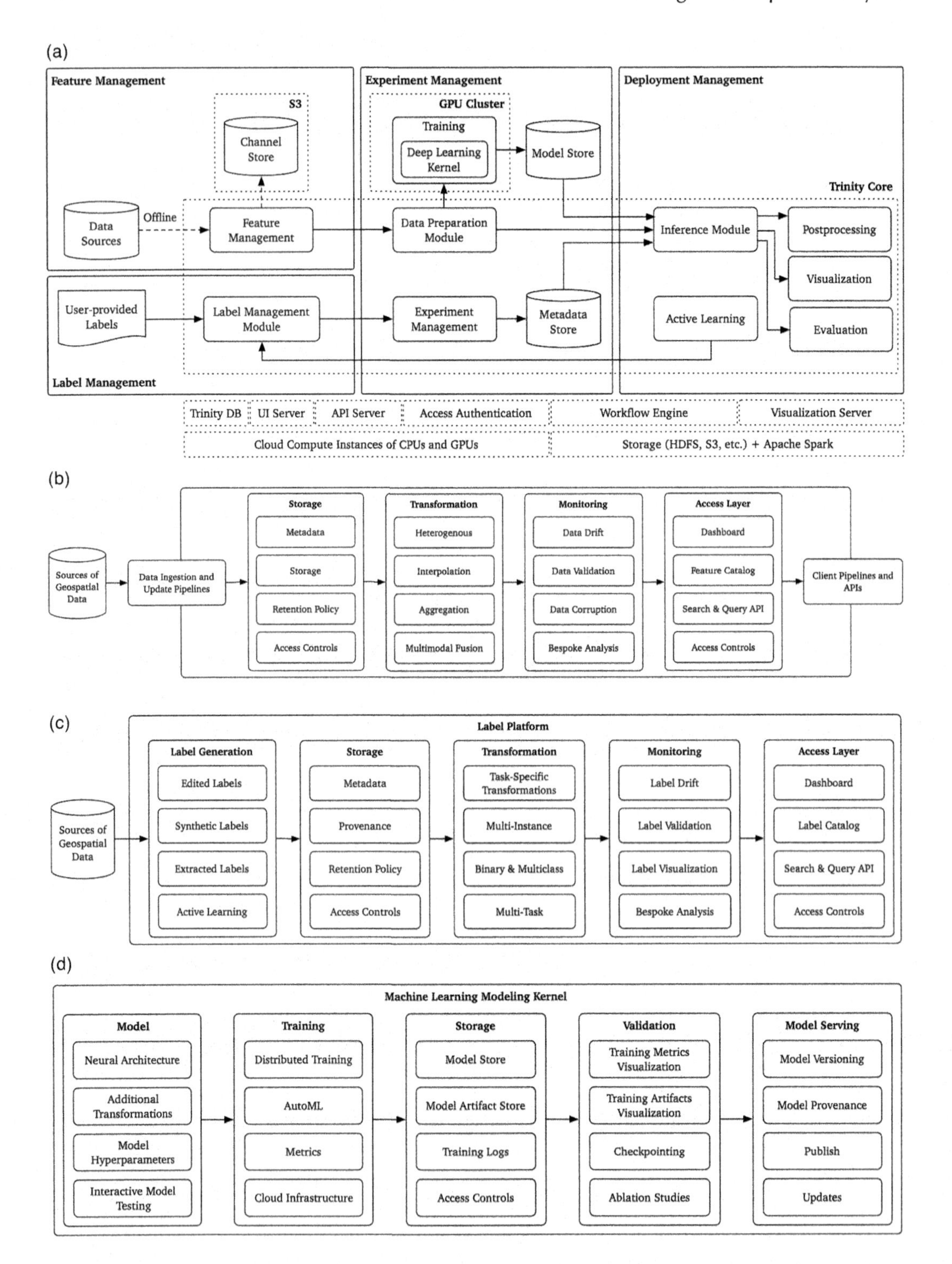

FIGURE 2.5 (a) A schematic of the various components of a data-centric geospatial AI platform like Trinity (Iyer et al., 2021). Schematic descriptions of the major components such as (b) the feature platform, (c) the label platform, and (d) machine learning kernel are also shown.

1. **Feature Generation and Ingestion**: Ease of feature generation and ingestion is one of the crucial aspects of any feature platform. A platform typically has a series of batch or streaming jobs that operates on some data sources in order to generate features that would be used in downstream machine learning applications. In the context of geospatial feature platforms, the features are usually (but not always) inherently spatial in nature. These upstream jobs can be simple ETL jobs (as is the case in platforms such as (Levine, 2020)). Alternatively, they may be streaming jobs as in (GO-JEK & Google, 2019). For instance in Trinity (Iyer et al., 2021), the geospatial features are computed with the keys as zoom-q tiles. The values for these tiles are feature representations as discussed in Section 2.4.
2. **Scalable and Persistent storage**: A scalable persistent storage is another crucial component of a feature platform. Platforms offer different storage decisions based on different considerations (see Section 2.6.3). Popular options for storage include S3-compatible stores like Sagemaker (Amazon, 2017), data lakes built on distributed filesystems (Databricks, 2021a) as well as spatial databases and key-value stores such as Feast (GO-JEK & Google, 2019). The choice of storage depends on the access patterns of the features. The storage layer handles not only the storage of the data but also of the metadata. Along with the storage of data, handling and cataloging the metadata is vital to ensure reproducibility of the features. Versioning of the features enables users of the platform to regenerate features. Thus, feature platforms need to store the lineage of the feature generation (also known as the provenance). As stated in Agrawal et al. (2019), tracking the lineage relationships between the different data versions and the model versions is very important to have a stable platform.
3. **Transformations**: A feature platform is expected to provide consistent ways of transforming features. One simple transformation is feature fusion, fusing features so as to enable stitching of disparate features representing multiple modalities together. Several techniques for fusing such features are covered in Section 2.4. Another transformation is aggregation across time. For example, if the traffic on a road segment is known for two consecutive days, the traffic for the 2 days combined can be deduced by a simple addition. However, for features generated with neural networks, special neural aggregators need to be designed. Furthermore, another type of transformation is to handle heterogeneous resolutions of the different features as we fuse them together. For example, a resolution of a few centimeters is very useful in the case of images but may result in sparsity issues when used with other mobility-based features if there isn't enough mobility data that is present in a given pixel for a chosen observation time interval. Hence the feature platform needs to provide ways to fuse these heterogeneous resolutions together.
4. **Monitoring and Feature Life Cycle Management**: A feature platform typically has modules to monitor the features at different layers of abstraction. Lifecycle management of features deals with validation of generated features, interpolation for populating missing features, and retirement of outdated features, among other aspects relevant to maintaining a given feature or set of related features. Over time, the statistical relevance of generated features for different downstream models changes (known as data drift) due to changes in the underlying data. The feature platform is expected to have ways to track the drift of a feature so as to be able to alert its downstream users if and when the drift occurs.
5. **Access Layer**: A feature platform is only useful if the users have the ability to explore and use the various stored features. Having an updated feature catalog that describes the features already present in the feature store is indispensable. In the context of a geospatial feature platform, a catalog also describes relevant aspects of the metadata of a feature such as geographic region of validity, time duration of underlying data, and spatial resolution of the feature. In addition, the access layer defines the access controls that cover both the

permissions of who has permissions to generate features and populate the feature store as well as the visibility of those features to other users. This is managed by an authentication mechanism at different levels of the feature store. Collaboration between stakeholders of a project can be facilitated by implementing efficient access control mechanisms.

2.6.3 Design Considerations for a ML Feature Platform

With the multitude of options available, there are several key design considerations that need to be taken into account and thought through with reference to business or project needs when designing or choosing an ML feature platform. Below, we highlight three main dimensions to guide this process that take into account available hardware or cloud computing infrastructure, intended cadence and latency of predictions, and model deployment.

1. **Online vs Offline**: A key consideration when designing or choosing a feature platform is whether it is expected to handle real-time (online) or batch queries. For an application with aggressive SLAs and needs to respond to queries interactively, an online feature platform is crucial. For a system that needs extensive aggregations and transformations and does not need to respond real time, an offline feature platform would be preferred.
2. **Point Queries vs Batch Queries**: Another dimension to be considered when designing a feature platform is the user access pattern. If the users are expected to use the platform in an interactive manner, the point queries would need to be supported by the platform. An example of such a point geospatial query would be to identify the location of a single point of interest (PoI). On the other hand, if the downstream ML algorithm requires the density of PoIs as a feature, that would necessitate the use of a batch query on the feature platform.
3. **Hosted/Cloud vs On-Premise Compute Infrastructure**: The other dimension of consideration is whether to use a hosted feature platform (such as Feast (GO-JEK & Google, 2019), Databricks (Databricks, 2021a) and Amazon Sagemaker (Amazon, 2017)) or to use a custom-built on-premises feature platform such as Overton (Re et al., 2020). This decision is based on the supported platforms, access patterns that are chosen to interact with the feature store, generalizability of the applications, and requirements of data privacy and security. For instance, some of the hosted general purpose feature platforms do not necessarily have support for geospatial queries due to which applications such as Trinity use a custom-built feature platform.

2.7 LABEL MANAGEMENT AND LABEL PLATFORM

Despite the promise of unsupervised and self-supervised models, most ML models in production are supervised and need labels to train (Oliver et al., 2018; Huyen, 2022). This makes the management of labels a core component of a ML system. The labeling platform is responsible for label generation and management. Labels used in a no-code platform can be represented according to the task at hand. For instance, for regression and object detection tasks, the labels are scalar values or tags, respectively. For segmentation tasks, the labels are geometries such as polygons or lines or buffered points. Objects like road centerlines, polygons of pedcrossings, and cycling segment tags (Wiki, 2022) can be cast as labels in a straightforward fashion while tasks such as turn restrictions need customized label representation. Although there are a variety of supervised ML models, in this section, we will be focusing on geospatial applications cast as semantic segmentation tasks. However, most of the points of discussion apply to other canonical computer vision tasks as well. One common way to generate labels for computer vision tasks from vector geometries is to transform the geometries into raster images. For example, in the raster, a pixel represents a zoom-24 tile and the raster label image has size 256×256 pixels—same as any of the input channels to the model.

2.7.1 Components of a Label Platform

Figure 2.5c describes the different components that are critical to a label platform irrespective of the specific implementations or business needs. Some key requirements and characteristics of these components are discussed below. For concreteness, we use a segmentation-based geospatial tasks as our target use case although the components mentioned here are generic enough to work for other types of tasks as well.

2.7.1.1 Label Generation and Editing

The most important component of a labeling platform is the label generation module. There are several ways of generating labels for a supervised learning task. Three key categories are manual, automatic, and synthetic label generation.

1. **Manual Labels**: One of the most common ways to generate labels is by using hand-generated labels and annotations. There are several labels ingest tools that make this task possible. Examples of commercial tools that help in label generation using hand labeling include ESRI's ArcGIS (ESRI, 2022) and iMerit (2021). Trinity uses an in-house labeling platform. Since the quantity and quality of labeled data are both critical, the label platform often has a crowdsourcing capability to enable quicker generation of labels. A popular example of such a platform is Amazon Sagemaker Data Labeling (Amazon, 2021) that integrates with Amazon MTurk Marketplace to enable fast generation of manual labels.
2. **Auto-Generated Labels**: For geospatial datasets, another source of labels is curated maps that already have the geospatial features coded in. For instance, for a sample task that aims to detect cycling road segments based on satellite imagery, one source of labels can be a coded map. An example of such an open-source map from which labels can be extracted is OpenStreetMap (OpenStreetMap, 2017).
3. **Synthetic Data Generation**: For some specific tasks, data can also be generated synthetically based on the tasks. An example of this in the medical domain can be seen in SinGAN-Seg (Thambawita et al., 2021). In the geospatial domain too, there have been several attempts at generating synthetic data. An example is Synthinel (Kong et al., 2020). One of the key reasons for generating synthetic data for labeling is to train anomaly detection models and to induce sufficient variability in the training data that may seldom be naturally occurring.

2.7.1.2 Label Visualization, Analysis, and Validation

Once the labels are generated in the label generation module as described in Section 2.7.1.1, the platform needs tools to visualize, analyze and validate the labels that are added. In Trinity, there is an in-house tool to perform visualization and analysis of the labels. General purpose machine learning platforms like TFX (Baylor et al., 2017) provide dedicated ways to visualize the statistics of the features partitioned by the label categories. In the context of geospatial platforms, there are tools like ArcGIS with similar capabilities. Trinity uses an in-house tool to visualize and analyze the labels that are generated and it is tightly integrated with the label generation process. The analysis of the labels is focused on four main aspects - correctness, consistency, completeness, and coverage.

1. **Correctness**: Correctness of labels answers the question of whether the labels capture exactly what is desired by the model and if all the labels are correct in the representation of the objective. For example, if the task requires the detection of road segments from satellite imagery, the labels should cover road segments, and not rivers or rail segments. Usually, correctness is handled by using redundancy. If there are human labelers, multiple users are required to label

the same instance and the aggregate labels are considered only if there is agreement between the labelers. If there is a disagreement beyond a specified threshold, the instance is bubbled up for review.

2. **Consistency**: Typically, labels are generated in a heterogeneous fashion by a large team of people. Since this is sometimes crowd-sourced, consistency ensures the labels are homogeneous in their representation. For instance, one human labeler may depict a road segment by a single line whereas another may depict the same by using a polygon with some buffer and width. Consistency in the labels is a crucial aspect to how well a downstream model can learn from the labels. Consistency can be ensured by proper instructions and clear guidelines provided to the labelers.
3. **Completeness**: A label is said to be complete if it does not omit any of the target features of interest in the provided area that is considered for labeling. Completeness along with correctness is critical for the downstream machine learning model to learn useful patterns for detection of target features.
4. **Coverage**: A label set is said to have good coverage if it covers all the different scenarios and distributions that would be encountered in the downstream applications. This ensures that the labeled data is unbiased and can lead to models that generalize well. In the context of Trinity, this is ensured by first stratifying the world into different categories called stratas as described in Section 2.7.1.4. Following this, the labels are generated so as to cover all or most of the different tiers.

2.7.1.3 Label Metadata and Catalog

In addition to validation and analysis of labels, it is imperative for the labels to also have key metadata attributes associated with them. For manually generated labels, metadata may include attributes such as the target region, the objective, the IDs of the annotators, and the reference datasets used to label. In the context of auto-generated or synthetic labels, metadata includes attributes such as the versions of the pipeline used to extract the label data, the constraints, and filters used to select the labels, and the versions of the source datasets. This metadata is crucial for users to be able to explore and understand the labels available on the ML platform and to be able to understand the implicit assumptions encoded in the generation of the labels to be able to leverage it for newer applications. Among other things, label analysis includes the ability to slice and dice the labels to verify the distribution of the labels by different attributes.

2.7.1.4 Stratification

Stratification is a process of dividing the surface of the earth into several tiers called stratas based on a variety of distinguishing attributes in order to have a lens of comparing different regions. A simple example of stratification is bucketing US counties based on population densities (Smith, 2017). The tiers can be computed at different granularities. For instance, Figure 2.6b shows the different tiers based on population density at a county-level granularity. In order to be globally uniform, one can use zoom-q tiles as the atomic unit at which the stratas are computed. Likewise, a platform can leverage different combinations of attributes to compute unbiased and global tiers. Using the different tiers as a common frame of reference, the training and prediction data distributions are measured for their similarities. If the training and test data distributions diverge, then there is an opportunity to retrain the model with updated labels in the tiers that are missing labels. This serves as an independent validation check when predicting if a pre-trained model would work effectively in a new region and to validate whether the data distribution of the test set matches the data distribution of the training set.

2.7.1.5 Active Learning

As elaborated in Cohn et al. (1996), Active learning facilitates a system to close the loop by selecting queries that enable systematic expansion of training dataset to progressively become better iteratively. In literature, active learning has been used with a variety of learning techniques from

Support Vector Machines (Tong & Koller, 2001) to Bayesian Convolutional Neural Networks (Gal et al., 2017). Rarely do generic ML platforms typically support active learning out of the box. However, in the context of the geospatial ML platforms, active learning assumes a greater role, since it provides for a systematic way to improve the quality of a downstream application model. Figure 2.6a provides an outline of an active learning loop. In Trinity, active learning is supported as a first-class feature of the platform. The active learning module in Trinity enables users to use their predictions as starting points to generate new labels. Once the user performs inference on a dataset, the active learning module selects the predictions that the current model was not confident on. These are candidates to be labeled by humans. This process iteratively improves the quality of the model over time as it enables labels to get better with in-line feedback from the users.

2.7.2 Design Considerations for a Label Platform

There are some key considerations that a labeling platform needs to handle in a typical machine learning application: representation of labels, class imbalance, and handling of labeling errors.

1. **Representation of Labels**: The labeling platform needs to have a representation that is compatible with the downstream supervised learning applications. In a generic machine learning-platform, labels can be strings representing tags or classes or sequences of strings or real numbers, in the case of regression tasks. In Trinity, as mentioned before, the labels are typically

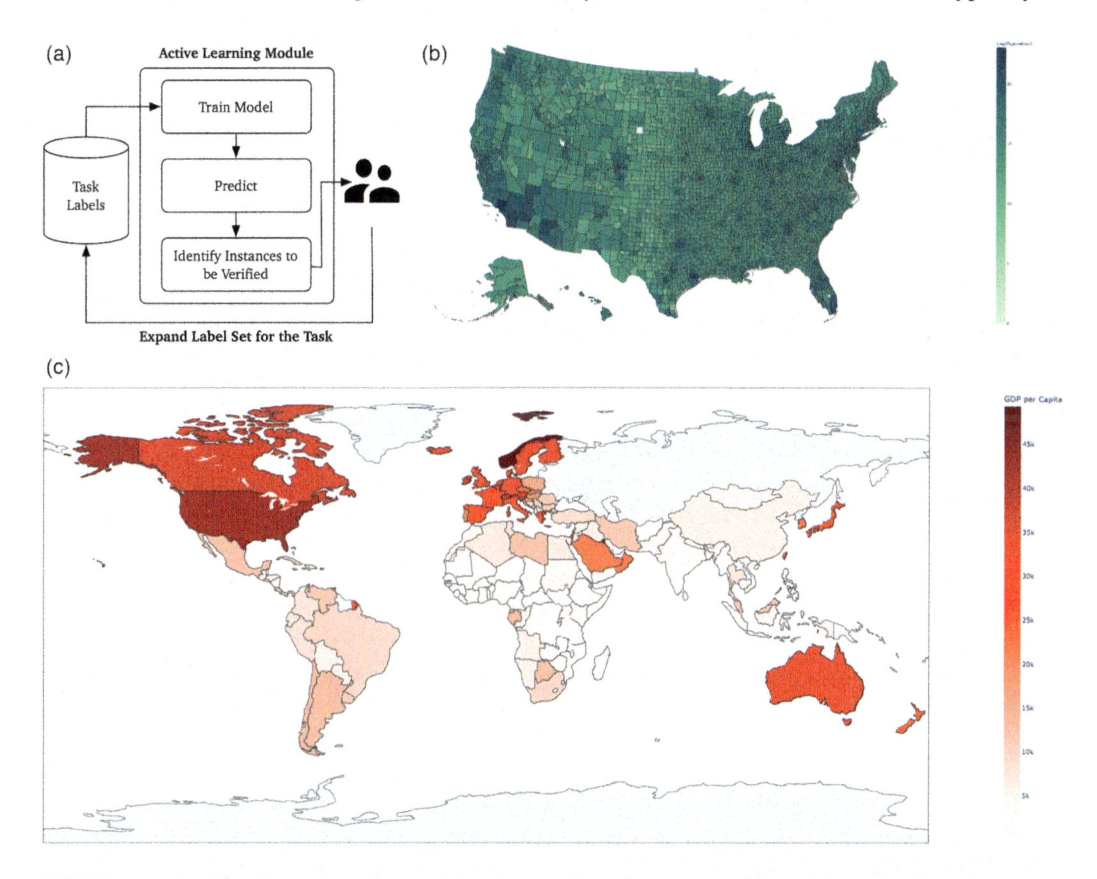

FIGURE 2.6 Active learning and stratified sampling are two methods for label improvement. (a) A schematic of the active learning loop. Shown are two examples of criteria for geographic stratification: (b) US counties based on logarithm of population (based on US Census 2010) and (c) Countries of the world based on GDP per capita in US dollars in 2007.

represented as annotated geometries for the purpose of training segmentation models. The usage and processing of labels in the downstream pipelines is elaborated upon in Section 2.8.

2. **Class Imbalance**: Class imbalance is a core issue that can heavily impact a machine learning model, especially in the geospatial domain. There are several techniques in literature to handle class imbalance (Galar et al., 2012), but the first step is for the platform to notify the users that there is indeed a class imbalance issue. Measuring data distributions independently using relevant stratification methodologies as reference is one possible solution. We delve into Stratification in the Section 2.7.1.4.
3. **Labeling Error Management Strategies**: The platform also needs to provide a strategy to identify and eliminate incorrect labels, so that they do not propagate in downstream applications. In literature, some of the techniques used to mitigate this effect are majority voting (Tao et al., 2018) and consensus labeling (Tang & Lease, 2011) using multiple labelers to tune out the bias.

2.8 MACHINE LEARNING INFRASTRUCTURE COMPONENTS

In order to implement the aforementioned modules, infrastructural pieces need to be engineered. In this section, we take a quick look at some of the commonly selected options. In Sections 2.8.1 and 2.8.2, we look at the some popular options and key considerations for data processing, orchestration, and storage infrastructure. These are not exhaustive and the reader is encouraged to take a deeper look into the associated references. In the context of Trinity, infrastructure comprises of several layers as described in Figure 2.7a.

2.8.1 Data Processing Framework

A typical large-scale AI platform needs pipelines to process, aggregate, transform and visualize large-scale data. There are several platforms and big data frameworks (Inoubli et al., 2018) that are used to process large-scale datasets. For example, TFX uses Tensorflow and Python with Google Cloud for compute, storage, and orchestration. In the context of a geospatial datasets, a popular option is ArcGIS (Johnston et al., 2001) that works on datasets that are small enough to fit in the RAM of a machine. There are also cloud variants (Kholoshyn et al., 2019) that provide ability to process larger data remotely. However, for much larger datasets, one of the more popular frameworks includes Apache Spark (Zaharia et al., 2016). Other potential frameworks include Apache Flink (Carbone et al., 2015). In Trinity, the large-scale data processing is done using Apache Spark, and orchestrated on proprietary workflow engines.

Selection of a data processing framework has some key considerations. The first consideration is the scale of data. For smaller datasets, frameworks such as ArcGIS would suffice. However, for datasets that are larger, usage of Spark is preferred (Lunga et al., 2020) for easy scalability. The second consideration is that of the storage ecosystem. A processing framework that can seamlessly interact with the storage platform is generally preferred. For instance, Spark on Yarn can seamlessly interact with labels, and features are stored in platforms based on S3 like Object stores or Hadoop. Third, the access patterns need to be considered to determine the best data processing framework. If the platform is intended to be more analytical, data access is done in batch, involves massive amounts of data and not in real time, then frameworks such as Spark may be a good candidate. On the other hand, for quicker, interactive, near real-time processing of small amounts of spatial data locally, tools such as the FME Workbench (Boyes et al., 2015) work well.

2.8.2 Storage, Compute, and Metadata Handling

In addition to the data processing layer, storage and compute are additional pieces in a geospatial AI platform. Storage decisions govern how the different artifacts are stored, versioned, retrieved,

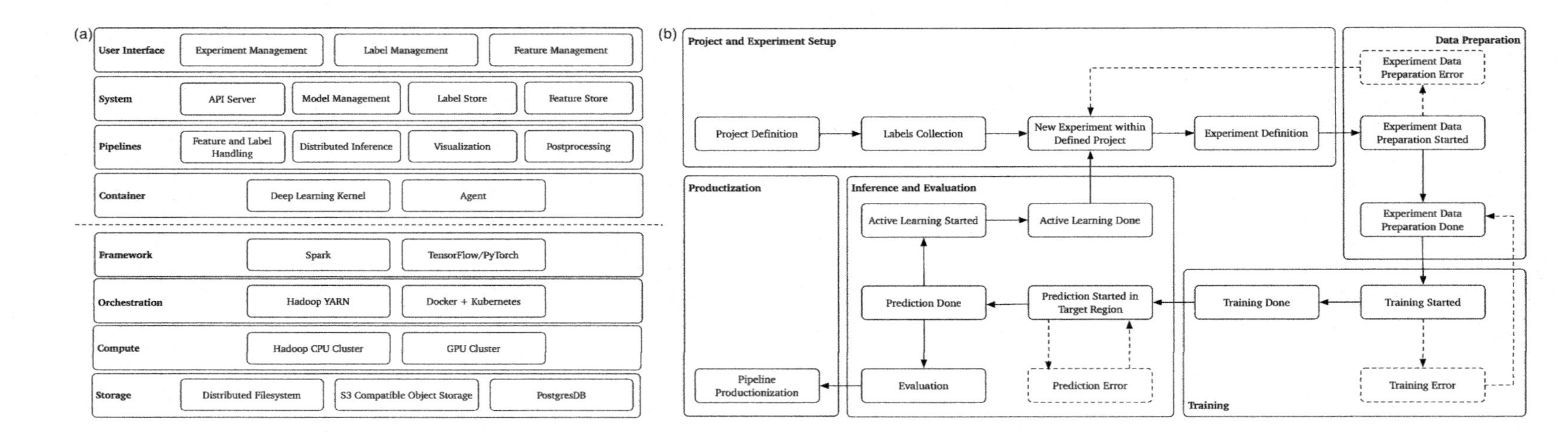

FIGURE 2.7 Schematic description of the (a) infrastructure layers for experiment management in Trinity (Iyer et al., 2021) and (b) the life cycle of Trinity experiments.

and managed while choice of the right compute infrastructure is crucial in meeting the desired SLAs. Data artifacts here include models, labels, and features, along with their respective metadata. Starting with the storage layer, Trinity is designed to work with multiple storage implementations. While its feature store is maintained in an S3 (Simple Server Storage) compatible storage, the intermediate data, inputs, and processed predictions are stored in a distributed file system such as HDFS (White, 2012)). For compute, Trinity uses GPU and CPU instances hosted on internal compute clusters. Third, the training is containerized using Docker and orchestrated by Kubernetes. Large-scale distributed predictions are carried out on CPU clusters orchestrated by Yarn and the jobs are implemented on Apache Spark. As of this writing, Trinity uses Tensorflow as the deep learning framework. Trinity uses Spark on Yarn for preprocessing of data, and channels and labels, as well as distributed inference, visualization of predictions, and post-processing of results for downstream applications. The experiment management module has an API server that runs on an internal cloud and offers end points to perform authentication, job submissions for training and predictions, and status checks. All of the experiment metadata, including the channels used, the network architecture used, predictions, and pointers to the resultant model files are stored in a database. The data flow pipelines are responsible for transformation of the labels and the channel data are described in Section 2.4.

2.9 MACHINE LEARNING MODELING KERNEL

The kernel is at the core of a typical ML platform and encapsulates neural net architectures, provides for model training, evaluation, handling of metrics, and inference. General purpose platforms typically use a dedicated kernel where users can code their ML models. For instance in TFX (Baylor et al., 2017), the kernel is implemented in Tensorflow. General purpose no-code AI platforms like Azure ML Studio (Chappell, 2015) provide the users with a plethora of standard ML algorithms for regression and classification. Figure 2.5d describes the key components of a typical machine learning kernel in a platform. The five pillars are model creation, training, storage of the models and artifacts, model performance validation and monitoring, and serving and update of the models. Most of the general purpose platforms take care of each of these aspects albeit with different specific implementations. For instance, for model creation, training, validation, hyperparameter optimization, and data transformations, TFX uses Tensorflow while Sagemaker uses MXNet in addition to specific model-building pipelines. Other options for scaling such infrastructure is by using Docker containers within a Kubernetes-based orchestration system. For data and model storage, TFX as well as Sagemaker provides a variety of options including S3-based stores such as Amazon S3 and Google Cloud. For serving and deployment of models, most platforms provide support for a microservice-based infrastructure where an endpoint is deployed on a container. One of the key choices in designing a geospatial ML platform is to identify and scope the types of tasks that it would support. In literature, some of the commonly addressed tasks in the context of geospatial computer vision include classification (Minetto et al., 2019), regression (Kumar et al., 2018), semantic segmentation and instance segmentation tasks (Su et al., 2019a), and object detection (Cheng et al., 2014) tasks.

In Trinity, the kernel houses deep learning models and is implemented in TensorFlow but is flexible enough to be modified for using other deep learning frameworks. The kernel provides multiple segmentation architectures such as FCNs (Long et al., 2015), SegNet (Badrinarayanan et al., 2017), UNet (Ronneberger et al., 2015), Tiramisu (Jegou et al., 2017), among many others. Users are able to select architectures from a catalog for any given task. This module is modular and extensible and new architectures can be added and evolved independent of the rest of the system. The platform is designed to support a variety of tasks including binary and multi-class segmentation. Users can formulate variety of use-cases as segmentation tasks. For instance, detection of the road centerlines (Figure 2.10b), detecting residential areas (Figure 2.10a), and detecting one-way roads (Figure 2.10c), among others (Figure 2.10) can be posed as binary segmentation tasks. Likewise, the multi-class tasks aim to detect different class tags of the target feature. For instance, type-of-road detection as depicted in Figure 2.10e that extracts different types of roads such as alleys, freeways, and ramps is a multiclass task. Additional examples are provided in the 10.5. Furthermore, the Trinity platform enables users to combine several binary and multiclass tasks into a single model. Such a task is called as multi-task segmentation (also called multi-label segmentation). Specifically, in such tasks, each pixel is associated with multiple labels to depict each task. As described in earlier sections, of these tasks are based on multi-modal learning for mixing different modalities of data. For instance, users can combine satellite imagery with additional signals to predict buildings, pedestrian crosswalks, and road signage more accurately. An extension of segmentation can be both instance-level (e.g., using Mask-RCNN) and semantic.

For the training models, the kernel code is packaged in a docker container and the training is launched on a GPU cluster. Different metrics on training and validation datasets such as per-task accuracy, precision, recall, loss, and fIoUs are logged during the training phase and visualized on a dashboard. Additionally, there's support for warm-starting training using pre-trained models for transfer learning. The kernel is packaged with all dependencies inside a virtual environment and runs on a Spark cluster during inference to leverage data locality and distributed inference using massive CPU-based computer resources. The inference phase is detailed in Section 2.10.3. The Trinity kernel also supports hyperparameter optimization, model selection, and channel selection as first-class features of the platform. It enables users to select the best neural architecture for the given task, perform ablation studies with different channels as well with the different hyperparameters such as learning rate, weight decay, and training batch size. Multiple batches of trials are launched by the system. The parameter selection is performed using Bayesian search to select the hyper-parameters that optimize the metric specified by the user. The rest of the section focuses on the serving of the models after they are trained and finalized.

2.9.1 Serving and Deployment of Trained Models

In a machine learning platform, model serving and deployment represent the stage where the models are taken to production. To begin with, there are several approaches that are generally followed in order to deploy a model. First, the Embedded model approach where the model is embedded within an app or a microcontroller that uses the localized model to perform inference on the fly (Warden & Situnayake, 2019). Next, there is a Model-as-a-Service (MaaS) approach where the model is typically served using a dedicated endpoint and deployed as a service, typically using an orchestration framework such as Kubernetes and containerization framework such as Docker and using a RESTful client. The advantage of this approach is that this caters to both real-time and batch queries and the service can be scaled by simply adding additional resources. Finally, there is a massively parallelized batch mode. In the batch mode, the model is embedded within a large-scale data processing framework such as Apache Hadoop or Spark and works on a large cluster of nodes and predicts on a massively large amount of data in a parallel fashion.

Different ML Frameworks adopt and encourage Trinity uses the batch model as the default approach to serving models although the models are flexible enough to be exported as a service for near-real-time applications. The active learning module ensures that the outputs generated can be used to generate new labels for training and the resulting models can be deployed again thereby providing a continuous deployment loop. This requires that the platform enable a A/B testing module to ensure that the new model does not suffer from data drift or model decay. The older models are then retired but still persisted for reproducibility, comparison, and project progress documentation.

2.10 TRINITY EXPERIMENT LIFECYCLE

We define the experiment lifecycle as the typical workflow that is followed by the user to go from an idea to a deployed product. Machine learning-based products are inherently different from the software engineering products. One key reason for that is that in machine learning, the data is also as important as code and needs to be tracked with the same diligence. Another key difference is that the ML workflows—geospatial or otherwise—typically involve a lot of experimentation and it is vital to be able to track, reproduce, and maintain these different experiments as part of the platform for effective deployment. There have been several open-source tools such as MLFlow (Zaharia et al., 2018) and FBLearner Flow (Dunn, 2016) geared toward managing the machine learning lifecycle better (Mackenzie, 2019).

The typical life cycle for a Trinity experiment (e.g., for detecting a geospatial feature) is described in Figure 2.7b. The users start with an objective geo-feature to be detected. Based on that, Trinity provides them with tools to collect or hand-annotate labels and then catalog them. Once the labels are generated, the user can define an experiment using an intuitive user interface by selecting the relevant channels and neural net architectures from a catalog using just a few clicks. The user can train the model once the experiment is defined. After the model is trained, the model can be used for several inference tasks from the platform. All the experiments and predictions made from Trinity are tracked and are always searchable and reproducible. The profiles, labels, metadata, models, and artifacts are persisted in their respective stores and are searchable via a catalog. The platform generates metrics for the training and predictions, and the active learning module helps the model use human expertise judiciously to obtained additional labels that can be used to iterate on the model. Finally, the model is exported to a product. Below we walk through each of these steps in further detail.

2.10.1 Project and Experiment Setup via the User Interface

A simple and intuitive user interface is crucial to the success of a machine learning platform and assumes a greater importance if the platform is designed as a no-code/low-code platform. In Trinity, this is implemented by presenting the users with a clear set of options for each of the three inputs that are required to start the experiment to initiate the experiment lifecycle described in Figure 2.7b. The three inputs are labels, choice of neural-net architecture, and choice of channels with clear descriptions of what the options are intended for. The data ingestion pipelines and the feature store perform alignment and fusion of different data modalities so that the details are abstracted away from the user. In other words, all of the design principles described in Sections 2.6, 2.7, and 2.9 pay rich dividends while interacting with platform users. Advanced options are also available for users with advanced knowledge of machine learning and deep learning techniques to perform power-user tasks such as channel selection, hyperparameter selection, addition of new neural architectures, and advanced settings for training objective optimization. A sample experiment view is depicted in Figure 2.8b. Upon the completion of training, with just a few clicks of buttons, the user can use the trained model to predict on any new region selected using a map tool as shown in Figure 2.8c and also a provide reference golden dataset in the prediction region that is used to generate various performance metrics (e.g., precision-recall curves as shown in Figure 2.8d, statistics such as f-IoUs, accuracy, and F1-scores) to evaluate the trained model. Evaluated models can be compared with each other directly thereby enabling faster model selection. All of the metrics are generated for each class for each task in the general case of a multi-class, multi-task trained model.

2.10.2 Data Preparation and Training

The data preparation module consists of pipelines for preparing data for training and inference phases. In Trinity, these pipelines are responsible for tasks such as label extraction and handling, channel extraction, data transformation, and image-like artifact creation. They are implemented using Apache Spark and run using an in-house workflow management engine. The end product of this module is a dataset that can be directly ingested by the kernel for training and prediction. The channel normalization and aggregation is done on the fly as part of the data preparation module. An

FIGURE 2.8 Screenshots of (a) the Trinity (Iyer et al., 2021) experiment definition UI, (b) selection of a target region for prediction using a trained model, and (c) an auto-generated precision-recall curve for a sample use-case.

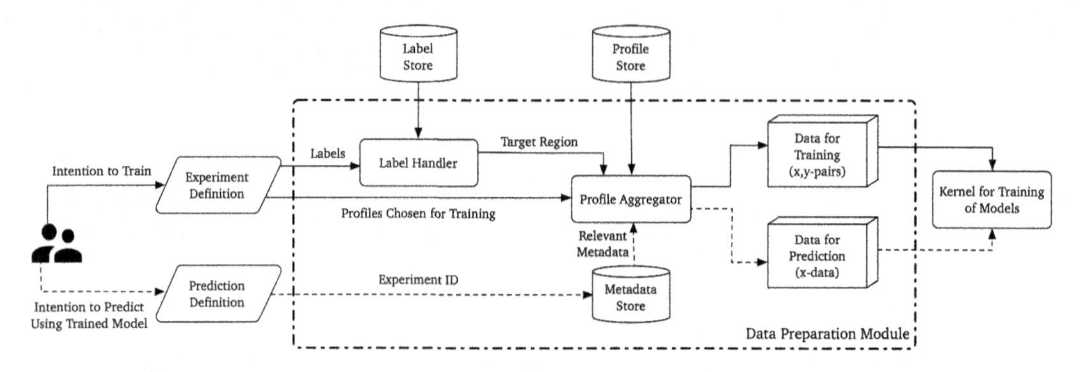

FIGURE 2.9 Data preparation module in Trinity (Iyer et al., 2021).

overview of this workflow is depicted in Figure 2.9 where the solid lines depict the training flow and the dashed lines represent the prediction flow.

Once the data preparation is done, the model training is carried out. The user is provided with options in the UI to select (either manually or automatically) the network architecture and other hyperparameters. The resultant model is persisted in a model store and used in all of the predictions based on this experiment. The training step also enables the users to perform hyperparameter optimization as well as perform ablation studies on the channels using Bayesian search (Wu et al., 2019). The training step also persists checkpoint models so as to be able to restart from an intermediate model. For every epoch, it also publishes metrics such as precision, recall, and IoU for every task and class for quick monitoring of the progress.

2.10.3 Scalable Distributed Inference

Once the models are trained, the user uses the prediction UI to specify the target region where the predictions are required. Multiple predictions on different areas can be launched from the same experiment simultaneously. Trinity uses distributed prediction with Apache Spark to use trained models on new and unseen datasets in a scalable fashion. Here, the natural partitioning of data and data locality is used to partition the data into several nodes. The inference is parallelized across the nodes using Tensorflow-based prediction code and the results are stored in a distributed file system. For a semantic segmentation task, this inference workflow generates a series of confidence heatmaps for each task.

2.10.4 Visualization and Evaluation of Predictions

Once the inference phase is complete as specified in Section 2.10.3, Trinity automatically generates visualizations of predictions as raster layers. These heatmaps serve as a quick sanity check to help the users check the quality of results. For example, Figure 2.10f shows the example of binary segmentation carried out for detecting pedestrian crossings. For multi-class segmentation tasks, the dominant class for each pixel is displayed in the visualization.

Along with the prediction, workflows for clustering and post-processing of results are also provided by the platform for easy deployment of the results. Several strategies for post-processing are supported including vectorization, which is the clustering of the predictions into vector geometries. In addition to vectorization, Trinity also supports density-based clustering algorithms such as weighted DBSCAN, as well as matching of predictions to the geo-features already on the map in a process called Map Matching (Chawathe, 2007). The prediction artifacts resulting from post-processing are used in downstream pipelines in different ways, viz. for feature detection, anomaly detection, and prioritization. Each of the above steps can be repeated for several different experiments, testing various hypotheses with the help of different channels,

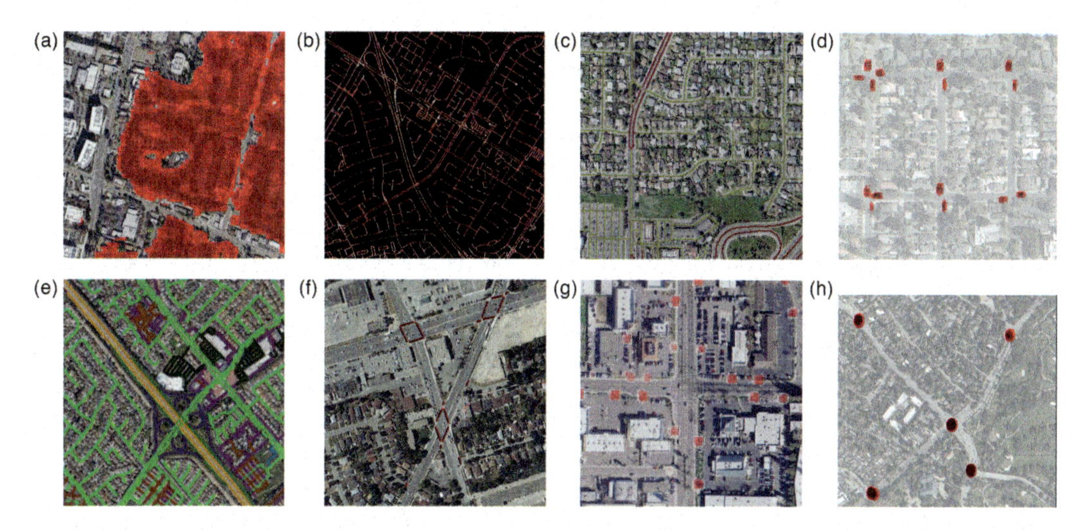

FIGURE 2.10 Visualizations of predictions from multimodal models trained on Trinity (Iyer et al., 2021) for eight different geospatial feature detection tasks. Predictions are visualized for (a) detection of residential areas using only mobility-based channels, (b) predicted road center lines using only mobility-based channels, (c) detection of one-way roads using only mobility-based channels, (d) detected stop sign locations using only mobility-based channels, (e) multiclass road type detection using only mobility-based channels, (f) detection of pedestrian crosswalks with imagery and mobility-based channels, (g) access point detection using only mobility-based channels, and (h) detected traffic lights locations using only mobility-based channels. In tasks that do not use satellite imagery, the imagery is shown only for reference.

label representations, and neural net in against golden datasets by the users using quantitative and qualitative metrics in order to select the best model for the given use case. This process is called as model selection. We use precision, recall, F1-score, and Intersection-Over-Union as the default metrics for comparison. Such metrics are associated with the model as model metadata and comprise part of the model's performance. MLFlow (Zaharia et al., 2018) affirms that one of the key aspects of MLOps in a platform is to enable tracking of experiment performance so as to enable reproducible results. In Trinity, model selection is enabled by using the multitude of metrics generated by the kernel for every model. The best model is then deployed as a product using one of the approaches detailed in Section 2.9.1. In Section 2.10.5, we take a look at the different types of products that are deployed and visualization of a few sample applications.

2.10.5 Product Types and Sample Applications

Several categories of geospatial products for web mapping applications that are possible based on the predictions depending on that goal at hand. First, we have reference layers of heatmaps that aid human editors with their decisions. Second, there are anomaly detectors where geo-feature prediction models trained on Trinity are used to detect anomalies in the existing map. The third category of products is prioritization filters to prioritize or filter other signals. For example, for certain applications, commercial areas may need to be prioritized warranting the use of a commercial area detection model. Fourth, the predictions are used to evaluate different data sources and vendor datasets. The visualizations of some reference models for 8 different selected sample applications are provided in Figure 2.10a–f.

2.11 CONCLUSIONS

This chapter presents an overview of the challenges and opportunities for joint multimodal modeling of various geospatial data modalities to solve important applications in geospatial analysis and remote sensing. It also motivates the need for a geospatial AI platform that can benefit domain experts, scientists, and engineers alike. Various strategies for modeling geospatial data modalities are discussed keeping in mind the needs of tasks that need unimodal inputs as well as multimodal models that need data representations that can be aligned and fused. The design of a geospatial AI platform is discussed with Trinity as a case study. This chapter lays out one possible template to build wide and deep geospatial AI platforms by narrowing down on suitable AI techniques after appropriate signal transformation and intelligent problem formulation, bringing them to the fold of canonical deep learning techniques in computer vision. This enables the platform to benefit from rapid advancement of computer vision tools and techniques in industry and academia. The progress in AI holds great promise for solving important problems in geospatial analysis and remote sensing. However, real democratization of AI happens when domain experts are technically empowered to solve intricate problems on their own without being encumbered by technical challenges. A platform like Trinity can help abstract out challenges with data processing, information and knowledge extraction, algorithms, software frameworks, and hardware dependencies thereby allowing wider penetration of AI-based solutions to geospatial applications.

NOTE

1 One can analogize this to the advantage of word embeddings in NLP produced by contextual language models like BERT (Devlin et al., 2018) over fixed embeddings from word2vec (Mikolov, 2013a, b), since a word's embedding is based on its location and context in a sentence.

REFERENCES

Acs, G., & Castelluccia, C. (2014). A case study: Privacy preserving release of spatio-temporal density in Paris. In *ACM SIGKDD* (pp. 1679–1688).

Agrawal, P., Arya, R., Bindal, A., Bhatia, S., Gagneja, A., Godlewski, J., … Wu, M.-C. (2019). Data platform for machine learning. In *ICMD* (pp. 1803–1816).

Ahas, R., Silm, S., Jarv, O., Saluveer, E., & Tiru, M. (2010). Using mobile positioning data to model locations meaningful to users of mobile phones. *JUT*, 17 (1), 3–27.

AirBnB. (2020). Zipline. Retrieved from https://airbnb.io/projects/.

Akbari, H., Yuan, L., Qian, R., Chuang, W.-H., Chang, S.-F., Cui, Y., & Gong, B. (2021). VATT: Transformers for multimodal self-supervised learning from raw video, audio and text. In *NeurIPS*, 34.

Alayrac, J.-B., Recasens, A., Schneider, R., Arandjelovic, R., Ramapuram, J., De Fauw, J., … Zisserman, A. (2020). Self-supervised multimodal versatile networks. In *NeurIPS*, 7.

Alwassel, H., Mahajan, D., Korbar, B., Torresani, L., Ghanem, B., & Tran, D. (2020). Self- supervised learning by cross-modal audio-video clustering. In *NeurIPS*, 33.

Amazon. (2017). Amazon sagemaker feature store. https://docs.aws.amazon.com/sagemaker/latest/dg/feature-store-getting-started.html.

Amazon. (2021). Amazon sagemaker data labeling. https://aws.amazon.com/sagemaker/data-labeling/.

Amer, M. R., Shields, T., Siddiquie, B., Tamrakar, A., Divakaran, A., & Chai, S. (2018). Deep multimodal fusion: A hybrid approach. *IJCV*, 126 (2), 440–456.

Ammour, N., Alhichri, H., Bazi, Y., Benjdira, B., Alajlan, N., & Zuair, M. (2017). Deep learning approach for car detection in UAV imagery. *Rem. Sen.*, 9 (4), 312.

Arandjelovic, R., & Zisserman, A. (2017). Look, listen and learn. In *ICCV* (pp. 609–617).

Arandjelovic, R., & Zisserman, A. (2018). Objects that sound. In *ECCV* (pp. 435–451).

Atrey, P. K., Hossain, M. A., El Saddik, A., & Kankanhalli, M. S. (2010). Multimodal fusion for multimedia analysis: A survey. *Multimedia Syst.*, 16 (6), 345–379.

Aytar, Y., Vondrick, C., & Torralba, A. (2017). See, hear, and read: Deep aligned representations. arXiv:1706.00932.

Badr, H. S., Du, H., Marshall, M., Dong, E., Squire, M. M., & Gardner, L. M. (2020). Association between mobility patterns and COVID-19 transmission in the USA: A mathematical modelling study. *Lancet Inf. Dis.*, 20 (11), 1247–1254.

Badrinarayanan, V., Kendall, A., & Cipolla, R. (2017). Segnet: A deep convolutional encoder-decoder architecture for image segmentation. *IEEE TPAMI*, 39 (12), 2481–2495.

Baevski, A., Hsu, W.-N., Xu, Q., Babu, A., Gu, J., & Auli, M. (2022). data2vec: A general framework for self-supervised learning in speech, vision and language. arXiv:2202.03555.

Ball, J. E., Anderson, D. T., & Chan Sr, C. S. (2017). Comprehensive survey of deep learning in remote sensing: Theories, tools, and challenges for the community. *JARS*, 11 (4), 042609.

Baltrusaitis, T., Ahuja, C., & Morency, L.-P. (2018). Multimodal machine learning: A survey and taxonomy. *IEEE TPAMI*, 41 (2), 423–443.

Barnes, J. A., & Harary, F. (1983). Graph theory in network analysis. *Social Networks*, 5 (2), 235–244.

Barnum, G., Talukder, S., & Yue, Y. (2020). On the benefits of early fusion in multimodal representation learning. arXiv:2011.07191.

Barthelemy, M. (2011). Spatial networks. *Phys. Rep.*, 499 (1–3), 1–101.

Barthelemy, M., & Flammini, A. (2008). Modeling urban street patterns. *Phys. Rev. Lett.*, 100 (13), 138702.

Baylor, D., Breck, E., Cheng, H.-T., Fiedel, N., Foo, C. Y., Haque, Z., ... Zinkevich, M. (2017). TFX: A Tensorflow-based production-scale machine learning platform. In *23rd ACM SIGKDD* (pp. 1387–1395).

Bayoudh, K., Knani, R., Hamdaoui, F., & Mtibaa, A. (2022). A survey on deep multimodal learning for computer vision: Advances, trends, applications, and datasets. *Visual Comput.*, 38, 2939–2970.

Bengio, Y., Courville, A., & Vincent, P. (2013). Representation learning: A review and new perspectives. *IEEE TPAMI*, 35 (8), 1798–1828.

Berke, A., Doorley, R., Larson, K., & Moro, E. (2022). Generating synthetic mobility data for a realistic population with RNNs to improve utility and privacy. In *ACM/SIGAPP SAC* (pp. 964–967).

Bhatti, U. A., Yu, Z., Yuan, L., Zeeshan, Z., Nawaz, S. A., Bhatti, M., ... Wen, L. (2020). Geometric algebra applications in geospatial artificial intelligence and remote sensing image processing. *IEEE Access*, 8, 1–14.

Boyes, G., Thomson, C., & Ellul, C. (2015). Integrating BIM and GIS: Exploring the use of IFC space objects and boundaries. In *Proceedings of the GISRUK*.

Briggs, C., Fan, Z., & Andras, P. (2021). A review of privacy-preserving federated learning for the Internet-of-Things. In M. Habib ur Rehman, & M. Medhat Gaber (Eds.), *Federated Learning Systems: Towards Next-Generation AI* (pp. 21–50). Cham: Springer.

Brown, T., Mann, B., Ryder, N., Subbiah, M., Kaplan, J. D., Dhariwal, P., ... Amodei, D. (2020). Language models are few-shot learners. In *NeurIPS*, 33 (pp. 1877–1901).

Bruna, J., Zaremba, W., Szlam, A., & LeCun, Y. (2013). Spectral networks and locally connected networks on graphs. arXiv:1312.6203.

Cadamuro, G., Muhebwa, A., & Taneja, J. (2019). Street smarts: Measuring intercity road quality using deep learning on satellite imagery. In *2nd ACM SIGCAS* (pp. 145–154).

Cai, H., Zheng, V. W., & Chang, K. C.-C. (2018). A comprehensive survey of graph embedding: Problems, techniques, and applications. *IEEE TKDE*, 30 (9), 1616–1637.

Calabrese, F., Colonna, M., Lovisolo, P., Parata, D., & Ratti, C. (2010). Real-time urban monitoring using cell phones: A case study in Rome. *IEEE TITS*, 12 (1), 141–151.

Campos-Cordobes, S., del Ser, J., Lana, I., Olabarrieta, I. I., Sanchez-Cubillo, J., Sanchez-Medina, J. J., & Torre-Bastida, A. I. (2018). Big data in road transport and mobility research. In F. Jimenez (Ed.), *Intelligent Vehicles* (pp. 175–205). New York: IEEE.

Cao, S., Lu, W., & Xu, Q. (2016). Deep neural networks for learning graph representations. In *AAAI*, 16 (pp. 1145–1152).

Cao, H., Sankaranarayanan, J., Feng, J., Li, Y., & Samet, H. (2019a). Understanding metropolitan crowd mobility via mobile cellular accessing data. *ACM TSAS*, 5 (2), 1–18.

Cao, H., Xu, F., Sankaranarayanan, J., Li, Y., & Samet, H. (2019b). Habit2vec: Trajectory semantic embedding for living pattern recognition in population. *IEEE TMC*, 19 (5), 1096–1108.

Carbone, P., Katsifodimos, A., Ewen, S., Markl, V., Haridi, S., & Tzoumas, K. (2015). Apache Flink™: Stream and batch processing in a single engine. *Bull. IEEE Comput. Soc. Tech. Committee Data Eng.*, 36 (4), p. 11.

Cardillo, A., Scellato, S., Latora, V., & Porta, S. (2006). Structural properties of planar graphs of urban street patterns. *Phys. Rev. E*, 73 (6), 066107.

Chappell, D. (2015). Introducing Azure machine learning. Microsoft.

Chawathe, S. S. (2007). Segment-based map matching. In *IEEE IVS* (pp. 1190–1197).

Chen, Y., Lin, Z., Zhao, X., Wang, G., & Gu, Y. (2014). Deep learning-based classification of hyperspectral data. *IEEE JSTAEORS*, 7 (6), 2094–2107.

Chen, C., Liao, C., Xie, X., Wang, Y., & Zhao, J. (2019). Trip2Vec: A deep embedding approach for clustering and profiling taxi trip purposes. *Springer PUC*, 23 (1), 53–66.

Chen, T., Kornblith, S., Norouzi, M., & Hinton, G. (2020a). A simple framework for contrastive learning of visual representations. In *ICML*.

Chen, Y.-C., Li, L., Yu, L., El Kholy, A., Ahmed, F., Gan, Z., ... Liu, J. (2020b). UNITER: Universal image-text representation learning. In *ECCV* (pp. 104–120).

Chen, J., Li, K., & Philip, S. Y. (2021). Privacy-preserving deep learning model for decentralized VANETs using fully homomorphic encryption and blockchain. In *IEEE TITS*.

Cheng, G., Han, J., Zhou, P., & Guo, L. (2014). Multi-class geospatial object detection and geographic image classification based on collection of part detectors. *ISPRS JPRS*, 98, 119–132.

Cho, K., Van Merrienboer, B., Bahdanau, D., & Bengio, Y. (2014). On the properties of neural machine translation: Encoder-decoder approaches. arXiv:1409.1259.

Closa, G., Maso, J., Zabala, A., Pesquer, L., & Pons, X. (2019). A provenance metadata model integrating ISO geospatial lineage and the OGC WPS: Conceptual model and implementation. *Trans. GIS*, 23 (5), 1102–1124.

Cohn, D. A., Ghahramani, Z., & Jordan, M. I. (1996). Active learning with statistical models. *JAIR*, 4, 129–145.

D'Zmura, M. (2020). Behind the scenes: Popular times and live busyness information (Google Blog post).

Databricks (2021a). Databricks feature store. https://docs.databricks.com/applications/machine-learning/feature-store/index.html.

Databricks (2021b). MLOps. Retrieved from https://databricks.com/glossary/mlops.

de Souza, E. N., Boerder, K., Matwin, S., & Worm, B. (2016). Improving fishing pattern detection from satellite AIS using data mining and machine learning. *PLoS One*, 11 (7), e0158248.

Defferrard, M., Bresson, X., & Vandergheynst, P. (2016). Convolutional neural networks on graphs with fast localized spectral filtering. In *NeurIPS*, 29.

Devlin, J., Chang, M.-W., Lee, K., & Toutanova, K. (2018). Bert: Pre-training of deep bidirectional transformers for language understanding. arXiv:1810.04805.

Di, L., Shao, Y., & Kang, L. (2013). Implementation of geospatial data provenance in a web service workflow environment with ISO 19115 and ISO 19115-2 lineage model. *IEEE TGRS*, 51 (11), 5082–5089.

Doersch, C., Gupta, A., & Efros, A. A. (2015). Unsupervised visual representation learning by context prediction. arXiv:1505.05192.

Donahue, J., & Simonyan, K. (2019). Large scale adversarial representation learning. arXiv:1907.02544.

Dong, Y., Chawla, N. V., & Swami, A. (2017). metapath2vec: Scalable representation learning for heterogeneous networks. In *23rd ACM SIGKDD* (pp. 135–144).

Doucette, P., Agouris, P., & Stefanidis, A. (2004). Automated road extraction from high resolution multispectral imagery. *ASPRS PERS*, 70 (12), 1405–1416.

Dunn, J. (2016). Introducing FBLearner Flow: Facebook's AI backbone. Facebook.

Dunnmon, J., Ganguli, S., Hau, D., & Husic, B. (2019). Predicting us state-level agricultural sentiment as a measure of food security with tweets from farming communities. arXiv:1902.07087.

Ebel, P., Gol, I. E., Lingenfelder, C., & Vogelsang, A. (2020). Destination prediction based on partial trajectory data. In *IEEE IVS* (pp. 1149–1155).

EPSG:3857. (2022). EPSG geodetic parameter registry. Official entry of EPSG:3857 spherical Mercator projection coordinate system. http://www.epsg-registry.org.

ESRI. (2022). ESRI Annotations. https://desktop.arcgis.com/en/arcmap/latest/manage-data/annotations/what-is-annotation.htm.

Frome, A., Corrado, G. S., Shlens, J., Bengio, S., Dean, J., Ranzato, M., & Mikolov, T. (2013). Devise: A deep visual-semantic embedding model. In *NeurIPS*, 26.

Gadzicki, K., Khamsehashari, R., & Zetzsche, C. (2020). Early vs late fusion in multimodal convolutional neural networks. In *2020 IEEE 23rd International Conference on Information Fusion (FUSION)* (pp. 1–6).

Gal, Y., Islam, R., & Ghahramani, Z. (2017). Deep Bayesian active learning with image data. In *ICML* (pp. 1183–1192).

Galar, M., Fernandez, A., Barrenechea, E., Bustince, H., & Herrera, F. (2012). A review on ensembles for the class imbalance problem: Bagging-, boosting-, and hybrid-based approaches. *IEEE TSMC C*, 42 (4), 463–484.

Ganguli, S., Dunnmon, J., & Hau, D. (2019a). Predicting food security outcomes using convolutional neural networks (CNNS) for satellite tasking. arXiv:1902.05433.

Ganguli, S., Garzon, P., & Glaser, N. (2019b). GeoGAN: A conditional GAN with reconstruction and style loss to generate standard layer of maps from satellite images. arXiv:1902.05611.

Ganguli, S., Iyer, C., & Pandey, V. (2021). Reachability Embeddings: Scalable self-supervised representation learning from Markovian trajectories for geospatial computer vi-sion. arXiv:2110.12521.

Gao, Q., Zhou, F., Zhang, K., Trajcevski, G., Luo, X., & Zhang, F. (2017). Identifying human mobility via trajectory embeddings. In *IJCAI*, 17 (pp. 1689–1695).

Geng, X., Li, Y., Wang, L., Zhang, L., Yang, Q., Ye, J., & Liu, Y. (2019). Spatiotemporal multi-graph convolution network for ride-hailing demand forecasting. In *AAAI Conference*, 33 (pp. 3656–3663).
Gharaee, Z., Kowshik, S., Stromann, O., & Felsberg, M. (2021). Graph representation learning for road type classification. *Pat. Rec.*, 120, 108174.
Gidaris, S., Singh, P., & Komodakis, N. (2018). Unsupervised representation learning by predicting image rotations. In *ICLR*.
Gift, N., & Deza, A. (2021). Practical mlops. "O'Reilly Media, Inc.".
GO-JEK, & Google. (2019). Feast feature store. https://feast.dev.
Goodfellow, I., Pouget-Abadie, J., Mirza, M., Xu, B., Warde-Farley, D., Ozair, S., ... Bengio, Y. (2014). Generative adversarial networks. arXiv:1406.2661.
Graves, A. (2013). Generating sequences with recurrent neural networks. arXiv:1308.0850.
Grill, J.-B., Strub, F., Altche, F., Tallec, C., Richemond, P., Buchatskaya, E., ... Valko, M. (2020). Bootstrap your own latent: A new approach to self-supervised learning. In *NeurIPS*.
Grover, A., & Leskovec, J. (2016). node2vec: Scalable feature learning for networks. In *22nd ACM SIGKDD* (pp. 855–864).
Guo, W., Wang, J., & Wang, S. (2019). Deep multimodal representation learning: A survey. *IEEE Access*, 7, 63373–63394.
H2O.ai. (2020). H2O.ai feature store. https://www.h2o.ai/feature-store/.
Hatchett, B. J., Benmarhnia, T., Guirguis, K., VanderMolen, K., Gershunov, A., Kerwin, H., ... Samburova, V. (2021). Mobility data to aid assessment of human responses to extreme environmental conditions. *Lancet P. Health*, 5 (10), e665–e667.
He, K., Zhang, X., Ren, S., & Sun, J. (2016). Identity mappings in deep residual networks. In *ECCV* (pp. 630–645).
Hermann, J., & Balso, M. D. (2017). Meet Michelangelo: Uber's machine learning platform. https://eng.uber.com/michelangelo-machine-learning-platform/.
Ho, S.-S., & Ruan, S. (2011). Differential privacy for location pattern mining. In *ACM SIGSPA-TIAL Workshop on Security and Privacy in GIS and LBS* (pp. 17–24).
Hochreiter, S., & Schmidhuber, J. (1997). Long short-term memory. *Neural Comput.*, 9 (8), 1735–1780.
Hopsworks. (2018). Feature stores for ML. https://www.featurestore.org.
Hopsworks. (2020). Hopsworks: The enterprise feature store. https://www.hopsworks.ai.
Hu, J., Guo, C., Yang, B., & Jensen, C. S. (2019a). Stochastic weight completion for road networks using graph convolutional networks. In *35th IEEE ICDE* (p. 1274–1285).
Hu, Y., Li, W., Wright, D., Aydin, O., Wilson, D., Maher, O., & Raad, M. (2019b). Artificial intelligence approaches. arXiv:1908.10345.
Huang, G., Liu, Z., Van Der Maaten, L., & Weinberger, K. Q. (2017). Densely connected convolutional networks. In *IEEE CVPR* (pp. 4700–4708).
Huang, J., Deng, M., Tang, J., Hu, S., Liu, H., Wariyo, S., & He, J. (2018). Automatic generation of road maps from low quality GPS trajectory data via structure learning. *IEEE Access*, 6, 71965–71975.
Huyen, C. (2022). Designing machine learning systems. O'Reilly.
Iguazio. (2021). Iguazio feature store. https://www.iguazio.com/feature-store.
iMerit. (2021). iMerit solutions. https://imerit.net/solutions/computer-vision/.
Inoubli, W., Aridhi, S., Mezni, H., Maddouri, M., & Mephu Nguifo, E. (2018). An experimental survey on big data frameworks. In *FGCS*, 86.
Isaacman, S., Becker, R., Caceres, R., Kobourov, S., Martonosi, M., Rowland, J., & Varshavsky, A. (2011). Identifying important places in people's lives from cellular network data. In *ICPC* (pp. 133–151).
Iyer, C. V. K., Hou, F., Wang, H., Wang, Y., Oh, K., Ganguli, S., & Pandey, V. (2021). Trinity: A no-code AI platform for complex spatial datasets. In *4th ACM SIGSPATIAL GeoAI* (pp. 33–42).
Jaegle, A., Gimeno, F., Brock, A., Vinyals, O., Zisserman, A., & Carreira, J. (2021). Perceiver: General perception with iterative attention. In *ICML* (pp. 4651–4664).
Jegou, S., Drozdzal, M., Vazquez, D., Romero, A., & Bengio, Y. (2017). The one hundred layers tiramisu: Fully convolutional densenets for semantic segmentation. In *IEEE CVPR* (pp. 11–19).
Jepsen, T. S., Jensen, C. S., & Nielsen, T. D. (2019). Graph convolutional networks for road networks. In *27th ACM SIGSPATIAL* (pp. 460–463).
Jepsen, T. S., Jensen, C. S., & Nielsen, T. D. (2020). Relational fusion networks: Graph convolutional networks for road networks. In *IEEE TITS*.
Jepsen, T. S., Jensen, C. S., Nielsen, T. D., & Torp, K. (2018). On network embedding for machine learning on road networks: A case study on the Danish road network. In *IEEE ICBD* (pp. 3422–3431).
Jiang, B., Seif, M., Tandon, R., & Li, M. (2021). Context-aware local information privacy. *IEEE TIFS*, 16, 3694–3708.

Jiang, X., de Souza, E. N., Pesaranghader, A., Hu, B., Silver, D. L., & Matwin, S. (2017). Trajectorynet: An embedded GPS trajectory representation for point-based classification using recurrent neural networks. arXiv:1705.02636.

Jin, X., & Davis, C. H. (2005). Automated building extraction from high-resolution satellite imagery in urban areas using structural, contextual, and spectral information. *EURASIP JASP*, 2005 (14), 1–11.

Johnston, K., Ver Hoef, J. M., Krivoruchko, K., & Lucas, N. (2001). *Using ArcGIS Geostatistical Analyst* (Vol. 380). Esri Redlands: ArcGIS.

Joshi, A. V. (2020). Amazon's machine learning toolkit: Sagemaker. In *MLAI* (pp. 233–243).

Karpathy, A., & Fei-Fei, L. (2015). Deep visual-semantic alignments for generating image descrip-tions. In *CVPR* (pp. 3128–3137).

Khan, A., & Hassan, Z. S. (2020). Doodash feature store. https://doordash.engineering/2020/11/19/building-a-gigascale-ml-feature-store-with-redis/.

Kholoshyn, I., Bondarenko, O., Hanchuk, O., & Shmeltser, E. (2019). Cloud ArcGIS online as an innovative tool for developing geoinformation competence with future geography teachers. arXiv preprint arXiv:1909.04388.

Kim, W., Son, B., & Kim, I. (2021). ViLT: Vision-and-language transformer without convolution or region supervision. In *ICML* (pp. 5583–5594).

Kingma, D. P., & Welling, M. (2013). Auto-encoding variational Bayes. arXiv:1312.6114.

Kipf, T. N., & Welling, M. (2016). Semi-supervised classification with graph convolutional net-works. arXiv: 1609.02907.

Klein, L. J., Marianno, F. J., Albrecht, C. M., Freitag, M., Lu, S., Hinds, N., … Hamann, H. F. (2015). PAIRS: A scalable geo-spatial data analytics platform. In *IEEE ICBD* (pp. 1290–1298).

Kong, F., Huang, B., Bradbury, K., & Malof, J. M. (2020). The Synthinel-1 dataset: A collection of high resolution synthetic overhead imagery for building segmentation. In *IEEE WACV*.

Korbar, B., Tran, D., & Torresani, L. (2018). Cooperative learning of audio and video models from self-supervised synchronization. In *NeurIPS* (pp. 7774–7785).

Krisp, J. M. (2010). Planning fire and rescue services by visualizing mobile phone density. *JUT*, 17 (1), 61–69.

Kumar, N., Velmurugan, A., Hamm, N. A., & Dadhwal, V. K. (2018). Geospatial mapping of soil organic carbon using regression kriging and remote sensing. *JISRS*, 46 (5), 705–716.

Kussul, N., Lavreniuk, M., Skakun, S., & Shelestov, A. (2017). Deep learning classification of land cover and crop types using remote sensing data. *IEEE GRSL*, 14 (5), 778–782.

Lazaridou, A., Pham, N. T., & Baroni, M. (2015). Combining language and vision with a multi-modal skip-gram model. arXiv:1501.02598.

Lee, M. A., Zhu, Y., Srinivasan, K., Shah, P., Savarese, S., Fei-Fei, L., … Bohg, J. (2019). Making sense of vision and touch: Self-supervised learning of multimodal representations for contact- rich tasks. In *ICRA* (pp. 8943–8950).

Levine, E. (2020). ML Lake: Building Salesforce's data platform for machine learning. https://engineering.salesforce.com/ml-lake-building-salesforces-data-platform-for-machine-learning-228c30e21f16.

Li, J., Selvaraju, R., Gotmare, A., Joty, S., Xiong, C., & Hoi, S. C. H. (2021). Align before fuse: Vision and language representation learning with momentum distillation. In *NeurIPS*, 34.

Liu, Y., Feng, X., & Zhou, Z. (2016). Multimodal video classification with stacked contractive autoencoders. *Signal Process.*, 120, 761–766.

Liu, Z. et al. (2017). Semantic proximity search on heterogeneous graph by proximity embedding. In *AAAI* (pp. 154–160).

Long, J., Shelhamer, E., & Darrell, T. (2015). Fully convolutional networks for semantic segmen-tation. In *IEEE CVPR* (pp. 3431–3440).

Lu, J., Batra, D., Parikh, D., & Lee, S. (2019). ViLBERT: Pretraining task-agnostic visiolinguistic representations for vision-and-language tasks. In *NeurIPS*, 32.

Lunga, D., Gerrand, J., Yang, L., Layton, C., & Stewart, R. (2020). Apache spark accelerated deep learning inference for large scale satellite image analytics. *IEEE J Sel. Top. Appl. Earth Obs. Remote Sens.*, 13, 271–283.

Lunga, D., Arndt, J., Gerrand, J., & Stewart, R. (2021). ReSFlow: A remote sensing imagery data-flow for improved model generalization. *IEEE J Sel. Top. Appl. Earth Obs. Remote Sens.*, 14, 10468–10483.

Mackenzie, A. (2019). From API to AI: Platforms and their opacities. *ICS*, 22 (13), 1989–2006.

Malpica, J. A., Alonso, M. C., Papı, F., Arozarena, A., & Martınez De Agirre, A. (2013). Change detection of buildings from satellite imagery and LiDAR data. *IJRS*, 34 (5), 1652–1675.

Marshall, S. (2016). Line structure representation for road network analysis. *JTLU*, 9 (1), 29–64.

Martelli, F., Renda, M. E., & Zhao, J. (2021). The price of privacy control in mobility sharing. *J. Urban Tech.*, 28 (1–2), 237–262.

Masucci, A. P., Smith, D., Crooks, A., & Batty, M. (2009). Random planar graphs and the London street network. *Eur. Phys. B*, 71 (2), 259–271.
Mayer, H. (1999). Automatic object extraction from aerial imagery: A survey focusing on buildings. *CVIU*, 74 (2), 138–149.
Meyer, J., Eitel, A., Brox, T., & Burgard, W. (2020). Improving unimodal object recognition with multimodal contrastive learning. In *IROS* (pp. 5656–5663).
Mikolov, T., Chen, K., Corrado, G., & Dean, J. (2013a). Efficient estimation of word representations in vector space. arXiv:1301.3781.
Mikolov, T., Sutskever, I., Chen, K., Corrado, G. S., & Dean, J. (2013b). Distributed representations of words and phrases and their compositionality. In *NeurIPS*, 26.
Minetto, R., Segundo, M. P., & Sarkar, S. (2019). Hydra: An ensemble of convolutional neural networks for geospatial land classification. *IEEE TGRS*, 57 (9), 6530–6541.
Morchhale, S., Pauca, V. P., Plemmons, R. J., & Torgersen, T. C. (2016). Classification of pixel-level fused hyperspectral and LiDAR data using deep convolutional neural networks. In *8th IEEE WHISPERS* (pp. 1–5).
Morvant, E., Habrard, A., & Ayache, S. (2014). Majority vote of diverse classifiers for late fusion. In *Joint IAPR International Workshops on Statistical Techniques in Pattern Recognition (SPR) and Structural and Syntactic Pattern Recognition (SSPR)* (pp. 153–162).
Ng, A. (2022). Data-centric AI. Retrieved from https://datacentricai.org.
Ngiam, J., Khosla, A., Kim, M., Nam, J., Lee, H., & Ng, A. Y. (2011). Multimodal deep learning. In *ICML*.
Nikolenko, S. (2021). *Synthetic Data for Deep Learning*. Berlin/Heidelberg, Germany: Springer.
Nisselson, E. (2018). Deep learning with synthetic data will democratize the tech industry. TechCrunch.
Noroozi, M., & Favaro, P. (2016). Unsupervised learning of visual representations by solving jigsaw puzzles. In *ECCV*.
NUMO, NABSA, & OMF. (2021). Privacy principles for mobility data. Retrieved from https:// www.mobilitydataprivacyprinciples.org/.
Oghaz, M. M. D., Razaak, M., Kerdegari, H., Argyriou, V., & Remagnino, P. (2019). Scene and environment monitoring using aerial imagery and deep learning. In *15th IEEE DCOSS* (pp. 362–369).
Oliver, A., Odena, A., Raffel, C. A., Cubuk, E. D., & Goodfellow, I. (2018). Realistic evaluation of deep semi-supervised learning algorithms. In *NeurIPS*, 31.
OpenStreetMap. (2017). OpenStreetMap annotations. https://www.openstreetmap.org.
Owens, A., & Efros, A. A. (2018). Audio-visual scene analysis with self-supervised multisensory features. In *ECCV* (pp. 631–648).
Owens, A., Wu, J., McDermott, J. H., Freeman, W. T., & Torralba, A. (2016). Ambient sound provides supervision for visual learning. In *ECCV* (pp. 801–816).
Panzarino, M. (2018). Apple is rebuilding maps from the ground up. TechCrunch.
Pennington, J., Socher, R., & Manning, C. D. (2014). Glove: Global vectors for word representation. In *EMNLP* (pp. 1532–1543).
Perez, A., Ganguli, S., Ermon, S., Azzari, G., Burke, M., & Lobell, D. (2019). Semi-supervised multitask learning on multispectral satellite images using Wasserstein generative adversarial networks (GANs) for predicting poverty. arXiv:1902.11110.
Perozzi, B., Al-Rfou, R., & Skiena, S. (2014). Deepwalk: Online learning of social representations. In *20th ACM SIGKDD* (pp. 701–710).
Poria, S., Cambria, E., Howard, N., Huang, G.-B., & Hussain, A. (2016). Fusing audio, visual and textual clues for sentiment analysis from multimodal content. *Neurocomputing*, 174, 50–59.
Potamianos, G., Neti, C., Gravier, G., Garg, A., & Senior, A. W. (2003). Recent advances in the automatic recognition of audiovisual speech. *Proc. IEEE*, 91 (9), 1306–1326.
Radford, A., Kim, J. W., Hallacy, C., Ramesh, A., Goh, G., Agarwal, S., ... Sutskever, I (2021). Learning transferable visual models from natural language supervision. In *ICML* (pp. 8748–8763).
Ramachandram, D., & Taylor, G. W. (2017). Deep multimodal learning: A survey on recent advances and trends. *IEEE Signal Process. Mag.*, 34 (6), 96–108.
Re, C., Niu, F., Gudipati, P., & Srisuwananukorn, C. (2020). Overton: A data system for monitoring and improving machine-learned products. In *10th CIDR*.
Reed, S., Akata, Z., Yan, X., Logeswaran, L., Schiele, B., & Lee, H. (2016). Generative adversarial text to image synthesis. In *ICML* (pp. 1060–1069).
Ronneberger, O., Fischer, P., & Brox, T. (2015). U-net: Convolutional networks for biomedical image segmentation. In *MICCAI* (pp. 234–241).
Ruderman, D. L. (1994). The statistics of natural images. *Network Comput. Neural Syst.*, 5 (4), 517.
Sculley, D., Holt, G., Golovin, D., Davydov, E., Phillips, T., Ebner, D., ... Dennison, D. (2015). Hidden technical debt in machine learning systems. In *NeurIPS*, 28.

Shumway, R. H., & Stoffer, D. S. (2000). *Time Series Analysis and Its Applications* (Vol. 3). Berlin/Heidelberg, Germany: Springer.

Simini, F., Gonzalez, M., Maritan, A., & Barabasi, A.-L. (2012). A universal model for mobility and migration patterns. *Nature*, 484 (7392), 96–100.

Singh, A., Hu, R., Goswami, V., Couairon, G., Galuba, W., Rohrbach, M., & Kiela, D. (2021). FLAVA: A foundational language and vision alignment model. arXiv:2112.04482.

Smith, D. A. (2017). Visualising world population density as an interactive multi-scale map using the global human settlement population layer. *J. Maps*, 13 (1), 117–123.

Smith, C. D., & Mennis, J. (2020). Incorporating GIS and technology in response to COVID-19. US CDC PDC, 17.

Smith, S., Patwary, M., Norick, B., LeGresley, P., Rajbhandari, S., Casper, J., … others (2022). Using DeepSpeed and Megatron to train Megatron-Turing NLG 530B, a large-scale generative language model. arXiv:2201.11990.

SnorkelAI. (2022). Tips for using a data-centric AI approach. Retrieved from https://snorkel.ai/tips-for-data-centric-ai-approach/.

Song, C., Koren, T., Wang, P., & Barabasi, A.-L. (2010). Modelling the scaling properties of human mobility. *Nat. Phys.*, 6 (10), 818–823.

Soto, V., & Frias-Martinez, E. (2011). Robust land use characterization of urban landscapes using cell phone data. In *1st WPUA* (Vol. 9).

Soto, V., Frias-Martinez, V., Virseda, J., & Frias-Martinez, E. (2011). Prediction of socioeconomic levels using cell phone records. In *UMAP* (pp. 377–388).

Srivastava, R. K., Greff, K., & Schmidhuber, J. (2015). Highway networks. arXiv:1505.00387.

Strano, E., Nicosia, V., Latora, V., Porta, S., & Barthelemy, M. (2012). Elementary processes governing the evolution of road networks. *Nat. Sci. Rep.*, 2 (1), 1–8.

Strickland, E. (2022). Andrew Ng: Unbiggen AI, Interview in IEEE Spectrum Magazine. Retrieved from https://spectrum.ieee.org/andrew-ng-data-centric-ai.

Su, H., Wei, S., Yan, M., Wang, C., Shi, J., & Zhang, X. (2019a). Object detection and instance segmentation in remote sensing imagery based on precise mask R-CNN. In *IEEE IGARSS* (pp. 1454–1457).

Su, W., Zhu, X., Cao, Y., Li, B., Lu, L., Wei, F., & Dai, J. (2019b). VL-BERT: Pre-training of generic visual-linguistic representations. arXiv:1908.08530.

Sutskever, I., Vinyals, O., & Le, Q. V. (2014). Sequence to sequence learning with neural networks. In *NeurIPS*, 27.

Taleb, A., Lippert, C., Klein, T., & Nabi, M. (2021). Multimodal self-supervised learning for medical image analysis. In *International Conference on Information Processing in Medical Imaging* (pp. 661–673).

Tan, H., & Bansal, M. (2019). LXMERT: Learning cross-modality encoder representations from transformers. arXiv:1908.07490.

Tan, M., & Le, Q. (2019). Efficientnet: Rethinking model scaling for convolutional neural networks. In *ICML* (pp. 6105–6114).

Tang, W., & Lease, M. (2011). Semi-supervised consensus labeling for crowdsourcing. In *SIGIR WCIR* (pp. 1–6).

Tao, D., Cheng, J., Yu, Z., Yue, K., & Wang, L. (2018). Domain-weighted majority voting for crowdsourcing. *IEEE TNNLS*, 30 (1), 163–174.

Thambawita, V., Salehi, P., Sheshkal, S. A., Hicks, S. A., Hammer, H. L., Parasa, S., … Riegler, M. A. (2021). SinGAN-Seg: Synthetic training data generation for medical image segmentation. arXiv:2107.00471.

Tong, S., & Koller, D. (2001). Support vector machine active learning with applications to text classification. *JMLR*, 2, 45–66.

Treveil, M., Omont, N., Stenac, C., Lefevre, K., Phan, D., Zentici, J., … Heidmann, L. (2020). Introducing mlops. O'Reilly Media.

Tsai, Y.-H. H., Bai, S., Liang, P. P., Kolter, J. Z., Morency, L.-P., & Salakhutdinov, R. (2019). Multimodal transformer for unaligned multimodal language sequences. In *ACL* (pp. 6558–6569).

United Nations. (2020). Future trends in geospatial information management: The five to ten year vision. United Nations Committee of Experts on Global Geospatial Information Management.

USGS. (2022). USGS Landsat 7 mission page. Retrieved from https://www.usgs.gov/landsat-missions/landsat-7.

Vaswani, A., Shazeer, N., Parmar, N., Uszkoreit, J., Jones, L., Gomez, A. N., … Polosukhin, I. (2017). Attention is all you need. In *NeurIPS*, 30.

Veer, R. V., Bloem, P., & Folmer, E. (2018). Deep learning for classification tasks on geospatial vector polygons. arXiv:1806.03857.

Velickovic, P., Cucurull, G., Casanova, A., Romero, A., Lio, P., & Bengio, Y. (2017). Graph attention networks. arXiv:1710.10903.

Venugopalan, S., Xu, H., Donahue, J., Rohrbach, M., Mooney, R., & Saenko, K. (2014). Translating videos to natural language using deep recurrent neural networks. arXiv:1412.4729.

Wang, D. et al. (2016). Structural deep network embedding. In *22nd ACM SIGKDD* (pp. 1225–1234).

Wang, D., Miwa, T., & Morikawa, T. (2020). Big trajectory data mining: A survey of methods, applications, and services. *Sensors*, 20 (16), 4571.

Wang, W., Bao, H., Dong, L., & Wei, F. (2021). VLMo: Unified vision-language pre-training with mixture-of-modality-experts. arXiv:2111.02358.

Warden, P., & Situnayake, D. (2019). Tinyml: Machine learning with Tensorflow lite on Arduino and ultra-low-power microcontrollers. O'Reilly.

Weir, N., Lindenbaum, D., Bastidas, A., Etten, A. V., McPherson, S., Shermeyer, J., ... Tang, H. (2019). Spacenet-MVOI: A multi-view overhead imagery dataset. In *ICCV* (pp. 992–1001).

White, T. (2012). Hadoop: The definitive guide. "O'Reilly Media, Inc.".

Wiki, O. (2022). Bicycle tags map. Retrieved from https://wiki.openstreetmap.org/wiki/Bicycle tags map.

Wu, S., Bondugula, S., Luisier, F., Zhuang, X., & Natarajan, P. (2014). Zero-shot event detection using multi-modal fusion of weakly supervised concepts. In *Proceedings of the IEEE CVPR* (pp. 2665–2672).

Wu, J., Chen, X.-Y., Zhang, H., Xiong, L.-D., Lei, H., & Deng, S.-H. (2019). Hyperparameter optimization for machine learning models based on Bayesian optimization. *JEST*, 17 (1), 26–40.

Xiao, X., Ganguli, S., & Pandey, V. (2020). VAE-Info-CGAN: Generating synthetic images by combining pixel-level and feature-level geospatial conditional inputs. In *13th ACM SIGSPATIAL IWCTS* (p. 1–10).

Xie, F., & Levinson, D. (2011). *Evolving Transportation Networks*. Berlin/Heidelberg, Germany: Springer.

Xu, K., Ba, J., Kiros, R., Cho, K., Courville, A., Salakhudinov, R., ... Bengio, Y. (2015). Show, attend and tell: Neural image caption generation with visual attention. In *ICML* (pp. 2048–2057).

Yeh, C., Perez, A., Driscoll, A., Azzari, G., Tang, Z., Lobell, D., ... Burke, M. (2020). Using publicly available satellite imagery and deep learning to understand economic well-being in Africa. *Nature*, 11 (1), 1–11.

Yu, B., Yin, H., & Zhu, Z. (2017). Spatio-temporal graph convolutional networks: A deep learning framework for traffic forecasting. arXiv:1709.04875.

Yu, B., Yin, H., & Zhu, Z. (2018). Spatio-temporal graph convolutional networks: A deep learning framework for traffic forecasting. In *27th IJCAI* (pp. 3634–3640).

Zaharia, M., Xin, R. S., Wendell, P., Das, T., Armbrust, M., Dave, A.,... Shenker, S. (2016). Apache Spark: A unified engine for big data processing. *Commun. ACM*, 59 (11), 56–65.

Zaharia, M., Chen, A., Davidson, A., Ghodsi, A., Hong, S. A., Konwinski, A., ... Zumar, C. (2018). Accelerating the machine learning lifecycle with MLFlow. *IEEE DEB*, 41 (4), 39–45.

Zehori, I. (2020). Bigabid feature store. https://www.bigabid.com.

Zhang, F., Du, B., & Zhang, L. (2015). Scene classification via a gradient boosting random convolutional network framework. *IEEE TGRS*, 54 (3), 1793–1802.

Zhang, J., Zheng, Y., & Qi, D. (2017). Deep spatiotemporal residual networks for citywide crowd flows prediction. In *31st AAAI Conference* (pp. 1655–1661).

Zhang, C., Yang, Z., He, X., & Deng, L. (2020). Multimodal intelligence: Representation learning, information fusion, and applications. *IEEE JSTSS*, 14 (3), 478–493.

Zhang, W., Jiang, B., Li, M., & Lin, X. (2022). Privacy-preserving aggregate mobility data release: An information-theoretic deep reinforcement learning approach. *IEEE TIFS*, 17, 849–864.

Zhao, L., Song, Y., Zhang, C., Liu, Y., Wang, P., Lin, T., ... Li, H. (2019). T-GCN: A temporal graph convolutional network for traffic prediction. *IEEE TITS*, 21 (9), 3848–3858.

Zheng, Y. (2015). Trajectory data mining: An overview. *ACM TIST*, 6 (3), 1–41.

Zheng, Y., Capra, L., Wolfson, O., & Yang, H. (2014). Urban computing: Concepts, methodologies, and applications. *ACM TIST*, 5 (3), 1–55.

Zheng, Y., Li, Q., Chen, Y., Xie, X., & Ma, W.-Y. (2008). Understanding mobility based on GPS data. In *10th ICUC* (pp. 312–321).

Zhou, J. et al. (2018). Graph neural networks: A review of methods and applications. arXiv:1812.08434.

Zhou, W., Klein, L. J., & Lu, S. (2020). Pairs Autogeo: An automated machine learning framework for massive geospatial data. In *IEEE ICBD* (pp. 1755–1763).

Zhu, J.-Y., Park, T., Isola, P., & Efros, A. A. (2017a). Unpaired image-to-image translation using cycle-consistent adversarial networks. In *IEEE ICCV* (pp. 2223–2232).

Zhu, X. X., Tuia, D., Mou, L., Xia, G.-S., Zhang, L., Xu, F., & Fraundorfer, F. (2017b). Deep learning in remote sensing: A comprehensive review and list of resources. *IEEE GRSM*, 5 (4), 8–36.

3 Temporal Dynamics of Place and Mobility

Kevin Sparks[1], Jesse Piburn[1], Andy Berres[2], Marie Urban[1], and Gautam Thakur[1]
Oak Ridge National Laboratory
Geospatial Science and Human Security Division[1]
Computational Sciences and Engineering Division[2]

CONTENTS

3.1 INTRODUCTION

The temporal patterns of our daily lives are in part defined by our circadian rhythms and influenced by a range of environmental and social aspects. Our daily temporal patterns (what we do, when we do it) can be split into aggregate sleep patterns and aggregate activity patterns, each influenced by their own set of complex factors. For instance, our aggregate sleep patterns, i.e., when we as a population tend to fall asleep and when we wake up, are determined by the solar clock (the natural light-dark cycle), our biological clock (circadian rhythms), and social clock (when school or work begins and whether the following day is a work day/school day) (Aschoff & Wever, 1976; Roenneberg et al., 2003, 2007; Monsivais et al., 2017). Our aggregate activity patterns, i.e., our daily activities we as a population perform when we are awake, are influenced by the places we live and the place types (e.g., Restaurant or Retail) we visit, which are in turn influenced by a complex set of interacting social, economic, cultural, religious, and environmental factors (Wu et al., 2014; McKenzie et al., 2015; Giuntella & Mazzonna, 2019; Janowicz et al., 2019; Sparks et al., 2020; Lískovec et al., 2022; Sparks et al., 2022). However, unlike our understanding of how the solar, biological, and social clocks determine our aggregate sleep patterns, there is limited understanding associated with what

DOI: 10.1201/9781003270928-5

the precise determinants of our aggregate activity patterns are, short of qualitative descriptions of social, cultural, and environmental factors.

There has been considerable effort spent in understanding what the determinants of sleep patterns are and how disruptions in aggregate sleep patterns affect health and economic outcomes (Aschoff & Wever, 1976; Duffy et al., 2011). Sleep deprivation leads to reduced daytime alertness and higher accident rates (Bonnet & Arand, 1995), hinders memory function, and negatively affects mood (Dinges et al., 1997). Day Light Savings Time (DST) has created an opportunity to assess a population's reaction to a sudden disruption between the alignment of the solar clock and our social clock. It results in social jetlag. It can take multiple days for people to re-adjust their wake-up times following the switch from Standard Time (ST) to DST (Monk & Folkard, 1976). A 2014 study identified changes in the rate of heart attacks associated with DST, beyond those expected by random or seasonal variation and found the Monday after the start of DST (losing an hour of sleep) was associated with a 24% increase in heart attacks, and the Tuesday following the end of DST (gaining an hour of sleep) was associated with a 22% reduction in heart attacks (Sandhu et al., 2014).

While we understand less about the determinants of activity patterns than we do about sleep patterns, we recognize that temporal patterns of places are likely influenced by (and influencing) a set of social, cultural, and environmental factors. When we are awake and active, we perform many complex actions, move through space, and interact with multiple places and people at different times of the day and on different days of the week. For example, a visit to a restaurant will be at different times than when we might visit a retail location or a place of worship. These differences illustrate temporal changes across place types. These temporal dynamics of place types also change across geographies, from city to city and country to country (McKenzie et al., 2015; Janowicz et al., 2019; Sparks et al., 2020); for instance, the time of day a restaurant becomes busy in Spain is different than when a restaurant becomes busy in the United States. The temporal dynamics of places become more complex when the time horizon extends to weekly and seasonal patterns. Temporal dynamics inform population modeling, estimations of energy use, and economic opportunities. The scientific question that continues to motivate a growing study in temporal dynamics of places is, what are the determinants of temporal patterns of places, and how do the temporal dynamics of places impact health and economic outcomes?

Beyond identifying that the temporal dynamics of places change across space, it has been challenging to identify the determinants of those temporal patterns due to the complex nature of what constitutes a population's aggregate activity patterns. Still, governments recognize the importance of guiding our daily temporal patterns through policy to impact citizens' economic and quality of life measurements. For example, Pakistan in 1997 shifted its work week from Sunday-Thursday with Friday-Saturday weekends to Monday-Friday with Saturday-Sunday weekends, citing economic reasons of foreign investment and trade opportunities (Cooper, 1997). The United Arab Emirates (UAE) in 2021 stated they were transitioning to a four-and-a-half-day working week (Monday-Friday afternoon), with weekends to consist of Friday afternoon, Saturday, and Sunday. This made the UAE the first nation to formalize a workweek shorter than 5 days, citing social well-being and economic competitiveness (Treisman, 2021). In March 2022, the United States Senate passed legislation to make DST permanent in 2023 (Broadwater & Nierenberg, 2022). These examples show a recognition by governments that seemingly small adjustments to the temporal dynamics of our daily and weekly activity patterns (e.g., when we work, how long we work, when we engage in leisure activities) are important enough to shape through policy for apparent positive health and economic outcomes.

Places are shaped by their own social, cultural, and environmental influences, each with its own unique behavioral patterns. Our aggregate sleep patterns are influenced by our social clock, the solar clock, and our biological clock, and our aggregate activity patterns are likely influenced by a set of social, environmental, and built environment influences. In the following paragraphs, we describe how the temporal patterns of places have bi-directional relationships with that place's (1) Social Norms, (2) Environmental Influences, and its (3) Built Environment.

3.1.1 Social Norms and Historical Contexts

Temporal patterns of places are influenced by social and cultural norms and political structures, including religious influences and perceptions of safety and outbreaks of violence. Recognized days of religious importance and worship often dictate times that some businesses close for the day and when weekends are defined. For example, businesses in Jerusalem, Israel are proportionally open less during Friday evenings and Saturdays. This is influenced by the Jewish tradition of Saturday as the day of worship, while businesses in predominantly Christian cities such as Barcelona, Spain are proportionally open less during Sunday, Christianity's traditional day of worship (Sparks et al., 2020). While the direct influence of religious-based activity patterns may change over time, there is anecdotal evidence of the historic influence remaining in many countries. Historic influences of cultural and social norms extend beyond religious days of worship and include the historic potential for outbreaks of violence. Altering our aggregate activity patterns in response to perceived safety and potential outbreaks of violence occur all over the world. Cities or neighborhoods that recognize a consistent potential for outbreaks of violence may regularly respond by closing businesses before sunset in an attempt to curb potential outbreaks of violence (Russell, 2020). This pre-mature closing may also occur in response to anticipated infrequent events such as natural hazards.

3.1.2 Environmental Influences and Mobility Disruptions

Living near a time zone border in the United States creates a misalignment between our social clocks and the solar clock. For people living on the eastern side of a time zone border (i.e., adjacent and to the right of a time zone border), it is one hour later than for people living on the western side of a time zone border (i.e., adjacent and to the left of a time zone border), even though both sets of people experience a nearly identical solar position. In these scenarios of social and solar clock misalignment, a study using biometric sleep tracking data found that counties in the United States on the eastern side of a time zone border are going to bed on average 19-minutes later than their corresponding adjacent county on the westerns side of the time zone border (Giuntella & Mazzonna, 2019). Whether this 19-minute difference translates to an average increase in activity patterns at commercial locations still needs to be confirmed by other data, but the finding creates a reasonable new hypothesis to test. This example illustrates the potential of longitudinal time zone effects, while latitudinal differences from place-to-place influence seasonal change and variation between the length of day depending on the time of the year. When considering the solar clock's influence on sleep patterns, our activity patterns likely change with the variation in the length of daylight, particularly at high-low latitudes. Daylight length and seasonal changes by latitude have shown a measurable impact on music listening behavior across countries and cultures (Park et al., 2019), however, whether this behavioral change extends to activity patterns varying by latitude would need to be empirically tested.

In response to large disruptions in overall mobility due to COVID-19, biometric tracking and cellphone-based GPS data illustrated how temporal activity patterns were changed from pre-pandemic behavior showing later wake-up times (Giuntella et al., 2021; Sparks et al., 2022). These environmental and mobility-based examples illustrate that if seemingly small temporal differences are systematic, over long-time horizons they add up to significant changes in temporal dynamics of places and human behavioral patterns.

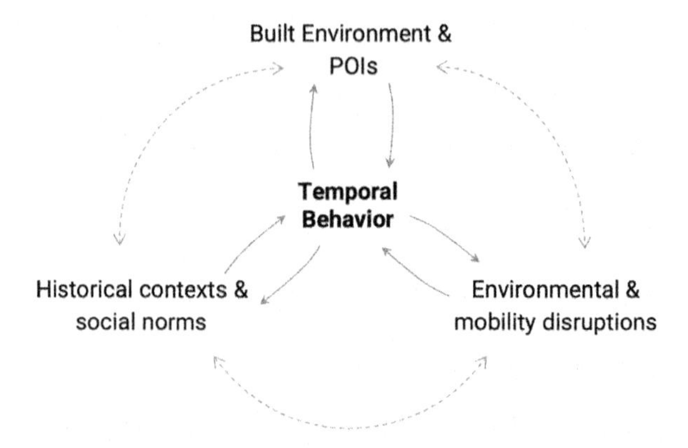

FIGURE 3.1 The influences and their relationships that determine the temporal patterns of a place.

3.1.3 Built Environment and Points of Interest

The land use of neighborhoods and districts within cities creates or discourages activity within them based on the usage types and the Points of Interest (POIs) that exist at those places. Jane Jacobs in her book *The Life and Death of Great American Cities* (Jacobs, 1961) talks about the temporal dynamics of Manhattan's Financial District.

> *"It is only necessary to observe the deathlike stillness that settles on the district after 5:30 and all-day Saturday and Sunday. This degree of underuse is a miserable inefficiency...", "... when a primary use is combined, effectively, with another that puts people on the street at different times, then the effect can be economically stimulating".*

While this comment about Manhattan's financial district was made in the 1960s, and the financial district has changed over time, the observation on single-use districts influencing when and how people spend time there remains highly relevant. If a place doesn't have amenities and POIs that draw people there after a certain time, then it won't be used during those time periods. Jacobs implies that the presence of certain place types and POIs acts as one of the determinants of temporal dynamics of places and suggests how an economic outcome may or may not be reached (i.e., multi-use districts can be economically stimulating).

To comprehensively identify the determinants of temporal dynamics of place, each one of the inputs discussed above should be considered. These inputs have complex, bi-directional relationships that change each other over time (Figure 3.1).

Research into temporal dynamics of places and human mobility has been motivated by the recent growth and availability of multi-modal geospatial, geosocial, transportation, and mobility data. This has led to a multitude of disciplines integrating temporal dynamics perspectives into their set of problems. As such, there have been siloed approaches to answering broader questions concerning aggregate temporal dynamics of places and human behavior. It is likely that no one discipline can solve the complex problems of the twenty-first century and there is a need and opportunity for cross-discipline sharing and collaboration (National Research Council, 2014). In the following sections, we detail the recent emergence of varying types of geosocial, mobility, and digital trace data and discuss how these data have prompted new problems in a variety of disciplines that often share common methods and interests.

3.2 DATA TYPES FOR TEMPORAL RESEARCH

There are multiple data types and representations that can help us understand temporal patterns of places and mobility. In the literature, there are place-based assessments that are concerned with activity in bucketed places as a spatial unit of measure over time (Wu et al., 2014; McKenzie et al.,

2015; Happle et al., 2020), and there are individual-based assessments that are concerned with trajectories of individual agents over time (Liu et al., 2010; Alessandretti et al., 2017, 2020; Schläpfer et al., 2021). "Places" in this context means everything from individual buildings up to countries and includes the transportation networks that connect them. Place-based temporal dynamics of aggregate activity patterns are our focus in this chapter.

Coupled with a growing amount of data, the advancement of methods to identify and forecast emerging temporal patterns within large time-series data will be necessary. Research to develop methods for time-series specific data involve time-series similarity measure such as Dynamic Time Warping (Keogh & Pazzani, 2001), adopting methods such as Non-Negative Matrix Factorization for finding emerging patterns with time-series data (Cichocki et al., 2009; Aledavood et al., 2022), and advancements in time-series forecasting (Taylor & Letham, 2018; Zhou et al., 2021). While we recognize the importance of developing methods to analyze temporal data, we chose to highlight a discussion around emerging data types due to the recent acceleration of new data becoming available.

Information regarding the temporal dynamics of places has historically only been available anecdotally. However, the growth of open spatiotemporal data now allows us to observe precise temporal patterns for places worldwide. In the following, we will discuss Probe-Based, Stationary Sensor, Placed-Based, and Human-centric data types relevant to studying spatiotemporal patterns.

3.2.1 Probe Data

We define *probe data* as data that tracks mobile sensors' locations as they move through space. The resulting data is a time-series of Global Positioning System (GPS) coordinates, which can have additional data for each timestep. Such fine-grained data can be collected through various types of handheld devices such as cellphones, trackers, and vehicles, among others. Probe data is typically owned by companies in the commercial sector and its use requires privacy considerations, which we will elaborate on in the discussion section.

Vehicles are one of the most common sources of probe data. Many modern vehicles have built-in GPS sensors and an integrated navigation system, which assists the driver in finding the best route to their destination, under consideration of current traffic conditions. Centralized systems collect information from vehicles within their system or other traffic information sources, in order to determine current traffic bottlenecks and recommend faster alternate routes. Probe data from vehicles is typically collected at very short intervals (from several times per second to once every few seconds), and it can include many other data fields, such as speed, fuel or electricity use, ignition status, acceleration, and braking. Examples of such data sources include TomTom, INRIX, and car manufacturers.

Cellphones are another ubiquitous source of probe data as today's cellphones are equipped with GPS sensors. However, unlike vehicles, there is much more variation in how cellphone data is collected. Cellphone data collection depends on which app the user has installed, and whether they have granted permission to share their device location. The most detailed GPS traces are available if the data is collected from navigation apps, which track locations at specific timesteps (from once per second to once every few seconds). Other apps may collect data only whenever the device has moved a specific distance, or when the dwell time in a location has reached a specific threshold. Depending on the level of detail, this data can be used similarly to vehicle data, but as cellphones typically travel with their owners, one can gain more insights into human behaviors, e.g., people stopping at a coffee shop on their way to the office. At an aggregate level, this type of data allows us to make inferences about the culture at a location, through information, such as the beginning and end of the workday, the time and duration of lunch breaks, or the visitation to parks or retail locations. Cellphone data has been used to create many opportunities for derivative data types, many of which we will discuss later.

Biometric Trackers and similar personal fitness devices can provide short-term insights into human activity and mobility. While the location information is often limited to a short duration, this data can provide insights into the population's activity levels (Giuntella et al., 2021), and popular times at parks, greenways, or other physical activity places.

Traffic Simulation can be an alternative to measured probe data in order to model and understand human mobility patterns. To set up a realistic simulation scenario, other data sources are required, such as population distributions, high-traffic times, and estimates of daily activity patterns. These are discussed in the following subsections. Beyond traffic planning, simulated traffic data have been used to plan emergency evacuations (Tranouez et al., 2012), and to determine building occupancy (Berres et al., 2019).

3.2.2 Stationary Sensor Data

Stationary sensors contrast with probe data and refer to any data stemming from sensors that have a fixed location.

Traffic Sensors have become very common in smart city environments. They are typically maintained by traffic agencies and placed in key locations for traffic management. They deliver information on traffic conditions to traffic operators. For instance, sensors are placed along the highway system to collect information such as vehicle counts and speeds, often with lane-level detail, and vehicle classification. Other sensors are placed at intersections, where they collect vehicle counts, speeds, and turn movements, as well as other information, such as traffic signal state. There are also semi-stationary sensors that work in groups which are placed along a stretch of road to determine travel times by tracking the arrival and departure of Bluetooth identifiers at each sensor. These sensors are usually only set up for a short time before they are moved to a new location, but they do not move during data collection. Traffic sensor data have been used to study building occupancy (Berres et al., 2022).

Aggregate Probe Data are derived from the collection of devices within a traffic system. While this data is not technically from one specific stationary sensor, but rather from a collection of probes, one could consider a road segment to be a virtual sensor, for which data is collected. The individual probes' information is aggregated in reference to a specific section of road, or a bigger area. No single data source includes all vehicles on the road, so it is important to understand the data source's so-called *penetration rate*, and the percentage of vehicles that share data with the source. Derived data typically focuses on information that is meaningful despite a small sample size, such as speed and travel time. Many data sources also include the number of probes that this data is collected from, which enables the data user to better understand data reliability.

Cellphone Towers manage the connectivity of all cellphones in their cell, and they coordinate handover with adjacent cells whenever a device crosses over into a neighbor's cell. The spatial precision of these data depends on the density of cellphone towers; however, it is sufficient to determine the approximate travel distances of different devices. They have been popular for the analysis of mobility in the United States during the early months of the COVID-19 pandemic (Gao et al., 2020; Thakur, 2020; Kupfer et al., 2021).

Traffic Analysis Zones (TAZ) are polygonal data that outline areas of interest to transportation officials to quantify the amount of traffic between different zones during different times of day (e.g., for commute). Cellphone data are a valuable resource for creating this data.

Call-Detail Records (CDRs) are the records telecommunications companies maintain for billing purposes. They contain details of the call and SMS records between anonymized mobile phone users and can be thought of as a derivative data product from the raw data generated by cellphone towers. The data contains the interaction between two user IDs that contain the duration of call or occurrence of an SMS and importantly IDs for the cellphone towers the interaction was routed through for both the sender and receiver. This type of data was one of the first data types used to measure large-scale human mobility interactions and infer place-based temporal dynamics

(González et al., 2008; Ahas et al., 2015). While the use of GPS-based mobility data has become increasing prominent over the last decade, CDRs are still an important source of insights being used recently for city-level building occupancy estimates (Barbour et al., 2019) as well as in developing areas where cell phone penetration is high, but smartphone-based GPS data are not (Pestre et al., 2020).

Wi-Fi Mac Traps (Access Point Logs): Each wireless access point (AP) or switch port (i.e., an aggregation of APs in a building) collects the wireless session record that contains individuals connecting to these access points. This record trace provides the individual's equipment (cellphone, laptops, etc.) Media Access Control (MAC) address, the start and end events for device associations (when they visited or left that specific AP), the date and time of those events, and the AP (or switch) IP and port numbers. A media access control address is a one-of-a-kind identifier assigned to a network interface controller for use as a network address in intra-network communications. From this information, we can deduce the association history (i.e., the location and time of user affiliation) for all MAC addresses. The MAC address of a device and its association to its owners is a unique combination and therefore can help in developing a digital trail of location-based visiting patterns of an individual holding the device.

3.2.3 Place-Based Data

We define place-based data as data that are focused on specific POIs such as gas stations, grocery stores, restaurants, and theaters. These data are often derivative products from probe-based cellphone data.

Activity or Popular Times data aggregate and process user location data to produce a measure of how busy a POI typically is throughout the week at an hourly resolution. While each product is produced through propriety processing and differs in user bases, the result is the creation of millions of individualized profiles representing the occupancy dynamics of real-world buildings and places across the globe. This data represents a notable increase in the resolution at which the impact of human dynamics can be incorporated into several different areas of science. Some sources that have demonstrated promise are SafeGraph and Google Popular Times (GPT) data. These data provide scaled estimates of how busy a POI is by the hour of the day and day of the week. These data are passively collected, which is more likely to be representative of the population than actively collected sources like Foursquare where a user has to "check-in" (McKenzie et al., 2020). The underlying data source for these estimates is passively collected cell phones. It has shown to be in line with other methods of estimating occupancies such as traffic counts (Bandeira et al., 2020) and Wi-Fi (Timokhin et al., 2020). Given Google's pervasiveness, it also has the largest spatial coverage of any source. The largest drawback with GPT is its black-box methodology. How the numbers are generated, which points of interest have their popular times data available, how often it changes over time, and the demographics of the users from which the data was collected can only be inferred because Google does not publish how the data is generated. All these drawbacks are noteworthy; however, all mobility data sources suffer from many if not all the same problems. We must acknowledge these data's shortcomings and progress with methods and conclusions that attempt to take them into account.

Geosocial data are from social network applications that allow users to "check-in" or tag locations as part of a post on the social network. Sources such as Foursquare, Twitter, and Facebook have been widely used in human mobility (Xu et al., 2022) and temporal dynamics research. Users must actively "check-in" or tag themselves to a location versus the passively collected activity or popular times data discussed previously. While interesting for social networking aspects, their application for understanding place-based activity patterns at large has been shown to be limited. Issues of self-selection and social desirability biases influence who, where, and when users include location-based information in their reported content.

Internet Protocol (IP) Network Flow Data is a unidirectional sequence of packets that transit through a network device and have certain standard features (e.g., IP address and port number source and destination) (e.g., router). It is used to understand individual activity patterns in online behavior. This equipment can be used to capture flow data. The acquired data provide fine-grained metering for thorough consumption analysis. The start and finish timings (or duration), source, and destination IP addresses, port numbers, protocol numbers, and flow sizes (in packets and bytes) are all included in flow records data. The destination IP address may be used to determine which websites were visited, while the port and protocol numbers can determine which application was utilized. The IP addresses are often geo-located, allowing fine-grained analysis of place-based activity patterns of individuals.

Buildings are essential to the human environment representing a critical spatial unit of measure when considering the temporal dynamics of places: according to the National Human Activity Pattern Survey (NHAPS), Americans spend 87% of their time in enclosed buildings. It is therefore important to understand where buildings are located, how big they are, and what they are used for. Microsoft has published building footprints for 129.6 million US buildings (Microsoft, 2018) and Oak Ridge National Laboratory (ORNL) has developed an authoritative building footprints dataset (Yang et al., 2018) for disaster and emergency response and recovery found in the Federal Emergency Management Agency (FEMA) website (https://disasters.geoplatform.gov/publicdata/Partners/ORNL/USA_Structures/). These footprints are 2-dimensional polygons of the building shape, but they do not include information on building type or building height, which are needed to determine the overall size of the buildings, and what they are used for. New et al. (2021) published building models for 122.9 million US buildings, which include estimated heights, number of floors, building type, and other parameters.

3.2.4 Human-Centric Data

Survey Data can provide valuable insights into the lives of a population. *Census* and Microcensus data can provide insights into the demographics of a particular area. Due to privacy concerns, the amount of detail available for public use varies. In the United States, the *Public Use Microdata Sample* (PUMS) provides 1- and 5-year aggregate data for Public Use Microdata Areas (PUMA), which contained up to 100,000 individuals at the time of definition. For each area, this data contains responses to the American Community Survey (ACS), such as age, family type, and the number of children, but also includes home and work PUMA, time of departure and arrival at home and at work, travel times, vehicle occupancy, and means of transportation to work, as well as very detailed demographic and health information. The *National Household Travel Survey* (NHTS) is performed by the U.S. Federal Highway Administration every couple of years. It contains detailed information on travel behavior, including demographics of each household, the number of drivers, vehicles, and typical weekly trips for each day of the week with the trip purpose (e.g., to/from work, school, grocery shopping), departure times, and arrival times.

Population Data and Modeling now incorporate locational and POI data as a refinement of population estimates at the building level due to the increased availability of building footprints extracted from high-resolution imagery discussed in the Place-Based Data subsection. For nearly a quarter of a century, ORNL has developed an annual update of ambient global population counts (Dobson et al., 2000) through the LandScan (https://landscan.ornl.gov) program. Each new global release consists of the country census and boundary updates where new data is available, high-resolution settlement or building footprints that effectively push people into the built environment, and POI or other volunteered or spatiotemporal data that identifies activity spaces over a 24-hour period for distribution of the data (Rose & Bright, 2014). There are several other global population datasets developed over the past decade involving other input datasets and modeling approaches. Leyk et al. (2019), review the available gridded global datasets and their "fitness of use" to guide the user community (from analysts, and researchers, to decision and policymakers) in the appropriate application of the data.

Focusing on activity spaces and exploiting the plethora of multimodal and high-resolution geospatial data within the United States, population estimates for the night (residential) and day are available at ~100 m^2 and developed through disaggregation of the block-level census and the use of dasymetric methods (intelligent areal interpolation methods) for development of the LandScan USA dataset (https://LandScan.ornl.gov). Building footprints are disaggregated into "pieces" (sub-building level in some cases) based on parcel and location or activity data (e.g., POI, OpenStreetMap, and Safegraph) (Moehl et al., 2021; Brelsford et al., 2021; Weber et al., 2018) and census data distributed. The method for population allocation for the daytime is much more complex and requires several other census datasets and methods to determine place of employment (day) movement from residential (night) (Bhaduri et al., 2007). Research by Sparks et al. (2022), defines different activity periods, morning, day, evening, and night and over several seasons and could be used to inform the gridded population estimates. There is also the opportunity to develop seasonal datasets over time for more dynamic and improved population estimates.

To maintain privacy, but quantify key social, cultural, and economic factors at a high spatial resolution, microdata collected at the PUMA level is decomposed through the PMEDM (Penalized Maximum-Entropy Dasymetric Modeling) algorithm to the block group (Nagle et al., 2014). The resulting probability estimates determine key demographic variables (e.g., ownership of a car, income level, and age) that typify individual or social vulnerability to various impacts such as energy and natural hazard/environment (Tuccillo & Spielman, 2021). Simulations of worker populations for dasymetric modeling and travel flows capture the dynamic movements of a daytime subpopulation for more accurate and relevant population estimates (Morton et al., 2017).

Outside the US, the data scene is a mixed bag of availability and either mirrors the United States in abundance (developed countries) whereas many developing countries around the world are data scarce and without a current or reliable census (Weber et al., 2018). The higher resolution population models flexibly adapt to the disparate data at the building level with resulting gridded population estimates for several countries around the world (https://LandScan.ornl.gov). This is accomplished either through disaggregation of census data (top-down approach) or the development of population estimates from the bottom up in lieu of reliable census data (Weber et al., 2018; Leasure et al., 2020).

Another novel global population modeling research focuses specifically on building occupancy (or activity spaces) through a Bayesian learning approach. The resulting data is an ambient probability estimate of people/1,000 ft^2 for the day, night, and episodic at the national level that accounts for uncertainty at the data input level (Stewart et al., 2016; Duchscherer et al., 2018). Night and day estimates are economically defined by POI hours of operation (Sparks et al., 2017, 2020).

3.3 WHAT'S GOING ON

Many data types detailed above have created an opportunity to empirically test previously held theories or created new hypotheses that are now testable across a range of disciplines. Below we briefly identify multiple disciplines with ongoing research problems related to temporal dynamics of places and mobility.

Transportation: In the field of Transportation, one of the primary concerns for short-term and long-term planning is to understand current and future demand on the traffic system. In the short term, some of the major challenges include optimization of traffic signals to allow for the fastest possible traffic flow and rerouting traffic along alternate routes in case of congestion. In the long term, these traffic planners make decisions about long-term investments, such as expanding existing roads (e.g., adding more lanes) or the addition of new roads (e.g., a bypass road to guide through traffic around their city. These goals require an understanding of how humans move through a city or between cities, and how this behavior may change in the future. The typical representation of these data is in the form of origin-destination matrices. These matrices contain information at a polygonal aggregate scale, which tells us how many people travel from each area to each other area, and at what time of day they travel. The reality, however, is much more complex, as most individuals have other trips throughout their week, to drop their children off at school, go grocery shopping, attend group gatherings, etc. Some of these insights can be obtained from survey data such as NHTS, or TAZ data, but these sources do not account for day-to-day variations, that are not planned for.

Urban Planning: Beyond traffic, population dynamics play an important role in city-scale urban planning. In areas with rapid population growth, a lack of urban planning can lead to excessive urban sprawl which can lead to a severe climate impact on the urban environment through the formation of urban heat islands (Myint et al., 2015). Another important task of urban planning is the zoning (or re-zoning) of urban areas. This determines how each neighborhood is structured, whether it is walkable, and how much community it will foster (Jacobs, 1961).

Geography: In human dynamics research occupancy signatures can be used to aid in POI classification and recommendation, and high-resolution population distributions. Spatial patterns of population and mobility exist on different timescales from daily (e.g., commute), to seasonal (e.g., vacations), and long-term (e.g., population growth). Over the past decade, the influx in social connectivity data, such as GPT and Foursquare, has contributed to increasingly detailed population modeling, measuring the distribution of temporal activities in cities and monitoring urban changes (Ahas et al., 2015). Furthermore, this type of data has been used to explore a place-based understanding of population dynamics by making inferences about a particular POI. The temporal visitation patterns of certain types of businesses differ widely depending on what function they serve. Accordingly, occupancy profiles of a POI can be used to classify the type of business of a POI (Silva & Silva, 2018; Milias & Psyllidis, 2021) as well as predict its expected temporal activity patterns (D'Silva et al., 2018; Timokhin et al., 2020). By combining POIs occupancy dynamics along with their spatial arrangement, one can characterize regional variability and produce estimates of a city's functional activity spaces (McKenzie et al., 2015; Janowicz et al., 2019; D'Silva 2020; Lískovec et al., 2022).

Computer Science: Data-driven approaches to design context-aware adaptive networking protocols in ad hoc and cellular networks are continuously evolving for the past decade (Thakur & Helmy, 2013). With the availability of finer-resolution spatiotemporal footprint of human mobility patterns and decades of theoretical research behind characterizing such pattern, the computer science community has added advantage to leverage this knowledge toward developing efficient, resilient, and robust wireless protocols that are adaptive to speed, occupancy, and activities.

National Security: Accurate estimates of where people are, for how long, and how many are significant to human and national security posture. Such estimates allow safe and secure evacuation planning, analysis of critical infrastructure, and allocation of resources such as energy and gas. Such mobility modeling characterization is essential for national security and offers evidence-based guidelines to government, non-profit, and private enterprise leaders on how to build the resilience of communities, ensure safe and inclusive patterns, develop predictability out of conflict, and widen national security discussions to include civilian methods to achieve these goals. These patterns studied over a course of duration inform migration that has major impacts on both the people and the places where they migrate which in turn informs measures to be taken toward national security assets.

Disaster Relief: Disaster relief and recovery, both natural and man-made, need a coordinated response including several federal and state government organizations. Accurate and timely assessments of the effect and amount of devastation are the pillars of any recovery operation in order to achieve optimal resource allocation and activation of first responders (Thakur et al., 2022). Human mobility and temporal patterns of places provide efficient planning and allocation as well as placement of much-needed resources in a timely manner.

3.4 DISCUSSION, CONCLUSION, AND OPPORTUNITIES

This section describes opportunities and constraints for cross-discipline collaborations and identifies challenges for advancing temporal dynamics research that focus on data scalability and privacy.

3.4.1 Common Grounds

Building occupancy has long been of interest to both the fields of Building Science and Human Dynamics albeit at different spatial and temporal scales. Various sources of mobility data however have brought closer their scales of study and over the last few years Building Science and Human Dynamics have started to overlap in both data and methods. This confluence presents many opportunities for cross-fertilization. Exploring this recent overlap from the building science perspective, Dong et al. (2021) provided a thorough review of the current state and research needs of occupant behavior modeling, comparing methods within the building science community to the data and methods used in human dynamics research. They conclude that a more accurate and detailed approach to modeling occupant behavior that considers demographic, behavioral changes, and social network analysis would "transform traditional building design and operation, and open up a new paradigm for future research" and suggest that emerging data sources and methodologies from research domains concerned with urban-scale human mobility such as transportation, location intelligence, and disaster management could potentially meet these needs.

Raw sensor data is noisy, incomplete, and difficult to move and model at scale. From a methods point of view, the patchy nature of data collection puts functional constraints on the methods and approaches you can apply. A sophisticated machine learning algorithm might work very well in areas with lots of data but not all areas have a lot of data. The best method in a data-rich area may not even be able to be applied in a data-poor area, much less still be the best method.

Using sensor data (mostly cellphone-based location data) to derive occupant dynamics for a POI has been an area of interest for building science and human dynamics research communities for the last several years and now has become of commercial interest.

3.4.2 Data Privacy

The combination of technological progress and the unfettered collection and exploitation of millions of people's personal data has brought with it numerous examples that have placed corporate and societal incentives in direct opposition. However, there are at least areas where corporate incentive has led to the production of data that has the potential to help solve pressing societal issues rather than creating new ones. Concerns of individual privacy have been increasing, leading many organizations both political and corporate to begin to put in place policies that limit the amount and type of data mobile phone apps can legally collect. Going forward, more research needs to investigate the viability of place-based activity patterns as a privacy-preserving alternative to current research using individual-based activity patterns.

3.4.3 Data Ownership

One of the major determining factors of how useful a dataset will be lies in its ownership. The major distinction lies between publicly owned data and commercial data. Publicly owned data has the advantage of a high longevity, as public entities are often on a tight budget and the purchase of sensors or other technology is seen as a long-term investment in the future of a city, state, or nation. As these entities are typically publicly funded, the data are often free to use for researchers (if sometimes protected under non-disclosure agreements on raw data due to privacy or legal concerns), and many cities have started freely sharing some of their data through open data portals. Commercially owned data, on the other hand, is often a lot richer in features than publicly owned data, as the commercial sector usually has more resources (financial as well as technical expertise) to dedicate to the development of new data products. However, the commercial sector is much more fast moving than the public sector, and data that is available today may not be available in the future.

3.4.4 Data Quality and Transparency

In 1933 Alfred Korzybski famously remarked "the map is not the territory." As sources of place data become more detailed and the questions we ask of them more nuanced, we must remember this advice and recall that more detailed is not synonymous with more accurate. The Commercial origins of most place data bring with it direct and indirect influences on what and how places are represented in the data or if a place is represented at all.

The lens of Activity or Popular times data is of course a distorted and sometimes opaque view on the true nature of human mobility. The types of places in which they maintain or ignore, how they decide to label them, where they can and cannot collect data, who the data is being collected from, and how they turn that data into a measure of temporal usage are all sources of distortion and must be taken into consideration when conclusions are being made. This is true however for all observational processes. Any phenomenon of interest exists within some larger environment through which it is to be observed. By deciding to observe one necessarily decides which aspects of that environment will be explicitly controlled for, which aspects will be acknowledged, and their influence mitigated, and which aspects, whether explicitly or implicitly, will be ignored. Activity and Popular Times data allow us an unprecedented view into how places are utilized around the world, yet they have obvious limitations that must be considered at all times.

Time is a limited resource. No matter how rich or poor any person, any town, or any city, we all have 24 hours in a day and 168 hours in a week. How we as individuals or as a collective choose to use that time tells us something about ourselves, our cities, our nations, and our businesses, that is not just an illustration of human behavior, but of economic processes, social processes, and the culture of places.

REFERENCES

Ahas, R., Aasa, A., Yuan, Y., Raubal, M., Smoreda, Z., Liu, Y., Ziemlicki, C., Tiru, M., & Zook, M. (2015). Everyday space–time geographies: Using mobile phone-based sensor data to monitor urban activity in Harbin, Paris, and Tallinn. *International Journal of Geographical Information Science*, 29(11), 2017–2039.

Aledavood, T., Kivimäki, I., Lehmann, S., & Saramäki, J. (2022). Quantifying daily rhythms with non-negative matrix factorization applied to mobile phone data. *Scientific Reports*, 12(1), 1–10.

Alessandretti, L., Sapiezynski, P., Lehmann, S., & Baronchelli, A. (2017). Multi-scale spatio-temporal analysis of human mobility. *PLoS One*, 12(2), e0171686.

Alessandretti, L., Aslak, U., & Lehmann, S. (2020). The scales of human mobility. *Nature*, 587(7834), 402–407.

Aschoff, J. & Wever, R. (1976). Human circadian rhythms: a multioscillatory system. *Federation Proceedings*, 35(12), 236–232.

Bandeira, J.M., Tafidis, P., Macedo, E., Teixeira, J., Bahmankhah, B., Guarnaccia, C., & Coelho, M.C. (2020). Exploring the potential of web based information of business popularity for supporting sustainable traffic management. *Transport and Telecommunication*, 21(1), 47–60.

Barbour, E., Davila, C.C., Gupta, S., Reinhart, C., Kaur, J., & González, M.C. (2019). Planning for sustainable cities by estimating building occupancy with mobile phones. *Nature Communications*, 10(1). doi: 10.1038/s41467-019-11685-w.

Berres, A., Im, P., Kurte, K., Allen-Dumas, M., Thakur, G., & Sanyal, J. (2019). A mobility-driven approach to modeling building energy, *2019 IEEE International Conference on Big Data (Big Data)*, 2019, pp. 3887–3895, doi: 10.1109/BigData47090.2019.9006308.

Berres, A., Bass, B., New, J.R., Im, P., Urban, M., & Sanyal, J. (2022). Generating traffic-based building occupancy schedules in Chattanooga, Tennessee from a grid of traffic sensors. *Building Simulation Conference 2021*. doi: 10.26868/25222708.2021.30744.

Bhaduri, B., Bright, E., Coleman, P., & Urban, M.L. (2007). LandScan USA: A high-resolution geospatial and temporal modeling approach for population distribution and dynamics. *GeoJournal*, 69(1), 103–117.

Bonnet, M.H. & Arand, D.L. (1995). We are chronically sleep deprived. *Sleep*, 18(10), 908–911.

Brelsford, C., Krapu, C., Tuccillo, J., McCarthy, M., McKee, J., & Singh, N. (2021). AI-improved resolution projections of population characteristics and imperviousness can improve resolution and accuracy of urban flood predictions. No. AI4ESP-1016. Artificial Intelligence for Earth System Predictability (AI4ESP) Collaboration (United States).

Broadwater, L. & Nierenberg, A. (2022). A groggy senate approves making daylight saving time permanent. *The New York Times*. https://www.nytimes.com/2022/03/15/us/politics/daylight-saving-time-senate.html.

Cichocki, A., Zdunek, R., Phan, A.H., & Amari, S.I. (2009). *Nonnegative Matrix and Tensor Factorizations: Applications to Exploratory Multi-way Data Analysis and Blind Source Separation*. John Wiley & Sons, Hoboken, NJ.

Cooper, K. (1997, March 1). Islamic Pakistan adopts Western-Style Weekends. *The Washington Post*. https://www.washingtonpost.com/archive/politics/1997/03/01/islamic-pakistan-adopts-western-style-weekends/06c24db6-0d81-4981-bd07-3d39f9773354/.

D'Silva, K. (2020). Modeling urban venue dynamics through Spatio-temporal metrics and complex networks. Doctoral dissertation, University of Cambridge.

D'Silva, K., Noulas, A., Musolesi, M., Mascolo, C., & Sklar, M. (2018). Predicting the temporal activity patterns of new venues. *EPJ Data Science*, 7, 1–17.

Dinges, D.F., Pack, F., Williams, K., Gillen, K.A., Powell, J.W., Ott, G.E., Aptowicz, C., & Pack, A.I. (1997). Cumulative sleepiness, mood disturbance, and psychomotor vigilance performance decrements during a week of sleep restricted to 4–5 hours per night. *Sleep*, 20(4), 267–277.

Dobson, J.E., Bright, E.A., Coleman, P.R., Durfee, R.C., & Worley, B.A. (2000). LandScan: A global population database for estimating populations at risk. *Photogrammetric Engineering and Remote Sensing*, 66(7), 849–857.

Dong, B., Liu, Y., Fontenot, H., Ouf, M., Osman, M., Chong, A., Qin, S., Salim, F., Xue, H., Yan, D., & Jin, Y. (2021). Occupant behavior modeling methods for resilient building design, operation and policy at urban scale: A review. *Applied Energy*, 293, 116856.

Duchscherer, S., Stewart, R., & Urban, M. (2018). Revengc: An R package to reverse engineer summarized data. *The R Journal*, 10(2), 114–123.

Duffy, J.F., Cain, S.W., Chang, A.M., Phillips, A.J., Münch, M.Y., Gronfier, C., Wyatt, J.K., Dijk, D.J., Wright Jr, K.P., & Czeisler, C.A. (2011). Sex difference in the near-24-hour intrinsic period of the human circadian timing system. *Proceedings of the National Academy of Sciences*, 108(3), 15602–15608.

Gao, S., Rao, J., Kang, Y., Liang, Y., & Kruse, J. (2020). Mapping county-level mobility pattern changes in the United States in response to COVID-19. *SIG Spatial Special*, 12(1), 16–26.

Giuntella, O. & Mazzonna, F. (2019). "Sunset time and the economic effects of social jetlag: Evidence from US time zone borders." *Journal of Health Economics*, 65, 210–226.

Giuntella, O., Hyde, K., Saccardo, S., & Sadoff, S. (2021). Lifestyle and mental health disruptions during COVID-19. *Proceedings of the National Academy of Sciences*, 118(9), e2016632118.

González, M.C., Hidalgo, C.A., & Barabási, A.-L. (2008). Understanding individual human mobility patterns. *Nature*, 453(7196), 779–782. doi: 10.1038/nature06958.

Happle, G., Fonseca, J.A., & Schlueter, A. (2020). Context-specific urban occupancy modeling using location-based services data. *Building and Environment*, 175, 106803.

Jacobs, J. (1961). *The Death and Life of Great American Cities*. Vintage Books, New York.

Janowicz, K., McKenzie, G., Hu, Y., Zhu, R., & Gao, S. (2019). Using semantic signatures for social sensing in urban environments. In: *Mobility Patterns, Big Data and Transport Analytics*. Elsevier: Amsterdam, Netherlands, pp. 31–54.

Keogh, E.J. & Pazzani, M.J. (2001). Derivative dynamic time warping. *In Proceedings of the 2001 SIAM International Conference on Data Mining*, Chicago, IL, Society for Industrial and Applied Mathematics, pp. 1–11.

Kupfer, J.A., Li, Z., Ning, H., & Huang, X. (2021). Using mobile device data to track the effects of the COVID-19 Pandemic on spatiotemporal patterns of national park visitation. *Sustainability*, 13(16), 9366.

Leasure, D.R., Jochem, W.C., Weber, E.M., Seaman, V., & Tatem, A.J. (2020). National population mapping from sparse survey data: A hierarchical Bayesian modeling framework to account for uncertainty. *Proceedings of the National Academy of Sciences*, 117(39), 24173–24179.

Leyk, S., Gaughan, A.E., Adamo, S.B., de Sherbinin, A., Balk, D., Freire, S., Rose, A., Stevens, F.R., Blankespoor, B., Frye, C., & Comenetz, J. (2019). The spatial allocation of population: A review of large-scale gridded population data products and their fitness for use. *Earth System Science Data*, 11(3), 1385–1409.

Lískovec, R., Lichter, M., & Mulíček, O. (2022). Chronotopes of urban centralities: Looking for prominent urban times and places. *The Geographical Journal*, 00, 1–11. Available from: doi: 10.1111/geoj.12426.

Liu, L., Andris, C., & Ratti, C. (2010). Uncovering cabdrivers behavior patterns from their digital traces. *Computers, Environment and Urban Systems*, 34(6), 541–548.

McKenzie, G., Janowicz, K., Gao, S. and Gong, L. (2015). How where is when? On the regional variability and resolution of geosocial temporal signatures for points of interest. *Computers, Environment and Urban Systems*, 54, 336–346.

McKenzie, G., Janowicz, K., & Keßler, C. (2020). Uncovering spatiotemporal biases in place-based social sensing. *AGILE: GIScience Series*, 1, 1–17.

Microsoft (2018). US Building Footprints. Retrieved from https://github.com/Microsoft/USBuildingFootpri nts.

Milias, V. & Psyllidis, A. (2021). Assessing the influence of point-of-interest features on the classification of place categories. *Computers, Environment and Urban Systems*, 86, 101597.

Moehl, J.J., Weber, E.M., & McKee, J.J. (2021). A vector analytical framework for population modeling. *The International Archives of Photogrammetry, Remote Sensing and Spatial Information Sciences*, 46, 103–108.

Monk, T.H. & Folkard, S. (1976). Adjusting to the changes to and from daylight saving time. *Nature*, 261(5562), 688–689.

Monsivais, D., Ghosh, A., Bhattacharya, K., Dunbar, R.I., & Kaski, K. (2017). Tracking urban human activity from mobile phone calling patterns. *PLoS Computational Biology*, 13(11), e1005824.

Morton, A., Piburn, J., Nagle, N., Aziz, H.M., Duchscherer, S.E., & Stewart, R. (2017). *A Simulation Approach for Modeling High-Resolution Daytime Commuter Travel Flows and Distributions of Worker Subpopulations*. Oak Ridge National Lab (ORNL), Oak Ridge, TN.

Myint, S.W., Zheng, B., Talen, E., Fan, C., Kaplan, S., Middel, A., Smith, M., Huang, H.P., & Brazel, A. (2015). Does the spatial arrangement of urban landscape matter? Examples of urban warming and cooling in Phoenix and Las Vegas. *Ecosystem Health and Sustainability*, 1(4), 1–15.

Nagle, N., Buttenfield, B., Leyk, S., & Spielman, S. (2014). Dasymetric modeling and uncertainty. *Annals of the Association of American Geographers*, 104(1), 80.

National Research Council. (2014). *Convergence: Facilitating Transdisciplinary Integration of Life Sciences, Physical Sciences, Engineering, and Beyond*. The National Academies Press, Washington, DC. doi: 10.17226/18722.

New, J., Adams, M., Berres, A., Bass, B., & Clinton, N. (2021). Model America-data and models of every US building. Oak Ridge National Lab (ORNL), Oak Ridge, TN. Oak Ridge Leadership Computing Facility (OLCF); Argonne National Laboratory (ANL) Leadership Computing Facility (ALCF). doi: 10.13139/ORNLNCCS/1774134.

Park, M., Thom, J., Mennicken, S., Cramer, H., & Macy, M. (2019). Global music streaming data reveal diurnal and seasonal patterns of affective preference. *Nature Human Behaviour*, 3(3), 230–236.

Pestre, G., Letouzé, E., & Zagheni, E. (2020). The ABCDE of big data: Assessing biases in call-detail records for development estimates. *The World Bank Economic Review*, 34(1), S89–S97. doi: 10.1093/wber/lhz039.

Roenneberg, T., Kumar, C.J., & Merrow, M. (2007). The human circadian clock entrains to sun time. *Current Biology*, 17(2), R44–R45.

Rose, A.N. & Bright, E. (2014). The LandScan Global Population Distribution Project: Current state of the art and prospective innovation. Oak Ridge National Laboratory.

Roenneberg, T., Wirz-Justice, A., & Merrow, M. (2003). Life between clocks: Daily temporal patterns of human chronotypes. *Journal of Biological Rhythms*, 18(1), 80–90.

Russell, A. (2020). Mapping Northern Ireland's Post-Brexit Future. *The New Yorker*. https://www.newyorker.com/news/letter-from-the-uk/mapping-northern-irelands-post-brexit-future.

Sandhu, A., Seth, M., & Gurm, H.S. (2014). Daylight savings time and myocardial infarction. *Open Heart*, 1(1), e000019.

Schläpfer, M., Dong, L., O'Keeffe, K., Santi, P., Szell, M., Salat, H., Anklesaria, S., Vazifeh, M., Ratti, C., & West, G.B. (2021). The universal visitation law of human mobility. *Nature*, 593(7860), 522–527.

Silva, L.A. & Silva, T.H. (2018). Extraction and exploration of business categories signatures. *Workshop on Big Social Data and Urban Computing*, Springer, Cham.

Sparks, K., Thakur, G., Urban, M., & Stewart, R. (2017). Temporal signatures of shops' and restaurants' opening and closing times at global, country, and city scales. *In GeoComputation 2017*, Leeds, UK.

Sparks, K., Thakur, G., Pasarkar, A., & Urban, M. (2020). A global analysis of cities' geosocial temporal signatures for points of interest hours of operation. *International Journal of Geographical Information Science*, 34(4), 750–776. doi: 10.1080/13658816.2019.1615069.

Sparks, K., Moehl, J., Weber, E., Brelsford, C., & Rose, A. (2022). Shifting temporal dynamics of human mobility in the United States. *Journal of Transport Geography*, 99, 103295.

Stewart, R.N., Urban, M., Duchscherer, S., Kaufman, J., Morton, A., Thakur, G., Piburn, J., & Moehl, J. (2016). A Bayesian machine learning model for estimating building occupancy from open source data, *Natural Hazards*, 81(3), 1929–1956.

Taylor, S.J. & Letham, B. (2018). Forecasting at scale. *The American Statistician*, 72(1), 37–45.

Thakur, G.S. & Helmy, A. (2013). COBRA: A framework for the analysis of realistic mobility models. *2013 Proceedings IEEE INFOCOM,* IEEE. doi: 10.1109/INFCOMW.2013.6562895.

Thakur, G., et al. (2020). COVID-19 joint pandemic modeling and analysis platform. *Proceedings of the 1st ACM SIGSPATIAL International Workshop on Modeling and Understanding the Spread of COVID-19.* doi: 10.1145/3423459.3430760.

Thakur, G., Sparks, K., Berres, A., Tansakul, V., Chinthavali, S., Whitehead, M., Schmidt, E., Xu, H., Fan, J., Spears, D., & Cranfill, E. (2022). Accelerated assessment of critical infrastructure in aiding recovery efforts during natural and human-made disaster. *Proceedings of the 29th International Conference on Advances in Geographic Information Systems.* doi: 10.1145/3474717.3483947.

Timokhin, S., Sadrani, M., & Antoniou, C. (2020). Predicting venue popularity using crowd-sourced and passive sensor data. *Smart Cities*, 3(3), 818–841.

Tranouez, P., Daudé, É., & Langlois, P. (2012). A multiagent urban traffic simulation. *Journal of Nonlinear Systems and Applications*, 98, 106.

Treisman, R. (2021, December 8). The UAE is adopting a 4.5-day workweek and a Saturday-Sunday weekend. NPR. https://www.npr.org/live-updates/morning-edition-2021-12-08#the-uae-is-adopting-a-4-5-day-workweek-and-a-saturday-sunday-weekend.

Tuccillo, J.V. & Spielman, S.E. (2021). A method for measuring coupled individual and social vulnerability to environmental hazards. *Annals of the American Association of Geographers*, 112(6), 1702–1725.

Weber, E.M., Seaman, V.Y., Stewart, R.N., Bird, T.J., Tatem, A.J., McKee, J.J., Bhaduri, B.L., Moehl, J.J., & Reith, A.E. (2018). Census-independent population mapping in northern Nigeria. *Remote Sensing of Environment*, 204, 786–798.

Wu, L., Zhi, Y., Sui, Z., & Liu, Y. (2014). Intra-urban human mobility and activity transition: Evidence from social media check-in data. *PLoS One*, 9(5), e97010.

Xu, S., Li, S., Huang, W., & Wen, R. (2022). Detecting spatiotemporal traffic events using geosocial media data. *Computers, Environment and Urban Systems*, 94, 101797.

Yang, H.L., Yuan, J., Lunga, D., Laverdiere, M., Rose, A., & Bhaduri, B. (2018). Building extraction at scale using convolutional neural network: Mapping of the united states. *IEEE Journal of Selected Topics in Applied Earth Observations and Remote Sensing*, 11(8), 2600–2614. doi: 10.1109/JSTARS.2018.2835377.

Zhou, H., Zhang, S., Peng, J., Zhang, S., Li, J., Xiong, H., & Zhang, W. (2021). Informer: Beyond efficient transformer for long sequence time-series forecasting. *In Proceedings of the AAAI Conference on Artificial Intelligence,* 35(12), 11106–11115.

4 Geospatial Knowledge Graph Construction Workflow for Semantics-Enabled Remote Sensing Scene Understanding

Abhishek Potnis, Surya S Durbha, Rajat Shinde, and Pratyush Talreja
Centre of Studies in Resources Engineering, Indian Institute of Technology Bombay, India

CONTENTS

4.1 INTRODUCTION AND MOTIVATION

Over the last few years, with the advancements in sensor technology, there has been a rapid rise in the use of remote sensing across multiple domains for a variety of applications. Due to this, there has been a meteoric rise in acquisition and availability of Earth Observation (EO) data. The volume of archived data in NASA's Earth Observing System Data and Information System (EOSDIS) has been documented to have reached 32.9 Petabytes in August 2019 (Wang and Yan, 2020). Moreover, it is estimated that the size of this data archive alone would exceed 246 Petabytes by 2025. Figure 4.1 depicts the estimated growth post-2021 in NASA's EOSDIS Data Archive. The archives of earth observation data captured throughout the years have huge potential in comprehending numerous natural as well as man-made phenomena. However, such archives of data would remain largely untapped, primarily due to the sheer volume and variety of the data. Thus, there is a dire need to semantically understand such heterogeneous remote sensing big data and derive actionable insights from it.

DOI: 10.1201/9781003270928-6

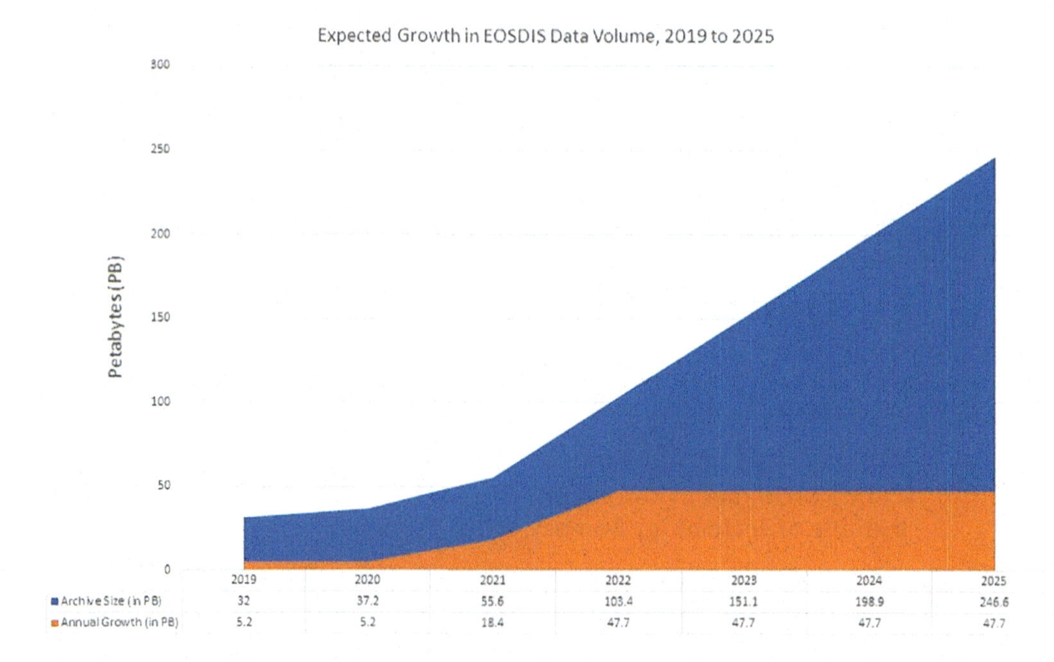

FIGURE 4.1 Estimated growth in data volume of NASA's EOSDIS data archive.[1]

Advent of different remote sensing platforms including spaceborne platforms such as satellites and airborne platforms such as drones and aircrafts has led to the challenge of complex heterogeneity and multi-modality while working with data products obtained from these platforms. In addition to the heterogeneity in platforms, there exists a wide variety of sensors for capturing relevant information depending on the applications. Multi-Spectral, Pan-Chromatic, Hyperspectral, and Microwave form the different remote sensing data products captured by various sensors mounted on different platforms. This further introduces heterogeneity in terms of the data processing and analysis due to the differences in the spatio-temporal as well as spectral resolution of such rich data products. This challenge of processing and analyzing voluminous, complex, and heterogeneous data to derive actionable insights calls for a need to focus on geospatial standards to ensure data interoperability among the data generated across these multimodal platforms.

4.1.1 Image Information Mining for Earth Observation (EO)

Numerous Earth Observation (EO) based applications such as monitoring of disasters such as forest fires, floods, earthquakes, and tsunamis require users to search and query remote sensing imagery based on their content. Such content-specific requirements for such crucial applications have fueled the inception and growth of the research area of Content-based Image Information Mining Systems over the last decade. Earlier, querying over an image was only restricted over the structural metadata—geographic coordinates, time of acquisition, sensor type, and acquisition mode. This restriction put severe constraints over operational users interested in the content of the remote sensing imagery. Over time, Content-based Image Retrieval (CBIR) Systems for Earth Observation data were researched upon. However almost all of those proposed systems were based on indexing of images in the feature space with most prominent features being analyzed were—texture, shape, and color. IBM QBIC System (Flickner et al., 1995), MIT Photobook System (Pentland et al., 1996), and Virage System are some of the examples of such CBIR System. Berkeley BlobWorld (Carson et al., 1999), UCSB Netra (Ma and Manjunath, 1999), and Columbia VisualSEEK (Smith and Chang, 1996) are the examples of Region-based Retrieval Systems that proposed to segment an image into regions followed by a similarity-based region retrieval approach. One of the earlier Image Information Mining systems for the EO domain was GeoBrowse (Marchisio et al., 1998). It leveraged distributed computing in addition to an object-oriented approach for relational databases toward data retrieval and storage. Geospatial Information Retrieval and Indexing System (GeoIRIS) (Shyu et al., 2007) facilitated identification of specific objects in satellite imagery using specialized descriptor features. Also, the GeoIRIS supported retrieval of satellite imagery using spatial configuration of identified objects. Self-organized maps (SOMs) were leveraged by the PicSOM system (Molinier et al., 2007) for optimization of content-based image retrieval operations.

As the research in this area of Image Information Mining progressed, the Knowledge-driven Information Mining (KIM) (Quartulli and Datcu, 2003) for EO data archives was proposed. Daschiel and Datcu (2005) proposed a system that extracted the primitive features of an image, followed by manual linking of the features with semantic interpretations. Hierarchical Segmentation was another approach that leveraged the intrinsic features of a region to develop heuristics for labeling of regions in satellite imagery. The Intelligent Interactive Image Knowledge Retrieval (I3K) (Durbha and King, 2005) used the concept of domain-dependent ontologies. The I3K System leveraged classical machine learning approaches for detecting and mapping object classes. Further deductive reasoning was leveraged to map the detected object classes to classes formalized in the ontology. The query processing and searching components were supported by description logic-based reasoning that facilitated implicit knowledge discovery from satellite imagery.

The Spatial Image Information Mining Framework (Kurte et al., 2017) inspired from the I3K System, uses the concept of domain-dependent ontologies. The SIIM framework largely focuses on formalizing the spatial-directional and topological relations among the regions in a remote sensing scene and consequently the development of Spatial Semantic Graph (SSG). Figure 4.2 depicts the generation of Spatial Semantic Graph (SSG) as proposed in the Spatial Image Information Mining (SIIM) Framework (Kurte et al., 2017). The Spatial Semantic Graph (SSG) is an RDF-based graph

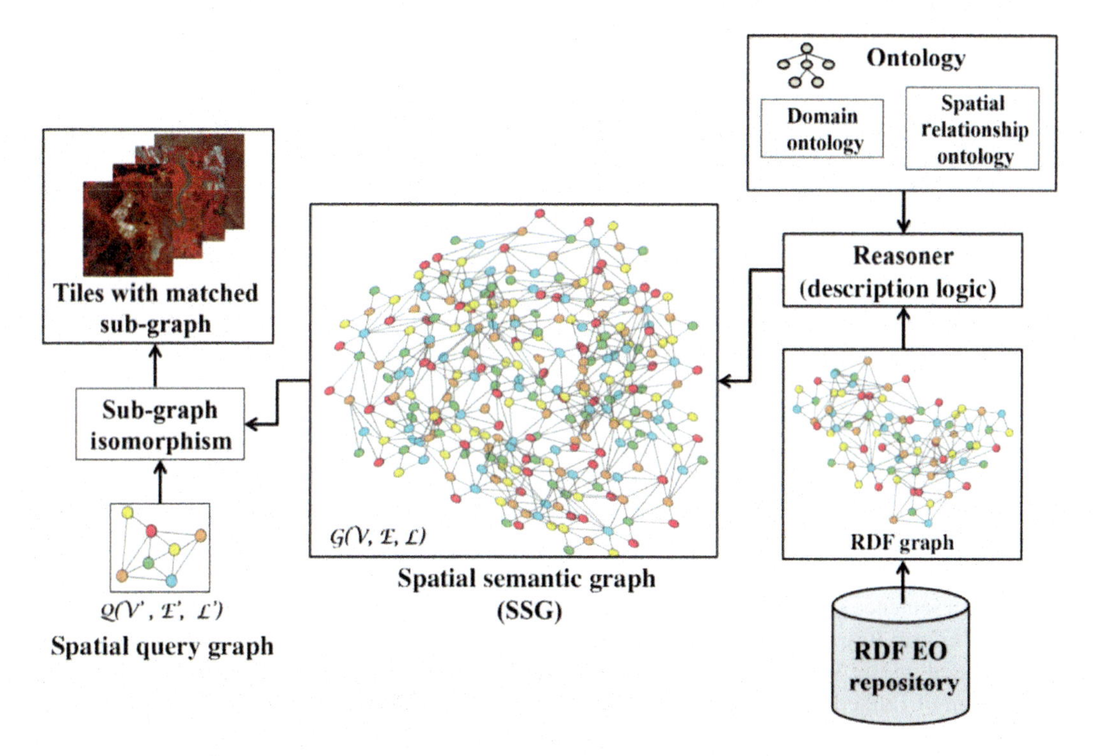

FIGURE 4.2 Generation and utility of spatial semantic graphs in content-based image retrieval as proposed by Kurte et al. (2017).

that enables effective description logic-based reasoning. The SIIM framework allows querying of content information, which none of other frameworks in this domain previously support. Its application is geared toward a post-flood scenario, where such content-based queries could prove to be great help for disaster management and mitigation. Thus, it strives to minimize the spatial semantic gap in remote sensing imagery. The SIIM also proposes an RDF-based data representation for modeling a remote sensing scene as Linked Data—mapping regions with relevant land use classes and their corresponding spatial relations with other regions. The proposed data model also facilitates representing the metadata such as geographical coordinates, and time of acquisition, along with CRS and other data lineage-related information. The Spatial Semantic Graph proposed in SIIM enables for efficient query answering and knowledge retrieval.

From the literature, it is evident that there is a paucity of EO-IIM Systems that support higher-level semantic abstraction, domain-specific spatio-contextual knowledge modeling, and complex spatial queries in addition to support intuitive natural language user interfaces. The paradigm of Geospatial Semantic Web aligns appropriately with this need to be able to seamlessly explore, navigate, integrate and make logical inferences over such heterogeneous dynamically evolving big data, thereby deriving actionable insights from it. This thus calls for a brief background on Semantic Web—Ontologies, Linked Data, and Geospatial Semantics.

4.1.2 Semantic Web

The web, in the nineteenth century, had been developed with the objective of sharing information through documents, intended to be read and understood by humans. However, as more and more people around the world have started using the web for a variety of needs, the complexity and the amount of heterogeneous data generated on the web each day has grown exponentially. This has demanded machines to be able to intelligently sift through and understand the huge amounts of data, thereby aiding humans to make decisions, across a wide spectrum of domains. It is this demand that has led computer scientists around the globe to explore this previously uncharted area of Semantic Web.

Semantic Web advocates structuring of the meaningful information in web pages (Berners-lee et al., 2001). It puts forth structural, syntactic, and semantic standardization of information on the web. Legacy search engines make use of keyword-based approaches to display the search results. This although effective for most queries does not prove fruitful when querying for complex information. For instance, searching for a word—'dog', the search engine will list all the web pages which contain the word 'dog' in them, without understanding the meaning of this word 'dog.' Semantic Web strives to eliminate this ignorance.

Knowledge-based systems have been studied by Artificial Intelligence researchers long before the advent of the World Wide Web. Automated Reasoning comprises Structured Data and Set of Inference Rules. Earlier Knowledge Representation Systems were centralized. The major challenge for knowledge representation has been to provide a standardized data model for representing data as well as domain-specific rules for reasoning over the data.

The Semantic Web Technologies – XML[2] (eXtensible Markup Language) and RDF[3] (Resource Description Framework) addresses this challenge by providing a standardized representation for modeling information. XML is used for syntactic standardization, while the RDF is used to convey the semantics. Information in the RDF representation is stored in the form of 'triples.' It is analogous to storing a sentence in the English language as Subject, Object, and Verb. To keep the information unique, Uniform Resource Identifiers bind concepts to unique definitions, which can be discovered and accessed on the Web. Thus, the paradigm of Semantic Web advocates storing of information in the form of Linked Data. This further ensures that the data are easily accessible, amenable for integration, and also compatible toward making logical inferences based on domain knowledge.

Linked Data is thus a method of publishing structured information on the web with an aim to interlink and enhance its value. It is intended to be read and "understood" by computers rather than humans. The delivery of such structured information on the web is facilitated by widely used web protocols such as HTTP and HTTPS. This term was first coined by Tim Berners Lee in 2006. Linked Open Data refers to the Linked Data that is open content—i.e., freely available to use and modify. DBpedia[4] and Freebase[5] are among the most popular Linked Open Data sets.

4.1.3 Ontologies and Reasoning

Tim Berners-lee et al. (Shadbolt et al., 2006) introduced the concept of Ontologies and highlighted its role in building the Semantic Web. An ontology is a document that formally anoints a taxonomy of classes and relations among the instances of the defined classes. Taxonomy is a representation of concepts in a hierarchical form, classifying concepts in a parent-child relationship for better understanding. An ontology thus helps represent different entities belonging to classes along with their relation among themselves, in addition to conforming to the concept of inheritance—enabling class entities to derive properties and relations. Ontologies contain equivalence relations to solve problems of similar concepts with different names. Ontologies can be consumed by programs or agents to achieve the desired results.

Formalizing domain-knowledge as ontologies can facilitate modeling of background information required for inferring the otherwise implicit knowledge from the incoming dynamically generated heterogeneous data. The process of deductive reasoning or inferencing in context to Semantic Web entails—(1) Classification of individuals to classes defined in an ontology. (2) Consistency checking of the individuals and their properties based on the axioms and rules defined in an ontology (3) Manifestation of otherwise implicit relationships and properties using the axioms formalized in the domain ontology. In that regard, the process of ontology-based deductive reasoning forms an integral part in making explicit the otherwise implicit knowledge, thus adding value to the existing knowledge base.

SWRL (Semantic Web Rule Language) was proposed by the W3C. The rule language, built on the Web Ontology Language (OWL) leverages the "antecedent→ consequent" structure to define axioms. It can easily assimilate axioms as a part of the defined schema, also facilitating enhanced expressivity at both the domain and application level.

The rule depicted in equation (4.1) is an example to infer the "hasAunt" relation:

$$\text{hasParent}(?x1,?x2) \wedge \text{hasSister}(?x2,?x3) \Rightarrow \text{hasAunt}(?x1,?x3) \tag{4.1}$$

[Source: http://www.w3.org/]

The "x1," "x2," and "x3" are referred to as *Named Individuals* in the ontological sense and are validated for the applicability of this rule. For example, if "x1" refers to a person named "John," "x2" refers to a person named "Mary" and "x3" refers to a person named "Jane." To establish the relationship "hasAunt" between John and Jane such that Jane is the aunt of John then the conditions—Mary being a parent of John and Jane being a sister of Mary would need to be met. Such rules represented in SWRL can be embedded in an ontology to aid in reasoning and inferencing the otherwise implicit classes and relationships.

4.1.4 Geospatial Knowledge Representation

Considering the diversity, volume, and richness of geospatial data sources, the concept of Semantic Web has been extended to the geospatial domain. The Geospatial Semantic Web constitutes domain-specific Geospatial Ontologies, Semantic Gazetteers, and standardized vocabularies toward describing and disseminating different geographic phenomenon in a standardized manner.

Sheth et al. (2008) have proposed a registry of domain-specific ontologies for modeling semantic sensor web. In that regard, a geospatial ontology to model the location-specific information, an ontology to represent the time-related information, and domain-specific ontologies to model the weather-related and sensor-related information have been developed. In this work, Sheth et al. have leveraged axioms encoded in SWRL, embedded in the ontologies to infer the otherwise implicit assertions from known instances.

Koubarakis et al. (2016) puts forth the life cycle of accessible Geospatial Linked Data. The cycle commences with ingestion of Earth Observation data followed by processes such as Context Extraction and Semantic Annotation to transform it into a Linked Data representation for storage, querying, reasoning, analyzing, and further publishing. Considering the numerous advantages of disseminating geospatial information in the form of Linked Open Data, there have been significant efforts by the geospatial research community at large to encourage the publishing of EO data as Geospatial Linked Data. This chapter thus advocates leveraging the Geospatial Linked Data Cloud to encourage standardization and interoperability, thereby envisaging a sustainable ecosystem of effective Geospatial Linked Data-based applications across different domains.

4.1.4.1 Ontology-Based Remote Sensing Image Analysis

Several research studies have leveraged Ontology-based knowledge representation to aid in remote sensing image analysis. The common strategy in these studies has been to embed in the ontological axioms—the conditions for assigning a particular class to a pixel based on the remote sensing indices values; and further perform reasoning to collate and generate a classified map result.

In Andrés et al. (2017), spectral rules have been embedded as Description Logic (DL) axioms to hierarchically infer complex concepts defined in Ontology. In addition to defining rules and concepts for relevant remote sensing indices, the authors also define contextual higher-order abstractions to relate with the more primitive features. Based on the contextual class definitions in the ontology, the image objects in the image are assigned labels. The higher-level semantic abstractions in the form of class labels such as "Vegetation," "ThickClouds," and "Shadow" have been defined using the 14 spectral rules (Baraldi et al., 2006). An example of the spectral rule for the "Vegetation," specific to the calibrated Landsat spectral bands is as follows:

$$\begin{aligned}\text{Vegetation}_{SR} = &(\text{ThematicMapper}_2 \geq 0.5*\text{TM}_1)\,\text{and}\\ &(\text{ThematicMapper}_2 \geq 0.7*\text{ThematicMapper}_3)\\ &\text{and}(\text{ThematicMapper}_3 < 0.7*\text{ThematicMapper}_4)\\ &\text{and}\left(\text{ThematicMapper}_4 > \max\begin{Bmatrix}\text{ThematicMapper}_1, \text{ThematicMapper}_2,\\ \text{ThematicMapper}_3\end{Bmatrix}\right)\\ &\text{and}(\text{ThematicMapper}_5 < 0.7*\text{ThematicMapper}_4)\\ &\text{and}(\text{ThematicMapper}_5 \geq 0.7*\text{ThematicMapper}_3)\\ &\text{and}(\text{ThematicMapper}_7 < 0.7*\text{ThematicMapper}_5)\end{aligned} \tag{4.2}$$

In equation (4.2), ThematicMapper_i denotes the pixel value of the ith band of Landsat Thematic Mapper (TM).

The spectral rule depicted in equation (4.2) has been embedded in the ontology using description logic axiom. The authors also proposed an alternate methodology to scale up reasoning capabilities in terms of the response time. The authors proposed to store the A-Box instances corresponding to individual pixels in an RDF Triple Store. It was found that reasoning over an RDF Triple Store greatly reduced the time taken for reasoning as depicted by the blue line in the graph.

Durand et al. (2007) proposed the use of spectral, spatial, and contextual concepts to identify image objects in remote sensing imagery. The remote sensing scene was proposed to be segmented, followed by feature extraction and then using the ontology-based deductive reasoner to infer labels for image objects. In addition to the spectral rules as leveraged by Andrés et al. (2017), the authors also leveraged spatial attributes such as diameter, area, perimeter, elongation, Miller Index, and Solidity Index; for image object identification. (Forestier et al., 2008) employed an unsupervised approach for clustering remote sensing imagery into segments followed by using low-level descriptors based on spectral and shape information for assigning class labels to individual segments.

Arvor et al. (2019) comprehensively justified the need for using ontologies for earth observation data analysis. The authors highlighted the challenges due to the sensory and semantic gap and elaborated on the advantages of leveraging an ontological knowledge base to address these challenges. Similar to (Andrés et al., 2017), the authors also proposed a hierarchical semantic enrichment of knowledge from primitive features and remote sensing indices to higher levels of abstractions in the ontology. The authors also advocated the use of ontologies considering the benefits in terms of knowledge sharing due to adherence to geospatial standards and interoperability.

Gu et al. (2017) employed a data-driven C4.5 (Quinlan, 1996) decision tree-based classification approach followed by a semantic knowledge-based approach for semantic classification high resolution remote sensing scenes. The authors proposed a three-level model including ontology modeling followed by initial classification based on decision tree algorithm followed by a re-classification of image objects using semantic rules defined in the ontology. The authors compared the results of the initial machine learning-based classification and the semantic network-based classification. It was found that the semantic network-based classification following the data-driven classification statistically outperformed the data-driven approach when compared in an isolated manner.

Arvor et al. (2013) comprehensively reviewed the advances in the research area of Object-based Image Analysis in Geospatial domain and subsequently the role of ontologies toward semantic interoperability, data publication, and automatic image interpretation. In addition to the benefits that the use of a knowledge-based approach entails, the authors examined the challenges in the form of scalability, updating, and processing performance issues envisaged as the research in this area progresses.

Alirezaie et al. (2019) proposed a novel fusion of deep neural network approach and a semantic knowledge-based approach for improving the performance of the semantic segmentation task over remote sensing imagery. The proposed 'semantic referee' conceptualizes the misclassifications through an ontological representation and provides relevant feedback to the neural network classifier to improve its performance. This work also proposed OntoCity—an ontology to model the structural, conceptual, and physical abstractions in the geospatial domain. The reasoner communicates with the deep learning classifier by providing feedback in the form of additional three channels to the original training data. The additional three channels containing estimation of (1) shadow, (2) height, and (3) uncertainty is appended to the existing RGB image to enhance the performance of the semantic segmentation deep learning model. The Auto-Encoder-based Deep Neural Network is assisted by the deductive reasoner by conceptualizing errors to improve the performance of the semantic segmentation operation. The uncertainty channel accounts for those pixels where the spatial relationship with the neighboring regions was found to be inconsistent by the ontology-based deductive reasoner.

The Land Cover Change Continuum (LC3) model as propounded in Harbelot et al. (2015) is a spatio-temporal semantic model to track changes in spatial entities over time and enable knowledge discovery. It employed an ontology-based system to model the geometric, attribute, and identity changes in geospatial entities. The LC3 model leveraged GeoSPARQL vocabulary to map spatio-temporal transitions of regions in remote sensing scenes. Another work (Ghazouani et al., 2019), on similar lines as (Harbelot et al., 2015), proposed a hierarchical semantic change interpretation approach for evolving earth observation data. The authors proposed a hierarchical enrichment model with (1) reasoning over pixels using visual primitives, (2) reasoning over semantic objects, (3) spatial reasoning, and (4) reasoning for inferring higher levels of abstraction in the form of semantic changes in the remote sensing scene. The work proposed different ontologies for multiple stages of enrichment. The top-most stage of inferencing in the proposed architecture dealt with rules for deducing changes in spatial relations between objects detected in the satellite imagery. It also

```
[] a rule : Expansion; rule : content """
IF {
?01 a DO : Scene Image.
?02 a DO : Scene Image.
?01 DO : has Identity ?id1.
?01 DO : hasTime ?t1.
?02 DO : hasIdentity ?id2.
?02 DO : hasTime ?t2.
?01 DO : inside ?02.
FILTER(?id1 = ?id2}
THEN { ?01 a DO : Expansion. } """
```

FIGURE 4.3 SWRL rule as proposed for the expansion event, indicating object "O1" expanded to "O2" in the remote sensing scene (Ghazouani et al., 2019).

formalized rules in Semantic Web Rules Language (SWRL) for change events such as expansion, shrinkage, urbanization, deforestation, and degradation. Figure 4.3 depicts the SWRL rule embedded in the ontology for the expansion event. In the SWRL rule, "O1" and "O2" refer to the remote sensing scenes being compared for change interpretation. "id1" and "id2" are the corresponding land use land cover (LULC) classes for the respective objects in the remote sensing imagery. The axiom enforces the conditions—if the land use classes of the two objects in their corresponding scenes are the same and if the first object has the topologically relation "inside" with the other object and both the scenes belong to different times ("t1" and "t2"), then it can be inferred that "O1" object expanded in the remote sensing scene captured at time "t2."

Although the research studies reviewed earlier propose their own methodologies and domain-dependent ontologies, there is still a need for a standardized upper-level ontology for a generic remote sensing data product. The upper-level ontology should ideally take into account the geospatial data intricacies in addition to focusing on the metadata. Geospatial metadata in the form of provenance and data lineage is a significant information source for various applications. Thus, an ideal upper-level ontology would facilitate extending itself to suit the needs of geospatial application developers worldwide.

4.1.4.2 Ontology-Based Approaches for Disaster Applications

Having understood the numerous advantages from an ontological representation of knowledge, there has been an increase in interest by remote sensing researchers toward leveraging ontologies for disaster-related geospatial applications. Furthermore, numerous Earth Observation (EO) services including wildfire monitoring, flood monitoring, earthquake monitoring, are reaping the benefits of geospatial linked data and semantic interoperability. In that regard, this section discusses some of the prominent research works in this area.

Kyzirakos et al. (2012) proposed the Wildfire Monitoring Service under the umbrella of the EU project—TELEIOS. The authors demonstrated the benefits of geospatial linked data by implementing and deploying the forest fire monitoring service for the National Observatory of Athens (NOA) under the Services and Applications For Emergency Response (SAFER) framework. Satellite images captured by ESA's MSG/SEVIRI are ingested into the system. Semantic annotation translates the images into a Linked Data representation. The annotated Linked Data is stored in the Strabon—Geospatial RDF Triple Store. The semantic annotation and reasoning are in accordance with the NOA ontology proposed in this work. The NOA Ontology reused some of the concepts from NASA's SWEET Ontology to ensure semantic interoperability. Classes such as "RawData," "Shapefile," and "Hotspot" were formalized in this ontology to aid in reasoning for forest fire monitoring. The developed web interface allows users to explore and visualize the input data in addition to supporting stSPARQL queries and visualization of the results.

In Alirezaie et al. (2017), the authors proposed a deductive reasoning framework for querying remote sensing imagery for monitoring disasters. The proposed SemCityMap framework provided the users with an interactive interface to run semantic queries involving high-level abstractions over the satellite image and visualized the results. The system leveraged GeoSPARQL capabilities in addition to proposing the OntoCity ontological model. The OntoCity ontology proposed in this work extended the GeoSPARQL Ontology in addition to the DOLCE+DnS Ultralite (DUL) upper-level ontology. It also supported path-finding between two points on the map using the ontological knowledge base and GeoSPARQL's routing functionalities. In Wang et al. (2017), the authors proposed the Hydrological Sensor Web Ontology. The HSWO has been developed and built over the W3C's SSN Ontology, W3C's Time Ontology, and OGC's GeoSPARQL Ontology. The proposed ontology formalizes concepts and relations that proliferate during a hydrological event such as HydrologicalStation, HydrologicalSensor, RunOffVolume, RainGuage, and defines rules to enable reasoning over these concepts and relations. The proposed ontology contains rules to identify different stages of floods from the perspective of hydrological sensors based on the values of water levels and precipitation.

In Gaur et al. (2019), the authors proposed the "emapthi"—a coarse-grained ontology that strives to integrate the emergency management and hazard planning activities with the data from heterogeneous sensors including social media to provide a holistic ontological knowledge base. The "empathi" ontology imports external ontologies such as FEMA, HXL, GeoNames, SKOS, and LODE to ensure semantic interoperability among concepts and relations. In Scheuer et al. (2013), the authors proposed an ontology targeted toward risk assessment of the flood disaster. In that regard, they formulate concepts as "SusceptibiltiyFunction," "HazardMap," "FloodExtentMap," "InundationDepth," in addition to modeling "StakeHolders" and "Authorities" toward capturing comprehensively the flood risk assessment knowledge and ensuring interoperability and adaptability.

In Sinha and Dutta (2020), the authors presented a systematic analysis of different flood ontologies—comparing and contrasting them using a parametric approach. The authors systematically identified and reviewed the ontologies available for the flood domain based on different parameters such as ontology type, phase of the flood focused on, and ontologies reused, thus striving to facilitate a concise yet holistic perspective of the developed ontologies. It is evident that most of the knowledge bases are primarily sourced from hydrological sensors. It is envisaged that a very high-resolution (VHR) remote sensing scene as a primary data product would enable enhanced situational awareness primarily due to the higher granularity of modeling spatial relations and objects in the imagery. Thus, there is a need to also explore in detail a high-resolution remote sensing imagery-based approach for flood knowledge modeling.

The Sem-RSSU framework proposed in Potnis et al. (2021) extends the Spatial Image Information Mining Framework (Kurte et al., 2017) focusing primarily on enhancing the machine as well as the user-level understanding of remote sensing imagery comprehensively, from a spatio-contextual perspective. The context of a post –flood scenario has been selected as the domain for this work. The proposed framework is focused mainly on the contextual capturing of the core essence of a satellite imagery at the semantic level and rendering it in ways that is machine as well as user understandable. Consequently, Sem-RSSU proposes the generic ontology – Remote Sensing Scene Ontology (RSSO) along with the domain-specific Flood Scene Ontology (FSO), to formalize earth observation data into knowledge graphs. The knowledge graph for remote sensing scenes, conceptualized as part of the research study (Potnis et al., 2021) is an extension of the Spatial Semantic Graph. The RSS-KG seeks to comprehensively describe a remote sensing scene from a contextual standpoint, preserving both the directional and spatial relationships. It thus enables one to transition from the image space to the semantic space, providing an opportunity to understand what the machine has "understood." Moreover, the linked data nature of the RSS-KG makes it amenable for seamless integration with data from other external sources, thus adding value to itself.

4.1.4.3 Knowledge Graphs

Although the term "Knowledge Graph" seems to have become popular in recent times, the use of graph-based forms of representing data is not new in the research area of computing. In context to Knowledge Representation in the research area of Semantic Web, Knowledge Graphs refer to databases that leverage the graph-based data model to represent concepts, entities, and their relations between them. Subsequently Knowledge Graphs help in representing heterogeneous data seamlessly in a node-edge form by leveraging URIs. The Resource Description Framework (RDF)—one of the de-facto data models to represent Knowledge Graphs (KGs) was thus conceptualized in the paradigm of Semantic Web, to aid in data interoperability and standardization.

Ontologies play an integral role in the Knowledge Graph ecosystem. Generally, Knowledge Graphs are required to conform to an Ontology, to ensure correctness and consistency. Thus the relation between a Knowledge Graph and an Ontology is analogous to the relation between a XML document and its corresponding XSD (XML Schema) document.

4.2 GEOSPATIAL KNOWLEDGE GRAPHS CONSTRUCTION

The Sem-RSSU framework proposed in Potnis et al. (2021) formalizes the Remote Sensing Scene Ontology (RSSO)[6] along with the domain-specific Flood Scene Ontology (FSO).[7] This book chapter builds over (Potnis et al., 2021) and discusses elaborately the Knowledge Graph Construction Workflow conceptualized in the Sem-RSSU framework.

Both the ontologies—RSSO and FSO have been developed in accordance with the Knowledge Engineering methodology prescribed by Noy and McGuinness (2001), depicted in Figure 4.4. The ontology development process in the Sem-RSSU framework involved:

1. **Determination of the Domain and Definition of the Scope for the Proposed Ontologies**: One of the primary objectives of the Sem-RSSU framework has been to model remote sensing scenes in a representation that is amenable to integration with other data sources and also compliant to inferencing. In that regard, the Remote Sensing Scene Ontology (RSSO) has been conceptualized for representing generic scenes acquired by a standard sensing platform. Similarly, the Flood Scene Ontology (FSO) has been conceptualized to formalize the contextual domain information of the flood disaster in a standardized manner.
2. **Re-use of Existing Ontologies:** The proposed Remote Sensing Scene Ontology (RSSO) reuses OGC's GeoSPARQL Ontology. Specifically, the topological vocabulary definitions from the GeoSPARQL Ontology have been extensively leveraged for the Sem-RSSU framework. The Flood Scene Ontology (FSO) imports and augments the Remote Sensing Scene Ontology (RSSO) thus demonstrating the extensible and interoperable features of the Sem-RSSU framework.
3. **Enumeration of Important Concepts and Class and Property Definitions**: The important concepts to model a generic remote sensing scene have been extensively brainstormed and put together to conceptualize both RSSO and FSO. Consequently, the classes, data

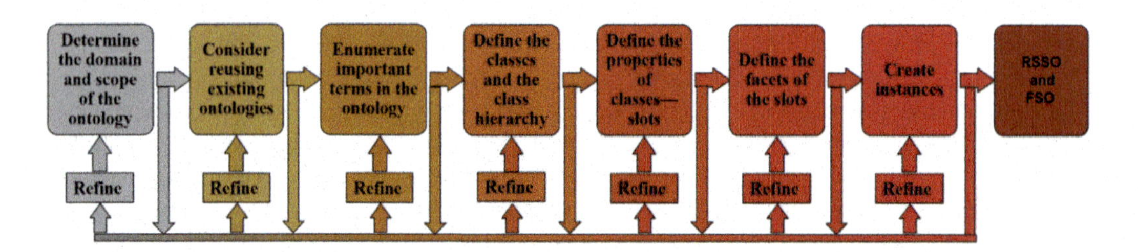

FIGURE 4.4 Ontology Engineering Methodology. (Adapted from Noy and McGuinness (2001).)

properties, and object properties have been defined. The most important atomic concept in a remote sensing scene has been identified to be a "region." The data properties and object properties have been planned and conceptualized accordingly.

4. **Instance Creation**: This process of instance creation in the ontologies for the Sem-RSSU framework has been conceptualized to be initiated on run-time. In other words, the instances for the ontologies—RSSO and FSO are generated dynamically based on the classified map generated by the Multi-class segmentation component in the proposed Sem-RSSU framework.

   During the course of development of the proposed Sem-RSSU framework, it was understood that the ontology development process is an iterative one. Thus, the ontologies – RSSO and FSO were iteratively enriched as the research on the framework progressed. Figure 4.5 depicts the taxonomy of classes that have been consolidated for representing the flood disaster information in a standardized format.

4.2.1 Knowledge Graph Construction Workflow

This section comprehensively describes the processes involved toward the construction of Knowledge Graphs for representing satellite imagery in a standardized manner. It elaborates on the self-contained structure of the workflow that enables a particular component to be upgraded without affecting the other components of the workflow.

Figure 4.6 depicts the end-to-end workflow for construction of Geospatial Knowledge Graphs. The workflow consists of four stages including Data Storage, Scene Ingestion, Data Mediation, and Semantic Enrichment.

The Data Storage layer includes a triple-store server that facilitates storage, retrieval, and updation of the Geospatial Knowledge Graphs generated from remote sensing scenes. It also facilitates storage of upper-level and domain-specific application ontologies for description-logic-based reasoning. The Scene Ingestion layer enables users to ingest satellite image scenes for knowledge graph generation.

4.2.1.1 Deep Learning-Based Multi-Class Segmentation

The workflow proposes to leverage a Deep Learning-based Multi-Class Segmentation Component for generating classification maps of the remote sensing scenes (Figure 4.7).

The classification map consists of class labels assigned to each pixel in the scene. The workflow advocates the use of multi-class segmentation as an initial step toward generation of Knowledge Graphs. The raster output serves as an input to the Geometry Shape Extraction component in the workflow.

4.2.1.2 Geometry Shape Extraction

WKT geometry is used to represent the multi-class segmentation maps based on the class labels predicted by the deep learning approach to each region in the scene. Noisy pixel labels are filtered by specifying a predefined threshold for area of a region in the knowledge graph. Python libraries—Shapely and Rasterio are utilized to implement this shape extraction from the raster classification maps. The WKT geometries obtained through the shape extraction are then encoded in a RDF form to represent the Linked Data Representation of the raw Knowledge Graph.

4.2.1.3 Resource Description Framework (RDF) Based Serialization

The classes' information from the Semantic Segmentation component serves as input to this component. This component generates RDF Triples containing the classes' information. This component is crucial as it serves as the initial step of transitioning from the image space to the semantic space. The above snippet depicts an example RDF graph generated from a part of a satellite imagery scene.

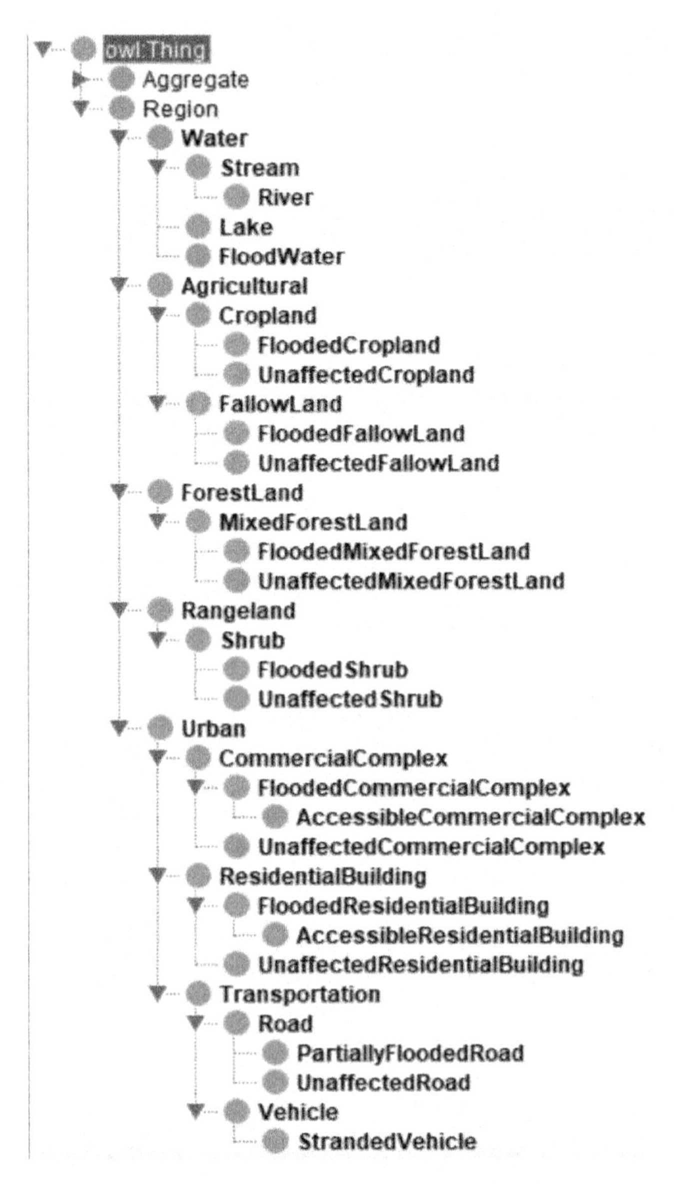

FIGURE 4.5 Visualization of Taxonomy of concepts that proliferate during the flood disaster as formalized in the Flood Scene Ontology (FSO) as proposed in Potnis et al. (2021).

The Linked Data Generator component generates graph representation of the land-use land-cover regions that conform to the Resource Description Framework (RDF).

The RDF-based Linked Data representation constitutes triples for each of the land-use land cover regions detected in the scene. Associated data properties are used to store the geometries and other spatial information conforming to the GeoRDF standard and the upper-level ontology.

Figure 4.8 depicts a snippet of the Linked Data representation in RDF demonstrating the mapping of the regions of type "Vehicle" and "ResidentialBuilding" in the classified remote sensing scene. The classified remote sensing scene on the right in Figure 4.8 is the output of the earlier discussed Multi-Class Segmentation component. This classified remote sensing scene is transformed to the Linked Data representation in the above manner with each region being identified with a unique URI and other relevant attributes.

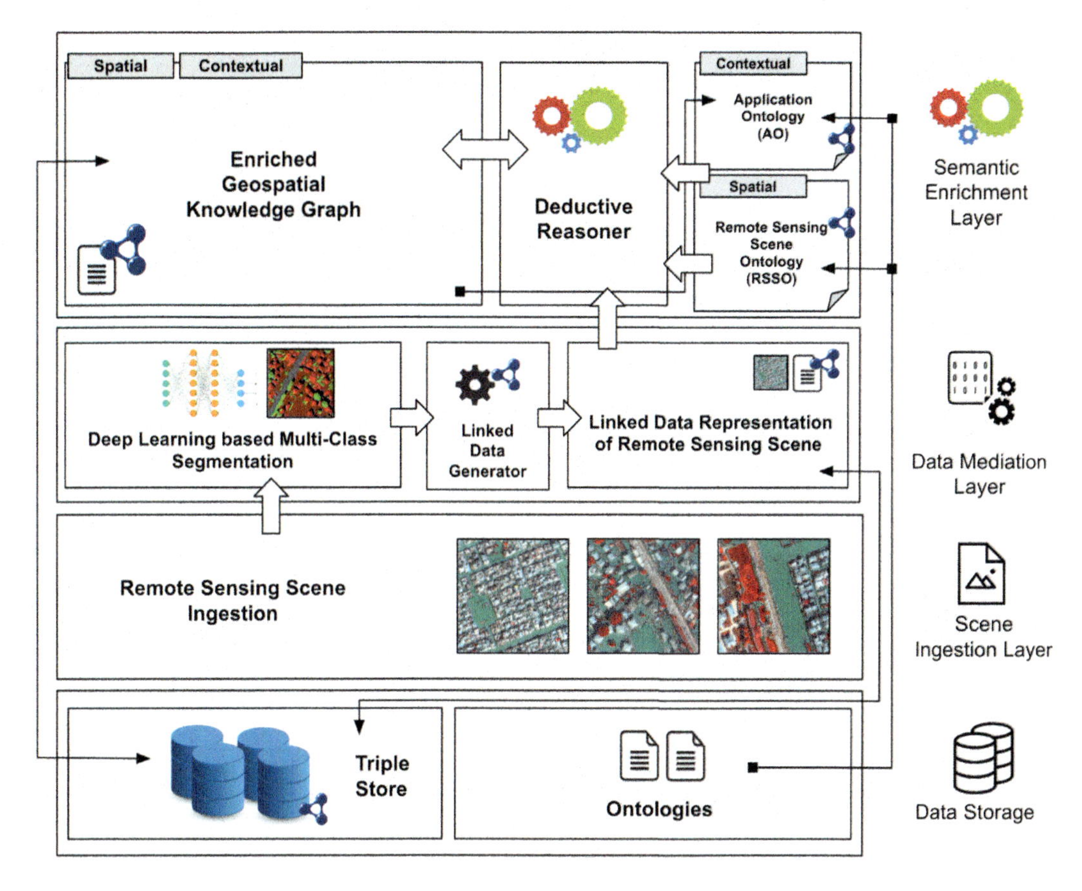

FIGURE 4.6 Geospatial knowledge graph construction workflow.

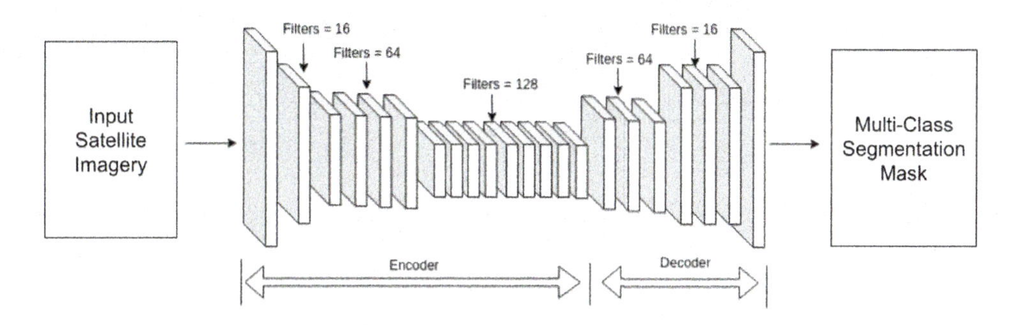

FIGURE 4.7 Deep learning-based multi-class segmentation.

4.2.1.4 Semantic Enrichment of Geospatial KG

The Semantic Enrichment layer in the workflow has been conceptualized to semantically enrich the linked data representation of the scene in a hierarchical manner. Figure 4.9 depicts the different sub-layers in the Semantic Enrichment layer of the workflow that translates the linked data representation of a remote sensing scene into an enriched geospatial knowledge graph.

The implicit spatial relations including topological and directional are inferrenced using the upper-level—Remote Sensing Scene Ontology (RSSO). This enables synthesizing relations between class entities resulting in the formation of a geospatial knowledge graph. The application ontology,

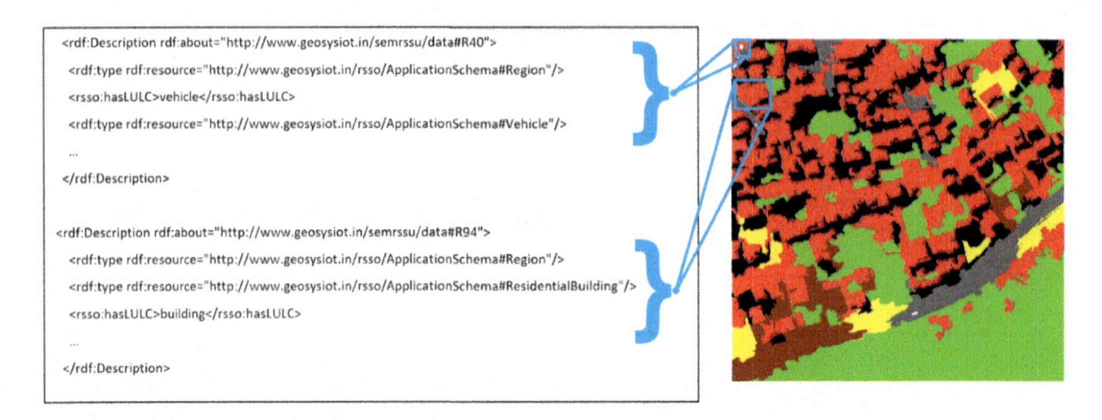

FIGURE 4.8 Snippet of the linked data representation in RDF and corresponding mapping between the regions of type "Vehicle" and "ResidentialBuilding" in the classified remote sensing scene.

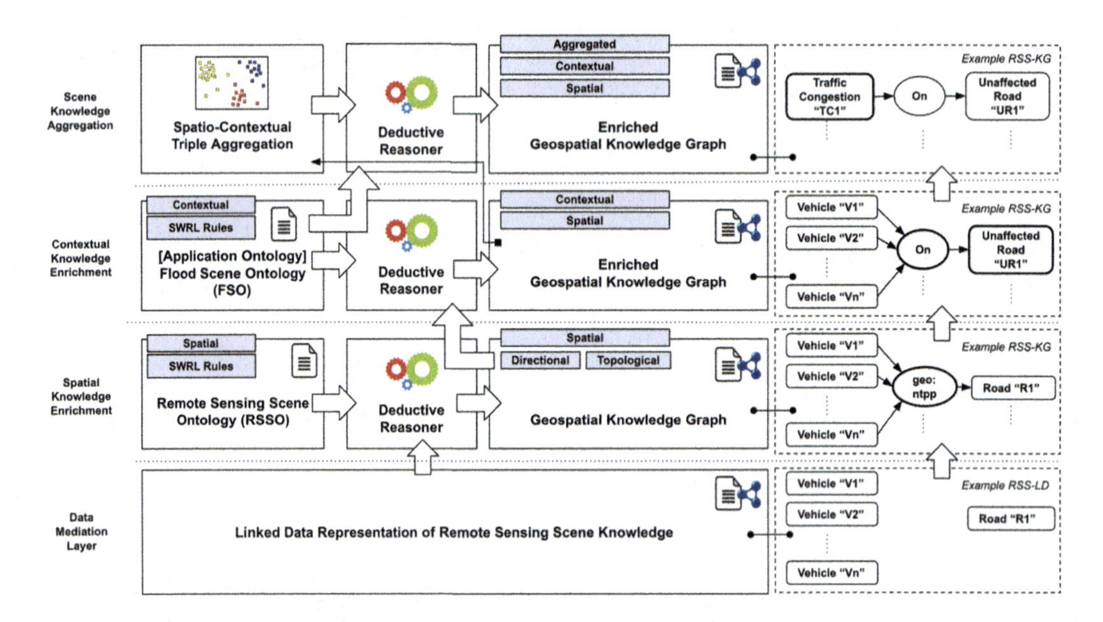

FIGURE 4.9 Hierarchical multi-level architecture for semantic enrichment in the reasoning layer as adopted from Potnis et al. (2021).

in this case—the Flood Scene Ontology, serving as the repository for flood-specific context, facilitates the deduction of concepts and relations that proliferate during the flood disaster. Thus the description-logic—based reasoner enables the synthesis and consequent knowledge inculcation in the knowledge graph. The multi-level strategy for knowledge enhancement in the knowledge graph through the upper-level and subsequent application ontological knowledge-base facilitates modularity and extensibility. This further encourages scalability and ease of integration with other heterogeneous sources of data for other applications.

An example to the right, in Figure 4.9, demonstrates the hierarchical knowledge enhancement in the knowledge graph as it traverses to the higher sub-layers in the Semantic enrichment layer. The linked data generator component synthesizes the "Vehicle" and "Road" instances from the remote sensing scene in the graph form, within the Data Mediation layer of the workflow.

Further the instances are enriched by inferring the spatial relation "geo:ntpp"—Non Tangential Proper Part using the upper-level ontology—RSSO. Subsequently, the relation is further specialized into "on" relation in addition to the "Road" instance being further specialized into "Unaffected Road." These specializations are a result of deductive reasoning using the application ontology—FSO. In the final stage of enrichment, the instances of "Vehicle" class are aggregated to "Traffic Congestion" using the aggregation concepts formalized in the application ontology.

4.3 APPLICATIONS AND USE CASES

The workflow for Geospatial Knowledge Graph Construction has been elaborated in this chapter as proposed as a part of the Sem-RSSU framework. The workflow aims for improved understanding of a remote sensing scene from a spatio-contextual perspective. For proving the effectiveness of the workflow and demonstrating real-world use, the flood disaster has been chosen and illustrated. However, the workflow discussed in this chapter can also be leveraged for numerous other applications consuming data products captured by different remote sensing platforms and other heterogeneous data sources.

Figure 4.10 depicts the grounded natural language scene description of an urban flood scene generated using the Sem-RSSU framework that leveraged the knowledge graph representation of remote sensing scenes. Similarly, several humanitarian applications including monitoring of natural calamities, human conflicts, and even research studies involving urban sprawl analysis, could potentially leverage and build upon, using the proposed workflow. It is envisaged that this workflow would instill confidence and encourage further research toward semantics-driven EO frameworks for comprehensive information extraction and subsequent for remote sensing applications. Figure 4.11 depicts the visualization of the enriched Geospatial Knowledge Graph for a satellite scene during a flood event. Intuitive visualizations using the node-edge graph form for entities and their corresponding relations are foreseen to help in improved situational awareness during disasters and extreme events.

Remote Sensing Scene

Remote Sensing Scene with Grounded Mappings Visualization

Grounded Spatio-Contextual Scene Description in Natural Language:
There is a road. There are accessibleBuildings along the road. There are floodedBuildings to the West and East direction of the road. There is traffic on the road. There are strandedVehicles to the West direction of the road. There is floodedVegetation to the West and East direction of the road.

FIGURE 4.10 Leveraging geospatial knowledge graphs to grounded remote sensing scene description using the semantics-driven remote sensing scene understanding framework (Potnis et al., 2021).

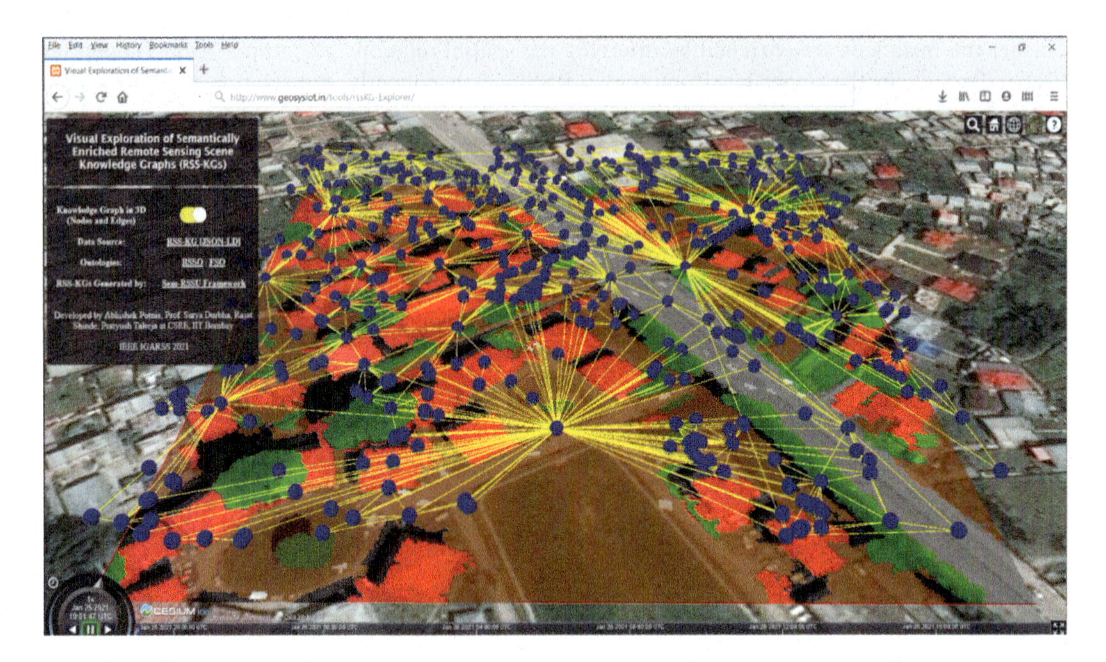

FIGURE 4.11 Visual exploration of the geospatial knowledge graph using cesium.[8]

4.4 SUMMARY

The workflow discussed in this chapter propounds leveraging both—a deep learning and a deductive reasoning-based approach for inferring the otherwise implicit information from EO data products. The workflow proposes the translation of EO imagery to geospatial knowledge graphs conforming to the upper-level ontology—Remote Sensing Scene Ontology (RSSO) that was developed as a part of this research. The upper-level ontology has been developed to represent a generic EO imagery in the form of a graph representation with entities and relations represented as triples. The ontology formalizes representation of land use land cover classes for each of the regions detected in the remote sensing imagery in addition to representing the metadata related to data lineage, CRS, and other ancillary information. The application ontology—Flood Scene Ontology (FSO) formalizes domain-specific concepts and relations that are relevant during the flood disaster. The ontology thus demonstrates workflow's malleability to adapt different remote sensing applications. The RSSO[9] and FSO[10] have been published on the internet, conforming to the open standards of Semantic Web, and can thus be consumed by linked data parsing software as well as web browsers alike. It is envisaged that the workflow proposed in this book chapter would form the basis of synthesizing Geospatial Knowledge Graphs for application scenarios such as disasters, surveillance of hostile territories, and urban sprawl monitoring, among numerous other remote sensing applications.

NOTES

1 https://www.earthdata.nasa.gov/learn/data-chats/data-chat-dr-christopher-lynnes.
2 https://www.w3.org/XML/.
3 https://www.w3.org/RDF/.
4 http://wiki.dbpedia.org/.
5 http://wiki.freebase.com/.
6 https://www.geosysiot.in/rsso/.
7 https://www.geosysiot.in/fso.

8 http://www.geosysiot.in/tools/rssKG-Explorer/.
9 https://www.geosysiot.in/rsso/.
10 https://www.geosysiot.in/fso.

REFERENCES

Alirezaie, M., Kiselev, A., Längkvist, M., Klügl, F., Loutfi, A., 2017. An ontology-based reasoning framework for querying satellite images for disaster monitoring. *Sens. Switz.* 17. doi: 10.3390/s17112545.

Alirezaie, M., Längkvist, M., Sioutis, M., Loutfi, A., 2019. Semantic referee: A neural-symbolic framework for enhancing geospatial semantic segmentation. *Semant. Web.* doi: 10.3233/SW-180362.

Andrés, S., Arvor, D., Mougenot, I., Libourel, T., Durieux, L., 2017. Ontology-based classification of remote sensing images using spectral rules. *Comput. Geosci.* 102, 158–166. doi: 10.1016/j.cageo.2017.02.018.

Arvor, D., Belgiu, M., Falomir, Z., Mougenot, I., Durieux, L., 2019. Ontologies to interpret remote sensing images: Why do we need them? *GISci. Remote Sens.* 56, 911–939. doi: 10.1080/15481603.2019.1587890.

Arvor, D., Durieux, L., Andrés, S., Laporte, M.A., 2013. Advances in geographic object-based image analysis with ontologies: A review of main contributions and limitations from a remote sensing perspective. *ISPRS J. Photogramm. Remote Sens.* 82, 125–137. doi: 10.1016/j.isprsjprs.2013.05.003.

Baraldi, A., Puzzolo, V., Blonda, P., Bruzzone, L., Tarantino, C., 2006. Automatic spectral rule-based preliminary mapping of calibrated landsat TM and ETM+ images. *IEEE Trans. Geosci. Remote Sens.* doi: 10.1109/TGRS.2006.874140.

Berners-lee, T., Berners-lee, T., Hendler, J., Hendler, J., Lassila, O., Lassila, O., 2001. The semantic web. *Sci. Am.* 284, 35–43. doi: 10.1007/978-3-642-29923-0.

Carson, C., Thomas, M., Belongie, S., Hellerstein, J.M., Malik, J., 1999. Blobworld: A system for region-based image indexing and retrieval. *Third International Conference on Visual Information Systems*, pp. 509–516. doi: 10.1007/3-540-48762-X_63.

Daschiel, H., Datcu, M., 2005. Information mining in remote sensing image archives: System evaluation. *IEEE Trans. Geosci. Remote Sens.* 43, 188–199. doi: 10.1109/TGRS.2004.838374.

Durand, N., Derivaux, S., Forestier, G., Wemmert, C., Gancarski, P., Boussaid, O., Puissant, A., 2007. Ontology-based object recognition for remote sensing image interpretation. *In 19th IEEE International Conference on Tools with Artificial Intelligence (ICTAI 2007)*, pp. 472–479. doi: 10.1109/ICTAI.2007.111.

Durbha, S.S., King, R.L., 2005. Semantics-enabled framework for knowledge discovery from Earth observation data archives. *IEEE Trans. Geosci. Remote Sens.* 43, 2563–2572. doi: 10.1109/TGRS.2005.847908.

Flickner, M., Sawhney, H., Niblack, W., Ashley, J., Huang, Q., Dom, B., Gorkani, M., Hafner, J., Lee, D., Petkovic, D., Steele, D., Yanker, P., 1995. Query by image and video content: The QBIC system. *Computer.* doi: 10.1109/2.410146.

Forestier, G., Wemmert, C., Gançarski, P., 2008. On combining unsupervised classification and ontology knowledge. *Int. Geosci. Remote Sens. Symp. IGARSS* 4, 1–4. doi: 10.1109/IGARSS.2008.4779741.

Gaur, M., Shekarpour, S., Gyrard, A., Sheth, A., 2019. Empathi: An ontology for emergency managing and planning about hazard crisis, *In 2019 IEEE 13th International Conference on Semantic Computing (ICSC)*, IEEE, pp. 396–403. doi: 10.1109/ICOSC.2019.8665539.

Ghazouani, F., Farah, I.R., Solaiman, B., 2019. A multi-level semantic scene interpretation strategy for change interpretation in remote sensing imagery. *IEEE Trans. Geosci. Remote Sens.* 57, 8775–8795. doi: 10.1109/TGRS.2019.2922908.

Gu, H., Li, H., Yan, L., Liu, Z., Blaschke, T., Soergel, U., 2017. An object-based semantic classification method for high resolution remote sensing imagery using ontology. *Remote Sens.* 9. doi: 10.3390/rs9040329.

Harbelot, B., Arenas, H., Cruz, C., 2015. LC3: A spatio-temporal and semantic model for knowledge discovery from geospatial datasets. *J. Web Semant.* 35, 3–24. doi: 10.1016/j.websem.2015.10.001.

Koubarakis, M., Kyzirakos, K., Nikolaou, C., Garbis, G., Bereta, K., Dogani, R., Giannakopoulou, S., Smeros, P., Savva, D., Stamoulis, G., Vlachopoulos, G., Manegold, S., Kontoes, C., Herekakis, T., Papoutsis, I., Michail, D., 2016. Managing big, linked, and open earth-observation data: Using the TELEIOS/LEO software stack. *IEEE Geosci. Remote Sens. Mag.* 4, 23–37. doi: 10.1109/MGRS.2016.2530410.

Kurte, K.R., Durbha, S.S., King, R.L., Younan, N.H., Vatsavai, R., 2017. Semantics-enabled framework for spatial image information mining of linked earth observation data. *IEEE J. Sel. Top. Appl. Earth Obs. Remote Sens.* 10, 29–44. doi: 10.1109/JSTARS.2016.2547992.

Kyzirakos, K., Karpathiotakis, M., Garbis, G., Nikolaou, C., Bereta, K., Sioutis, M., Papoutsis, I., Herekakis, T., Michail, D., Koubarakis, M., Kontoes, C., 2012. Real time fire monitoring using semantic web and linked data technologies. *In the 11th International Semantic Web Conference (ISWC 2012)*, Boston, MA, pp. 7–10.

Ma, W.Y., Manjunath, B.S., 1999. NeTra: A toolbox for navigating large image databases. *Multimed. Syst.* 7, 184–198. doi: 10.1007/s005300050121.

Marchisio, G.B., Li, W.H., Sannella, M., Goldschneider, J.R., 1998. GeoBrowse: An integrated environment for satellite image retrieval and mining. *Int. Geosci. Remote Sens. Symp. IGARSS* 2, 669–673. doi: 10.1109/igarss.1998.699546.

Molinier, M., Laaksonen, J., Häme, T., 2007. Detecting man-made structures and changes in satellite imagery with a content-based information retrieval system built on self-organizing maps. *IEEE Trans. Geosci. Remote Sens.* 45, 861–874. doi: 10.1109/TGRS.2006.890580.

Noy, N.F., McGuinness, D.L., 2001. Ontology development 101: A guide to creating your first ontology. *Stanf. Knowl. Syst. Lab.* 25. doi: 10.1016/j.artmed.2004.01.014.

Pentland, A., Picard, R.W., Sclaroff, S., 1996. Photobook: Content-based manipulation of image databases. *Int. J. Comput. Vis.* 18, 233–254. doi: 10.1007/BF00123143.

Potnis, A.V., Durbha, S.S., Shinde, R.C., 2021. Semantics-driven remote sensing scene understanding framework for grounded spatio-contextual scene descriptions. *ISPRS Int. J. Geo-Inf.* 10, 32. doi: 10.3390/ijgi10010032.

Quartulli, M., Datcu, M., 2003. Information fusion for scene understanding from interferometric SAR data in urban environments. *IEEE Trans. Geosci. Remote Sens.* 41, 1976–1985.

Quinlan, J.R., 1996. Improved use of continuous attributes in C4.5. *J. Artif. Intell. Res.* doi: 10.1613/jair.279.

Scheuer, S., Haase, D., Meyer, V., 2013. Towards a flood risk assessment ontology: Knowledge integration into a multi-criteria risk assessment approach. *Comput. Environ. Urban Syst.* 37, 82–94. doi: 10.1016/j.compenvurbsys.2012.07.007.

Shadbolt, N., Hall, W., Berners-Lee, T., 2006. The semantic web revisited. *IEEE Intell. Syst.* 21, 96–101. doi: 10.1109/MIS.2006.62.

Sheth, A., Henson, C., Sahoo, S.S., 2008. Semantic sensor web. *IEEE Internet Comput.* 12, 78–83. doi: 10.1109/MIC.2008.87.

Shyu, C.R., Klaric, M., Scott, G.J., Barb, A.S., Davis, C.H., Palaniappan, K., 2007. GeoIRIS: Geospatial information retrieval and indexing system - Content mining, semantics modeling, and complex queries. *IEEE Trans. Geosci. Remote Sens.* 45, 839–852. doi: 10.1109/TGRS.2006.890579.

Sinha, P.K., Dutta, B., 2020. A systematic analysis of flood ontologies: A parametric approach. *Knowl. Organ.* 47, 138–159. doi: 10.5771/0943-7444-2020-2-138.

Smith, J.R., Chang, S.-F., 1996. VisualSEEk: A fully automated content-based image query system. *Proc. Fourth ACM Int. Conf. Multimed.* 96, 87–98. doi: 10.1145/244130.244151.

Wang, L., Yan, J., 2020. Stewardship and analysis of big Earth observation data. *Big Earth Data.* doi: 10.1080/20964471.2020.1857055.

Wang, C., Chen, N., Wang, W., Chen, Z., 2017. A hydrological sensor web ontology based on the SSN ontology: A case study for a flood. *ISPRS Int. J. Geo-Inf.* 7, 2. doi: 10.3390/ijgi7010002.

5 Geosemantic Standards-Driven Intelligent Information Retrieval Framework for 3D LiDAR Point Clouds

Rajat C. Shinde, Surya S Durbha, Abhishek V. Potnis, and Pratyush V. Talreja
Centre of Studies in Resources Engineering,
Indian Institute of Technology Bombay, India

CONTENTS

DOI: 10.1201/9781003270928-7

5.1 INTRODUCTION AND MOTIVATION

Earth observation (EO) refers to the exploration of our earth as an individual *system* or a *system of systems* and its associated features and properties. It constitutes an analysis of physical, chemical, and biological cycles and their interrelation, such as biogeochemical, biophysical, or geochemical cycles. Any phenomenon around you, related to you and your surroundings, and in general, happening on the Earth can be defined using these cycles, making the EO research significant and impactful. For example, the photosynthesis effect, atmospheric carbon cycle, water cycle, perspiration, precipitation, climate change are a few examples. Over the years, EO datasets have been growing in size and volume attributed to multiple factors such as—rapid prototyping, increased involvement in research and industrial development, innovation in small-scale sensor manufacturing, rapid processing, and high-performance computing using GPUs and TPUs. With the plethora of EO data, it becomes critical to have a standardized framework and approaches for using these datasets. It is required for continued reusability of the data sources and addressing the heterogeneity arising because of multiple stakeholders involved in the process. Additionally, standardization is significant in achieving interoperability between datasets acquired by sensors of the same mode of operation and different modes of operation. For example, interoperability encourages that satellite imagery by different providers to follow similar syntactic and semantic structures; Also, interoperability encourages that satellite imagery and imagery acquired using airborne platforms (drones, UASs) to follow a consistent syntactic and semantic structure subject to the justification of fundamental conditions (particular geometric structure, for instance).

LiDAR stands for Light Detection and Ranging and is an active remote sensing data acquisition technique capturing the 3D surrounding accurately. Although LiDAR sensors are used for various applications, including self-driving cars and photonics, in terms of remote sensing, they are used for mapping and surveying. Typically, LiDAR datasets can be acquired onboard spaceborne, airborne (ALS—Airborne LiDAR Scanning), ground stations and mobile mapping (TLS—Terrestrial LiDAR Scanning), and handheld platforms. Regarding mode of usage, LiDAR datasets find uses as raw point clouds, meshes, 3D surfaces, waveforms, or multiview 2D images. Generally, point clouds are considered the most accurate 3D representation, with the point being the fundamental unit of the captured object. According to the latest reports and releases, the LiDAR market is expected to touch $2.8 billion by 2025 [1]. With these advancements, it becomes significant to design an ecosystem for adopting LiDAR technology and related stack to be used in a standardized and interoperable manner. The motivation behind this chapter is to explore the concept of interoperability in LiDAR data processing. The chapter addresses the need for interoperability and standardization for geospatial and geosemantic standards, focusing on LiDAR point clouds and 3D scenes. It also touches upon the design principle of the pipeline for knowledge discovery or information extraction from raw 3D LiDAR point clouds. Additionally, to automate and develop an intelligent information extraction framework using 3D LiDAR point clouds. The chapter focuses on presenting a step-by-step approach for intelligently and automatically extracting knowledge from a raw 3D LiDAR point cloud by presenting a case study of question-answering over airborne 3D LiDAR point clouds for the urban domain. In principle, the center of attention is to address questions like—

- How many trees have a height of more than 40 m? (useful in profiling forests)
- What is the canopy cover of the forests? (useful in biomass and carbon estimation)
- How many buildings above 20 m are next to the main road? (useful in urban planning)
- How much material is required to fill this pothole/manhole? (useful in road management)

We envisage that the chapter would make the readers capable of the current status of working with airborne 3D LiDAR point clouds. Although our focus is on airborne LiDAR scanning (ALS) but with relevant preprocessing, the concept as a whole can be employed for other LiDAR sources as well. After reading this, the readers should be equipped with the tools and technologies required

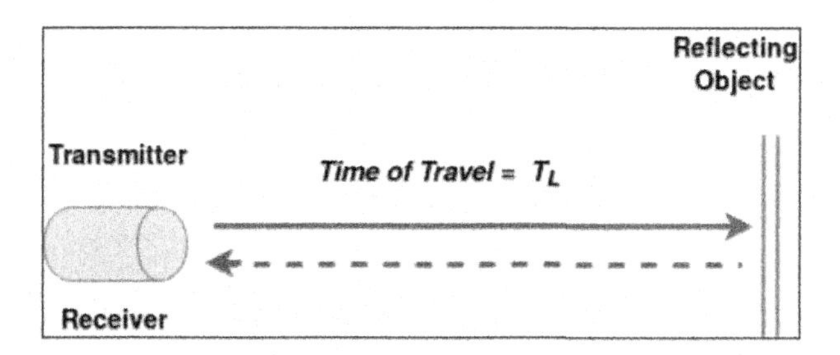

FIGURE 5.1 Illustration of working principle of LiDAR-based on Time of Travel (ToT).

for developing a scalable, interoperable, and standards-driven LiDAR processing framework for answering the above-mentioned questions.

5.2 LiDAR—LIGHT DETECTION AND RANGING

LiDAR (**Light Detection and Ranging**) is an optical active remote sensing data acquisition technique generating highly accurate 3D x, y, z measurements. LiDAR works on the principle of Time of Travel (ToT) from the sensor to the target object for accurately capturing the x, y, and z measurements. It also requires applying some mathematical transformations to consider the sensor alignment and orientation. The LiDAR scanning principle is based on Electronic Distance Measuring Instrument (EDMI), where the receiver captures the energy reflected from the object, and the transmitter fires the laser pulse (Refer to Figure 5.1). The ToT for the laser pulse determines the distance between the transceiver and the object.

With recent development in sensor manufacturing and solid-state technology, LiDAR technology has been used as a primary alternative for mapping and surveying applications to traditional surveying techniques due to its cost-effective and efficient implementation. LiDAR uses a laser for densely sampling the region of interest (RoI) by mapping the target in the form of points based on ToT.

5.2.1 Types of LiDAR Data Sources

The LiDAR types are differentiated based on the types of LiDAR scanning platforms and the mode of operation of the scanning. The different types (Refer to Figure 5.2 for illustrations) are described as follows:

- Spaceborne LiDAR Scanning
- Airborne LiDAR scanning (ALS)
- Ground-based Scanning—Terrestrial LiDAR scanning (TLS) and Mobile Mapping

In Spaceborne LiDAR scanning, the sensor is deployed onboard a space-based platform. Some examples are—ICESat, GEDI, LVIS, CALIPSO, ADM-Aeolus, ICESat-2. Typically, spaceborne LiDAR datasets are used for atmospheric and vegetation profiling over the region of study. In an ALS system, the scanner is installed onboard an aircraft, Unmanned Aerial Vehicle (UAV), or helicopter. The laser waves are emitted from the transmitters on the scanner toward the ground, and the receptors receive the reflected waves. All the reflected waves are mapped to a point on the ground, based on the time of flight, thus forming a massive 3D point cloud.

The LiDAR point cloud data storage usually follows LAS file specification. The word "LAS" stands for Log ASCII Format. LAS is a binary file format that is efficient for processing and analysis using computers. The LAS file contains metadata of the survey area in a header file. Each point in a LAS file stores x, y, z coordinate information, GPS timestamp, return number, scan angle, scan line

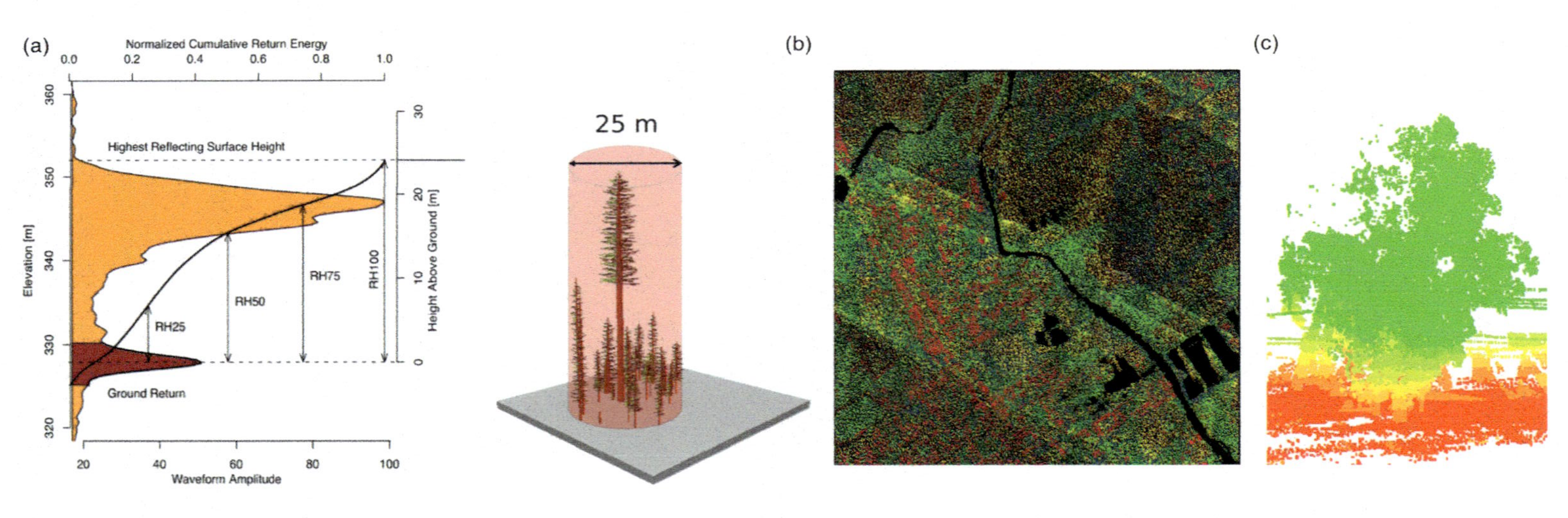

FIGURE 5.2 Illustration of types of LiDAR datasets - (a) spaceborne (Source: GEDI[1]), (b) airborne, and (c) ground-based (Terrestrial LiDAR scans—TLS).

TABLE 5.1
List of Remote Sensing LiDAR Datasets

Dataset Name	Details	Domain of Research
• Hessigheim 3D (H3D) benchmark (Kölle et al., 2021)	• High-resolution 3D point clouds and textured meshes from UAV LiDAR and multi-view-stereo	• Urban environment
• Dayton Annotated Laser Earth Scan (DALES) (Varney et al., 2020)	Eight classes; half-billion points	• Urban, sub-urban, rural and commercial scenes
• Semantic 3D net dataset (Hackel et al., 2017)	Eight classes including, villages and soccer fields; more than 4 billion points	• Urban environment
• IEEE Data Fusion Contest 2015[2]	• Point density: 65 points/m^2	• Urban and harbor area
• Open Topography[3]	• Hosts high-resolution topography dataset	• Urban, forests, disasters
• Vaihingen 3D (V3D) dataset[4] (Cramer, 2010)	• ISPRS 3D semantic labeling benchmark	• Urban environment

direction, and id as attributes. The points stored in a point cloud are invariable to the permutation of the sequence in which points are stored. This is why LiDAR point clouds are unstructured, unlike images that follow gridded representation and do not store the coordinates.

5.2.2 List of Remote Sensing-Based Open LiDAR Datasets

Artificial Intelligence research involving automated data processing prominently requires accurately labeled datasets and ground truth for validation. In remote sensing research, the accuracy of the ground truth data is of significance because of the remote data acquisition and challenges in capturing ground truth in geographically constrained regions. LiDAR data sources with accurate ground truth information have been scarce for a long time. Only recently have more open LiDAR data sources been available for research attributed to progress in open data policy throughout the globe. Table 5.1 presents a brief study of available remote sensing-based open airborne LiDAR data sources categorized according to the research domain.

For an extensive catalog of the available 3D LiDAR point cloud datasets, the readers are recommended to follow (Guo et al., 2021; Hu et al., 2021).

5.3 INTEROPERABILITY AND GEOSEMANTICS STANDARDIZATION FOR LiDAR

The rapid surge in manufacturing of sensor technology and research and development in the small-scale ICs (integrated chips)—VLSI, MSI, have increased the interest of new stakeholders in the LiDAR sensor manufacturing. Similarly, new tools and libraries are constantly being developed and enhanced on the processing side. The emergence of new datasets poses a huge challenge in processing regarding heterogeneity, diversity, and quantity.

5.3.1 Need for Interoperability in LiDAR

The heterogeneity in using LiDAR data arises due to the involvement of various stakeholders. This could be due to differences in the—(1) file format used for storing LiDAR data, (2) method of compression used, (3) algorithm used for processing, and (4) mode of acquisition. This heterogeneity primarily leads to difficulty in amalgamating sensors, raw data, processed results, services, and products in the LiDAR data processing ecosystem. To achieve this integration, *interoperability* is pursued by the scientific community to address the challenges. As defined by IEEE, 1991, "*Interoperability is the ability of two or more systems or components to exchange information and to use the information that has been exchanged.*" Although various global bodies have proposed multiple definitions of interoperability, this definition seems relevant for LiDAR sensors and data processing. Additionally, the heterogeneity can be further classified at different levels of interoperability such as technical (mode of operation of sensors), syntactic (data storage format), semantic (e.g., *{x, y, z}* or *{latitude, longitude, elevation}*), conceptual (variation in concept for photonics and urban) and organizational.

5.3.2 Geospatial Standardization

The way forward to achieving the various levels of interoperability as described above is by defining standards for the individual levels and for the system. Now a question might arise—What exactly is a standard and who defines it? The standard can be defined[5] as follows—A standard is a documented agreement between providers and consumers, established by consensus, that provides rules, guidelines, or characteristics ensuring materials, products, and services are fit for purpose.

Typically, the above definition seems relevant in the real world as well and justifies some of the tasks we do on a day-to-day basis. Some examples are—buying railway tickets, checking food licenses for our favorite packed snacks, and paying for our television subscription. Majorly the Standards Development Organizations (SDOs) develop these international standards following an iterative vetted and consensus-driven approach by referring documents and previous policies for reference.

5.3.2.1 International Bodies for Geospatial Standardization

Some of the active international bodies working toward standardization [6] are highlighted as follows:

- **IEEE**[7]: Institute of Electrical and Electronics Engineers
- **ITU**[8]: International Telecommunication Union
- **OGC**[9]**:** Open Geospatial Consortium
- **ISO/TC**[10]: International Standards Organization/Technical Committee
- **IHO**[11]: International Hydrographic Organization
- **W3C**[12]: World Wide Web Consortium

The ISO/TC 211 Geographic information/Geomatics is a working group of the ISO/TC working in the domain of geospatial standards.

5.3.2.2 Geospatial Standards for 3D LiDAR Data

Achieving interoperability is desired to encourage increased engagement of the developers, service providers, and users to enhance accessibility significantly. This can be achieved by following the recent cloud services' developments-Infrastructure-as-a-Service (IaaS), Software-as-a-Service (SaaS), and Platform-as-a-Service (PaaS), thus making the ecosystem accessible in a modular manner over the web and mobile environments. This introduces collaboration and development of community practices and is described using the continuum of data to knowledge model in Figure 5.3. It illustrates the transition of raw 3D LiDAR data about surroundings and their components linked together and provides a deeper understanding of a problem domain.

Figure 5.4 illustrates some of the 3D LiDAR standards and APIs (making it convenient to use the standards) proposed by the international standardization bodies for achieving this trajectory from data to geospatial knowledge. The standardization is important for reusability and accessibility in research and development. Moreover, it defines a benchmark for constant improvement over time instead of reinventing the wheel. On the other hand, standardization is an iterative process as it requires a constant participation of the scientific community to improve the existing standards.

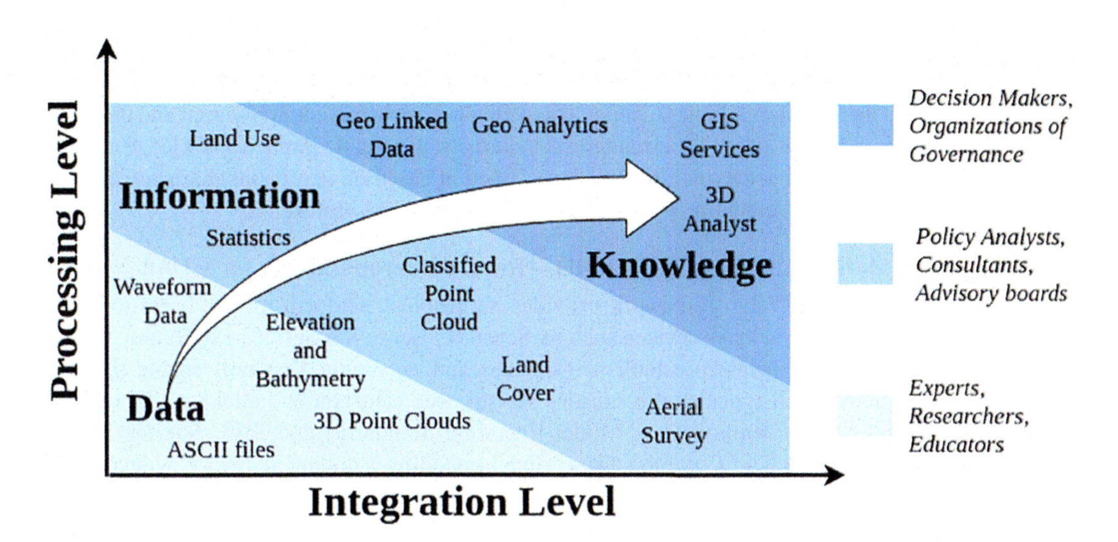

FIGURE 5.3 Data to knowledge continuum with 3D LiDAR point clouds as a reference. (Adapted from "A Guide to the Role of Standards in Geospatial Information Management" by OGC, ISO/TC 211, and IHO.)

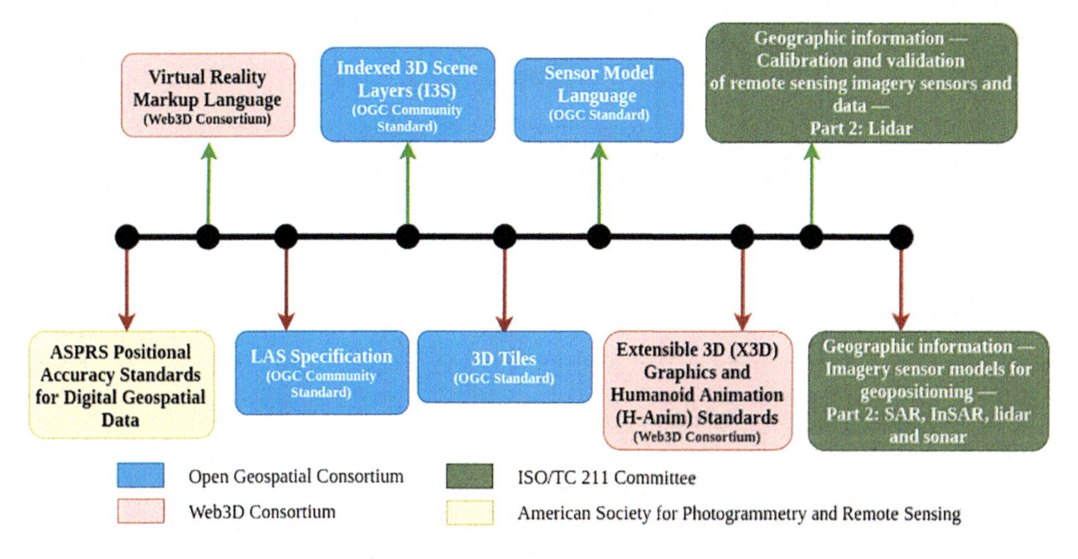

FIGURE 5.4 Various LiDAR and 3D specifications proposed by international standardization bodies—OGC, Web3D consortium, ISO/TC 211 committee, and ASPRS.

5.3.3 Designing a Geosemantic Standards-Driven Framework for 3D LiDAR Data

The Geosemantic framework builds upon the geospatial Linked Data and RESTFul APIs (For example, OGC API 3D GeoVolumes) to achieve interoperability of heterogeneous resources. The Linked Data approach uses the linking of resources over the internet that is both machine and human processible. It enables the service providers to provide a way for clients (mobile or web) to work with resources from diverse resources. RDF or Resource Description Framework is a W3C standard for data interchange over the web and is a widely used standard for creating and defining Linked Data. The geospatial Linked Data is the core of the Semantic Web, which aims at interconnecting and making the resources accessible, thus allowing reasoning and mining implicit semantic relationships from the data. With the rise in the number of sensors and a variety of available datasets, there is a need to standardize LiDAR data acquisition and transmission. OGC standards such as Sensor Web Enablement and SensorML[13] came into existence to realize this objective. LiDAR data standardization is also imperative in developing the airborne LiDAR data processing ecosystem in a collaborative and modular approach. Collaborative development paves the way for multiple stakeholders to enhance participation in research and development at multiple stages. Also, the 3D geosemantic standards would enable implicit knowledge discovery and spatio-temporal and contextual information extraction based on geospatial characteristics. In this section, we focus on implementing interoperability in LiDAR data acquisition.

5.3.3.1 LiDAR Markup Language (LiDARML)—Toward Interoperability for LiDAR

Sensor Web Enablement (SWE) framework provides an efficient platform for the users to interact with the sensors using various services such as Sensor Observation Service (SOS) and Sensor Planning Service (SPS). It helps to establish interfaces and encodings that will enable the services and applications of all types to the sensor over the web (Durbha & Joglekar, 2021). SWE typically comprises SWE Common Data Model Encoding Standard[14] and SWE Service Model Implementation Standard.[15] The Common Data Model encoding standard defines low-level data models for exchanging sensor-related data between various sensor nodes of the SWE framework. On the other hand, the Service Model Implementation Standard defines packages for common use for various services. Standardizing LiDAR data acquisition units helps observe the acquired data over the internet. The data acquisition unit could be deployed over moving platforms such as drones or vehicles, and in such a case, it would be difficult to visualize the data in real time.

Snippet 5.1: Code snippet showing the subunits defined in the LiDAR data acquisition[16] unit – Range Finder[17] and the Dynamic Position Locator[18] using the SensorML standard

```
<sml:components>
<sml:ComponentList>
<sml:component name="RangeFinder" xlink:title="urn:csre:sensors:2018"
xlink:href="https://geosysiot.in/liskg/RangeFinder.xml">
<sml:component name="DynamicPositionLocator" xlink:title="urn:accord:sensors:gps" xlink:href="https://www.
  geosysiot.in/liskg/DynamicPositionLocator.xml"/>
</sml:ComponentList>
</sml:components>
```

The SensorML focuses to provide a robust and standardized means of defining the components, processes, and features related to observations and measurements. It includes the sensors, actuators that trigger based on the sensor values, and the pre and post-event operations. In the context of LiDAR data acquisition units, a standardized format could help combine various point cloud datasets for specific applications over the internet, thus enabling interoperability among the existing infrastructure and sensor networks. The proposed standard, named *LiDARML* (LiDAR Markup Language) for data acquisition unit defined using the XML (eXtensible Markup Language), is intended for a LiDAR data acquisition system consisting of a rangefinder (LiDAR scanner) and a location sensor. The data acquisition unit comprises a range-finder that calculates the distance of objects based on time of flight and a GPS-based dynamic position locator. Since the rangefinder is proposed to be deployed on a moving platform, its location would be changing dynamically. Thus visualizing and analyzing it over the internet would require its location not only in the local reference system but also in a defined projected spatial reference system.

5.4 DEVELOPMENT OF A SCALABLE LiDAR INFORMATION MINING FRAMEWORK: A SYSTEMS PERSPECTIVE

Automated information retrieval from raw data has been desirable and is also the genesis for modern computer vision developments enhancing machine comprehension and understanding. Although

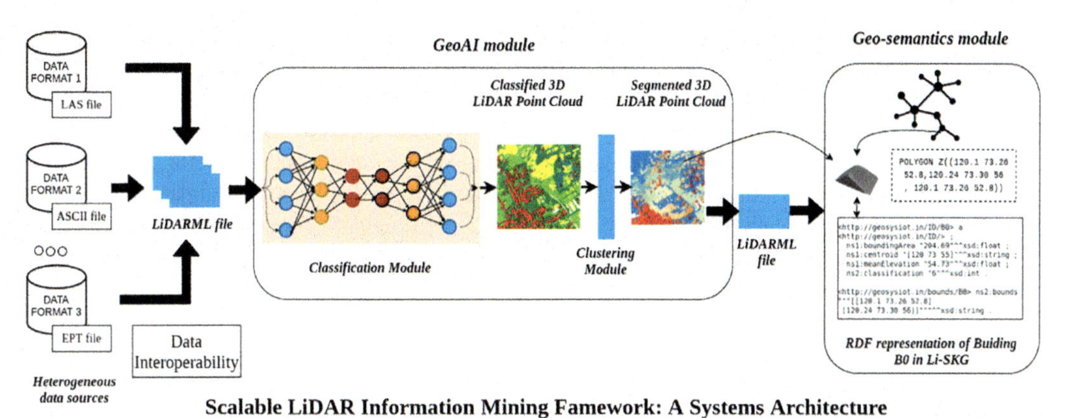

FIGURE 5.5 Illustration of a general architecture for designing a scalable LiDAR information mining framework.

comprehensive research and development are ongoing, this still seems to be a challenging problem. In research, this task is defined as information retrieval, knowledge extraction, knowledge discovery, or information mining. Recently, considerable research has been made to enhance machine comprehension and understanding using deep learning for classification problems. (Wijmans et al., 2019) proposed extraction of behavioral patterns from the social networking sites for perceiving real-world inputs in near real-time. In EO and remote sensing research, image information mining from remote sensing datasets has been widely researched for satellite images (2D dataset) over the years. Some of the examples of such systems are—GeoBrowse (Tusk et al., 2003), GeoIRIS (Shyu et al., 2007), I^{3}KR (Durbha & King, 2005), KEO (Datcu et al., 2010), and Deepti (Maskey et al., 2020). We recommend the readers follow the review of EO image information mining (IIM) for a comprehensive understanding of the evolution in IIM frameworks (Quartulli & Olaizola, 2013).

Figure 5.5 illustrates the general architecture for designing a scalable LiDAR information mining framework. The unclassified LiDAR point cloud is passed to the GeoAI module (explained in detail in the coming section). For many applications, the output of the GeoAI module, usually a segmented LiDAR point cloud, is also a significant result, but it is application-agnostic. It is not useful for decision-makers, which usually count on the knowledge extracted from the classified point cloud. The overall objective seeks to implement an end-to-end semantic information mining framework for achieving semantic interoperability based on already-defined standards. The semantic interoperability is achieved by using RDF or Linked Data representation and building upon the existing geosemantic standards.

5.4.1 Geo-Artificial Intelligence (GeoAI) Module for 3D LiDAR Point Cloud Processing

The GeoAI module of the proposed LiDAR information mining framework comprises point classification and segmentation algorithms. Recently, a new research trend could be observed for automatic feature extraction for segmentation and classification using deep learning and machine learning techniques. Deep learning networks use stacked layers of convolutional, activation, pooling, normalization layers, etc., for learning high-level features implicitly and hierarchically. Due to irregular structure, it is challenging to use raw point cloud data as input to the neural networks. Earlier this issue was addressed by using 3D voxel grids for representing the raw point cloud data (Maturana & Scherer, 2015a; Wu et al., 2015). The voxels (similar to pixels in 2D images) represent the shape of the objects and provide structure to the unstructured point cloud data. On the other hand, there have been researches exploring the conversion of 3D point clouds into multi-view images for implementing 2D Convolutional Neural Networks (CNN) architectures, thus making it easier for implementation.

A new approach for 3D object classification and scene segmentation was introduced by Qi et al. (2017a), where the authors propose passing the unordered point clouds as input without any preprocessing. The point cloud is represented as a set of 3D points where each point is a vector of its (x, y, z) coordinates. The point attributes represent the dimensions of a particular point and can be increased by adding extra feature channels such as color and surface normals. Segmentation and clustering of point cloud data have been studied for particular applications such as forest inventory and airplane landing zone detections (Maturana & Scherer, 2015b; Hu & Xie, 2016, p. 3). Since point cloud data is attributed to high spatial resolution and accuracy, the small footprint provided by the raw point cloud can generate promising results after segmenting the raw data and then clustering them for similar classes. SegCloud (Tchapmi et al., 2018) is a semantic segmentation pipeline for the point clouds that voxelized the point cloud and uses 3D Fully Convolutional Network to perform voxel predictions.

PointNet++ (Qi et al., 2017b) incorporated local geometrical features to improve the results proposed in PointNet. Li et al. (2020) propose a geometric graph CNN for 3D point cloud segmentation, named TGNet. In Thomas et al. (2019), the authors propose a flexible convolution operator for 3D point clouds, named KPConv. In Wu et al. (2019), the authors propose PointCONV, a deep convolutional network for 3D point clouds. Also, augmenting CNN with hand-crafted features has been widely researched in domain-specific applications, including deep learning for remote sensing and Earth Observation. Based on this, LiDARNet and LiDARNet++ (Shinde et al., 2021a) have been proposed, using a 3D feature stack and the learned convolutional features. For an extensive overview and comprehensive understanding of deep learning techniques for point cloud classification, the readers are recommended to follow (Guo et al., 2021).

In our proposed system, any of those mentioned above deep learning-based approaches for point cloud segmentation can be used. This flexibility is attributed to the modularity and differentiability associated with implementing deep learning models. Also, deep learning approaches can be used for processing big data making it scalable in implementation. The output of the GeoAI module generates a segmented 3D point cloud where each segment represents a cluster of points belonging to the same class.

5.4.2 GeoSemantics Module: Toward Semantically Enriched LiDAR Knowledge Graph

The Geosemantics module generates a semantically enriched knowledge graph from the segmented point cloud. Each point in the segmented point cloud has "Cluster Id" and "Classification value" as additional attributes. Each cluster forms an entity in the knowledge graph and is represented as a Unique Resource Identifier (URI). The spatio-contextual relationships between the entities such as "*touching*," "*is externally connected to*," "*contains*," are also stored in the knowledge graph using object properties. Additionally, the geometry of each cluster is stored instead of storing all the points forming that cluster to represent the objects efficiently. In Kurte et al. (2017), the authors propose using the entities' bounding box as a geometric representation. Potnis et al. (2021) propose to incorporate object geometry in Well-Known Text (WKT) format. In the proposed system for representing 3D geometries, WKT with the "Z" dimension is proposed. For example, If a polygon geometry in 2D is represented as POLYGON (30 10, 40 40, 20 40, 10 20, 30 10) in WKT format, then we can represent the 3D structure corresponding to the same polygon as POLYGON Z(30 10 5, 40 40 10, 20 40 3, 10 20 6, 30 10 5) where the third entry in each array represents the elevation values.

The semantic enrichment also comprises defining semantic *rules* and *axioms* describing certain conditions on the entities and their relationships. This semantic enrichment is achieved by reasoning the knowledge graph using semantic reasoners or reasoning engines capable of inferring logical consequences from a set of predefined facts and axioms. Once the inferred knowledge graph is generated, it can be queried using a query language such as GeoSPARQL.[19] The overall system forms the basis for implementing a Knowledge Base Question-Answering framework for 3D LiDAR data and is presented in the following section.

5.5 CASE STUDY: KNOWLEDGE BASE QUESTION-ANSWERING (KBQA) FRAMEWORK FOR LiDAR

Question-answering (QA) systems enable querying or asking questions to the machine which provide answers based on the processed information. In Visual Question-Answering (VQA), the processing requires extraction of meaningful information relevant to the input questions from the corresponding images based on the objects present in the images. VQA was introduced to model an agent for predicting answers given a question describing the image (Agrawal et al., 2017; Goyal et al., 2019; Shrestha et al., 2019) and the visual features of the image. The processing agent in VQA implicitly captures spatial and topological relationships and is widely studied in machine comprehension, machine intelligence, or machine understanding. Semantic labeling of the question-answer pairs along with the corresponding images is a big challenge for implementing VQA due to the subjectivity in defining the questions. For EO and remote sensing, in Lobry et al. (2019), the authors propose an approach to automatically annotate the questions using OpenStreetMap imagery for the region of interest. The annotations are used for training a deep learning model capable of performing question-answering (Lobry et al., 2020a, b). The authors in Shinde et al. (2021b) attempted to develop a deep learning-based 3D question-answering framework to model airborne LiDAR point clouds and associated questions to predict the answer based on 3D topological relationships. This approach is designed as a binary question-answering schema, i.e., the answering schema is restricted to *yes/no* as answers. The challenge with the above-mentioned studies lies in interpretability and explainability due to the white-box nature of deep learning-based approaches. In this case study, the focus is to equip readers in designing a Knowledge Base Question-Answering framework (KBQA) based on a proposed LiDAR Scene Knowledge Graph (LiSKG).

5.5.1 Dataset Details

The authors have used the LiDAR data openly available with the UP DREAM (Andaya et al., 2015) (Disaster Risk and Exposure Assessment for Mitigation) program for the study. The dataset comprised ~60 point cloud scenes with each scene having more than 100 million points from multiple urban locations geographically distributed over the Philippines.

5.5.2 Problem Formulation: Knowledge Base Question-Answering (KBQA) for LiDAR

The KBQA problem can be described as follows. The problem primarily explores the usability of a knowledge graph for question-answering to achieve enhanced 3D perception and interactivity. The knowledge graph is generated using the 3D LiDAR point cloud. Typically, it can be categorized in the following steps:

1. Classify the input LiDAR point cloud and segment it to generate unique semantic clusters. Each cluster is uniquely represented using URIs.
2. Represent the entire LiDAR point cloud as a knowledge graph in RDF format comprising entities (clusters) and their associated spatio-contextual relationships.
3. Framing the natural language questions as GeoSPARQL queries for querying over the knowledge graph.

5.5.3 Generating LiDAR Scene Knowledge Graph—LiSKG

Figure 5.6 illustrates the dataflow representing the translation of an unclassified point cloud to LiSKG. In general, question-answering is subjective and open-ended because there are multiple

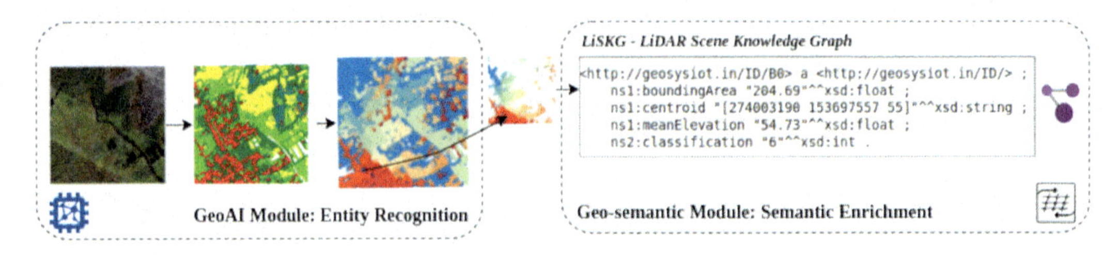

FIGURE 5.6 Illustration of high-level dataflow representing the translation of an unclassified point cloud to LiSKG.

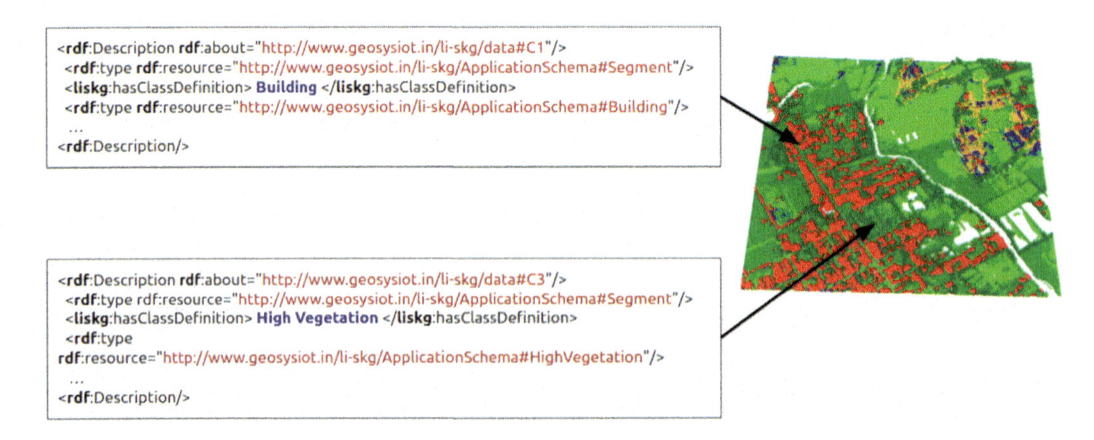

FIGURE 5.7 Illustration of the LiSKG as Linked Data showing segments of Building and High Vegetation classes in the Resource Description Format (RDF).

ways of phrasing a question or describing a scene in question-answer pairs. This makes the model training challenging and poses the need for an exhaustive and rich set of question-answer pairs. To overcome this issue, researchers have been trying to accomplish this objective in a closed-world domain by restricting the domain of application or the type of question and answers. KBQA, on the other hand, can be implemented in an open-world domain attributed to the implicit knowledge associated in the form of entities.

LiDARNet++ (Shinde et al., 2021a) followed by clustering constitutes the GeoAI (Refer to Figure 5.5) module for the knowledge base question-answering framework. In this case, the clusters form the entities. Each entity and corresponding relationships are represented using a unique uniform resource identifier (URI) and are stored as LiDAR Scene Knowledge Graph (LiSKG) in the RDF Linked data format.

In Figure 5.7, the 3D point cloud transformed to RDF Linked Data is illustrated. It refers to a segment with ID as "C1" and has a Polygon geometry. It has a class definition as "Building" and is defined by the rule "hasClassDefinition." Similarly, the segment with ID "C3" has a Polygon geometry and has a class definition of "High Vegetation." The LiSKG Linked Data is stored in a Graph database such as GraphDB,[20] AllegroGraph,[21] and NEO4J,[22] and can further be queried for information retrieval using the query language syntax as discussed in the upcoming section.

5.5.4 Natural Language to GeoSPARQL in Knowledge Base Question-Answering (KBQA)

Conventionally, querying over a Knowledge base is performed using a semantic query language such as SPARQL.[23] Based on SPARQL, OGC standardized GeoSPARQL[24] as a geographic query language for RDF data to retrieve and manipulate data stored in RDF format. It extends the SPARQL functions for geospatial domains and enables querying involving geometrical features. Typically, while implementing a KBQA, natural language questions are required to be translated to corresponding GeoSPARQL queries. Although recently, there has been the development of other query languages corresponding to the databases, such as the Cipher query language for NEO4J. Moreover, there has been research in implementing approaches like Named Entity Recognition (NER) to extract the semantics and contextual features from the natural language questions and form standardized template-based queries.

In this case study, our focus is on translating natural language (English) questions to GeoSPARQL queries for querying the LiSKG. The questions are designed to include "*has/is*" or possession relationships, and "*touches/adjacent to/surrounded,*" i.e., spatial relationships. Figure 5.8a illustrates a visualization of the 3D LiDAR point cloud and Figure 5.8b shows examples of the questions. These questions are then translated to equivalent GeoSPARQL queries. Table 5.2 illustrates some examples of natural language questions, translated GeoSPARQL queries, and the corresponding responses in WKT format along with the visualization of the geometry. The domain-specific 3D spatial and topological relationships can also be mapped and queried using BimSPARQL (Zhang et al., 2018), a domain-specific functional SPARQL extension.

The queries use the spatial predicates defined in GeoSPARQL such as '*geof:sfContains*' and '*geof:sfTouches*' for describing the spatial relationships between resources. For example, '*geof:sfContains*' maps the spatial relation representing a geometry *geom1* containing or surrounding another geometry *geom2*. This approach using GeoSPARQL querying opens up research dimensions for querying spatial and topological relationships (RCC - Region Connection Calculus and Egenhofer relationships) for qualitative reasoning. The responses (refer to Column "WKT Response and 2D Geometry Visualization" of Table 5.3) of the executed queries could be *geometries* or spatial features comprising Point, LineString, Polygon, PolyhedralSurface, and GeometryCollection. Usually, such vector geometry objects are represented using Well-known Text (WKT) or a compact, non-human-readable, and efficient computer processing Well-Known Binary (WKB) format. The 3D geometry is represented in WKT format by appending "**Z**" as "**POLYGON Z**" in the "**POLYGON**" representation used for the 2D geometry.

Recently, one of the research works prepends a Natural Language Processing engine for generating the discussed GeoSPARQL queries directly from Natural Language questions defined in English by using Neural Machine Translation (Potnis et al., 2020). This opens up new research dimensions and enables the technology to be useful for users without prior knowledge and understanding of geospatial databases and geospatial technology; essentially policymakers, students, and researchers.

(b) Snippet 1: Examples Illustrating Question-Answer pairs

Question	Answer
Is there a building?	***Yes***
Are there multiple buildings?	***No***
Does the scene contain a building touching the road?	***No***
Is the area of ground more than the vegetation?	***No***
Is there a region with high elevation?	***Yes***

FIGURE 5.8 Visualization of the 3D LiDAR point cloud—(a) topographical visualization. (b) Examples illustrating natural language questions.

TABLE 5.2
Natural Language Question, Corresponding GeoSPARQL Query and Response after Querying over LiSKG (LiDAR Scene Knowledge Graph)

Natural Language Question	GeoSPARQL Query	WKT Response and 2D Geometry Visualization

• Is there a building in the scene?	• PREFIX liskg: <http://www.geosysiot.in/liskg>
	select distinct ?area1
	where {
	?area1 liskg:hasClassDefinition
	liskg:Building .
	}
		• POLYGON Z(30 10 0, 40 40 5, 20 40 10, 10 20 10, 30 10 0)

• Does the scene contain a building surrounded by high vegetation?	• PREFIX geof:<http://www.opengis.net/def/function/geosparql/>
	PREFIX liskg:<http://www.geosysiot.in/liskg>
	select distinct?area1?area2
	where {
	?area1 liskg:hasClassDefinition
	liskg:HighVegetation.
	?area2 liskg:hasClassDefinition
	liskg:Building.
	?area1 liskg:hasGeometry?geom1.
	?area2 liskg:hasGeometry?geom2.
	filter (geof:sfContains
	(?geom1,?geom2))
	}
		• POLYGON Z((35 10 0, 45 45 5, 15 40 10, 10 20 5, 35 10 0), (20 30 0, 35 35 2, 30 20 2, 20 30 0))

5.6 SUMMARY AND FUTURE TRENDS

LiDAR is emerging as a reliable and efficient technology in academic research and industrial development. This has led to progressing innovations across various domains ranging from low-cost and high-accuracy sensor manufacturing to rapid data acquisition and processing. Moreover, it finds immense potential in application domains such as accurate mapping and surveying, forest inventory management, building asset monitoring, and autonomous vehicles. With increased diversity and heterogeneity, interoperability is pursued to support reusable research. Interoperability is achieved by defining standards, and various international bodies are constantly working on drafting and publishing these standards with community support. Although there have been constant developments in publishing new standards and iteratively improving the existing ones in the geospatial standards domain, a research gap exists in the LiDAR processing ecosystem. This chapter focuses on introducing concepts for standardizing LiDAR data processing. It highlights an example of LiDARML for standardizing LiDAR scanning units based on the OGC SensorML standard. The chapter touches on the Geosemantics framework and presents how the LiDAR point cloud can be used in a REST-based architecture, thus making it accessible as linked data over the web. Also, the chapter covers the existing drafts and published standards for LiDAR visualization over the web.

The LiDAR point cloud dataset is inherently big and multidimensional and requires high computing power. With the recent advancements in high-performance computing and deep learning, scalable implementation of frameworks has been possible. Additionally, following the success of deep learning approaches such as Convolutional Neural Networks (CNNs), Recurrent Neural Networks (RNNs), and Generative Adversarial Networks (GANs) for vision tasks, and deployment of the developed model in production using MLOps, it has been possible to design an end-to-end data pipeline. In this regard, the chapter comprehensively describes the state of deep learning for 3D LiDAR point clouds and presents results for 3D LiDAR point cloud classification. The 3D LiDAR point cloud classification forms the processing core for translating the input raw point cloud to a classified point cloud. This helps design the proposed LiSKG—LiDAR Scene Knowledge graph, which completes the data-to-knowledge continuum. LiSKG enables knowledge discovery and 3D information mining, thus forming the foundation for 3D information retrieval by building on the concepts of Linked Data and the semantic web. Finally, the chapter presents a case study on 3D LiDAR question answering for urban environments.

The primary objective of the chapter is to present a holistic discussion on the amalgamation of conventional AI (artificial intelligence) technologies arising from knowledge extraction and semantic web with the relatively recent development of 3D computer vision, specifically concerning 3D LiDAR point clouds. We envisage that this chapter would open the grounds for an understanding of achieving interoperability for 3D LiDAR point clouds. Moreover, it is targeted at upcoming researchers and learners to explain the developments in GeoAI for 3D LiDAR data processing. The case study for 3D question answering shall equip the learners to develop a thought process in implementing a standards-based framework for 3D information retrieval.

In future directions, we foresee that there shall be further developments for standards-based 3D information retrieval and processing tasks. It is challenging due to existing developments and constantly evolving research, but it also sets a foundation for cementing a standard development approach at different levels. It is envisioned that the existing LiDAR datasets shall be made available for the users by following RESTFul microservices, which eventually will open up a plethora of research dimensions and developments.

5.7 WHERE TO LOOK FOR FURTHER INFORMATION

1. Web3D Consortium Standards Strategy. Online: https://www.web3d.org/strategy
2. A Guide to the Role of Standards in Geospatial Information Management. Online: https://ggim.un.org/documents/Standards%20Guide%20for%20UNGGIM%20-%20Final.pdf
3. HydroComplexity-Geosemantics Framework. Online: http://hcgs.ncsa.illinois.edu/#introduction
4. Newcomers Earth Observation Guide. Online: https://business.esa.int/newcomers-earth-observation-guide#ref_8

ACKNOWLEDGMENTS

The authors of this chapter thank the reviewers and editors for allowing them to present their thoughts. The authors are thankful to the Taylor and Francis Group, LLC/CRC Press for publishing a book on - *Advances in Scalable and Intelligent Geospatial Analytics: Challenges and Applications*, which is a very relevant topic in the current scenario. The authors would like to thank the OpenTopography Facility (with support from the National Science Foundation) for providing the open lidar data. The authors are thankful to the Google Cloud Research Programs team for awarding the Google Cloud Research Credits with the award GCP19980904 to use the high-end cloud computing platforms.

NOTES

1 Global Ecosystem Dynamics Investigation (GEDI): https://gedi.umd.edu/mission/mission-overview/.
2 2015 IEEE GRSS Data Fusion Contest. Online: http://www.grss-ieee.org/community/technical-committees/data-fusion.
3 OpenTopography: https://opentopography.org/.
4 ISPRS Semantic Labeling Contest (3D): https://www.isprs.org/education/benchmarks/UrbanSemLab/results/vaihingen-3d-semantic-labeling.aspx.
5 A Guide to the Role of Standards in Geospatial Information Management. By OGC, ISO/TC 211, and IHO. Online: https://ggim.un.org/documents/Standards%20Guide%20for%20UNGGIM%20-%20Final.pdf.
6 ISPRS webpage on Standardization sites: https://www.isprs.org/education/standardization.aspx.
7 Institute of Electrical and Electronics Engineers. Online: https://www.ieee.org/.
8 International Telecommunication Union. Online: https://www.itu.int/.
9 Open Geospatial Consortium. Online: https://www.ogc.org/.
10 International Organization for Standardization. Online: https://www.iso.org/technical-committees.html.
11 International Hydrographic Organization: Online: https://iho.int/.
12 World Wide Web Consortium. Online: https://www.w3.org/.
13 OGC SensorML Standard - https://www.ogc.org/standards/sensorml.
14 OGC SWE Common Data Encoding Model - http://www.opengeospatial.org/standards/swecommon.
15 OGC SWE Service Model Implementation Standard - http://www.opengeospatial.org/standards/swes.
16 LiDARML description for LiDAR Data Acquisition unit: https://www.geosysiot.in/liskg/LidarDataAcquisitionUnit.xml.
17 LiDARML description for Range Finder unit: https://geosysiot.in/liskg/RangeFinder.xml.
18 LiDARML description for Dynamic Position Locator unit: https://www.geosysiot.in/liskg/DynamicPositionLocator.xml.
19 OGC GeoSPARQL - A Geographic Query Language for RDF Data: https://opengeospatial.github.io/-ogc-geosparql/geosparql11/spec.html.
20 GraphDB: https://graphdb.ontotext.com/documentation/standard/index.html.
21 AllegroGraph: https://allegrograph.com/.
22 Neo4J: https://neo4j.com/.
23 SPARQL 1.1 Query Language: https://www.w3.org/TR/sparql11-query/.
24 OGC GeoSPARQL - A Geographic Query Language for RDF Data: https://opengeospatial.github.io/ogc-geosparql/geosparql11/spec.html.

REFERENCES

Agrawal, A., Lu, J., Antol, S., Mitchell, M., Zitnick, C. L., Parikh, D., & Batra, D. (2017). VQA: Visual question answering: www.visualqa.org. *International Journal of Computer Vision*, *123*(1), 4–31. doi: 10.1007/s11263-016-0966-6.

Andaya, K. J., Alviar, J., Mars, P., Cruz, C., Sarmiento, C. J., Balicanta, L., & Paringit, E. (2015). Airborne LiDAR surveying in the Philippines: Data acquisition of the Nationwide Disaster Risk and Exposure Assessment For Mitigation (DREAM) program in 18 Major River Basins. *ACRS 2015–36th Asian Conference on Remote Sensing: Fostering Resilient Growth in Asia, Proceedings,* Asia.

Cramer, M. (2010). The DGPF-test on digital airborne camera evaluation overview and test design. *Photogrammetrie - Fernerkundung - Geoinformation*, *2010*(2), 73–82. doi: 10.1127/1432-8364/2010/0041.

Datcu, M., King, R. L., & D'Elia, S. (2010). Introduction to the special issue on image information mining: Pursuing automation of geospatial intelligence for environment and security. *IEEE Geoscience and Remote Sensing Letters*, *7*(1), 3–6. doi: 10.1109/LGRS.2009.2034822.

Durbha, S. S., & Joglekar, J. (2021). *Internet of Things*. Oxford University Press, Oxford.

Durbha, S. S., & King, R. L. (2005). Semantics-enabled framework for knowledge discovery from Earth observation data archives. *IEEE Transactions on Geoscience and Remote Sensing*, *43*(11), 2563–2572. doi: 10.1109/TGRS.2005.847908.

Goyal, Y., Khot, T., Agrawal, A., Summers-Stay, D., Batra, D., & Parikh, D. (2019). Making the V in VQA matter: Elevating the role of image understanding in visual question answering. *International Journal of Computer Vision, 127*(4), 398–414. doi: 10.1007/s11263-018-1116-0.

Guo, Y., Wang, H., Hu, Q., Liu, H., Liu, L., & Bennamoun, M. (2021). Deep learning for 3D point clouds: A survey. *IEEE Transactions on Pattern Analysis and Machine Intelligence, 43*(12), 4338–4364. doi: 10.1109/TPAMI.2020.3005434.

Hackel, T., Savinov, N., Ladicky, L., Wegner, J. D., Schindler, K., & Pollefeys, M. (2017). Semantic3D.Net: A new large-scale point cloud classification benchmark. *ISPRS Annals of the Photogrammetry, Remote Sensing and Spatial Information Sciences, 4*(1W1), 91–98. doi: 10.5194/isprs-annals-IV-1-W1-91-2017.

Hu, X., & Xie, Y. (2016). Segmentation and clustering of 3D forest point cloud using mean shift algorithms. *2016 4th International Conference on Machinery, Materials and Computing Technology (ICMMCT 2016)*, Hangzhou, pp. 1275–1279.

Hu, Q., Yang, B., Khalid, S., Xiao, W., Trigoni, N., & Markham, A. (2021). Towards Semantic Segmentation of Urban-Scale 3D Point Clouds: A Dataset, Benchmarks and Challenges (arXiv:2009.03137). http://arxiv.org/abs/2009.03137.

Kölle, M., Laupheimer, D., Schmohl, S., Haala, N., Rottensteiner, F., Wegner, J. D., & Ledoux, H. (2021). The Hessigheim 3D (H3D) benchmark on semantic segmentation of high-resolution 3D point clouds and textured meshes from UAV LiDAR and multi-view-stereo. *ISPRS Open Journal of Photogrammetry and Remote Sensing, 1*, 100001. doi: 10.1016/j.ophoto.2021.100001.

Kurte, K. R., Durbha, S. S., King, R. L., Younan, N. H., & Vatsavai, R. (2017). Semantics-enabled framework for spatial image information mining of linked earth observation data. *IEEE Journal of Selected Topics in Applied Earth Observations and Remote Sensing, 10*(1), 29–44. doi: 10.1109/JSTARS.2016.2547992.

Li, Y., Ma, L., Zhong, Z., Cao, D., & Li, J. (2020). TGNet: Geometric graph CNN on 3-D point cloud segmentation. *IEEE Transactions on Geoscience and Remote Sensing, 58*(5), 3588–3600. doi: 10.1109/TGRS.2019.2958517.

Lobry, S., Murray, J., Marcos, D., & Tuia, D. (2019). Visual question answering from remote sensing images. *IGARSS 2019-2019 IEEE International Geoscience and Remote Sensing Symposium*, 4951–4954. doi: 10.1109/IGARSS.2019.8898891.

Lobry, S., Marcos, D., Kellenberger, B., & Tuia, D. (2020a). Better generic objects counting when asking questions to images: A multitask approach for remote sensing visual question answering. *ISPRS Annals of the Photogrammetry, Remote Sensing and Spatial Information Sciences, 2*, 1021–1027. doi: 10.5194/isprs-annals-V-2-2020-1021-2020.

Lobry, S., Marcos, D., Murray, J., & Tuia, D. (2020b). RSVQA: Visual question answering for remote sensing data. *IEEE Transactions on Geoscience and Remote Sensing, 58*(12), 8555–8566. doi: 10.1109/TGRS.2020.2988782.

Maskey, M., Ramachandran, R., Ramasubramanian, M., Gurung, I., Freitag, B., Kaulfus, A., Bollinger, D., Cecil, D. J., & Miller, J. (2020). Deepti: Deep-learning-based tropical cyclone intensity estimation system. *IEEE Journal of Selected Topics in Applied Earth Observations and Remote Sensing, 13*, 4271–4281. doi: 10.1109/JSTARS.2020.3011907.

Maturana, D., & Scherer, S. (2015a). VoxNet: A 3D convolutional neural network for real-time object recognition. *IEEE International Conference on Intelligent Robots and Systems.* doi: 10.1109/IROS.2015.7353481.

Maturana, D., & Scherer, S. (2015b). 3D convolutional neural networks for landing zone detection from LiDAR. *2015 IEEE International Conference on Robotics and Automation (ICRA)*, 3471–3478. doi: 10.1109/ICRA.2015.7139679.

Potnis, A. V., Shinde, R. C., & Durbha, S. S. (2020). Towards natural language question answering over earth observation linked data using attention-based neural machine translation. *IGARSS 2020-2020 IEEE International Geoscience and Remote Sensing Symposium*, 577–580. doi: 10.1109/IGARSS39084.2020.9323183.

Potnis, A. V., Durbha, S. S., & Shinde, R. C. (2021). Semantics-driven remote sensing scene understanding framework for grounded spatio-contextual scene descriptions. *ISPRS International Journal of Geo-Information, 10*(1), 32. doi: 10.3390/ijgi10010032.

Qi, C. R., Su, H., Mo, K., & Guibas, L. J. (2017a). PointNet: Deep learning on point sets for 3D classification and segmentation. *Presented in the IEEE Conference on Computer Vision and Pattern Recognition*, Honolulu, HI.

Qi, C. R., Yi, L., Su, H., & Guibas, L. J. (2017b). PointNet++: Deep hierarchical feature learning on point sets in a metric space. *NIPS'17: Proceedings of the 31st International Conference on Neural Information Processing Systems,* Long Beach, CA, pp. 5105–5114.

Quartulli, M., & Olaizola, G. I. (2013). A review of EO image information mining. *ISPRS Journal of Photogrammetry and Remote Sensing*, *75*, 11–28. doi: 10.1016/j.isprsjprs.2012.09.010.

Shinde, R. C., Durbha, S. S., & Potnis, A. V. (2021a). LiDARCSNet: A deep convolutional compressive sensing reconstruction framework for 3D airborne LiDAR point cloud. *ISPRS Journal of Photogrammetry and Remote Sensing*. doi: 10.1016/j.isprsjprs.2021.08.019.

Shinde, R. C., Durbha, S. S., Potnis, A. V., Talreja, P., & Singh, G. (2021b). Towards enabling deep learning-based question-answering for 3D LiDAR point clouds. *2021 IEEE International Geoscience and Remote Sensing Symposium IGARSS*, pp. 6936–6939. doi: 10.1109/IGARSS47720.2021.9553785.

Shrestha, R., Kafle, K., & Kanan, C. (2019). Answer them all! Toward universal visual question answering models. *2019 IEEE/CVF Conference on Computer Vision and Pattern Recognition (CVPR)*, pp. 10464–10473. doi: 10.1109/CVPR.2019.01072.

Shyu, C.-R., Klaric, M., Scott, G. J., Barb, A. S., Davis, C. H., & Palaniappan, K. (2007). GeoIRIS: Geospatial information retrieval and indexing system: Content mining, semantics modeling, and complex queries. *IEEE Transactions on Geoscience and Remote Sensing*, *45*(4), 839–852. doi: 10.1109/TGRS.2006.890579.

Tchapmi, L., Choy, C., Armeni, I., Gwak, J., & Savarese, S. (2018). SEGCloud: Semantic segmentation of 3D point clouds. *Proceedings - 2017 International Conference on 3D Vision, 3DV 2017*. doi: 10.1109/3DV.2017.00067.

Thomas, H., Qi, C. R., Deschaud, J. E., Marcotegui, B., Goulette, F., & Guibas, L. (2019). KPConv: Flexible and deformable convolution for point clouds. *Proceedings of the IEEE International Conference on Computer Vision*, pp. 6410–6419. doi: 10.1109/ICCV.2019.00651.

Tusk, C., Koperski, K., Aksoy, S., & Marchisio, G. (2003). Automated feature selection through relevance feedback. *Proceedings of 2003 IEEE International Geoscience and Remote Sensing Symposium (IGARSS 2003) (IEEE Cat. No.03CH37477)*, vol. *6*, pp. 3691–3693. doi: 10.1109/IGARSS.2003.1295239.

Varney, N., Asari, V. K., & Graehling, Q. (2020). DALES: A large-scale aerial LiDAR data set for semantic segmentation (arXiv:2004.11985). http://arxiv.org/abs/2004.11985.

Wijmans, E., Datta, S., Maksymets, O., Das, A., Gkioxari, G., Lee, S., Essa, I., Parikh, D., & Batra, D. (2019). Embodied question answering in photorealistic environments with point cloud perception. *2019 IEEE/CVF Conference on Computer Vision and Pattern Recognition (CVPR)*, pp. 6652–6661. doi: 10.1109/CVPR.2019.00682.

Wu, W., Qi, Z., & Fuxin, L. (2019). PointCONV: Deep convolutional networks on 3D point clouds. *Proceedings of the IEEE Computer Society Conference on Computer Vision and Pattern Recognition*, pp. 9613–9622. doi: 10.1109/CVPR.2019.00985.

Wu, Z., Song, S., Khosla, A., Yu, F., Zhang, L., Tang, X., & Xiao, J. (2015). 3D ShapeNets: A deep representation for volumetric shapes. *Proceedings of the IEEE Computer Society Conference on Computer Vision and Pattern Recognition*. doi: 10.1109/CVPR.2015.7298801.

Zhang, C., Beetz, J., & de Vries, B. (2018). BimSPARQL: Domain-specific functional SPARQL extensions for querying RDF building data. *Semantic Web*, *9*(6), 829–855. doi: 10.3233/SW-180297.

6 Geospatial Analytics Using Natural Language Processing

Manimala Mahato
Department of Computer Engineering
Shah and Anchor Kutchhi Engineering College

Rekha Ramesh
Department of Artificial Intelligence and Data Science
Shah and Anchor Kutchhi Engineering College

Ujwala Bharambe
Computer Engineering Department
Thadomal Shahani Engineering College

CONTENTS

DOI: 10.1201/9781003270928-8

6.1 INTRODUCTION

Geospatial Information is embedded in most of the unstructured data that we come across. If there are entities in any data, those entities exist in one or more locations. It incorporates values across time and dynamic location and is referred to as geotext (Hu et al., 2019). The availability of massive training data from the web and tremendous progress in machine learning algorithms has been a catalyst in the growth of NLP field. The NLP techniques are now increasingly being investigated in the geospatial domain. This section will include a brief overview in the area of Geospatial Analytics using NLP.

6.1.1 Geospatial Analytics

A geospatial analysis is used to discover patterns and trends among entities in data using information from GPS, social media, mobile devices, location sensors, satellite imagery, etc. Geospatial data describes objects, events, and phenomena that are located. In some cases, the location may be static (e.g., the location of a building, an accident event), while in other cases, it may be dynamic (e.g., migrating birds, a forest fire spreading). Location data (usually coordinates on Earth), attribute data (the characteristics of the object or event), and time data (the length of time the object, event, or phenomenon existed) are all variables within geospatial data (Hu et al., 2019).

It has been a decade since geotextual, or spatio-textual, data proliferated rapidly. There is geospatial information associated with most web pages. Few examples are geo-tagged photographs uploaded to social photo-sharing services (namely, Flickr and Instagram) containing both description tags and geographic information; geo-tagged microblog posts (namely, geo-tagged tweets) containing text with geographic location information; reviews of local businesses (namely, Yelp and Amazon Reviews) containing both text and location tags; access to location-based social networks (namely, Foursquare); and text-based local news sources (Chen et al., 2021).

A geotext can also be derived from texts that don't explicitly mention place names (Wing and Baldridge, 2014). There are non-spatial words that are geo-indicative, such as beach, sunshine, and traffic. The amount, velocity, and variety of the geospatial data from these unstructured texts are often large. They are organized according to a general geographic representation theory (Goodchild et al., 2007), geotext data can be formulated (Hu, 2018) as in equation (6.1)

$$\langle L,[T],D\rangle \tag{6.1}$$

where

L is the geographic location of any place names mentioned in geotext data.
T is optional. This is the timestamp associated with the data record.
D is the textual description of geotext data.

6.1.1.3 Sources of Geotext

Geotext is available from different sources (Cong et al., 2016) as shown in Figure 6.1.

- **Labels of Vector and Raster Data** are metadata or non-spatial data associated with a location on a digitized map such as the name of the city, population, race, housing, etc. The knowledge associated with a geographic location in the form of metadata can be exploited to convert them into some actionable items by the concerned authorities. Raster land-use maps are correlated with land-use categories, such as 'Commercial' and 'Agricultural'. Satellite images, geotagged photographs, and maps, as well as Trip Advisor, can be used to supplement spatial data with additional information to aid geospatial analytics.
- **Location Based Web Data** includes geotagged Wikipedia pages, travel blogs, and news articles mentioning any place names. Several sources, such as newspapers, historical

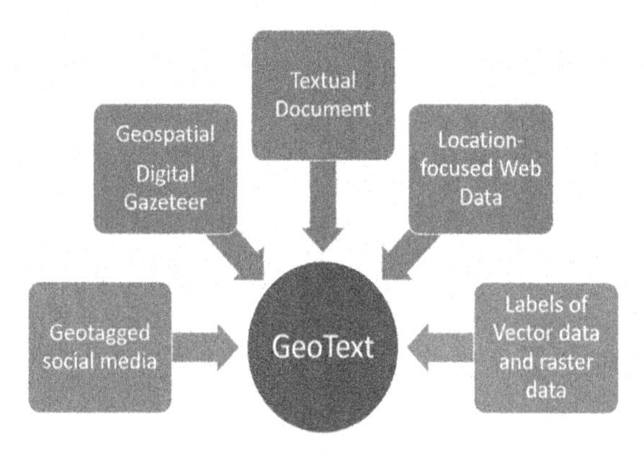

FIGURE 6.1 Sources of geotext.

archives, or Web sites, can give us geospatial information that was previously only available in unstructured texts.

- **Textual Documents** such as historical narratives, archaeological writings, and stories are another source of geotext.
- **Digital Georeferenced Gazetteers** link place names to geographic footprints. Numerous georeferenced records can be found in library catalogs, museums, and bibliographic indexes.

  These data sources can be merged with geospatial data using the digital georeferenced gazetteer. Navigation systems, for example, will be able to have voice-activated user interfaces and provide richer route-finding data. An "event gazetteer" could show the progressive and generalized footprints of named events, such as hurricanes, tornadoes, and migrations (Hill and Goodchild, 2000).
- **Geotagged Social Media** shows the location of articles posted on social networks such as Twitter, Facebook, and Instagram. Among the many sites that use location geotagging are Twitter, Facebook, Instagram, and even Google.

6.1.2 Introduction to Natural Language Processing

One of the most rapidly developing and widely used technologies in the field of AI is NLP. The NLP algorithms convert human speech into expressions that computers can understand. It offers a plethora of techniques for analyzing unstructured or semi-structured text and extracting relevant features from it (Bitters, 2011). The involvement of NLP techniques may range from basic preprocessing methods to any machine learning algorithms or very advanced techniques that can be utilized to infer and interpret semantic information from a large volume of text (refer to Figure 6.2) (Daniel and James, 2019).

A lexical analyzer analyses words into their linguistic components (morphemes). It further finds the relation between these morphemes and converts the word into its root form. It also assigns the possible Part-Of-Speech (POS) to the word. The lexical units can be individual characters, words, sequences of words (n-grams), sentences, paragraphs, or documents. Syntax Analysis ensures the correctness of sentence structure adhering to the rules of the formal grammar. It is used to find dependency among tokens and interpret the grammatical structure of sentences by constructing parse trees. Semantic analyzer assigns meanings to the syntactical units identified in previous steps based on the contextual knowledge. Sense disambiguation is the major part of this step. Discourses are collections of coherent sentences. This stage focuses on identifying and resolving coreferences among the sentences in a document. For example, consider a set of sentences "*Mohini visited National flower show in Bangalore. It had a wide variety and exotic collection of flowers.*

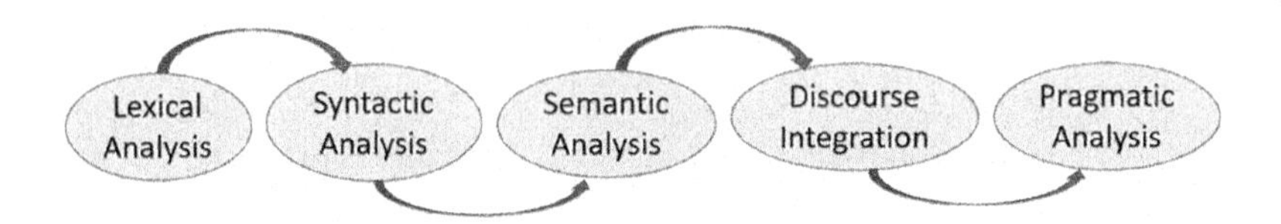

FIGURE 6.2 Stages of NLP.

She always loves to visit such shows." Here, "Mohini," "National *flower show*" and "Bangalore" are possible entities. "She" and "It" are references to the entities "Mohini" and "flower show." Pragmatics Analysis is the study of relation between utterances (words, phrases, and sentences used for communication) and context. It tries to define the rules that govern their interpretation.

6.1.3 Geospatial Data Meets NLP

Texts from a variety of sources are analyzed and computational methods are applied using NLP. In the geospatial domain, NLP can be utilized for spatial understanding of where events, places, or people may be related to a given phenomenon based on big data or large datasets (Hu, 2018). In most cases, NLP is used to extract meaning in an automated fashion from large collections of corpora, often by combining statistical or artificial intelligence techniques, using web scraping or document searching to obtain the data.

In the wider research literature, we are now beginning to see methodologies and techniques developed to understand a variety of topics within spatial data gathering and analytical frameworks. It may be possible for natural conversation to reveal patterns regarding the places in the world that people are interested in or converse about during everyday speech, or to reveal patterns from data recorded such as tweets, blogs, or websites. Now that data is available, it can be processed to provide new insight into where event patterns occur, how they relate to other events, such as natural disasters, and what may follow. NLP can be employed to identify parts of speech, sentence structure patterns, word frequency patterns, or even local dialects or slang terms. It is also possible that geographical references in text can be vague (e.g., a city rather than a specific city), where machine learning techniques can be used to determine what city or area the vague reference may refer to.

Many text documents describe various types of movements by individuals, groups, animals, vehicles, and other moving objects. It is possible to map movement descriptions in text to a number of component elements, including the source of the movement, the destination, the route, the direction, the distance, the start and end time, and the duration. Using natural language, this set of path components describes the spatiotemporal characteristics of the path of a moving object (e.g., migratory patterns of birds and animals).

6.1.3.1 Researchers Interest in Geospatial Data from Text Using NLP

Geospatial information being embedded into unstructured data can be in many forms such as coordinates of locations, proper place names or addresses ("The train starts from Bombay Central station to Pune at 10.00am") or locations are given relative to other locations ("The Infotech park is down the street from his house"). Similar information about different geographical areas, historical information, etc., is quite often passed on from person to person through word of mouth and hence may not be available in the literature. There has been a lot of interest among researchers in geotext.

Geospatial research areas also exploit social media posts such as Twitter text messages (tweets) using NLP. For example, a model is presented for building type classification from tweets by implementing geospatial text mining methods (Häberle et al., 2019). For example, annotations of OpenStreetMap (https://www.openstreetmap.org) objects can be enriched by tweets (Chen et al., 2018). Furthermore, by combining Twitter data with the remote sensing imagery, population density and the quantity of social media usage in slums in Mumbai (Klotz et al., 2017), extraction of insightful information about urban poverty (Hannes et al., 2018) has been studied.

6.2 OVERVIEW OF NLP TECHNIQUES IN GEOSPATIAL ANALYTICS

Geospatial information extraction and identification from unstructured data can be improved significantly using NLP. The process of information extraction and identification consists of various NLP tasks such as spatial and temporal information and relation extraction, named entity recognition, event extraction, location extraction, place extraction, and temporal expression extraction. This section will demonstrate some of the standard NLP tasks used for geospatial analytics for information extraction with example application.

6.2.1 Event Extraction

The event extraction task aims at identification and labeling of events from unstructured plain texts into a structured form, which mostly describes "who, when, where, what, why" and "how" of real-world events that happened. This task is useful for many NLP applications including information retrieval, question answering, and summarization (Li et al., 2021b). For example, the following news headline has been marked up for event references.

A red alert was <event> sounded </event> for Mumbai and neighbouring districts in Maharashtra as the monsoon <event> arrived </event> a day ahead of schedule.

6.2.2 Parts-of-Speech (POS) Tagging

POS tagging is a very powerful and effective NLP technique for the extraction of geographic named entities from unstructured data efficiently (Jurafsky and Martin , 2019). POS tagging is the task of assigning the best possible tags such as verbs, nouns, pronouns, adjectives, adverbs, preposition, conjunction, and their sub-categories to the words based on the functions of each word in a given text. This process converts a sentence to a list of words with their appropriate tags. Table 6.1 shows the POS tagging of the sentence, "*How many national highways are there in Maharashtra?*"

6.2.3 Temporal Information Extraction

Temporal information is defined as the information used to measure the durations or intervals of events and to order events. Recently, researchers pay more attention to the extraction of temporal information from documents. Considering the following query is being asked to question-answering (QA) system, "*Who was the Prime Minister of India five years ago?*" QA system will be able to answer correctly if the temporal information about when the question being asked is known to QA. Extraction of temporal information is a necessary task for applications very useful such as information retrieval (IR) systems, knowledge base (KB) construction, and question-answering (QA) systems.

In general, temporal information, such as temporal points, intervals, and durations, is represented through *Temporal expression* (Lim et al., 2019). Temporal expressions can be represented using five reference forms as defined in Figure 6.3. It is possible to create timeline through the

TABLE 6.1
Example of POS Tagging

Words	Tags	Words	Tags
How	WRB- WH adverb	are	VBP - Verb
Many	JJ-Adjective	there	EX
National	JJ-Adjective	in	IN
Highways	NNS-Noun plural	Maharashtra	NNP - Proper noun, singular

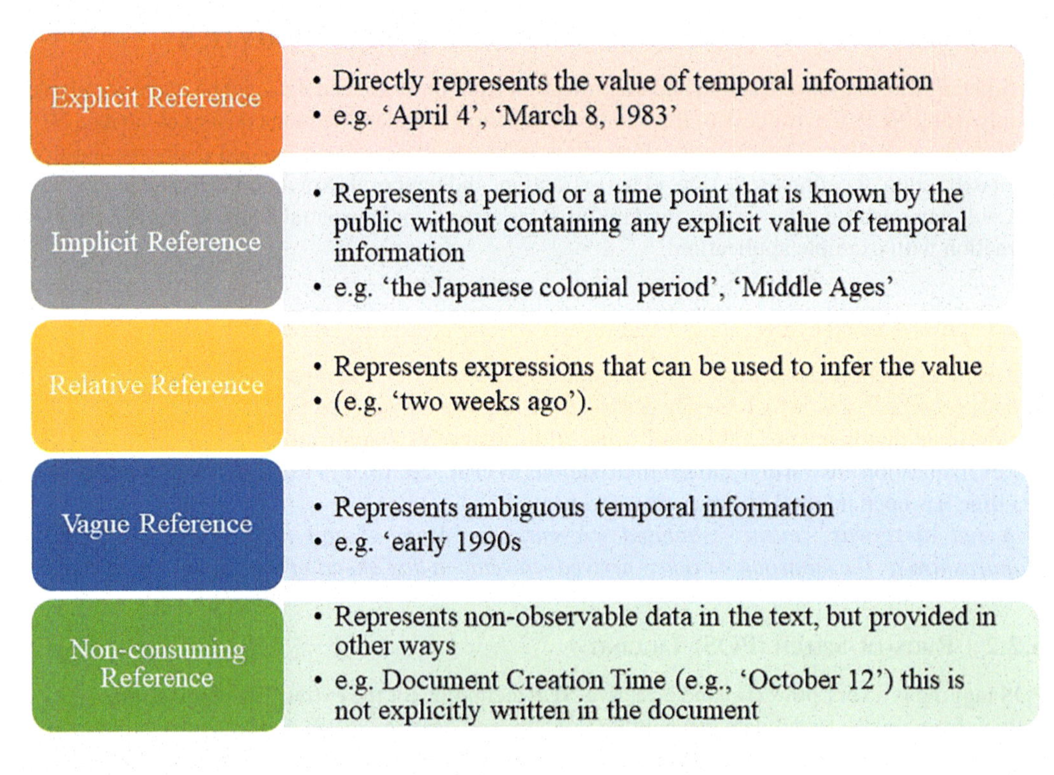

FIGURE 6.3 Reference forms for temporal information representation.

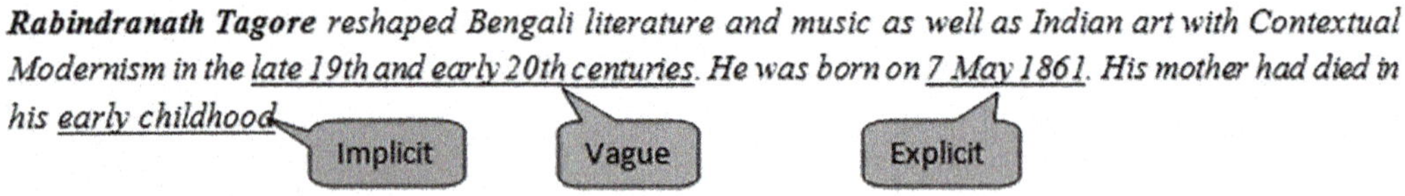

Rabindranath Tagore *reshaped Bengali literature and music as well as Indian art with Contextual Modernism in the late 19th and early 20th centuries. He was born on 7 May 1861. His mother had died in his early childhood*

Implicit Vague Explicit

FIGURE 6.4 Example of temporal references.

identification and labeling of time expressions. In order to use the temporal information for further analysis, it is necessary to convert the reference forms into a more structured form which includes the position of the temporal expression within the document, the value of the expression (e.g., "2022–04–11," "1 month") and any additional information available in the expression. This conversion process is known as temporal information extraction. Figure 6.4 shows an example of temporal references.

6.2.4 Spatial-Temporal Relationship Extractions

In geospatial analytics, extraction of semantic relationships between the entities (e.g., person, organization, location) is an important task important tasks for various applications such as traffic management, navigation, robotics, and question-answering systems. Relationships between two or more entities can be spatial or temporal. For example, question-answering system may involve answering queries about the location of any entity with respect to other entities. A sentence in Figure 6.5 shows many spatial relations embedded in it. Identifying spatial relations is a challenging task. In equation (6.1), the preposition "on" is an indication of spatial relation. Such indications are context specific. Ambiguity in spatial information can be resolved using statistical machine learning models effectively.

Given text	Spatial Relations
The Taj Mahal is located on the right bank of the Yamuna River in a vast Mughal garden that encompasses nearly 17 hectares, in the Agra District in Uttar Pradesh"	(i) Taj Mahal *is located on right bank of* Yamuna
	(ii) Taj Mahal *is in* Mughal Garden
	(iii) Mughal Garden *is* very vast (nearly 17 hectares)
	(iv) Taj Mahal *is in* Agra District
	(v) Agra *is in* Uttar Pradesh

FIGURE 6.5 Example of spatial relations.

Temporal relation, represented using temporal expressions, provides the order or duration of events mentioned in a given text. Temporal relations can be represented by expressions such as {equals, before, starts, meets, overlaps, during, since, finishes}. For example, "*The patient has a history of* ***2 months*** **<**temporal information**>** *of* ***Diarrhea***<Event>." NLP techniques, specifically, syntactic analysis are applied to identify the temporal and event expressions and then determine the temporal relations between them. If enough annotated sentences are available, then machine learning algorithms can be used to automatically extract the temporal relationships.

6.2.5 Named Entity Recognition (NER)

NER is one of the most important data pre-processing tasks in NLP. Any noun phrases, verb phrases, or both of a given sentence are defined as Entities. NER model is used to identify and classify entities into predefined categories such as the names of any person, company product event, organizations, any geographic locations, expressions of times, expression of dates, quantities, monetary values, percentages, etc. Any NER model takes any text as input and produces annotated text as output.

In general, entities are extracted from a given text using the following three steps namely, (1) *Noun Phrase Identification*: noun phrases are extracted using dependency parsing and POS tagging, (2) *Phrase Classification*: extracted noun phrases from step (1) are classified into their respective classes. For example, "New Delhi is the capital of India." Here, the noun phrases are "New Delhi, capital and India." Among them "New Delhi" and "India" are geographic entities which represent physical places on a map. (3) *Entity Disambiguation*: Resolving the ambiguities of misclassified entities by creating a validation layer (Jurafsky and Martin, 2019). There three major types of tools are being used in NLP application; (1) The Stanford Named Entity Recognizer (SNER)[1] is a JAVA tool developed by Stanford University for entity extraction using Conditional Random Fields (CRFs). It is able to recognize people, organizations, geographical locations, and other entities. (2) SpaCy library[2] is a Python framework known for being fast and very easy to use. The library also has a powerful statistical system that can be used to develop customized NER extractors. (3)The Natural Language Tool Kit (NLTK)[3] Python library includes more than 50 corpora and lexical resources and has a classifier to recognize named entities called ne_chunk, but also allows NER tags to be used within Python.

For example, "The delegation, which included the commander of the <ORG> U.N. </ORG> troops in <LOC> Basnia </LOC> <PERS> Lt. Gen. Sir Michael Rose </PERS> reached <LOC> Sarajevo </LOC> or <TIME> 13th October </TIME>"

6.3 APPLICATIONS OF NLP IN GEOSPATIAL ANALYTICS

In spite of the fact that NLP isn't strictly geospatial, human language is often rooted in geographical space and time and pertains to inherently geospatial themes like culture. Geospatial phenomena can be better understood with the help of NLP techniques, which offer a more comprehensive view that may not be accessible through traditional analysis. Therefore, NLP offers superior spatial and temporal analysis techniques compared to traditional geospatial methods. As a result, it provides an in-depth understanding of geospatial phenomena.

Geospatial relationships, regional hierarchies, and geographic laws and theories can be integrated with a range of advanced NLP methods and applications, some of which are already in use. This section discusses different NLP techniques with specific reference to applications in geospatial analytics. Our goal is to demonstrate the use of NLP in geospatial domain and talk about some of the challenges therein. The field of Geospatial analytics makes use of NLP techniques to analyze data to provide better insights, including identifying places, activities, and events, along with linguistic patterns associated with them, which may not be attainable through traditional techniques for spatiotemporal analysis. Such applications can be basically categorized as

1. Geoparsing and toponym disambiguation.
2. Geospatial semantics and natural language processing.
3. Geospatial information analysis.
4. Geospatial text analysis from social media.

NLP techniques such as geospatial part of speech tagging, noun phrase extraction, NER, Co-reference resolution, etc., are extensively applied in applications. This section will provide an introduction of some such applications.

6.3.1 Geoparsing and Toponym Disambiguation

Over the years, the way we represent places in text has been the same: we still use toponyms. Toponyms are place names, as their etymology states: topo (place) and nym (name). A toponym can be found in almost every bit of information on the Web and in the digital world: almost every news story references some place on Earth, explicitly or implicitly. The major problems in identifying a geographical entity by using toponyms are ambiguity, synonymy, and the fact that the meaning of toponyms changes over time.

Natural language processing (NLP) is dependent on the ability to deal with ambiguities that are inherent in the language used by people. When it comes to toponyms, ambiguity may take many forms: a proper name may refer to different classes of entities ('Amerawati' can be used as a name for either the writer 'Umbravati' or a city in Maharashtra, or may be used as a name for different examples of the same class: for example, 'Amravati' is also the name of a city in Andhra Pradesh. Recognition of 'Amaravati' as a place name and identification of which Amravati is located in Maharashtra or Andhra Pradesh are aspects of geoparsing which are examined in an upcoming section.

6.3.1.1 Geoparsing

Geoparsing is the technique of extracting place names from texts and matching them with proper nouns and spatial coordinates in an unambiguous manner (Refer Figure 6.6). It is important to many applications, such as Geospatial Information Extraction (GIE) and Geospatial Information Retrieval (GIR). Generally, two steps are involved in geoparsing; (1) Toponym Recognition (2) Toponym Resolution.

Toponym Recognition: The goal of toponym recognition is to find words and phrases that can be used to represent place names (or toponyms). In Table 6.2, we present a comparison of various toponym recognition systems. Furthermore, a toponym is a single or multiple-word name for a location that refers to a geographically placed detail and a group of people that use it. Toponyms

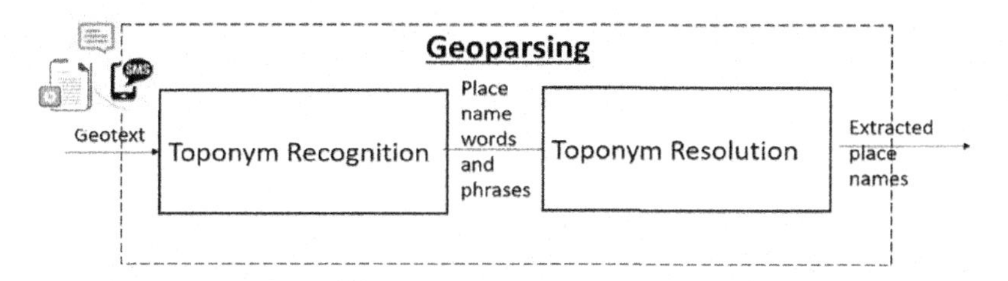

FIGURE 6.6 Steps of Geoparsing.

TABLE 6.2
Comparison of Toponym Recognition System and Its Application Domain

Source: Twitter-based hashtag syntax (Starbird and Stamberger, 2010), **Application Domain:** Disaster	Survey of various Twitter-based analysis methods
Source: Twitter data (Crooks et al., 2013), **Application Domain:** Disaster	The study used signal-to-noise ratio SNRTs for filtering social media sensor observations
Source: Twitter and traditional media (Murthy et al., 2013), **Application Domain:** Flood	Location aware Twitter Data analysis for Urdu language
Source: Twitter datasets(geo-tagged) (Yu et al., 2019), **Application Domain:** Sandy, Hurricanes, Irma, and Harvey	Deep learning-CNN
Source: Social media, Twitter NeuroTPR, (Wang et al., 2020), **Application Domain:** NER in context of location from social media messages	Neural Network, RNN
Source: 2015 Nepal earthquake and 2017 Jiuzhaigou (Wang and Hu, 2019), **Application Domain:** Earthquake	The spatial logistic growth model is used to estimate impact areas. Applying the spatial statistical analysis to the citizen sensor data to estimate the earthquake impact area based on the spatial growth trend.
Source: Social media, Twitter GSP (Nizzoli et al., 2020), **Application Domain:** Structured geographic information is used to enhance text documents	Regression Model with Knowledge Graph

differentiate between populated and deserted areas, as well as places with hills, rivers, and highways, and micro-toponyms like building names.

The Cartographic Location and Vicinity Indexer, Edinburgh Geoparser, GeoTxt, and TopoCluster are a few of the many geoparsing applications that have been developed, including locating disaster-related areas from social media messages (Wang and Hu, 2019). There were initial efforts to develop a web-based API-based Geoparsers (GeoTxt, TopoCluster, etc.). This was followed by NLP-based toponym recognition and resolution that took advantage of tools such as OpenNLP and StanfordNER. A growing number of deep learning-based geoparsers have been developed recently (Refer Figure 6.7).

Toponyms are included as named entities(NE) in the field of NLP. Multiple challenges are associated with traditional NER tools, which have limited capabilities to process geotext to identify toponyms. Geoname and W3C geo vocabulary are two of the most popular gazetteers for toponym recognition. In recent times, people have been using deep learning to recognize the toponyms for different languages such as Chinese (Wang and Hu, 2019).

Toponym Disambiguation: Disambiguation of toponyms refers to the process of determining the appropriate location that is indicated by a place name. The toponym resolution process in the geospatial domain entails identifying the relationship between a toponym (the mention of a place) and its physical footprint (the location of that place).

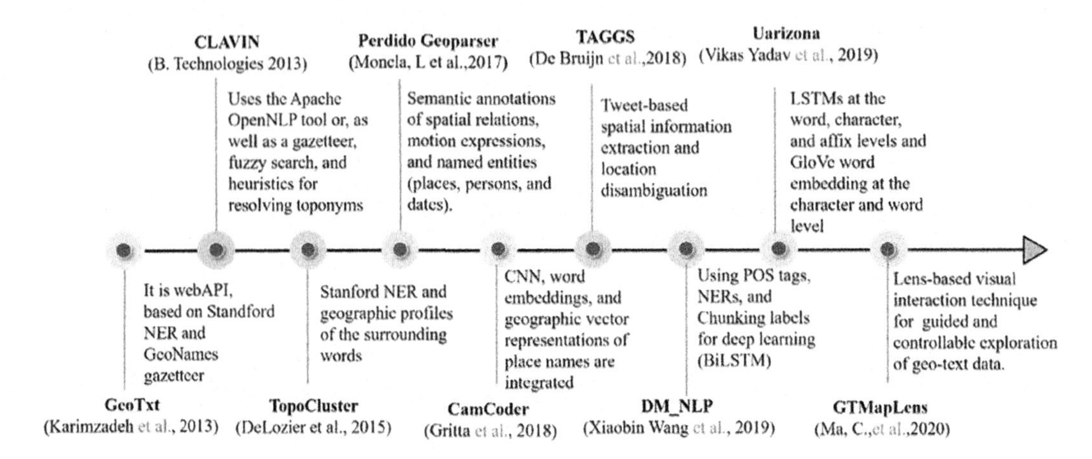

FIGURE 6.7 Timeline of available Geoparsers.

In the past, people have used the same geospatial names to identify their places, leading to referential ambiguity of place names. In some cases, the original name is modified (as in "Bombay" vs. "Mumbai"). A name is often reused without modification ("Aurangabad" in Maharashtra, India vs. "Aurangabad" in Bihar, India). In order to correlate latitude/longitude coordinates or polygons with place names or toponyms found in a document, disambiguation is necessary. An algorithm for toponym resolution converts toponyms to spatial footprints automatically.

In general, toponym disambiguation approaches are categorized into three categories: map-based, data-driven, and knowledge-based. Using maps-based methods, it is possible to identify ambiguous toponyms by using the context of other toponyms in the same text. A data-driven approach is based on machine learning, while a knowledge-based approach uses sources of knowledge (gazetteers, ontologies, etc.). For the disambiguation of toponyms, some approaches use topology-based methods, while most of them are based on knowledge graphs.

6.3.2 Geospatial Geosemantic in Natural Language

Semantic analysis is the process of understanding and interpreting the meaning of text considering the context, grammar, and logical structuring of sentences. It helps to extract invaluable information while reducing manual efforts.

Semantic Analysis can be of two types: Lexical Semantic Analysis, which involves understanding the dictionary meaning of each word of the text individually and Compositional Semantics Analysis, which involves understanding the meaning of the text by combining the individual words. Using semantic analysis to analyze geospatial concepts and relations, also known as geospatial semantics, we can simplify the design of geographical information systems (GIS) by enhancing the interoperability between distributed systems and developing richer user interfaces. Geospatial semantics has been studied from several perspectives. There are many perspectives for modern geospatial analysis, one of which includes the use of natural language processing, which is a new direction since large amounts of unstructured geotext are being generated nowadays. This section will provide a systematic review and discussion on the existing geospatial semantic research areas including semantic interoperability, geospatial Semantic Web, place semantics, and cognitive geographic concepts (Hu, 2017).

6.3.2.1 Role of NLP in Geosemantic

In Geospatial semantics, semantic technologies for comparing, linking, and accessing geospatial data are considered to be the best means to semantically model geospatial concepts. Ontologies have the advantage of making the process easier and more efficient. Ontologies used in the geospatial domain are generally called geo-ontologies and geographic ontologies. Geospatial ontology can be applied in many different ways. Traditionally, it has been used for enabling semantic interoperability and geospatial modeling technique for knowledge representation. There are several examples of Ontology-Driven GIS (ODGIS), such as those proposed by (Fonseca, 2001) and Fallahi et al. (2008). Kuhn (2003) proposed semantic reference systems (SRS) to facilitate semantic interoperability, which serves as an ontology-based equivalent to the existing spatial and temporal reference systems. Additionally, geo-ontologies are used for geospatial modeling in various geospatial domains (for example, cultural and agricultural infrastructures, wetlands, and transportation) where they can be used as knowledge bases.

Ontology contains the following elements: (1) **Individuals**: Things that can be named in the data (e.g., Mahatma Gandhi Road, LBS Road, etc.) (2) **Classes**: A collection of individuals, also known as concepts (e.g., Road) Properties: These form a connection between individuals and values (e.g., Road has a road length of 22 km) (3) **Relationships**: The relationship between two individuals (Road is composed of RoadSegment, i.e., Road → compose by → RoadSegment). (4) **Axioms**: A crucial component of ontologies, they help us to derive assumptions from data and draw conclusions.

The '*label*' has been used to describe all the above components. Labels represent a human-understandable entity that is expressed in natural language (mostly in English). This enables access to the machine for further processing through natural language processing. There are two ways that NLP is used in geosemantics: First, for automatically generating geo-ontologies, otherwise known as ontology learning. Secondly, geo-ontologies can be used for a variety of applications, such as ontology-based geospatial information retrieval and domain model representation that can be used for geospatial information processing.

Geo-ontology learning is the process by which geo-ontology is constructed from geotext (Kokla and Guilbert, 2020). This process involves learning all geo ontology entities from geotext using NLP. There are a variety of different types of concepts/classes that can be learned according to the requirements of the application. As an example, if it is a flood response ontology, then the terms are extracted and learned from the geo-ontology, such as place, year, city, damaged area, date, flood type, etc. Learning this is accomplished using different approaches, including machine learning (Qiu et al., 2019), deep learning (Zhang et al., 2018), dictionary learning, etc. The extraction of relational and property information from geo text is also an important aspect of geospatial extraction for geo ontology. A machine learning method (support vector machine) was used to extract geospatial relations, which is often used in literature. Recently, deep learning was used (LSTM; Long Short-Term Memory, GRU; Gated Recurrent Units).

6.3.2.2 Geosemantic Similarity Using NLP

Similarity plays a significant role in many geospatial applications. When geotext based similarly is involved, it is more complicated, for example, the similarity between the word "City" and "Town" is synonymous. However, in other cases, it is quite complex and perhaps subjective. For example, train and car are intuitively similar as they both are means of transport while car and road, which are also related but are not similar (i.e., they often occur together but each has a different role). This approach can be expressed to geosemantic relatedness following an adaptation of the Tobler's first law of geography (Montello et al., 2003). "Every term is geo-semantically related to all other terms, but terms that co-occur with specifiable geographic relations are more related than other terms."

There are multiple NLP-based approaches available in the literature for identifying this geosemantic relatedness. Kavouras and Kokla (2002) presented a methodology for exploring and identifying semantic information within geospatial ontologies. By applying NLP techniques, a set of semantic relations is extracted in preparation for the integration process. Semantic relations are extracted from the source such as homonyms, hyponyms, is-part-of, has-parts, and adjacent to relations with help of external auxiliary source WordNet Tversky similarity measures count the number of common relations among the two categories to determine the similarity between them. Spatial location similarity can generally be determined through comparisons of spatial footprints derived from textual sources and digital gazetteers, such as Thesaurus of Geographic Names (TGN) and the Alexandria Digital Library (ADL).

6.3.3 Geospatial Information Analysis

The process of analyzing geospatial information involves combining, visualizing, and combining different types of geospatial data, and we are analyzing it within the framework of geotext. Geospatial information was extracted, geospatial information was retrieved and geospatial questions were answered in this information analysis (Refer Table 6.3). Geospatial information utilizes and analyzes both structured and unstructured data from documents and combines it with geographic data. In order to find relevant knowledge from heterogeneous geospatial information, we need to have a deep understanding of their semantic meaning and develop strategies for efficiently collecting, analyzing, and using them with minimal human intervention.

Geospatial information analysis involves systems that index both structured and unstructured text as well as incorporating geographic information into these indices. Identifying relevant knowledge among heterogeneous geospatial information requires understanding the semantic meaning of data and establishing efficient strategies for collecting and analyzing data with minimal human intervention. Textual data can be associated with spatial and temporal information by using NLP techniques, such as named entity recognition, relation extraction, semantic similarity, and word sense disambiguation.

6.3.3.1 Geospatial Information Extraction (GIE)

Information extraction is a method for automatically processing unstructured or semi-structured natural language texts and retrieving specific forms of data that are relevant to the task at hand while ignoring other data types. Geospatial entities can be natural (e.g., Ocean Island Beach), physical (e.g., vegetable field), socioeconomic (e.g., University School District Public School Private School Museum), or manmade entities (e.g., Weather Station Astronomical Observatory). GIE also focuses on extracting relationships between entities. Among the relationships observed between entities are Mereological Relationships, Topological Relationships, Spatial Distance Relations, Qualitative Distance Relations, Temporal Distance Relations, and Temporal Properties. A general pipeline for geospatial information extraction is shown in Figure 6.8, which identifies geospatial entities from different sources.

TABLE 6.3
Set of Geospatial Information Analysis Tasks

Task1: Geospatial Information Extraction	Task 2: Geospatial Information Retrieval	Task 3: Geospatial Question Answering
Input: Geotext **Output:** Geospatial concepts such as cities rivers monuments, geographical feature types (e.g., "village," "dam"), geographical coordinates, addresses (postal codes), geographical expressions (involving spatial prepositions and toponyms), Geospatial Relation Extraction such as "Part Of," "Near." **Major NLP Task Required**: Named entity recognition and classification, word sense disambiguation	**Input:** Query (set of keywords) with geographical constraints, e.g., "seaside mall in Mumbai" **Output:** Set of relevant documents **Major NLP Task Required**: (1) Recognition and disambiguation of toponyms and indexing and searching thematic and spatial information (geospatial concept and relation extraction), (2) Spatial relevance measures, (3) Ranking of documents using both thematic and spatial relevance (geosemantic similarity measures)	**Input**: Question with geographical terms, e.g., What is the highest mountain in the world **Output**: Answer or set of answers **Major NLP Task Required**: Part-of-speech taggers, named entity recognition, syntactic parsers and semantic analysis

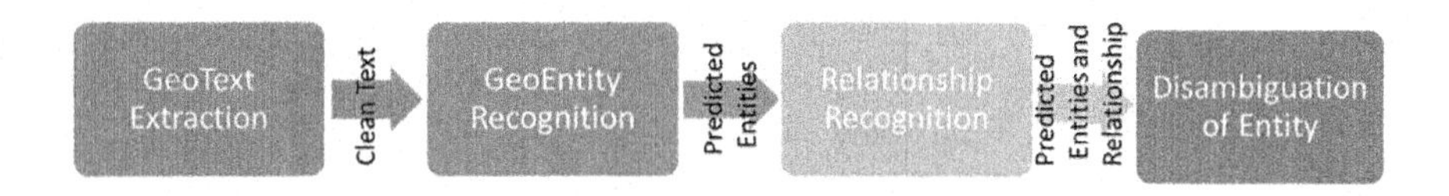

FIGURE 6.8 General pipeline of geospatial information extraction.

The NLP base extraction model proposed by Tahrat et al. (2012) extracts geospatial relations (such as rivers, bodies of water, cities, and towns) from free texts. Two types of spatial features were extracted in this work. There is an absolute feature such as 'place' and a relative feature such as 'north of Mumbai.' As part of this study, geo texts are first tokenized, after which lexical analysis is conducted for extracting named entities. Later, morpho-syntactic analysis is carried out for retrieving word types and semantic relationships.

Chasin et al. (2014) developed a method in order to identify events, temporal relations, and named entities contained in geotexts, and disambiguate these locational and named entities. A support vector machine-based algorithm is used to identify important events. Several features refer to the characteristics of the event word in the sentence (part of speech, grammatical aspect, separation from a named entity, etc.) or the phrase by itself (length). This approach uses the Tarsqi Toolkit (TTK)[4] to extract temporal information from text newswire articles. TTK is a set of processing components for extracting temporal information from news wire articles. In addition to extracting time expressions, events, and temporal links, TTK ensures consistency of temporal information. Furthermore, it makes use of the Heidel Time tool, which is based on regular expressions for extracting temporal relationships.

A study by Scott et al. (2021), for instance, matched Antarctic species to the appropriate toponyms mentioned in various journal articles by extracting Antarctic species from texts. Based on online travel blogs, Wikipedia articles, Geonames toponyms, and Twitter posts, Adams and McKenzie (2018) developed a neural network-based model to extract places from crowdsourced data.

An ontology-based extraction of information (OBIE, Wimalasuriya, and Dou, 2010) is a process by which domain knowledge is formally described and then extracted into concepts, properties, and instances. Geospatial-oriented semantic information extraction methods can be applied to various tasks, including the specialization of text data, the exploration of textual descriptions of space, and the retrieval of geospatial information. In these approaches, natural language texts are examined with the goal of extract information such as places (O'Hare & Murdock, 2013), locative expressions (Liu et al., 2014), activities (Hobel and Fogliaroni, 2016), and events (Wang and Stewart, 2015). As part of approaches to semantic information extraction from natural language texts, gazetteers are commonly used to identify place names and relationships among them. To extract types of places, events, or activities, vocabularies and taxonomies are often employed. However, ontologies, which represent domain knowledge, can also assist in semantic information extraction. According to Ballatore and Adams (2015), emotion vocabulary was used to analyze travel blog posts to extract place nouns of natural and built places. An instance-based learning approach is proposed by Stock and Yousaf (2018) for interpreting natural language descriptions of location. Two ontologies are used to model spatial relations among features and characteristics for representing context of geospatial features using a point, line, and area ontology.

6.3.3.2 Geospatial Information Retrieval

The geospatial information retrieval (GIR) method uses traditional text-based searches alongside location-based queries, such as place names or maps, to search the web, enterprise documents, or local documents. Geospatial information is added to both structured and unstructured documents through GIR systems. A GIR system is designed to reliably answer queries in which the geospatial aspect is present, such as "What wars were fought in Ukraine?" or "Where are the best restaurants in

Delhi?" In many GIR systems, word sense disambiguation and semantic similarity are determined using natural language processing or other metadata. Several GIR systems exist in the literature which has components like spatial indexing and spatial queries. GIPSY (Georeferenced Information Processing System) (Larson, 1996), Spatially Aware Information Retrieval on the Internet (SPIRIT) (Purves et al., 2007) and Franken place (Adams and McKenzie, 2012) are the most popular GIR. In GIR, spatial and time dimensions are added to the classic retrieval problem. It is necessary to make both a thematic and a spatial match in order to match the data with a query such as "restaurant in Delhi." GIR considers the following steps, according to Purves et al. (2007).

- As a first step, geographic references must be recognized and extracted from a query or a document through methods such as named entity recognition and geo-parsing.
- GIRs must determine which meaning the user intends with place names, since they are not unique.
- There are many examples of vague geospatial footprints and vague names used to refer to geographic locations; typically, vernacular names ("a restaurant in Delhi") are used as examples. When a restaurant query is made, the GIR system must select the correct boundaries.
- Finally, and in contrast to classical IR, documents must also be indexed according to specific geographical regions.

In GIR systems, the following stages are typically performed: geoparsing, indexing of text and geospatial data, storage of data, ranking of geospatial relevance on the basis of geographic queries, and browsing of results via maps. Natural language texts in GIR are analyzed in two ways: (1) extracting geographic information from NL texts to build knowledge bases, and (2) determining the geospatial scope of Web documents. Among the purposes of GIR is to build knowledge bases from geospatial information found in natural language texts, as well as to determine the geographical boundaries of Web documents. The former allows the construction and populating of geo-ontologies, while the latter allows the matching and retrieval of relevant documents relating to geospatial queries.

6.3.3.3 Geospatial Question Answering

Question-answering systems search external data sources to answer questions (information retrieval based) or search internal databases to match them against answers (knowledge based). Figure 6.9 shows an example of a question-answering tool that can answer inquiries about location, durations, dates, events, and even relative times ("when will Ukraine war end?," "How many years were between World Wars I and II?").

Geospatial Question Answering (GQA) refers to methods and algorithms designed to help users get answers to geospatial questions (Domènech, 2017). GQA is a technique that uses different kinds of information sources to answer geographic questions, such as text, geodatabases, and spatially-enhanced knowledge bases. Answers to these questions can be expressed in natural language, structured into tables and graphs, or visualized as maps. A GQA architecture typically aims to resolve

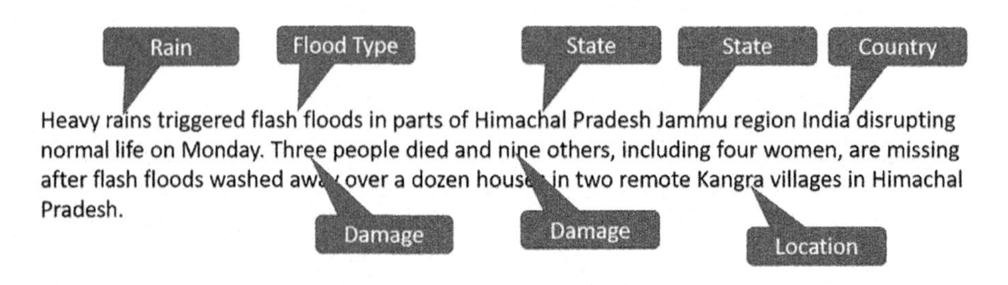

FIGURE 6.9 Example of entity extraction from news article.

three tasks: (1) question classification and intent analysis, (2) searching for relevant sources, and (3) extracting answers from the sources (Domènech, 2017).

A geospatial question includes information about a wide range of topics including geospatial objects (such as location), fields (such as temperature), events (such as floods), and networks (such as transportation). Obtaining answers to these questions can be accomplished through everyday communication, web searches, or geospatial information systems. GQA analyzes geospatial questions in two ways: (1) by classifying the question and (2) by recognizing its intent. There are two kinds of queries. The first kind is questions that include a spatial relationship "in" (including "at" and "from"), as in "flood in Mumbai"; the other kind is queries that are located relatively (such as "near," "close"). And they each have a type of answer which describes what should be explained in the answers (Domènech, 2017).

The selection of relevant responses is dependent on the information sources used to answer the questions (Hamzei, 2021). The answers to geographic questions are often extracted from a single snippet of a document using traditional geospatial information retrieval techniques (e.g., Mai et al., 2019). In conjunction with NLP techniques and word representation models (e.g., Bag-of-Words, Term Frequency-Inverse Document Frequency, Word embedding, etc.) these technologies train a model based on annotated datasets for extracting answers. The majority of GQA studies, in terms of question types, ask what/which questions about geographic places (e.g., Scheider et al., 2020). The process of answering complex questions such as where-questions and how-to-get-there questions consists either of retrieving stored coordinates or of selecting part of the textual information without adapting to the question context. A deep learning-based question-answering system was recently proposed by Li et al. (2021a). Based on a few individual words (e.g., named entities), this method provides the answer. It does so by using recursive neural networks (RNN) to model the compositional properties of geotext.

6.3.4 Spatiotemporal/Geospatial Text Analysis from Social Media

Recent developments in social media have shifted the focus of geospatial applications toward identifying events, places, and times of disasters or sentiment indicators. Adding to the complexity, there is a hefty volume of ambiguous text in the form of unstructured social media from specific regions that change over time. It is often difficult to extract useful information from the text in social media posts due to the unstructured nature of these posts (Refer Figure 6.9). Among the textual content of a post may be information on the topic of discussion, the sentiment of the user who posted the message, where the message was written, opinions on certain arguments, or risk prevention. The following example shows how one news post is analyzed to extract geospatial entities such as Rain, Flood type, state, country, and damage.

There have been a number of reviews that discussed the use of social media data to research relevant topics. There is a systematic review of the benefits and risks of analyzing disaster data from social media by Wiegmann et al. (2020), as well as a review of social media analytics for hospitality and tourism by Mirzaalinet et al. (20190. The use of social media in the response to natural disasters was examined by Muniz-Rodriguez et al. (2020), while the use of social media in park management was examined by Wilkins et al. (2021).

The data from social media is often ad hoc and unstructured. Majority of work of social media analytic in geospatial domain is in disaster area. It is heavily used for Biological, Climatological, Geophysical, Hydrological, Meteorological, Industrial disaster, etc. Information can be extracted for disaster analysis from this data. In this analysis, geospatial data will be gathered analyses, visualized a about the type and location of an event (disaster type, common name verification status date, time, and location), type-specific information that includes the impact on the population (magnitude, depth, severity, damage caused, deaths, injuries, displacements), Causes and effects of the event(-trigger, follow-up), description of the episode (narrative) and responses to the event (reaction actions taken, lessons learned).In disaster management. It is generally used to (1) disseminate information,

(2) plan disasters and provide disaster training, (3) solve collaborative problems and make decisions, and (4) gather information.

One of the common applications of social media analysis is sentiment analysis. Incoming text messages are analyzed to determine their underlying sentiment (e.g., neutral, negative, or positive) using this tool. The spread of Coronavirus, for instance, prompted people to share their fears, reactions and sentiments on social media. Malhan et al. (2022) collected unstructured data from Twitter and extracted opinions including sentiments of people on COVID-19 from all over the world by analyzing the tweets and the location, date, and time of tweets as well. Malhan et al. (2022) represented the countries with the highest number of tweets in ascending order, such as India, United Kingdom, United States, and Ireland, and displayed the countries with the most positive tweets related to Coronavirus. Multiple challenges remain to be addressed in geospatial analytics for social media data, including

- Extracting geospatial information for location-based services and applications that is contextually aware and personalized.
- Analyzing the spatiotemporal relationships between spatial tracking and the spatial-temporal trajectory of the geospatial data.
- Integrating social media data with automated map readers that can handle a variety of map features.
- Estimating demographic parameters and their spatial distribution within social media.
- Analyzing the spatial and temporal spread of social sentiment and detection using unstructured data.
- The analysis of the spatial and temporal spread of social sentiment and its detection, as well as utilizing social media semantics to evaluate public opinion dynamics over time and space.

6.4 FUTURE SCOPE OF NLP IN GEOSPATIAL ANALYTICS

A number of challenges are unique to geospatial domains of natural language processing, many of which are listed above. Researchers are examining how to use natural language processing to interpret finely-grained spatial relationships in geotext. There are many current NLP methods that are able to recognize entities, concepts, and relationships within natural language, but relatively few explicitly address spatial relationships. There is a long way for addressing spatial processes with natural language processing. Spatial autocorrelation is one such concept that is fundamental in geospatial domains, but it is rarely incorporated into natural language understanding.

Geospatial analytic has also been heavily influenced by advances in deep learning. When recognizing place names from texts, deep networks like bidirectional recurrent neural networks are helpful. The use of metonymies in texts may be able to be identified by a new class of deep learning-based NLP methods (Hu et al., 2019). Most approaches to toponym resolution still resolve place names based only on points, as countries, rivers, and other geospatial features can have better spatial footprints represented as polylines, polygons, or polyhedrons (in a 3D space). Despite many geoparsers, it is not possible to compare their performances directly due to the lack of open and annotated corpora.

A spatial-temporal pattern can also be identified with natural language processing. The goal is not just to analyze the individual entities/toponyms in the text, but also to identify broader themes or relationships between them. For example, when providing location recommendation services, you have to take into account not only the location but also the spatiotemporal relationship between locations.

6.5 SUMMARY AND CONCLUSIONS

Geospatial data can be found within a variety of natural language texts, including news articles, news articles, Wikipedia pages, social media posts, historical archives, travel blogs, and so forth. There are many sources of data available today and these sources are continually increasing over time. Consequently, there is a need for natural language processing techniques to process the geodata and this has a substantial impact on geospatial analytics.

Every geotext has a spatial footprint, an optional timestamp associated with the data record and textual content. Geospatial Analytics needs to extract spatial and temporal knowledge from explicitly or implicitly hidden in the geotext. Coupled with the inherent ambiguity present in the natural language text, this becomes an extremely challenging task. NLP offers a plethora of techniques that will solve these problems to a great extent. Role of NLP in geospatial analytics spans from simple preprocessing techniques to semantic interpretation of a large body of text. The combination of these two areas has created a lot of interest among researchers in the geospatial field. A geospatial analytics application cannot function without toponym recognition and disambiguation, also known as geoparsing. The recognition and resolution of place names in text, geoparsing enables the extraction of structured geospatial information from unstructured text. This will be handled by natural language processing along with geospatial information analysis through social media. The main focus of this chapter is to present the role of natural language processing in the overall geospatial information analysis process.

NOTES

1 https://nlp.stanford.edu/software/CRF-NER.shtml.
2 https://spacy.io/usage/linguistic-features#named-entities.
3 https://www.nltk.org/.
4 https://github.com/tarsqi/ttk.

REFERENCES

Adams, B., & McKenzie, G. (2012). Frankenplace: An application for similarity-based place search. In *Proceedings of the International AAAI Conference on Web and Social Media* (Vol. 6, No. 1, pp. 616–617)), Toronto, Ontario, Canada.

Adams, B., & McKenzie, G. (2018). Crowdsourcing the character of a place: Character-level convolutional networks for multilingual geographic text classification. *Transactions in GIS*, *22*(2), 394–408.

Ballatore, A., & Adams, B. (2015, May). Extracting place emotions from travel blogs. *In Proceedings of AGILE* (Vol. 2015, pp. 1–5).

Bitters, B. (2011, July). Geospatial reasoning in a natural language processing (NLP) environment. In *Proceedings of the 25th International Cartographic Conference* (Vol. 253).

Chasin, R., Woodward, D., Witmer, J., & Kalita, J. (2014). Extracting and displaying temporal and geospatial entities from articles on historical events. *The Computer Journal*, *57*(3), 403–426.

Chen, X., Vo, H., Wang, Y., & Wang, F. (2018). A framework for annotating OpenStreetMap objects using geotagged tweets. *Geoinformatica*, *22*(3), 589–613.

Chen, Z., Chen, L., Cong, G., & Jensen, C. S. (2021). Location-and keyword-based querying of geo-textual data: A survey. *The VLDB Journal*, *30*(4), 603–640.

Cong, G., Feng, K., & Zhao, K. (2016, May). Querying and mining geo-textual data for exploration: Challenges and opportunities. In *2016 IEEE 32nd International Conference on Data Engineering Workshops (ICDEW)* (pp. 165–168). IEEE Helsinki, Finland, May 16-20, 2016.

Crooks, A., Croitoru, A., Stefanidis, A., & Radzikowski, J. (2013). # Earthquake: Twitter as a distributed sensor system. *Transactions in GIS*, *17*(1), 124–147.

Daniel, J., & Martin James, H. (2019). *Speech and Language Processing: An Introduction to Natural Language Processing, Computational Linguistics and Speech Recognition*, 3rd Edition Draft. Prentice Hall, Hoboken, NJ.

Domènech, D. F. (2017). Knowledge-based and data-driven approaches for geographical information access. Doctoral dissertation, Universitat Politècnica de Catalunya (UPC).

Fallahi, G. R., Frank, A. U., Mesgari, M. S., & Rajabifard, A. (2008). An ontological structure for semantic interoperability of GIS and environmental modeling. *International Journal of Applied Earth Observation and Geoinformation*, *10*(3), 342–357.

Fonseca, F. T. (2001). Ontology-driven geographic information systems. The University of Maine.

Goodchild, M. F., Yuan, M., & Cova, T. J. (2007). Towards a general theory of geo-graphic representation in GIS. *International Journal of Geographical Information Science*, *21*(3), 239–260.
Häberle, M., Werner, M., & Zhu, X. X. (2019, July). Building type classification from social media texts via geo-spatial textmining. In *IGARSS 2019-2019 IEEE International Geoscience and Remote Sensing Symposium* (pp. 10047-10050). IEEE.
Hamzei, E. (2021). Place-related question answering: From questions to relevant answers. Doctoral dissertation. The University of Melbourne
Hannes Taubenbock, Jeroen Staab, Xiao Xiang Zhu, ¨ Christian Geiß, Stefan Dech, and Michael Wurm, "Are the Poor Digitally Left Behind? Indications of Urban Divides Based on Remote Sensing and Twitter Data," ISPRS International Journal of Geo-Information, vol. 7, no. 8, pp. 304, Aug. 2018.Hill, L. L., & Goodchild, M. F. (2000). Digital gazetteer information exchange (DGIE). Final Report of Workshop Held, October 12–14, 1999.
Hobel, H., & Fogliaroni, P. (2016, July). Extracting semantics of places from user generated content. In *Proceedings of the 19th AGILE International Conference on Geographic Information Science* Helsinki (Finland).
Hu, Y. (2017). Geospatial semantics. arXiv preprint arXiv:1707.03550.
Hu, Y. (2018). Geo-text data and data-driven geospatial semantics. *Geography Compass*, *12*(11), e12404.
Hu, Y., Mao, H., & McKenzie, G. (2019). A natural language processing and geospatial clustering framework for harvesting local place names from geotagged housing advertisements. *International Journal of Geographical Information Science*, *33*(4), 714–738.
Jurafsky Daniel and Martin James H.. 2019. *Speech and Language Processing: An Introduction to Natural Language Processing, Computational Linguistics and Speech Recognition.* 3rd Edition Draft. Prentice Hall, New Jersey.
Kavouras, M., & Kokla, M. (2002). A method for the formalization and integration of geographical categorizations. *International Journal of Geographical Information Science*, *16*(5), 439–453.
Klotz, M., Wurm, M., Zhu, X., & Taubenböck, H. (2017, March). Digital deserts on the ground and from space. In *2017 Joint Urban Remote Sensing Event (JURSE)* (pp. 1-4). IEEE.
Kokla, M., & Guilbert, E. (2020). A review of geospatial semantic information modeling and elicitation approaches. *ISPRS International Journal of Geo-Information*, *9*(3), 146.
Kuhn, W. (2003). Semantic reference systems. *International Journal of Geographical Information Science*, *17*(5), 405–409.
Larson, R. R. (1996). Geographic information retrieval and spatial browsing. Geographic information systems and libraries: patrons, maps, and spatial information, *Papers Presented at the 1995 Clinic on Library Applications of Data Processing*, April 10–12, 1995.
Li, H., Hamzei, E., Majic, I., Hua, H., Renz, J., Tomko, M., ... & Baldwin, T. (2021a). Neural factoid geospatial question answering. *Journal of Spatial Information Science*, *23*, 65–90.
Li, Q., Li, J., Sheng, J., Cui, S., Wu, J., Hei, Y., ... & Yu, P. S. (2021b). A compact survey on event extraction: Approaches and applications. arXiv preprint arXiv:2107.02126.
Lim, C. G., Jeong, Y. S., & Choi, H. J. (2019). Survey of temporal information extraction. *Journal of Information Processing Systems*, *15*(4), 931–956.
Liu, F., Vasardani, M., & Baldwin, T. (2014, November). Automatic identification of locative expressions from social media text: A comparative analysis. In *Proceedings of the 4th International Workshop on Location and the Web* (pp. 9–16)). Shanghai, China.
Mai, G., Yan, B., Janowicz, K., & Zhu, R. (2019). Relaxing unanswerable geographic questions using a spatially explicit knowledge graph embedding model. *In International Conference on Geographic Information Science*, Springer, Cham (pp. 21–39) Mérida, Yucatán, México.
Malhan, Y., Saxena, S., Mala, S., & Shankar, A. (2022). Geospatial modelling and trend analysis of coronavirus outbreaks using sentiment analysis and intelligent algorithms. In *Artificial Intelligence in Healthcare*, Springer, Singapore (pp. 1–19).
Montello, D. R., Fabrikant, S. I., Ruocco, M., & Middleton, R. S. (2003, September). Testing the first law of cognitive geography on point-display spatializations. In *International Conference on Spatial Information Theory*, Springer, Berlin, Heidelberg (pp. 316–331).
Muniz-Rodriguez, K., Ofori, S. K., Bayliss, L. C., Schwind, J. S., Diallo, K., Liu, M., ... & Fung, I. C. H. (2020). Social media use in emergency response to natural disasters: A systematic review with a public health perspective. *Disaster Medicine and Public Health Preparedness*, *14*(1), 139–149.
Murthy, D. and Longwell, S.A., 2013. Twitter and disasters: The uses of Twitter during the 2010 Pakistan floods. Information, Communication & Society, 16(6), pp.837-855.

Nizzoli, L., Avvenuti, M., Tesconi, M., & Cresci, S. (2020). Geo-semantic-parsing: AI-powered geoparsing by traversing semantic knowledge graphs. *Decision Support Systems, 136*, 113346.
O'Hare, N., & Murdock, V. (2013). Modeling locations with social media. *Information Retrieval, 16*(1), 30–62.
Purves, R. S., Clough, P., Jones, C. B., Arampatzis, A., Bucher, B., Finch, D., ... & Yang, B. (2007). The design and implementation of SPIRIT: A spatially aware search engine for information retrieval on the Internet. *International Journal of Geographical Information Science, 21*(7), 717–745.
Qiu, Q., Xie, Z., Wu, L., Tao, L., & Li, W. (2019). BiLSTM-CRF for geological named entity recognition from the geoscience literature. *Earth Science Informatics, 12*(4), 565–579.
Scheider, S., Meerlo, R., Kasalica, V., & Lamprecht, A. L. (2020). Ontology of core concept data types for answering geo-analytical questions. *Journal of Spatial Information Science, 20*, 167–201.
Scott, J., Stock, K., Morgan, F., Whitehead, B., &Medyckyj-Scott, D. (2021). Automated georeferencing of antarctic species. In *11th International Conference on Geographic Information Science (GIScience 2021)-Part II. Schloss Dagstuhl-Leibniz-Zentrum für Informatik.*
Starbird, K., & Stamberger, J. A. (2010, May). Tweak the tweet: Leveraging microblogging proliferation with a prescriptive syntax to support citizen reporting. In *ISCRAM.*
Stock, K., & Yousaf, J. (2018). Context-aware automated interpretation of elaborate natural language descriptions of location through learning from empirical data. *International Journal of Geographical Information Science, 32*(6), 1087–1116.
Tahrat, S., Roche, M., & Teisseire, M. (2012, August). Extraction of geospatial information from documents. In *Geographic Information Retrieval Tutorial-Panel discussion.* AGILE Workshop (p. 2).
Wang, J., & Hu, Y. (2019, November). Are we there yet? Evaluating state-of-the-art neural network based geoparsers using EUPEG as a benchmarking platform. In *Proceedings of the 3rd ACM SIGSPATIAL International Workshop on Geospatial Humanities* (pp. 1–6).
Wang, W., & Stewart, K. (2015). Spatiotemporal and semantic information extraction from Web news reports about natural hazards. *Computers, Environment and Urban Systems, 50*, 30–40.
Wang, J., Hu, Y., & Joseph, K. (2020). NeuroTPR: A neuro-net toponym recognition model for extracting locations from social media messages. *Transactions in GIS, 24*(3), 719–735.
Wiegmann M, Kersten J, Klan F, Potthast M, Stein B (2020) Analysis of detection models for disaster-related tweets. In: Proceedings of the 17th international conference on information systems for crisis response and management. ISCRAM '20. ISCRAM
Wilkins, E.J., Wood, S.A. and Smith, J.W., 2021. Uses and limitations of social media to inform visitor use management in parks and protected areas: A systematic review. Environmental Management, 67(1), pp.120-132.
Wimalasuriya, D. C., & Dou, D. (2010). Ontology-based information extraction: An introduction and a survey of current approaches. *Journal of Information Science, 36*(3), 306–323.Wing, B., & Baldridge, J. (2014, October). Hierarchical discriminative classification for text-based geolocation. In *Proceedings of the 2014 Conference on Empirical Methods in Natural Language Processing (EMNLP)* (pp. 336–348).
Yu, M., Huang, Q., Qin, H., Scheele, C. and Yang, C., 2019. Deep learning for real-time social media text classification for situation awareness–using Hurricanes Sandy, Harvey, and Irma as case studies. International Journal of Digital Earth, 12(11), pp.1230-1247.
Zhang, X., Ye, P., Wang, S., Du, M. (2018). Geological entity recognition method based on deep belief networks. *Acta Petrologica Sinica, 34*, 343–351.

Section III

Scalable Geospatial Analytics

7 A Scalable Automated Satellite Data Downloading and Processing Pipeline Developed on AWS Cloud for Agricultural Applications

Ankur Pandit, Suryakant Sawant, Rishabh Agrawal, Jayantrao Mohite, and Srinivasu Pappula
TCS Research and Innovation

CONTENTS

7.1 INTRODUCTION

The agricultural sector has been using satellite imagery for land use and weather predictions ever since NASA's "Landsat" program started taking images of earth from space in the early 1970s. But their utilization in the agricultural domain remained limited due to issues like inadequate resolution, sparse frequency, and limited integration with other data sources. Nowadays, with the increased deployment of "Earth Observation Satellites" that acquire images and other forms of remote sensing observations of the earth from relatively low altitudes, we have an abundance of data coming from the sky. Satellite-based Earth Observation (EO) sensors in space are continuously monitoring the globe at high frequency and with high resolution. Remote sensing has significantly contributed to the agriculture sector by providing multi-sensor data of larger regions on a temporal basis with a good frequency of revisits and resolution. The primary application of remote sensing in the agriculture domain involves the estimation of spatially distributed time-series vegetation health maps [1–3], soil moisture estimation [4–6], crop yield estimation [7,8], crop classification [9,10], etc. Most of these parameters can be estimated from the indices such as normalized difference vegetation index

DOI: 10.1201/9781003270928-10

(NDVI) [11], soil-adjusted vegetation index (SAVI) [12], normalized difference water index (NDWI) [13], Leaf Chlorophyll Index (LCI) [14], and ratio vegetation index (RVI) [15], which can be generated using the band combination of satellite datasets, e.g., Sentinel-2 or Landsat. The traditional method of calculating such indices (for both geographical and temporal analysis) using a Geographic Information System-based tool is only practical for the smaller region that may be covered by a single or few satellite scenes. Deriving numerous indices for different parameter estimations at a big scale (such as a district, state, or nation) using the standard technique, on the other hand, takes time and resources. This is due to the large quantity of satellite data that must be downloaded and analyzed. A traditional technique eventually becomes inadequate, especially in commercial organizations where stakeholders demand faster and more decisive outcomes at the state or country level. Therefore, an automated satellite data downloading and processing pipeline would be a more optimal and efficient option than the traditional technique. However, there are certain challenges in handling geospatial data on in-house local/on-premise systems because big geospatial data exceed the capacity of commonly used computing systems as a result of 5 V's (volume, velocity, variety, value, veracity) [16].

Cloud computing [17], which is the on-demand distribution of information technology resources over the Internet, can help with the difficulty of handling huge amounts of satellite data. It operates on a pay-as-you-go approach, in which users can ingress technology services such as storage, databases, and processing power, from cloud providers on an as-needed basis, rather than purchasing, operating, and maintaining data centers and servers physically. Cloud services provide the highest level of data security and privacy, allowing businesses to focus on their core competencies. Furthermore, cloud computing gives customers more flexibility by allowing them to access data from any web-enabled device, such as laptops, notebooks, and other mobile devices. Cloud service platforms enable the customer to rent more processing power as needed without having to purchase expensive servers that cost a million dollars. Many businesses are increasingly interested in moving their applications to cloud environments rather than spending a ton of money on hardware, software, license, and renewal charges.

Nowadays, major cloud-based platforms such as Google Earth Engine (GEE) [18], Amazon Web Services (AWS) [19], and Microsoft Planetary Computer [20] are offering services to use terabytes to petabytes of data acquired on a temporal basis by some of the selected satellites. These cloud platforms use different storage systems, access interfaces, and abstractions for satellite data sets. Table 7.1 provides a comparative analysis of cloud-based platforms that are capable of handling (i.e., data storage and processing) a huge amount of satellite data. Such platforms help derive meaningful satellite-based outputs (raster and zonal statistics) that are useful for various agricultural applications. Apart from the above-mentioned major cloud-based platforms, Sentinel-hub [21], Open Data Cube (ODC) [22], System for Earth Observation Data Access, Processing and Analysis for Land Monitoring (SEPAL) [23], OpenEO [24], Joint Research Centre Earth Observation Data and Processing Platform (JEODPP) [25], and pipsCloud [26] are also available for the same purpose. These platforms, however, employ various abstractions, access interfaces, and storage methods for satellite data sets.

Tata Consultancy Services (TCS) "Digital Farming Initiative" (DFI) [27] platform combines open-source satellite imagery with the richness of application development and machine learning capabilities of AWS and domain expertise to deliver intelligent capabilities to the agricultural industry. We have developed and deployed an automated satellite data downloading and processing pipeline (ADDPro) on the AWS cloud environment. The main goal of developing the ADDPro pipeline is to gradually get rid of the organization's dependency on available geospatial processing services such as GEE or Sentinel-hub, which are free for research, education, and non-profit use but chargeable for commercial or operational applications. The availability of an in-house pipeline helps in reducing operational costs and increasing service personalization based on the needs of the client(s). We would like to highlight that in the ADDPro pipeline, we have implemented multi-sensor data fusion technique to generate cloud-free crop health products on a temporal basis. This feature is very useful during the Kharif cropping season (southwest summer monsoon season spanning from June to October) which is mostly cloudy and through optical satellites, it is difficult to obtain the ground information from the sky. This particular feature can be implemented as per the requirement. Currently, our

TABLE 7.1
Comparative Analysis of Cloud-Based Platforms

Particular	Google	Amazon	Microsoft
Platform name	Google earth engine	Earth on AWS	Planetary computer
Data source	Open source commercial partners	Open source commercial partners	Open source commercial partners
Data store	Google cloud storage	Amazon S3 buckets	Azure cloud storage
Who maintains the data?	Google and third party	AWS and third party	Microsoft
Has high update/sync frequency?	Yes	Yes	Yes
Has data transfer cost?	No. Data is not migrated unless explicitly required	Yes. Requester pay bucket (minimal cost)	Unknown
Has an active community?	Yes	Limited support. Very few examples	Unknown
Periodic addition of new open-source data sets	Yes	Yes	Unknown
Requires API to access data?	Yes. Available in JavaScript, Python, R, Go	Yes. AWS SageMaker S3 API call	Yes. Azure data store call
Platform offering commercial/open source	Open source (mainly for research, Govt., non-profit organizations), commercial on Google cloud platform	Commercial	Commercial
Compute restriction	~$10E^{13}$ pixels for on-the-fly processing	Depends on the SageMaker/Docker/Lambda Instance compute	Yes. Depends on the instance size selected in Hub
Has simplified data processing logic?	Yes	No. Need to use native Python or R modules	No. Need to use native Python or R modules
How user code base is managed?	Using git like a native repository. Users are also provided with Python API for remote access	Using Git like native repository on AWS cloud platform	Git-Hub

in-house developed ADDPro pipeline is commercially operational for various customers in India as well as abroad. The overall pipeline was developed in the Python 3.6 programming language.

The pipeline is capable of generating raster outputs and zonal statistics of Sentinel-2 derived indices (i.e., NDVI, NDWI, SAVI, LCI, and RVI) at the plot level. Furthermore, apart from mentioned indices (i.e., NDVI, NDWI, SAVI, LCI, and RVI), many other vegetation indices procedures that are accessible on the Sentinel-hub platform can also be added to the ADDPro pipeline as per the requirement. The pipeline is efficient in generating temporal indices outputs from the number of plots (of varying shape and size) spread across any given geography.

7.2 SATELLITE IMAGERY RESOLUTIONS

Data from satellite (Figure 7.1) based EO sensors differ as follows:

- The spatial resolution determines how detailed the image is and specifies what is the size of 1 pixel on the ground (e.g., 1 pixel equals to 10-m by 10-m area).
- The temporal resolution, which specifies how often an area is imaged by a sensor from the same viewing angle, is determined by how long it takes for a satellite sensor to revisit an area (e.g., Sentinel 2 and Landsat 8 satellite revisit the same location after every 10 and 16 days respectively).

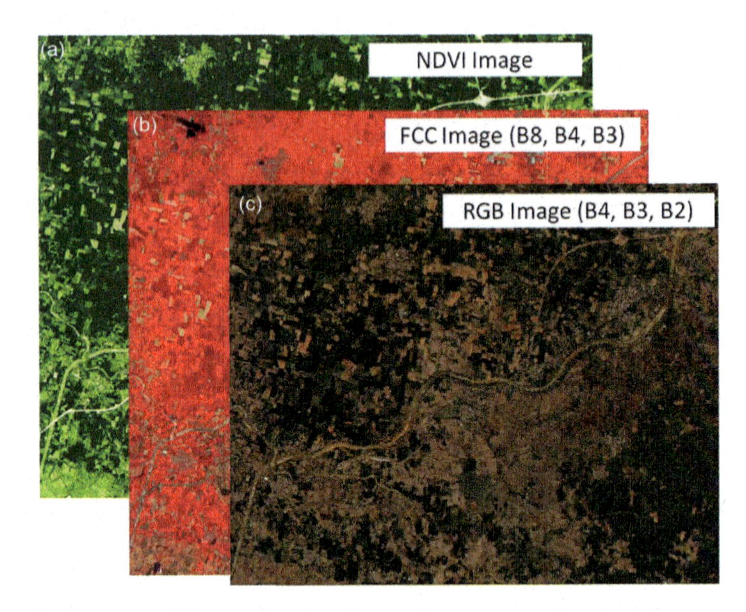

FIGURE 7.1 Satellite data (a) RGB image bands Red – 4, Green – 3 and Blue – 2, (b) FCC image band NIR – 8, Red – 4, Green – 3, and (c) NDVI image calculated from bands 8 and 4. The data has a temporal resolution of 10 days and the spectral resolution of 13 spectral bands. (European Space Agency.)

- The spectral resolution determines how finely a sensor can determine the electromagnetic wavelength range of observation and is an important parameter when looking for data acquired from specific regions of the electromagnetic spectrum including non-visible wavelength regions (e.g., multi-spectral Sentinel 2 satellite has 13 wavelength regions, Landsat 8 has 11 wavelength regions).
- The radiometric resolution of the sensor determines the ability of the sensor to distinguish received energy in different grey scale values, often referred to as bit depth (e.g., radiometric resolution of Sentinel-2 satellite is 12-bit, Landsat-8 is 12-bit and Landsat-9 is 14-bit). For a 12-bit, sensor received energy can be stored in integer values from 1 to 4096 integer values whereas for 14-bit it's from 1 to 16,383 integer values.

Satellite sources vary widely when it comes to the various resolutions they provide and the frequency at which data is generated has increased dramatically in recent years. The acquisition of data from these sources has grown so fast that the traditional methods of data processing cannot keep up with the rate of data acquisition. Also, multiple steps are involved from the acquisition of earth observations to the interpretation of acquired images. Stakeholders from various domains like agriculture, natural resources monitoring, and management are constantly struggling with the volume of EO data to obtain meaningful insights. Multi-spectral EO at high temporal observations is difficult to process in a resource-constrained environment and requires flexible computing, storage, and high network throughput to process.

7.3 APPLICATION OF TECHNOLOGY IN MONITORING CROP HEALTH

Crop health is a measure of how healthy the cultivation of plants is under stress from biotic factors caused by living organisms like bacteria and viruses, and abiotic factors like drought, extremes in temperature, and the chemical composition of the soil. Crop health is measured by metrics related to crop appearances, crop growth, crop production, damage from pests, competition by weeds, etc. Details about Sentinel-2 derived indices used for crop health monitoring are given in Table 7.2.

TABLE 7.2
Details about Sentinel-2 Derived Indices Used for Crop Health Monitoring

S. No.	Index	Sentinel-2 Bands Used	Formula
1.	NDVI	Band 04 (RED) & Band 08 (NIR)	(NIR – RED)/(NIR + RED)
2.	NDWI	Band 03 (GREEN) & Band 08 (NIR)	(GREEN – NIR)/(GREEN + NIR)
3.	SAVI	Band 04 (RED) & Band 08 (NIR)	[(NIR – RED)/(NIR + RED + L)] * (1 + L), where L = 0.5
4.	LCI	Band 04 (RED), Band 05 (REDE) & Band 08 (NIR)	(NIR – REDE)/(NIR + RED)
5.	RVI	Band 04 (RED) & Band 08 (NIR)	(NIR)/(RED)

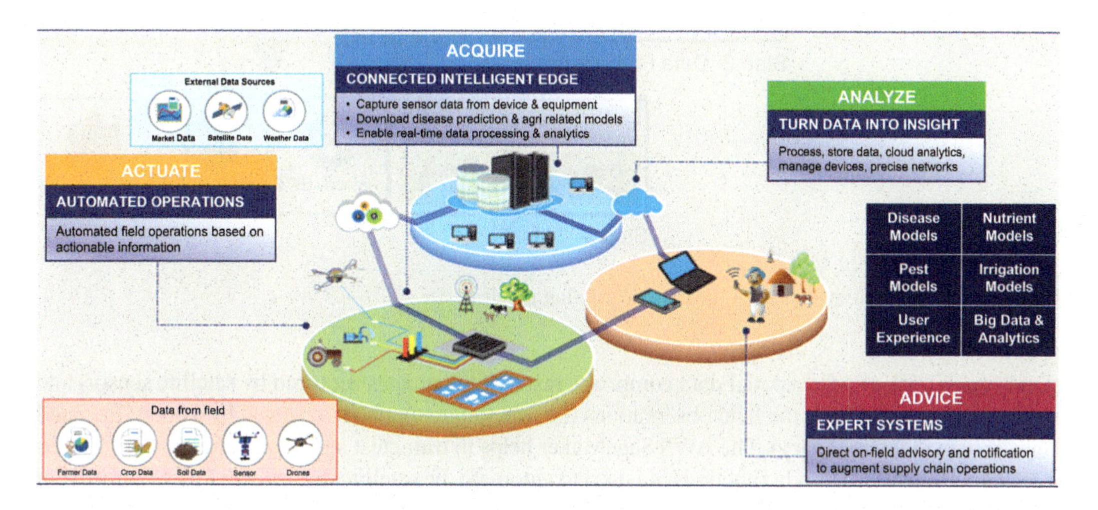

FIGURE 7.2 DFI Sky Earth Integration (TCS Digital Farming Initiatives, 2018).

Traditional field-level crop health monitoring is a resource and time-intensive process often limited to a smaller area and a limited number of times in a season. But with earth observations from satellites and drones, it is easy to assess the crop health over a large area in a short period. Also using the Internet of Things (IoT) enabled sensors, farmers can monitor crop health and environmental conditions continuously. Utilizing technology for crop health monitoring enables farmers to do monitoring at scale, and in real-time and helps them to make well-informed decisions at the right moment. The time saved to generate crop health information can be used to undertake crop management activities thereby reducing the crop failure risk.

Figure 7.2 represents a high-level view of how data from satellites and IoT sensors can be combined and analyzed on a digital platform to provide cutting-edge capabilities to farmers and other stakeholders in the agricultural industry.

7.4 HIGH-LEVEL SOLUTION—CROP HEALTH MONITORING USING SATELLITE DATA

We have integrated open-source satellite observations and TCS DFI capabilities for field-level crop health and canopy moisture monitoring. The geo-coordinates of the field being monitored are gathered via TCS's DFI solution's "Ground Truthing API" (Not to be confused with AWS SageMaker's Ground Truth). The field data is used to pull in the right data set from the open-source, Sentinel-2, satellite imagery dataset. TCS's proprietary machine learning models hosted in Amazon SageMaker analyze the images and generate the output. The machine learning models such as Random Forest and Support Vector Machines are used for crop classification, crop yield estimation, generation of

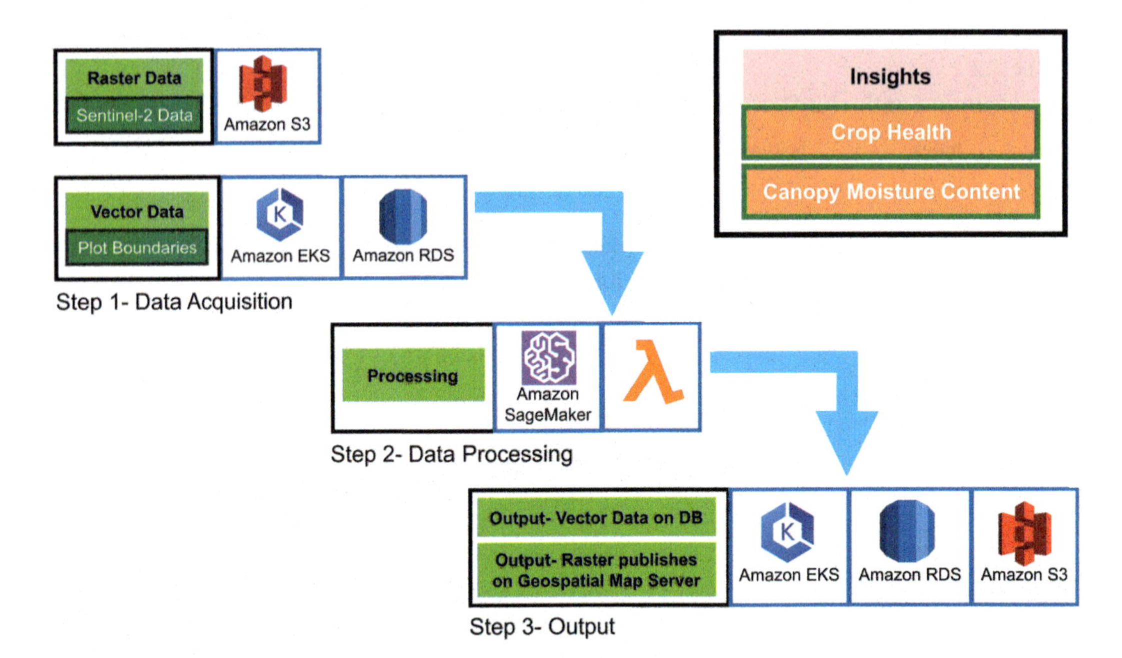

FIGURE 7.3 A high-level view of the data processing pipeline.

cloud-free NDVI, etc. Geospatial data comprises rasters (i.e., images acquired by satellite sensor) and vector data obtained from the field observations (i.e., vector geometry represented by a point, line, and polygon, e.g., field boundary). The AWS SageMaker helps to train, test and deploy the models as a service endpoint. AWS Lambda functions are used to automate the satellite data processing pipeline from data aggregation to generating the results using the model endpoints set using SageMaker. Finally, the model output raster and vector data are stored in the relational database and S3 bucket for consumption on the UI. The High-level view of the complete data processing pipeline is given in Figure 7.3.

7.5 AWS COMPONENTS USED IN THE SOLUTION

AWS's broad and deep capabilities in infrastructure, machine learning, data storage, and application development provide the perfect platform to develop such a solution. Figure 7.4 illustrates different components of AWS used in our solution. Some of the key services from AWS are:

- **Registry of Open Data** from AWS makes it simple to locate datasets made publicly available through the AWS services. This solution uses the Sentinel-2 dataset to make it sharable through S3 buckets, allowing the platform team to focus more on data analysis rather than data acquisition.
- **Amazon SageMaker** provided a platform to train, test, and host the proprietary ML model of TCS. SageMaker's ability to work with 3rd party models ensured that the DFI platform had a flexible ML platform to deploy and operate their model efficiently and at the same time, keep their IP protected at all times.
- **Amazon Elastic Kubernetes Service (EKS)**, a managed platform for Kubernetes applications, helped DFI to host its application logic in a standard way. EKS enabled the team to migrate a big part of their Kubernetes-based application to AWS with minimal changes.
- **AWS Lambda**, the serverless computing service lets the application run workflows without having to provision or manage any server. AWS Lambda was used to integrate the application with the ML endpoints hosted in SageMaker as well as create the data processing workflow.

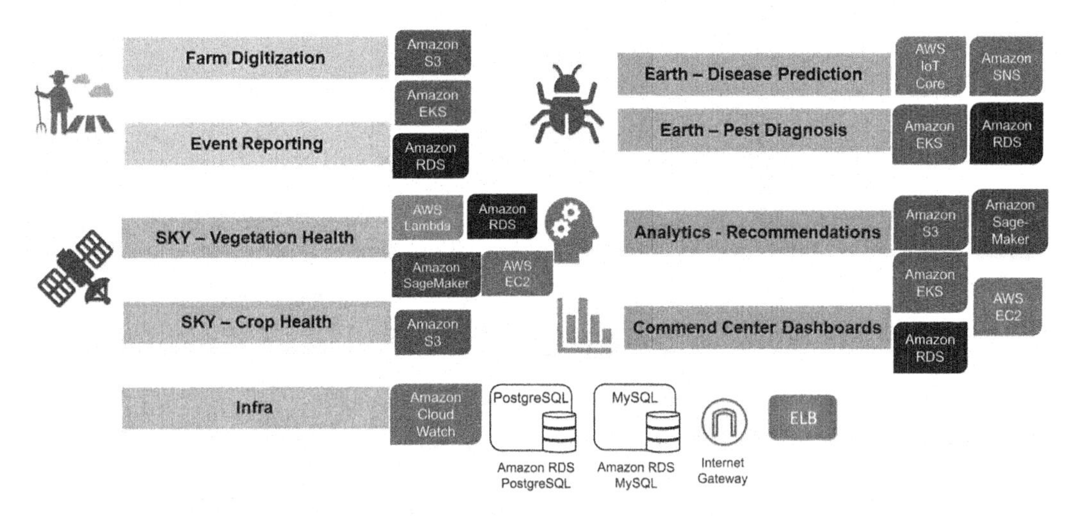

FIGURE 7.4 AWS components used in developing the overall crop health monitoring solution.

- **Amazon S3** is an object storage service offering industry-leading scalability, data availability, security, and performance.

7.6 SOME OF THE KEY ADVANTAGES OF HAVING AWS BASED DATA PIPELINE WERE

- **Efficient Data Processing**: No need to download satellite images with all bands. *The registry of open data* from AWS has stored spectral bands separately in the S3 bucket. This approach helped to get only the required bands (e.g., Band 4 and Band 8 and Quality band) which reduced the cost of data transfer and local storage. The total size of one granule (100 km^2 area) is ~800 MB, but with this approach, only about 100 MB of data is transferred.
- **Near Real-Time Analytics:** As soon as satellite data was made available the, S3 event notifications ensured that the data processing could start immediately, thereby reducing the delay in the acquisition and pre-processing.
- **Scaled-Up Processing:** On-demand provisioning of high-end computing resources helped process a large amount of satellite data to get insights over large areas, for instance across the entire watershed or country to understand the vegetation health status.
- **Use of Managed Services Helps Focus on the Problem:** Use of managed services like Amazon SageMaker, Lambda, and EKS ensured that the researchers and engineers could focus on the ML model and platform capabilities while AWS took care of the management of the underlying infrastructure.

7.7 DETAILED SOLUTION

This solution collects field geo-coordinates using a Smartphone application (developed by TCS DFI) and relays them into an application hosted in Amazon EKS, which in turn stored the data in Amazon RDS. A field boundary API exposed the field geo-coordinates to the SageMaker instance. For the respective field boundaries, the satellite data is pulled from the AWS S3 bucket. Finally, the satellite data is processed in an AWS SageMaker instance and crop health is estimated for each field. The output of the crop health algorithm in a raster and vector format is then relayed to a private (i.e., TCS DFI) S3 bucket. Further, the raster output is published as a map service compatible

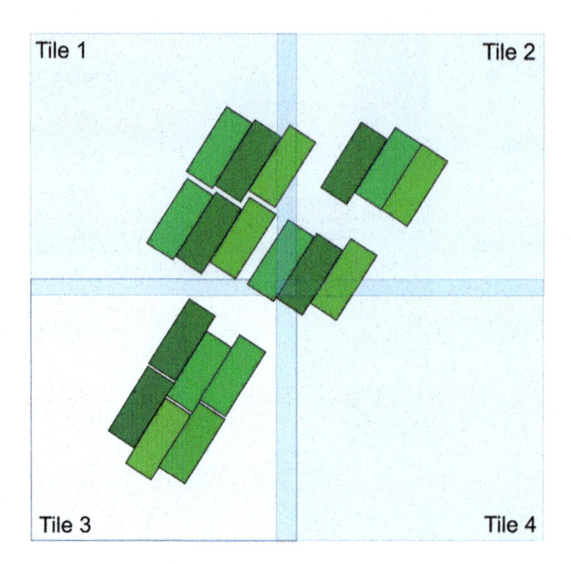

FIGURE 7.5 Detailed process flow diagram of ADDPro pipeline.

with Open Geospatial Consortium (OGC) standards and the vector data is stored in the private (i.e., TCS DFI) AWS relational store. The entire data processing pipeline is scheduled using a Lambda function.

7.7.1 Key Steps of the ADDPro Pipeline

Figure 7.5 provides a detailed flowchart of the complete ADDPro pipeline. Following key steps are involved in the ADDPro pipeline-

a. The pipeline application must be given the necessary settings shown in Table 7.3 to start the process. The top five parameters mentioned in Table 7.3 must be given for searching and generating a list of available Sentinel-2 data products from the Copernicus Open Access Hub for a specific geometry (i.e., *n*th farm of a given shapefile). There are two possible scenarios depending on the date range, the size and shape of the farms, and other search parameters: (1) numerous Sentinel-2 products may be available for a particular farm or (2) a single Sentinel-2 product accommodates multiple farms residing in a given shapefile. Figure 7.6 depicts the placement of the farms and the accompanying tiles for demonstrative purposes.
b. Upon running the pipeline, initially, the process accesses the field boundary from field boundary API in geojson format.
c. Next, the process performs searching of Sentinel-2 data for geometries (or farms) from Copernicus Open Access Hub and generates a list of available datasets. For each Sentinel-2 data, a request string has been created as per the AWS format.
d. Afterward, the process download selected bands (refer to Table 7.2 for bands involved in index calculation) of respective Sentinel-2 data (present in the list) in instance storage from the AWS S3 bucket.
e. Later, the process generates various index tiles (e.g., NDVI, NDWI, LCI, SAVI, RVI) using band combinations (refer to Table 7.2 for index calculation formula).
f. Next, the process performed a subset operation to generate farm-level rasters from the respective index tile generated in the previous step (d).
g. All the subsetted rasters from different tiles available for the particular farm(s) were mosaiced and farm-level statistics were generated.

TABLE 7.3
List of Parameters Required for ADDPro Process Initiation

S. No.	Parameter/{Description}	Sample Input
1	S2_PROC_LEVEL / {Level-1C (top of the atmosphere), Level-2A (bottom of the atmosphere)}	[Level-1C or Level-2A]
2	FROM_DATE (format: mmdd) / {mention process start date}	'1201'
3	TO_DATE (format: mmdd) / {mention process end date}	'1210'
4	CLOUD_MAX_PER_ALLOWED / {mention percentage of cloud cover in Sentinel-2 data}	10
5	SHAPEFILE_PATH / {mention path of a shapefile consist of various farms}	['./India_shp/country.shp']
6	S2_INDEX_LIST / {mention list of indexes need to calculate}	['NDVI', 'NDWI', 'LCI', 'SAVI', 'RVI']
7	SEASON_ID / {mention customized ID for a crop season}	'Karif_2021'
8	S2_YEAR_ LIST_TO_PROCESS (Format: yyyy) / {mention list of years need to process}	['2021', '2020', '2019', '2018']
9	S2_ PROC_REQUIRED / {1 = download as well as process data, 0 = Only download the data}	1
10	S2_ROOT_PROCESS_DIRECTORY / {mention path of root directory}	'./Sentinel-2'
11	S3_PREPROD_DIRECTORY_NAME / {mention directory name of AWS S3 bucket for publishing raster}	'poc.sky.preprod'
12	S3_BACKUP_DIRECTORY_NAME / {mention directory name of AWS S3 bucket for raster backup}	'poc.sky.backup'
13	CLOSE_INSTANCE_AFTER_PROC / {0- do not close the instance, 1- close the instance after completion of process}	1
14	CREATE_MOSAICED_RASTER / {0- execute raster mosaicing, 1- do not execute raster mosaicing}	1

h. As per the need, the pipeline saves the output raster (geospatial image) and field-level statistics in vector (CSV format) to the Amazon EBS at each stage.
i. From Amazon EBS, all farm rasters and zonal statistics have been transferred to the project-specific AWS S3 bucket. Raster data has subsequently been uploaded to the GeoServer. For business, policy, and vegetation growth/health monitoring, the output rasters and zonal data have been used.

Each iteration of the ADDPro pipeline must run for 10 days (e.g., 01–10 January 2022; 11–20 January 2022; 21–31 January 2022) to cover the entire Indian geography if the farms are spread out throughout several states in India. The execution term (i.e., start and finish dates) may vary depending on the size of the target region where the farms are located. For example, if all of the study farms are close together and in small geographies such as Kerala and Goa (Indian states), the iteration period of 5 days would be sufficient. A distinct log file is produced during each iteration, and it provides details about each stage of the processing pipeline. Logging assists with application troubleshooting and provides us with information about how our programs are performing on each of the several processing components.

7.8 SAMPLE ANALYSIS FOR FIELD-LEVEL CROP HEALTH

Figure 7.7 shows field boundary, raw satellite data covering all the fields, estimated crop health, and field-level crop health with statistics. The field boundary is acquired using the Smartphone mobile application or using the web application or by uploading the area of interest in geospatial data

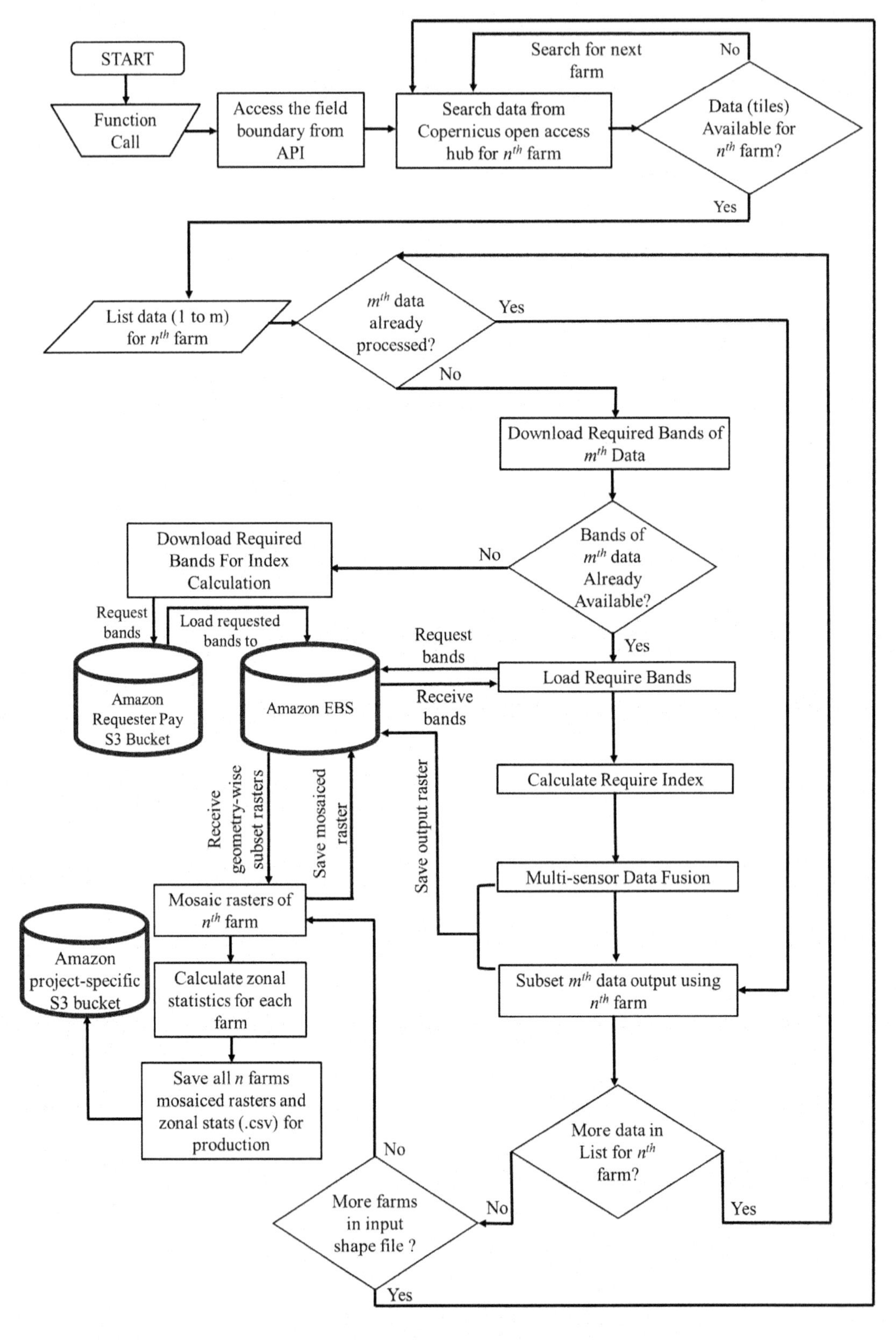

FIGURE 7.6 Geometry (or farms) and its corresponding Sentinel-2 tiles.

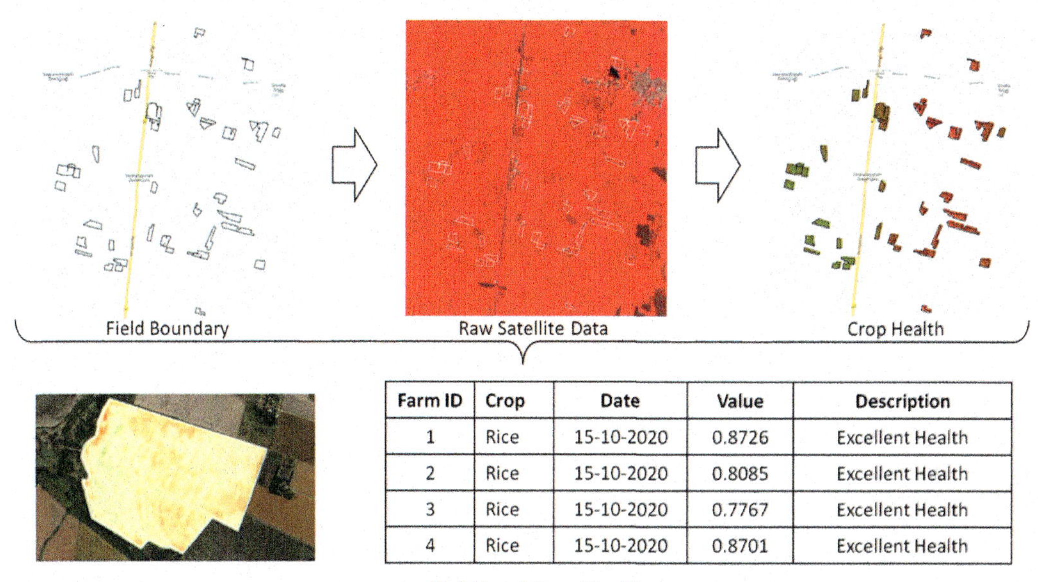

Farm ID	Crop	Date	Value	Description
1	Rice	15-10-2020	0.8726	Excellent Health
2	Rice	15-10-2020	0.8085	Excellent Health
3	Rice	15-10-2020	0.7767	Excellent Health
4	Rice	15-10-2020	0.8701	Excellent Health

FIGURE 7.7 Example of field-level crop health analysis.

format (e.g., Shapefile, GeoJSON). Raw satellite image represents surface spectral reflectance in specific wavelength regions, e.g., Red and NIR regions. The crop health image shows the estimated crop health information for the selected region. The field-level crop health image shows within-field variations in crop health in the Green-Yellow-Red color scheme. The healthy areas are represented by green, normal health in yellow and stressed areas in red color. Tabular data shows the date of the estimate, field-level aggregation (mean, minimum, maximum, etc.) of the crop health information, date of satellite pass over the region, and description of crop health.

7.9 TIME SERIES OF SATELLITE DATA AND CROP CONDITION INFORMATION

Using the time series of satellite observations the crop condition and status can be determined. Figure 7.8 shows various aspects of the crop (1) start of the season (i.e., planting), (2) peak of the season (i.e., crop maturity), (3) end of the season (i.e., harvest), and (4) fallow or within season fallow. Each dot represents the date of satellite pass and zonal (here field level) variability in terms of minimum, mean and maximum vegetation index. The small vertical lines are the box plots with the first quartile, mean and third quartile. The outliers are shown in the dots above and below. For each crop, the growth pattern for the selected vegetation index varies. The weather condition, crop type, crop variety, and growing conditions affect the overall duration of a cropping season. For example, the growing season of wheat in a tropical region is different than in a temperate region. Overall the choice of crops and cropping conditions govern the number of cropping seasons. In Figure 7.8, we can observe two crops grown in a year. One of the crops is short duration (wheat) and another is the long duration (cotton). In some regions with favorable weather conditions, three or four crops are also grown. In harsh weather conditions, only one crop is grown in one year. Using a combination of satellite-based observations (Sky), local weather (Earth), and in-situ crop information (Earth) useful insights are generated. The insights such as crop type, crop health, duration of abiotic stress, cropping pattern, cropping intensity, and average crop duration help toward the localization of crop management decisions and yield predictions.

Considering the dynamic nature of agricultural systems the satellite-based observations help to derive spatial and temporal insights at scale. The intersection of domain knowledge from Agronomy,

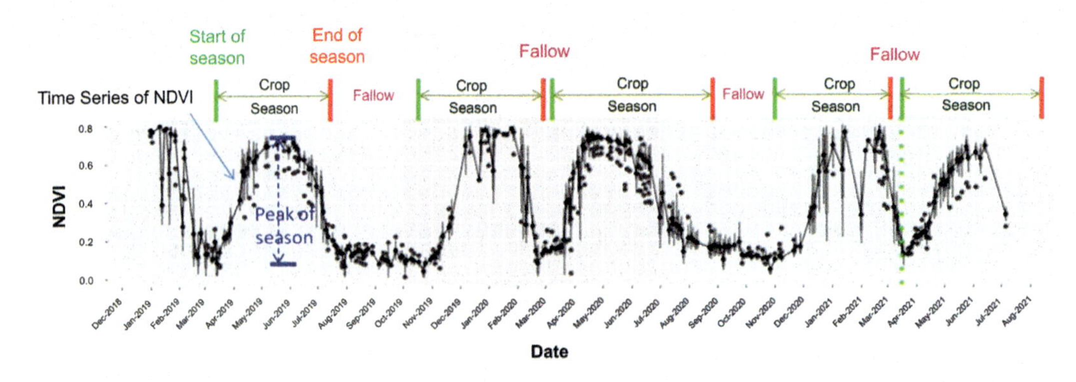

FIGURE 7.8 Time series of satellite-based vegetation index.

Remote Sensing, Geographic Information Systems, and Data Sciences is needed to generate usable insights. At TCS DFI, we use the convergence of the Sky-Earth continuum to generate near-real-time insights on crop conditions for actionable decision-making.

7.10 CONCLUSION

Geospatial data has become an invaluable tool for creating value in the agriculture business, eliminating reliance on field data collection at a larger scale. Extensive utilization of voluminous geospatial data empowers organizations to analyze the crop conditions at the farm level and take necessary actions based on near-real-time feedback. In this context, the automated satellite data downloading and processing pipeline has emerged as an efficient tool for handling terabytes of satellite data and generating decisive insight in near-real-time with the help of advanced machine learning algorithms. Thanks to AWS for providing a complete platform through which the overall pipeline has been effectively developed, deployed, and executed on a temporal basis. The geospatial map server (e.g., GeoServer, MapServer) makes the raster outputs available for visualization and analysis. Additionally, the zonal statistics produced by output rasters are very beneficial for monitoring vegetation dynamics, productivity, and changes in distribution at the regional scale. The entire ADDPro pipeline has been moved to the AWS cloud computing platform, which has eliminated the expense of maintaining local infrastructure. This pipeline is easy to scale up and deployable for various geographies around the world, however, it requires standardization of input data. Currently, the ADDPro pipeline is commercially operational for multiple clients in India as well as abroad. Through this pipeline, we are delivering outputs (rasters and zonal statistics) on a temporal basis at a farm level. This in-house product eliminates the dependency on third-party applications such as GEE or Sentinel hub for some of the most crucial use cases in the agricultural sector. Continuous feedback helps us in fulfilling customer expectations with the product features. Also, an interactive user interface and visualization are put forward.

ACKNOWLEDGMENT

The collection of Sentinel-2 datasets that are available for various geographies during each iteration at the pan-India scale has been provided by the Copernicus Open Access Hub, for which we are grateful. Additionally, we appreciate Tata Consultancy Services Limited for funding our study.

REFERENCES

1. Erener, A. Remote sensing of vegetation health for reclaimed areas of Seyitömer open cast coal mine. *International Journal of Coal Geology* 2011, 86(1), 20–26.

2. Kundu, A.; Dwivedi, S.; Dutta, D. Monitoring the vegetation health over India during contrasting monsoon years using satellite remote sensing indices. *Arabian Journal of Geosciences* 2016, 9(2), 1–15.
3. Lausch, A.; Bastian, O.; Klotz, S.; Leitão, P.J.; Jung, A.; Rocchini, D.; Schaepman, M.E.; Skidmore, A.K.; Tischendorf, L.; Knapp, S. Understanding and assessing vegetation health by in situ species and remote-sensing approaches. *Methods in Ecology and Evolution* 2018, 9(8), 1799–1809.
4. Price, J.C. The potential of remotely sensed thermal infrared data to infer surface soil moisture and evaporation. *Water Resources Research* 1980, 16, 787–795.
5. Das, N.N.; Mohanty, B.P. Root zone soil moisture assessment using remote sensing and vadose zone modeling. *Vadose Zone Journal* 2006, 5(1), 296–307.
6. Ahmad, S.; Kalra, A.; Stephen, H. Estimating soil moisture using remote sensing data: A machine learning approach. *Advances in Water Resources* 2010, 33(1), 69–80.
7. Ferencz, C.; Bognar, P.; Lichtenberger, J.; Hamar, D.; Tarcsai, G.; Timár, G.; Molnar, G.; Pasztor, S.Z.; Steinbach, P.; Szekely, B.; Ferencz, O.E.; Ferencz-Árkos, I. Crop yield estimation by satellite remote sensing. *International Journal of Remote Sensing* 2004, 25(20), 4113–4149.
8. Awad, M.M. Toward precision in crop yield estimation using remote sensing and optimization techniques. *Agriculture* 2019, 9(3), 54.
9. Dhumal, R.K.; Vibhute, A.D.; Nagne, A.D.; Rajendra, Y.D.; Kale, K.V.; Mehrotra, S.C. Advances in classification of crops using remote sensing data. *International Journal of Advanced Remote Sensing and GIS* 2015, 4(1), 1410–1418.
10. Ji, S.; Zhang, C.; Xu, A.; Shi, Y.; Duan, Y. 3D convolutional neural networks for crop classification with multi-temporal remote sensing images. *Remote Sensing* 2018, 10(1), 75.
11. Myneni, R.B.; Hall, F.G.; Sellers, P.J.; Marshak, A.L. The interpretation of spectral vegetation indexes. *IEEE Transactions on Geoscience and Remote Sensing* 1995, 33(2), 481–486.
12. Huete, A.R. A soil-adjusted vegetation index (SAVI). *Remote Sensing of Environment* 1988, 25(3), 295–309.
13. Gao, B. C. NDWI: A normalized difference water index for remote sensing of vegetation liquid water from space. *Remote Sensing of Environment* 1996, 58(3), 257–266.
14. Zebarth, B.J.; Younie, M.; Paul, J.W.; Bittman, S. Evaluation of leaf chlorophyll index for making fertilizer nitrogen recommendations for silage corn in a high fertility environment. *Communications in Soil Science and Plant Analysis* 2002, 33(5–6), 665–684.
15. Major, D.J.; Baret, F.; Guyot, G. A ratio vegetation index adjusted for soil brightness. *International Journal of Remote Sensing* 1990, 11(5), 727–740.
16. Li, Z. Geospatial big data handling with high performance computing: Current approaches and future directions. In W. Tang; S. Wang (Eds.), *High Performance Computing for Geospatial Applications* (pp. 53–76), 2020, Springer, Cham.
17. Qian, L.; Luo, Z.; Du, Y.; Guo, L. Cloud computing: An overview. In *IEEE International Conference on Cloud Computing 2009*, Berlin, Germany, Springer, 2009, 626–631.
18. Gorelick, N.; Hancher, M.; Dixon, M.; Ilyushchenko, S.; Thau, D.; Moore, R. Google Earth Engine: Planetary-scale geospatial analysis for everyone. *Remote Sensing of Environment* 2017, 202, 18–27.
19. Mathew, S.; Varia, J. Overview of amazon web services. Amazon Whitepapers 2014. Retrieved from https://docs.aws.amazon.com/whitepapers/latest/aws-overview/aws-overview.pdf, Accessed on 10 Feb 2022.
20. Luers, A.L. Planetary intelligence for sustainability in the digital age: Five priorities. *One Earth* 2021, 4(6), 772–775.
21. Sinergise. Sentinel Hub by Sinergise, 2020. Retrieved from https://www.sentinel-hub.com/.
22. Killough, B. Overview of the open data cube initiative. In *IEEE International Geoscience and Remote Sensing Symposium*, Valencia, Spain, 2018, 8629–8632.
23. FAO. SEPAL, a big-data platform for forest and land monitoring 2021. Retrieved from https://www.fao.org/3/cb2876en/cb2876en.pdf, Accessed on 15 Jan 2021.
24. Pebesma, E.; Wagner, W.; Schramm, M.; Von Beringe, A.; Paulik, C.; Neteler, M.; Reiche, J.; Verbesselt, J.; Dries, J.; Goor, E.; et al. OpenEO- A common, open source interface between earth observation data infrastructures and front-end applications; Technical Report, 2017; Technische Universitaet Wien, Vienna, Austria.
25. Soille, P.; Burger, A.; De Marchi, D.; Kempeneers, P.; Rodriguez, D.; Syrris, V.; Vasilev, V. A versatile data-intensive computing platform for information retrieval from big geospatial data. *Future Generation Computer Systems* 2018, 81, 30–40.

26. Wang, L.; Ma, Y.; Yan, J.; Chang, V.; Zomaya, A.Y. pipsCloud: High performance cloud computing for remote sensing big data management and processing. *Future Generation Computer Systems* 2018, 78, 353–368.
27. TCS Digital Farming Initiatives. The TCS PRIDE™ Model- Empowering Farmers! 2018. Retrieved from https://www.itu.int/en/ITU-D/Regional-Presence/AsiaPacific/SiteAssets/Pages/E-agriculture-Solutions-Forum-2018/TCS%20Digital%20Farming%20Initiatives_Shankar%20Tagad_ESF%202018_v0.2.pdf. Accessed on 21 March 2022.

8 Providing Geospatial Intelligence through a Scalable Imagery Pipeline

Andrew Reith,[a] *Jacob McKee,*[b] *Amy Rose,*[a]
Melanie Laverdiere,[c] *Benjamin Swan,*[a] *David Hughes,*[a]
Sophie Voisin,[a] *Lexie Yang,*[a] *Laurie Varma,*[d]
Liz Neunsinger,[d] *and Dalton Lunga*[a]
[a]Geospatial Science and Human Security Division,
Oak Ridge National Laboratory
[b]Descartes Labs
[c]NV5 Geospatial
[d]Communications Division, Oak Ridge National Laboratory

CONTENTS

DOI: 10.1201/9781003270928-11

8.1 GEOSPATIAL INTELLIGENCE R&D CHALLENGES

On December 10, 2021, a devastating tornado swept through western Kentucky, displacing hundreds of people and killing 56. A few days later, the White House tweeted a photo of President Biden receiving a briefing on the affected areas and response efforts from officials with the Federal Emergency Management Association (FEMA) and US Department of Homeland Security (DHS). A map of Mayfield, Kentucky, sat on an easel behind the president, showing damage to homes, stores, schools, religious facilities, and other buildings—telling an impactful story of lives and landscapes quickly changed.

The map was created with data from USA Structures, a database developed by Oak Ridge National Laboratory's (ORNL's) Geospatial Science and Human Security Division to supply data on the built environment to FEMA and others for any place in the United States. Maps like this, built from satellite imagery, are vital to planning for and responding to emergencies, developing public health campaigns, and undertaking urban planning. Advanced data analytics enable us to create maps like USA Structures, combing through the mounds of imagery and data produced by satellites and other remote sensing technologies after data inputs are made analysis ready. The extensive process of gathering, processing, and refining satellite imagery for analysis is integral to transforming raw data into usable geospatial intelligence. At the core of ORNL's ability to deliver geospatial intelligence to decision-makers is a workflow called the Imagery Parallel Integration and Processing Engine (PIPE). PIPE processes Level 1B images, resulting in Level 3A imagery[1] adjusted for terrain, sensor perspective to deliver high-resolution imagery and metadata for knowledge discovery and decision-making.

This chapter describes ORNL's contributions to imagery preprocessing for geospatial intelligence research and development (R&D) in four sections. First, we discuss challenges involved in building an effective imagery preprocessing workflow and the world-class high-performance computing (HPC) resources at ORNL available to process petabytes of imagery data. Second, we highlight how we developed imagery preprocessing tools over three decades while paving the way for our current cutting-edge machine learning and computer vision algorithms that are impacting humanitarian and disaster response efforts. Third, we discuss how PIPE modules work together to turn raw images into analysis-ready datasets. Fourth, we look toward the future and discuss planned advancements to PIPE and computing trends that will affect geospatial intelligence R&D.

8.1.1 Challenges to Advancing Geospatial Intelligence

Many challenges confront geospatial scientists as they work to deliver usable imagery and metadata for decision-making. Raw satellite imagery, available in abundance but not immediately useful, must be preprocessed to improve resolution and visibility. What was once a manual process tethered to desktop storage and applications now uses the power of HPC, offering the power of modern data management and analysis. As with all scientific R&D, there is always room for improvement, and ORNL continues to address challenges associated with compute power and startup costs, scalability, speed and resolution, and privacy and security.

8.1.1.1 Compute Power and Startup Costs

The increasing amounts of imagery available from satellites and remote sensing technologies used for disaster response, public health campaigns, and urban planning must be preprocessed efficiently and stored for use on demand. Furthermore, preprocessing methodologies used in PIPE increase image file size by 10–12 times due to many intermediate outputs and various storing formats. Compute power needs to be robust enough to manage terabytes of data through the process of ingesting, preprocessing, and storing imagery and can leverage the resulting metadata to create maps and models.

Building the computing capabilities needed to develop geospatial intelligence tools at any scale involves the prohibitive costs associated with acquiring hardware and software architecture, conducting data analysis, and maintaining data storage. Costs are also incurred in developing in-house talent expertise and acquiring (i.e., hiring) new talent. ORNL is the first national lab to develop an imagery workflow like PIPE, which resulted in high initial investments in infrastructure and expertise development. Naturally, the trial-and-error nature of R&D also increased costs that later entrants may not incur.

8.1.1.2 Scalability

Scalability applies the best information and solutions over numerous situations and geographic areas, thus increasing the usability of imagery and data. To date, many efforts are applicable only to limited areas of interest or geographic regions, requiring the same data-gathering and processing steps to be undertaken for each new incident or research question. Advances must increase generalizability to deliver efficient efforts and effective solutions. Data processing steps must scale up to meet the demands of the voluminous data now available for research and decision-making. In addition, connecting new methods to an existing workflow may not be as easy as plug-and-play, and the need to fit future methods to current design will impact scalability.

8.1.1.3 Speed and Resolution

Advances in algorithms have driven our ability to process giant caches of images in shorter amounts of time with higher resolution, putting the highest-quality data into the hands of decision-makers faster. The current drive toward cloud computing, edge computing, and GeoAI necessitates we take time to consider which of these emerging modalities will benefit geospatial intelligence R&D activities.

8.1.1.4 Data Privacy and Security

Images and their metadata need to be protected based on agreements under which the images were obtained. Agreements through commercial licenses or government contracts may dictate how images are collected, used, and stored. Permissions to access libraries of postprocessed images may be restricted to specific researchers, analysts, and administrators.

8.1.2 ORNL High-Performance Computing Resources

ORNL has become a world leader in HPC since the inception of our computing and computational sciences program in the mid-1980s. Staff members and external users alike can access world-class computing resources for data processing and storage, benchmarking, and simulations. ORNL's HPC resources allow scientists to make breakthroughs by employing more precise calculations and developing increasingly complex models that reduce time to insight and deliver high-fidelity solutions. We have used Summit, an IBM AC922 system introduced in 2018 that solves 200 quadrillion calculations per second and is the nation's fastest supercomputer, to benchmark algorithms, demonstrate proof of concept, and test the speed with which mapping, building extraction, and other geospatial intelligence R&D activities can be achieved.

The Compute and Data Environment for Science (CADES)[2] at ORNL supports scientists with compute and data infrastructure, including a suite of computing and analytic tools, dedicated data storage, high-speed data transfer, and scalable HPC capabilities. In our work associated with PIPE, we utilize a dedicated moderate-level[3] CADES cluster of 11 nodes, some with multiple CPUs, that provide the compute power of 17. Nine nodes have 256 GB of RAM and 36 CPUs, and two have just over 1 TB of RAM and 128 CPUs. This configuration enables us to run 17 matrix operations on imagery concurrently—1 operation on each of the first nine nodes and 4 operations on the remaining two. CADES also provides 3.7 petabytes of long-term storage organized as JBODs, discrete drives that allow us to add on data storage easily and cost-effectively as our needs increase.

8.2 PUSHING THE BOUNDARIES OF GEOSPATIAL INTELLIGENCE

Geospatial intelligence has grown in importance since the 9/11 attack, pivoting quickly in the following years from its stated mission in 2001 of tightening homeland security to responding to natural disasters such as 2005s Hurricane Katrina (Stair, 2001; *ORNL Review*, 2006). Today, users of geospatial intelligence seek data-driven methods for locating people in need, dispersing supplies, and improving future outcomes related to human health and security. PIPE builds on advancements contributed by ORNL over the last 30 years that modernize geospatial intelligence R&D methodologies.

ORNL's first efforts in high-resolution geospatial intelligence involved developing settlement maps. Black-and-white imagery was downloaded to desktops, and rudimentary computer programs processed 1 km^2 areas. One of our initial efforts was the Settlement Mapping Tool, a GUI-supported tool that leveraged NVIDIA K80 GPUs. It used user-generated training data to extract areas of likely human settlement at a resolution of 0.25 arc-seconds from very high-resolution (typically a half-meter) satellite imagery. Although the tool was capable of processing imagery very quickly (~500 km^2/minute), the training process was labor intensive and, for this reason, analysts initially acquired view-ready (processed to Level 2A) imagery strip by strip, as needed. Later, hard drives of raw imagery were downloaded to a server and orthorectified at ORNL on a dedicated Intel Xeon CPU E5-2687W processor with 20 cores, and the postprocessed imagery was made available to customers. By processing the imagery in house, ORNL researchers found that they were able to achieve more control over the preprocessing steps, with better results. As computing power increased and algorithms for detecting settlement improved, we shifted our focus from extracting coarse settlement areas to extracting building-level data.

New capabilities and the desire to address the need for improved estimates of population for consequence assessment led ORNL to our first effort in high-resolution geospatial intelligence, LandScan Global, which is widely considered the gold standard of population and mapping data in the United States (Lakin, 2020). The introduction of GPUs in the mid-2000s and growing capabilities in remote sensing and imaging technology—along with advancements in computer vision, machine learning, and algorithm development—led to ORNL improvements in feature extraction, image segmentation, and object detection, which have been leveraged in PIPE.

Today, we routinely work with the National Geospatial-Intelligence Agency (NGA), FEMA, DHS, and the Gates Foundation to provide data and models to understand and address human and health security needs of the US and world populations. The pre-event imagery contained in USA Structures allowed us, for example, to model damage to commercial, residential, government, historical, and industrial buildings in the aftermath of flooding west of Nashville in August 2021. We also provided USA Structures data that was foundational to crowdsourcing initiatives to characterize building damage following Hurricanes Harvey and Maria in 2017 and the Carr Fire and Kilauea Lava Flow in 2018. To support recovery efforts in Texas following Hurricane Harvey, ORNL processed 2,000 images covering 26,000 $miles^2$ in 24 hours. In 2013, ORNL began a multiyear collaboration with the Global Polio Eradication Initiative to locate previously unmapped populations in Nigeria to increase vaccine coverage among children. We used a support vector machine algorithm to create settlement maps and machine learning to extract buildings as a benchmark dataset. The dataset helped identify chronically missed settlements in Nigerian polio vaccination campaigns, and the initial project has been expanded to support other world health missions in sub-Saharan Africa and South Asia (ORNL, 2017; Rose et al., 2020).

ORNL's Imagery PIPE lies at the heart of our ability to produce high-quality data for decision-making and planning. Originating as an orthorectification workflow in 2014, PIPE completes a set of preprocessing steps that deliver analysis-ready images and metadata for downstream compute and analysis. We then use our ISOSCELES workflow to select representative training samples and our SCRIBE workflow to extract building feature data. ReSFlow, another workflow we have developed, partitions training data into "buckets" of images with similar building features important

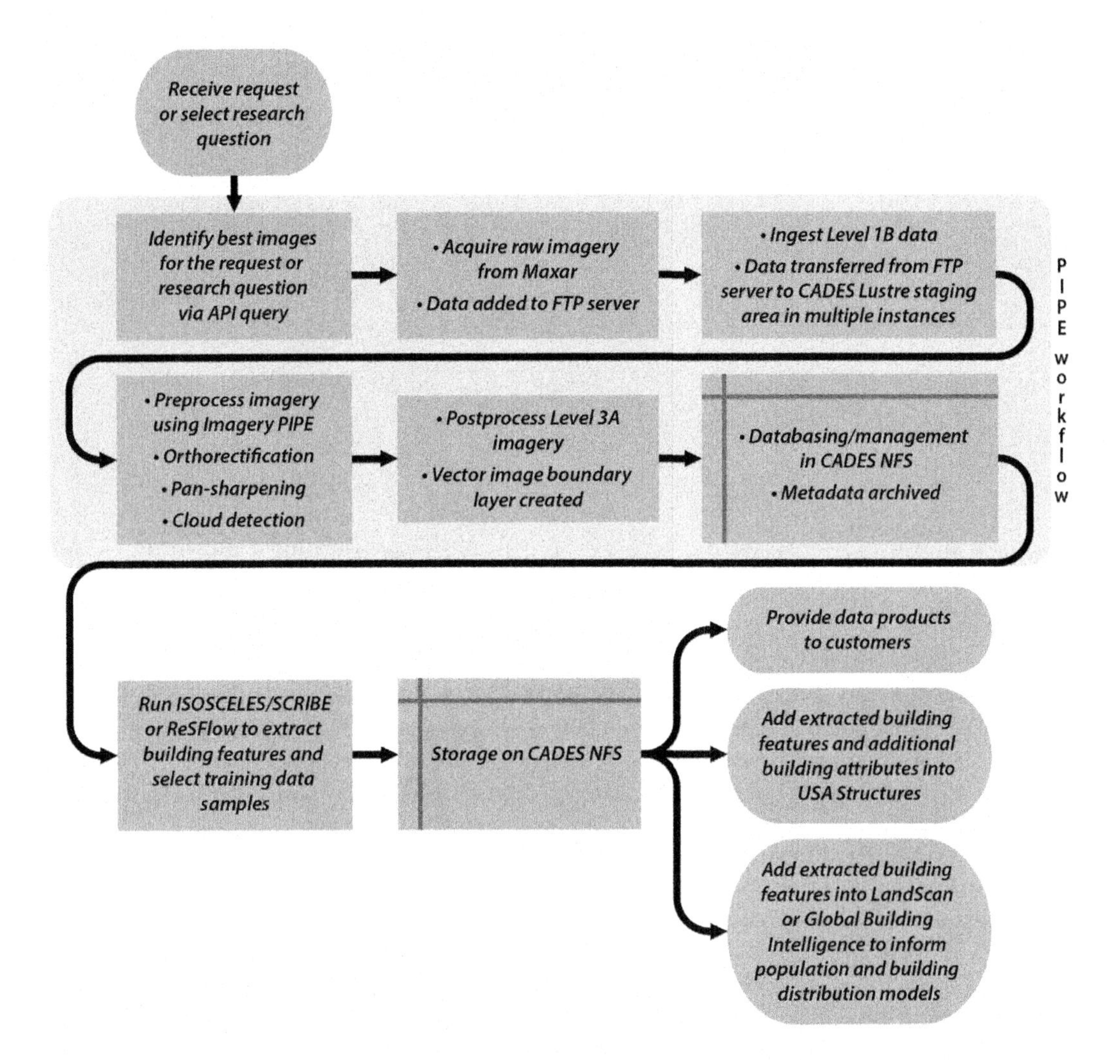

FIGURE 8.1 ORNL's imagery processing pipeline. (Source: ORNL.)

to the research question to reduce the complexity of data analysis. After imagery has been made analysis ready, we can plug resulting data into modeling and mapping tools such as our LandScan suite, USA Structures, and Global Building Intelligence. Figure 8.1 provides an overview of how our workflows and mapping and modeling applications work together.

8.2.1 Enabling Research

We use ISOSCELES (**I**terative **S**elf-**O**rganizing **SCE**ne **LE**vel **S**ampling) (Swan et al., 2021), an unsupervised machine learning program and workflow designed to efficiently sample very large image datasets for effective training data. ISOSCELES employs a multilevel sampling approach, first selecting highly representative images from a whole target image domain, then highly representative subsets of those images to be labeled and used for model training. This method allows for inclusion of both broad, image-level variables, such as viewing angle and time of day, and complex, within-image variables, such as local spectral texture. As an unsupervised and data-driven approach, ISOSCELES can be quickly employed on large target datasets without prior knowledge of the data or any ancillary or reference data. Additionally, it can be readily adapted to work with data from a wide variety of sensors.

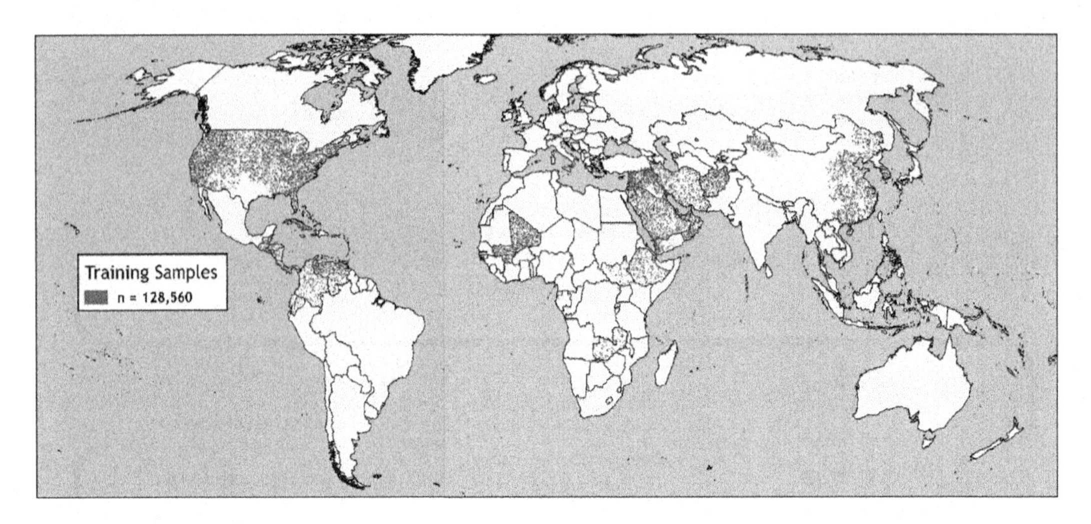

FIGURE 8.2 Distribution of ORNL's training data library. (Source: ORNL.)

SCRIBE (**SC**alable High-**R**esolution **I**magery-Based **B**uilding **E**xtraction) (Laverdiere et al., 2019) provides a generic and scalable approach to large-scale building extraction. Powered by ORNL's ISOSCELES and PIPE (discussed in "Building the Imagery Pipeline") workflows, SCRIBE performs building extraction to quickly and accurately define the built environment in high resolution using data on the location of buildings. SCRIBE's convolutional neural network (CNN) framework delivers better generalization of features and leverages recent advances in deep feature learning for pattern recognition and classification tasks.

Like SCRIBE, **ReSFlow** (**Re**mote **S**ensing Data **Flow**) (Lunga et al., 2021) provides a workflow for automated scalable object detection and segmentation of remote sensing data for building extraction and training data definition. ReSFlow overcomes poor generalization of models trained with highly variable data—resulting from heterogeneous remote sensing technologies, variations, and climactic conditions in the scenes gathered by satellite imagery—and algorithmic and model limitations. By identifying images with similar characteristics and assigning them to "buckets" via a two-step partitioning approach, ReSFlow reduces instrument, collection, environment, and scene content variation and maps both satellite imagery and deep learning models to the buckets.

Over the time ORNL has been working in geospatial intelligence R&D, we have developed a robust training data library containing ~128,000 training data samples representing data points from all over the world (Figure 8.2). To build the library, we identified representative image samples via ISOSCELES and digitized corresponding structures. This diverse, high-quality large dataset enables us to generate highly generalizable machine learning models that perform well on new target domains where no training data exists.

8.2.2 Mapping

ORNL's Geospatial Science and Human Security Division has contributed to developing highly resolved mapping of the built environment. Most notably, **USA Structures** forms a dataset of the location and description of US buildings that has been used as a starting point for damage assessments following extreme weather events (Laverdiere et al., 2022). After SCRIBE is leveraged to create a building outline dataset, additional data inputs are leveraged to describe each building including Census housing unit data, Homeland Infrastructure Foundation Data, and LightBox parcel data for occupancy type classifications (e.g., residential, commercial). USA Structures has a consistent metadata and schema across all states and territories, includes unique building identifiers, and incorporates light detection and ranging (LiDAR)-derived structures from the NGA's 133 cities dataset.[4] USA Structures uses the SCRIBE workflow to develop a robust and generalizable model trained with nearly 60,000 manually annotated training samples to create a baseline inventory of buildings for the United States and its territories, "current" as of August 2021, that can be referenced for disaster response.

8.2.3 Large-Scale Modeling

ORNL entered the geospatial intelligence R&D field in 1997 with the LandScan Program,[5] which was initiated to address the need for improved estimates of population for consequence assessment. The unpredictable nature of natural and manmade disasters and the risks to populations across the globe drove our development of population distribution models to capture where people might be found throughout the day, rather than relying on traditional sources such as housing data contained in censuses. Further, it was critical to develop highly resolved population estimates to evaluate events across multiple geographic scales. ORNL's initial effort, **LandScan Global**, a 30 arc-second (~1 km) global regular grid of estimated population counts, has been updated annually since the first version was released in 1998.

The ability to create a new version of this dataset each year requires the knowledge and capture of changes in the nature and extent of global human habitation over time. To support this, we began pioneering developments in machine learning and computer vision specifically to identify anthropogenic signals apparent in overhead imagery. This R&D eventually enabled the rapid, large-scale detection of human settlements and building features from high-resolution imagery and became the basis for early efforts to develop an improved-resolution (3 arc-seconds, ~100 m) population distribution known as LandScan HD.

LandScan HD incorporates current land use and infrastructure data from a variety of sources, applies occupancy estimates from ORNL's Population Density Tables project (Stewart et al., 2016), and leverages novel image processing algorithms developed at ORNL to rapidly map building structures (Yang et al., 2018) and urban structural units (Arndt and Lunga, 2021) using HPC. In this way, LandScan HD is developed using a "bottom-up" approach where high-resolution population estimates are not dependent on a recently conducted, high-quality census. This approach is particularly useful for parts of the world that regularly experience large changes in population distribution due to rapid growth, natural hazards, or conflict. The first country-scale LandScan HD dataset was created in 2014, and a continuous stream of new country-scale datasets have been developed ever since.

Building on the modeling approach developed for LandScan Global—and taking advantage of higher-quality data available for the United States—we improved both spatial and temporal resolution with our first release of **LandScan USA** in 2004. LandScan USA aimed to capture the diurnal variation of population that is critical for a variety of analyses and actions including emergency preparedness and response. In 2016, the original LandScan USA model was reengineered to incorporate advances in geospatial technology, machine learning approaches, and new input data sources including the USA Structures dataset. We continue to make annual improvements to the underlying model and release a new version of the LandScan USA dataset each year.

ORNL's **Global Building Intelligence** (GBI) modeling platform (Rose, 2021) uses a Bayesian learning framework to automate prediction of building characteristics (e.g., construction materials) and building use for any building on the planet when data is unavailable. Still under development, GBI harmonizes and links diverse geospatial data at multiple spatial and temporal scales. The graph network model powering GBI integrates capabilities including automated feature extraction, occupancy modeling, and height derivation as connected nodes in the graph. GBI will strengthen location intelligence across multiple domains including humanitarian assistance and disaster relief; energy security and reliability; collateral damage assessments; and emergency readiness, response, recovery, and resiliency.

8.3 BUILDING THE IMAGERY PIPELINE

Projects analyzing available data to define the location and extent of human population often begin with generating a foundational layer delineating the built environment where people are likely to live and work (Cheriyadat et al., 2007). ORNL's PIPE image processing workflow helps create that foundational knowledge, which can then be input to mapping and modeling applications to support solutions. PIPE has been developed over several years, adding on new capabilities and increasing the precision of data inputs.

PIPE grew organically as we leveraged ORNL HPC resources to do geospatial intelligence work. First, ORNL acquired an account with the NGA's vendor Maxar Technologies[6] that allowed us to download raw imagery. Our next task was to acquire sufficient storage—essentially, a "staging area" where imagery could be quickly downloaded and efficiently preprocessed. We procured ~300 TB of storage on a Lustre-distributed file system and 11 compute nodes through ORNL's CADES. Lustre efficiently distributes jobs to nodes, making it ideal for our work. With the proper short-term storage infrastructure in place, our next task was to establish the processing piece of the workflow infrastructure. We set up 7 (later expanded to 11) Intel Xeon E5-2698 processors. Jobs were allocated to these nodes for orthorectification, pan-sharpening, and generation of image statistics and valid image pixels. Jobs related to cloud detection, the final step of the workflow, were allocated to the nodes as well. Long-term storage was also established.

Once an information request is received or a research question is selected, we perform a manual search on desired features via API query to Maxar's database. Relevant imagery catalog IDs are sent to NGA for approval and forwarding to the vendor. Level 1B imagery obtained from Maxar is transferred from an FTP server and ingested to a CADES Lustre staging area in multiple instances, and then preprocessed in CADES scalable HPC condos using the PIPE workflow. Postprocessing steps including creating a vector image boundary layer and incorporating cloud-masked image areas into the vector image boundary layer also take place in the CADES condos. After processing, the resulting Level 3A imagery is stored on the CADES JBODs, which connect to NVIDIA DGX machines where the main machine learning and GeoAI-related activities are conducted. With such a computing configuration, the PIPE processes imagery for general use through the streamlined connection to CADES cluster so data can be accessed on demand for further analyses.

8.3.1 Imagery Ingest

The process of imagery ingest involves getting the right imagery to an environment where it can be processed. The first step in this process is identifying the imagery needed based on the final use case. For the purposes of population estimation, this generally requires prioritizing image recency, ground sample distance (GSD), cloud cover, and other characteristics that may increase the accuracy and precision of the building feature extraction. Attention is also paid to minimizing the area covered by numerous overlapping image strips. Therefore, based on the area of interest, our goal is generally to obtain full coverage of the newest, cloud-free, or low-cloud images with the highest spatial resolution.

At ORNL, once images are identified, the selected Level 1B imagery is downloaded by Maxar via an FTP server. This download includes several important files. The main file is the raw image transferred in national imagery transmission format (.ntf), which contains the remotely sensed data. The second most crucial file is the image metadata file (.imd), which contains valuable information about the image, including sensor type, satellite angles, sun angles, GSD, and estimated cloud cover. Other files include the attitude file (.att), describing the position and orientation of the sensor relative to the Earth; the ephemeris file (.eph), describing the location of the sensor in space at the time of image capture; and the geometric calibration file (.geo), which contains photogrammetric information such as the principal distance, camera attitude, and perspective center. Two final files transmitted with the imagery include the rational polynomial coefficient file (.rpb) and tile map file (.til), which are used for orthorectification. The preprocessed image (.tif), image metadata file (.imd), and accompanying files (.xml files for the image statistics, .shp for the image masks) are copied to long-term storage. It should be noted that, for a four-band image, ORNL receives two sets of these files—one for the panchromatic image and one for the multispectral image.

The imagery ingest process is admittedly archaic given its reliance on FTP, which is slow, and given that it is made up of multiple human-in-the-loop steps. Because it is so crucial to ensure that the imagery ingest step does not impede preprocessing and building feature extraction, among other later steps, we optimized the process by spawning multiple instances of the command to download each raw image strip, which better utilizes compute resources and ensures that imagery is quickly available for the next steps in the PIPE workflow.

8.3.2 Orthorectification

Generating digital orthophotos is the foundation for any remote sensing pipeline. In essence, digital orthophotos contain both the image characteristics of a photograph and the geometric properties of a map. Thus, the image acquired via satellite must be transformed from image space to map space (Zhou et al., 2005), resulting in a geometrically corrected image that removes the effects of tilt and terrain. For our model of the Earth, we use EGM96 (Earth Gravitational Model, 1996), SRTM (Shuttle Radar Topography Mission), and ASTER (Advanced Spaceborne Thermal Emission and Reflection Radiometer). EGM96 represents the vertical datum, and SRTM and ASTER are our digital elevation models. SRTM delivers a spatial resolution of 30 m available from 60° N to 60° S. For geographic regions beyond 60°, ASTER is used as a substitute. To tie the images to their respective geographic locations, we use the rational polynomial coefficients provided in the metadata for each image. Output files from this process include the orthorectified image (.tif). As this process can be computationally intensive, we have written our own version in C++ and enabled the use of GPUs. Figure 8.3 illustrates the orthorectification process for remote sensing imagery.

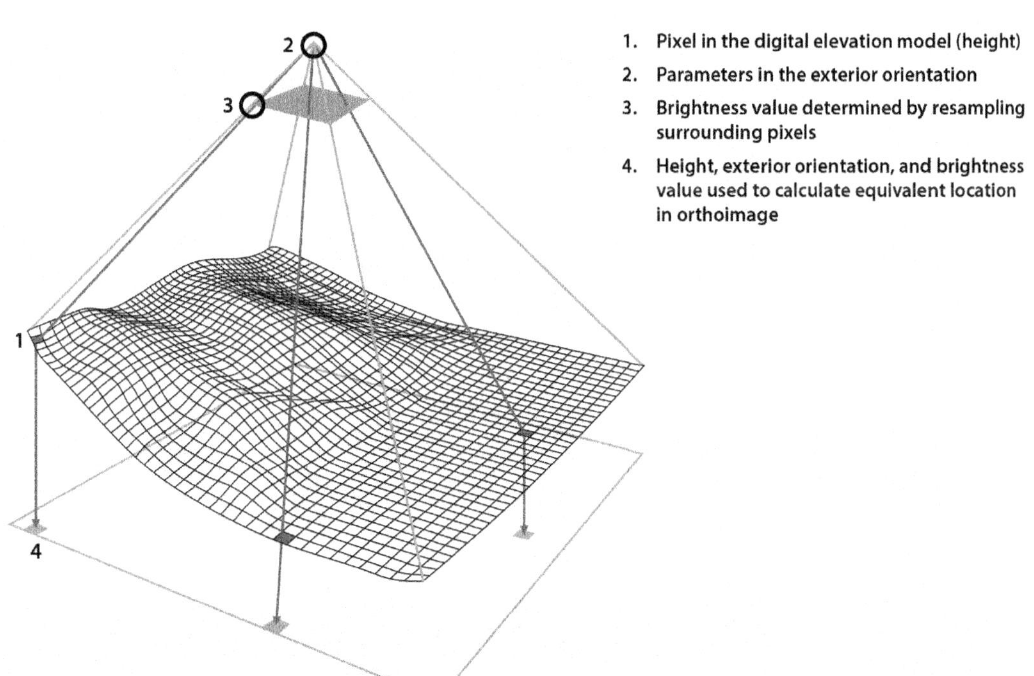

FIGURE 8.3 Typical orthorectification process for remote sensing imagery. (Source: ORNL.)

8.3.3 Pan-Sharpening

Most imagery available for processing is not inherently high resolution to allow high-density outputs. While multiple super-resolution approaches—from the multispectral, multiframe, and single-frame families—exist to enhance the resolution of a given image, one should be very cautious to avoid generating high-resolution details that are not there. Pan-sharpening, as a multispectral approach, involves leveraging a high-resolution panchromatic image that covers a large spectrum to enhance a low-resolution image of a narrower spectral band, resulting in a high-resolution multispectral image (.tif output file). The methodology assumes a correlation between the larger spectrum and narrow spectral band that allows the details from the high-resolution panchromatic image to be transferred to the low-resolution band image. Although this limits the spectral band to the same resolution as the panchromatic image, the main advantage of this multimodal image fusion is the production of outputs more robust to hallucination phenomenon.

Multiple algorithms can be used to achieve pan-sharpening. As we did for the orthorectification step, we have written our own version of the NNDiffuse approach (Sun et al., 2014) in C++ and CUDA to leverage ORNL HPC resources for scalability and speed.

8.3.4 Cloud Detection

ORNL's cloud detection capability is, conceptually, a preprocessing and normalization step followed by a relatively simple CNN (Johnston et al., 2017). In PIPE, cloud detection is run on orthorectified and pan-sharpened imagery; the output is a pixelwise cloud salience score between 0.0 and 1.0, as well as a binary (i.e., thresholded) cloud mask. PIPE's cloud detection module is prepared only for three-band visible (i.e., RGB) or single-band near-infrared band.[7]

Due to the possible presence of fog, haze, and other partially transmissive (i.e., scattering) phenomena—all of which we refer to as simple "haze"—cloud detection is not quite a simple binary classification problem. Because haze thickness varies along a continuum, it is difficult to label imagery as cloud/not-cloud while maintaining consistency in how haze is labeled. To ensure consistency, we manually annotate our training, validation, and evaluation data with three classes: cloud, not-cloud, and haze. Only regions with completely opaque atmosphere are labeled as cloud, and only regions without noticeable atmospheric effects are labeled as ground. Areas labeled as haze are ignored during training, and performance scores are reported separately for each class during evaluation.

The threshold used to generate the binary mask from the salience image is configurable; the default is chosen to have a low false-positive rate, resulting in an accuracy of 97.94 assuming balanced classes (96.3% true positive [TP], 0.43% false positive [FP]). When optimizing the threshold for accuracy alone, the algorithm achieves 98.6% accuracy (98.4% TP, 1.2% FP). Figure 8.4 demonstrates the cloud detection capability, showing an example input image along with its corresponding salience image and thresholded mask.

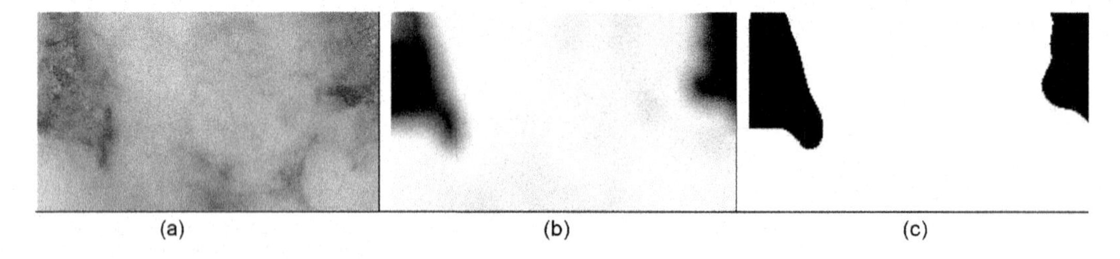

FIGURE 8.4 Images illustrating ORNL's cloud detection capability. (a) Single-band near infrared input image, (b) detection image, (c) thresholded image. ((a) Copyright © 2010 DigitalGlobe NextView License; (b and c) image developed by ORNL using DigitalGlobe imagery.)

Our current model, trained on WorldView-2 imagery, consists of three convolutional layers interspersed with two max-pooling layers and followed by a final affine layer and softmax. The model is fully convolutional, so the dimensions of the input image can vary. However, because remote sensing imagery is often very large, running the model on a full-size image frequently exceeds GPU memory. Our deployment application supports arbitrarily large images by cropping the input image into smaller tiles and then stitching the prediction results. This tiling process allows multiple GPUs to be used in parallel to perform prediction on a single, large image; reading and preprocessing the next subimage runs on the CPU, while prediction with the CNN is accomplished in parallel on the GPU.

After PIPE runs the cloud detection module, the binary mask is converted to a polygon (.shp), which is then buffered for use in the training sample selection workflow. The preprocessed image (.tif), image metadata file (.imd), and accompanying files (.xml files for image statistics,.shp for image masks) are copied to long-term storage.

The CNNs, as well as the application that deploys it, have undergone continuous development. Initial research efforts used Python and Caffe, while the model was deployed to production using a custom application written in C++. To have a single codebase, efficient for both research and deployment, we now research, develop, and deploy cloud detection using the Julia (Bezanson et al., 2017) and Flux (Innes et al., 2018) frameworks.

8.3.5 Postprocessing and Output

Once the preprocessing modules outlined in the previous sections are completed, the imagery is prepared for analysis. Specific to large-scale building extraction as defined in the SCRIBE workflow, postprocessing steps may include creating a vector image boundary layer and incorporating cloud-masked image areas into the vector image boundary layer. This boundary layer indicates where usable image pixels exist within a given image scene. In this case, usable image pixels are locations where ground features are visible, excluding cloudy or hazy regions that obscure the ground. This image boundary layer is critical for identifying which portions of an image are optimal for more accurate building feature extraction (Figure 8.5). Once, machine learning techniques have been applied to imagery to identify building features, the resulting raw raster output is converted to vector in parallel. Vector-building features are ideal for other applications in support of projects such as the LandScan Program, USA Structures, and GBI.

8.4 FUTURE CONSIDERATIONS

We face an interesting dilemma in the 2020s. Advancing technology—Earth-observing satellites streaming images every 30 seconds, and new trends in crowdsourcing information during

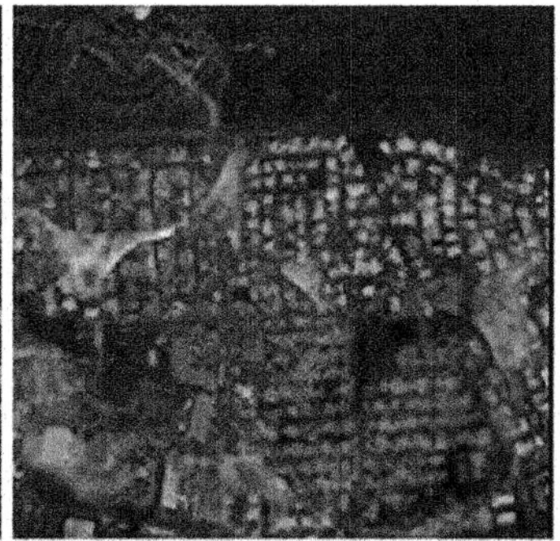

FIGURE 8.5 Sample image boundary layer. (Copyright © 2010 DigitalGlobe NextView License.)

natural disasters and smart devices—provides researchers and analysts with voluminous amounts of imagery and other data of vastly improved quality. The geospatial intelligence R&D field's main challenge now is, as it is for other domains using data science to solve problems, to harness the power of supercomputing to efficiently process and synthesize available information. The long time frames associated with manual data extraction are unworkable at global scale, and automated technologies deliver more consistent processes and therefore more accurate insights at a level of efficiency that is far beyond what humans can do alone, for a level of understanding of our planet that currently does not exist. The initial impetus for developing PIPE—enabling the large-scale mapping of anthropogenic signals on Earth—is still our guiding motivation today. As we attempt to map and describe the changing physical world, we must develop algorithms that add new capabilities to PIPE to optimize existing modules. We must look for ways to leverage the evolving HPC compute and infrastructure landscape and rearchitect the existing technology stack. As we look toward the future, we see opportunities to add new capabilities to PIPE, to harness the power of cloud computing in our work, and to adapt PIPE to other data types and applications.

8.4.1 Adding Atmospheric Compensation to the Pipeline

The ability of deep neural networks to learn complex representations with high-level features has allowed for the development of machine learning models that can generalize over large image datasets with an accuracy that was not possible with previous state-of-the-art techniques (Rasti et al., 2020; Zhang et al., 2016; Zhu et al., 2017). However, even models trained with a large and diverse set of samples can fail to produce acceptable results when applied to novel image domains. While domain adaptation using a strategically chosen set of samples from the target image domain is often a successful means of achieving suitable results, this method is costly, time-consuming, and prone to delays.

One avenue for improving model generalizability without using additional samples involves applying techniques to compensate for atmospheric effects on the radiant energy reaching the image sensor. Ozone thickness, total water vapor, and type and concentration of aerosols can produce complex changes in the apparent reflectivity of surface objects through attenuation and backscattering. Common machine learning practices for data standardization and normalization are inadequate for addressing these large shifts in the distribution of apparent reflectivity.

Prior to the relatively recent advent of using deep learning in remote sensing, most machine learning models were trained for only one image or a handful of images, so it has not been standard practice to expend the time and processing effort to employ atmospheric compensation. The limited literature on this topic generally shows accuracy gains with the method, although the magnitude varies considerably (Elmahboub et al., 2009; Lin et al., 2015; Pu et al., 2015; Rumora et al., 2020; Vanonckelen et al., 2013). However, because most of these studies involved only a single image, the question of how the method impacts generalizability over large image sets remains unanswered.

Challenges to deploying atmospheric compensation include automation of key parameter settings, availability of ancillary data on atmospheric conditions, and the potential requirement for additional quality control steps to ensure uniformity of outputs. Given the challenges and to fill a gap in the literature, we have launched an effort to quantify the impact of atmospheric compensation methods on building extraction tasks. Our initial experiments have shown promising performance gains, and we are building on this work to obtain a statistically robust comparison of the effectiveness of multiple atmospheric compensation methods for a diverse set of geographic and land cover contexts. Next steps will include examining whether these methods can be effectively scaled up to be compatible with the SCRIBE workflow, both in terms of computational complexity and the ability to maintain accuracy when implemented in a fully automated system.

8.4.2 Leveraging Cloud Computing to Advance Our Imagery Processing Capabilities

Future iterations of our imagery processing pipeline could include containerization of PIPE and deployment within a cloud environment. Containerization, such as the use of Docker, would allow for a code base to be neatly packaged alongside the varying dependencies at the finest granularity, including the operating system specifications. Containerization also lends itself to transferability and reproducibility when deployed on different compute environments, such as when deploying to another workstation or migrating to the cloud. While the compute resources utilized by PIPE are rarely idle, transitioning to a cloud platform such as AWS, AZURE, or GCP would have immediate tangible benefits. Instead of physical machines located on-premises, configuring a K8s (Kubernetes) compute cluster in the cloud would bring more flexibility to our operations. For example, if the physical hardware utilized by PIPE is idle, we will continue to incur costs associated with the physical space it occupies, electrical consumption, and maintenance. In contrast, if the compute resources in a cloud environment are not being utilized, the end user does not incur any cost (i.e., you only pay for what you use). This type of model also lends itself to scalability as resources are easily scaled up or down depending project needs. On the other hand, acquiring additional compute nodes to scale on-premises hardware is both expensive and time-consuming. Even after the needs of the project begin to subside, on-premises hardware will still be incurring costs.

8.4.3 Adapting PIPE to Other Applications

PIPE currently enables ORNL to deliver timely, high-quality information to multiple customers, as well as high-resolution data inputs to a range of ORNL projects. This contrasts with the tendency for commercial tools to focus narrowly on one use case or purpose. Our in-house infrastructure ensures that customer requests receive a faster response, and the development of PIPE reflects ORNL's focus on working toward solutions to real-world problems. It is important to note that the computing configurations and optimization strategies used in PIPE for preprocessing of high-resolution satellite imagery are also applicable to other source data types, including synthetic aperture radar, LiDAR, and stereo pairs, and we are committed to exploring ways to develop PIPE capabilities for use with nonimagery data inputs. While the specific preprocessing modules used may vary depending on data type, the same compute, storage, and input/output considerations come into play. Leveraging PIPE for other input data types will provide ample opportunities to pursue multimodal geospatial data analytics and develop novel data fusion approaches at scale, as well as to broaden the research and response applications of our work. Our future work will endeavor to provide "anytime, anywhere" answers to humanitarian problems that are local to global in scale.

As with many new technologies, our understanding of the knowledge, technology, and methodologies needed to advance geospatial intelligence R&D grew organically. While hindsight will always be clearer than the view researchers get while working on new methods or technologies, it is useful to reflect on the lessons learned during that process, to assist others who are advancing the field or blazing paths in wholly new research areas. The temptation with the ever-growing compute power we have experienced over the course of our work—and that we can only expect to accelerate as we move forward—is to build a bigger system that can compute more data, faster. In our experience, this is only part of the equation. Efficiency, control, and attention to scaling are also crucial. Bottlenecks we have experienced in imagery processing point to the need to ensure the individual steps of the PIPE workflow sync up so that when one module completes its work, the next one is ready to move the data forward. Achieving this level of control is, of course, a principal challenge facing HPC today, and all the research it supports. While it is impossible to account for technological advances that occur over the time anyone spends working in a new research area, we have found that early thought on scaling is key to efficient computing processes. Likewise, the data volumes that advancing computing technology creates calls for careful thought on data storage.

Over our three decades of research, we have matured how images are acquired, ingested, preprocessed, stored, and applied to problems that require timely and high-resolution data for human health and security response and planning. We will apply the lessons we have learned while building PIPE to our continuing work on imagery processing and health and human security, public health, and urban planning projects. We will continue to investigate the benefits of exploiting ORNL HPC resources for building extraction, object detection, classification, and data fusion strategies. We are working toward better methods for selecting representative samples; increasing the generalizability of models; and improving the quality and consistency of structure detection, automated model construction, and change detection at scale. We also will explore the efficacy of leveraging emerging remote sensing data modalities including unmanned aerial vehicle images, and new avenues in artificial intelligence approaches.

ACKNOWLEDGMENTS

The authors would like to recognize Devin White and Andrew Hardin, currently with Sandia National Laboratories, who served as developers of PIPE's orthorectification module, as well as Tyler McDaniel, currently with Amazon Web Services, who served as developers for PIPE's pan-sharpening module. We would like to recognize Don March, who served as the lead developer for PIPE's cloud detection module, made significant contributions to integrating the cloud detection capability with PIPE's other modules, and provided valuable suggestions during the chapter revision. We would like to recognize Jeanette Weaver, currently with WS LLC, who contributed to PIPE's pan-sharpening and orthorectification capabilities. We would like to recognize Chris Davis of ORNL for his assistance maintaining PIPE's orthorectification and pan-sharpening capabilities. We would like to recognize support provided by CADES, including group leaders Elton Cranfill and Brian Zachary and staff members Steve Moulton, Ryan Prout, Greg Shutt, and Chris Layton (currently with ORNL's Information Technology Services Division. CADES staff helped set up and currently help maintain a range of DGX and Condo computational resources and network storage for deploying PIPE and feature extraction workflows. We also would like to recognize Jacob Arndt of ORNL, who served as data scientist and co-developer for the ReSFlow framework and made significant contributions to developing and deploying the machine learning algorithms.

NOTES

1. See https://earthdata.nasa.gov/collaborate/open-data-services-and-software/data-information-policy/data-levels for standard definitions of imagery levels.
2. https://cades.ornl.gov.
3. Based on definitions in Committee on National Security Systems (2009).
4. The "133 Cities dataset" is an outgrowth of the Nunn-Lugar-Domenici Domestic Preparedness Program (Office of Justice Programs, 2001) started in 1997 with funding from the US Department of Defense (DoD). (Funding was later transferred to the US Department of Justice.) The program identified 120 of the nation's most populous cities and facilitated customized assessment, training, and equipment support for domestic terrorism initiatives. In 2010, the National Imagery and Mapping Agency, housed in the DoD, undertook a new mission to map 133 "top priority cities," most of which had been identified by the Nunn-Lugar-Domenici program (Patterson, 2010).
5. https://LandScan.ornl.gov.
6. In August 2021, Colorado-based Maxar Technologies was awarded a 5-year contract with NGA to continuing providing support for a data analytic system that exploits hundreds of data feeds and leverages multiple geospatial tools for US Department of Defense and Intelligence Community users (Maxar Technologies, 2021).
7. The cloud detection algorithm expects that imagery is from a distribution similar to the training data, all of which has daytime illumination conditions and near-nadir viewing geometries. The algorithm has not been proven to detect clouds near dusk, dawn, or nighttime. Additionally, the algorithm is not currently trained on examples with ice or snow.

REFERENCES

Arndt, J.; Lunga, D. Large-scale classification of urban structural units from remote sensing imagery. *IEEE Journal of Selected Topics in Applied Earth Observations and Remote Sensing* 2021, *14*, 2634–2648. doi: 10.1109/JSTARS.2021.3052961.

Bezanson, J.; Edelman, A.; Karpinski, S.; Shah, V. Julia: A fresh approach to numerical computing. *SIAM Review* 2017, *59*(1), 65–98.

Cheriyadat, A.; Bright, E.; Potere, D.; Bhaduri, B. Mapping of settlements in high-resolution satellite imagery using high performance computing. *GeoJournal* 2007, *69*, 119–129. doi: 10.1007/s10708-007-9101-0.

Committee on National Security Systems. Security Categorization and Control Selection for National Security Systems, version 1. CNSS Instruction no. 1253. Fort Meade, MD: National Security Agency, October 2009.

Elmahboub, W.; Scarpace, F.; Smith, B. A highly accurate classification of TM data through correction of atmospheric effects. *Remote Sensing* 2009, *1*(3), 278–299. doi: 10.3390/rs1030278.

Innes, M.; Saba, E.; Fischer, K.; Gandhi, D.; Rudilusso, M. C.; Joy, N. M.; Karmali, T.; Pal, A.; Shah, V. Fashionable modelling with flux. *CoRR* 2018, abs/1811.01457.

Johnston, T.; Young, S.; Hughes, D.; Patton, R.; White, D. Optimizing convolutional neural networks for cloud detection. *Proceedings of the Machine Learning on HPC Environments* 2017. doi: 10.1145/3146347.3146352.

Lakin, M. "GIS: LandScan Goes Public." *ORNL News,* May 5, 2020. https://www.ornl.gov/news/gis-landscan-goes-public.

Laverdiere, M.; Yang, L.; Swan, B. SCRIBE workflow. *Internal Presentation during Sponsor Visit,* May 2019.

Laverdiere, M.; Hauser, T.; Swan, B.; Yang, L.; Whitehead, M.; Moehl, J.; Schmidt, E. USA structures: High resolution feature extraction and attribution for the United States & territories. *Presentation at the OpenStreetMap US Government Committee Monthly Meeting,* January 24, 2022.

Lin, C.; Wu, C.-C.; Tsogt, K.; Ouyang, Y.-C.; Chang, C.-I. Effects of atmospheric correction and pansharpening on LULC classification accuracy using WorldView-2 imagery. *Information Processing in Agriculture,* 2015, *2*(1), 25–36. doi: 10.1016/j.inpa.2015.01.003.

Lunga, D.; Arndt, J.; Gerrand, J.; Stewart, R. ReSFlow: A remote sensing imagery data flow for improved model generalization. *IEEE Journal of Selected Topics in Applied Earth Observations and Remote Sensing* 2021, *14*, 10468–10483.

Maxar Technologies. Maxar awarded big data analytics contract from NGA. *Press release,* August 23, 2021. https://www.maxar.com/press-releases/maxar-awarded-big-data-analytics-contract-from-nga.

Oak Ridge National Laboratory (ORNL). *A 21st Century Planning Tool Built on AI.* Success Story. ORNL: Oak Ridge, TN, 2017.

Office of Justice Programs, US Department of Justice. Nunn-Lugar-Domenici Domestic Preparedness Program: Program Overview. NCJ no. 193756. Office of Justice Programs: Washington, DC, September 2001. https://www.ojp.gov/ncjrs/virtual-library/abstracts/nunn-lugar-domenici-domestic-preparedness-program-program-overview (accessed February 16, 2002).

ORNL Review. Preparing for the threats. *ORNL Review* 2006, *39* (1), 12–18.

Patterson, D. NIMA's shared data, secret data. Government Technology, July 27, 2010. https://www.govtech.com/security/nimas-shared-data-secret-data.html (accessed February 16, 2002).

Pu, R.; Landry, S.; Zhang, J. Evaluation of atmospheric correction methods in identifying urban tree species with WorldView-2 imagery. *IEEE Journal of Selected Topics in Applied Earth Observations and Remote Sensing* 2015, *8*(5), 1886–1897. doi: 10.1109/JSTARS.2014.2363441.

Rasti, B.; Hong, D.; Hang, R.; Ghamisi, P.; Kang, X.; Chanussot, J.; Benediktsson, J. A. Feature extraction for hyperspectral imagery: The evolution from shallow to deep: Overview and toolbox. *IEEE Geoscience and Remote Sensing Magazine* 2020, *8*(4), 60–88. doi: 10.1109/MGRS.2020.2979764.

Rose, A. Global building intelligence factsheet, July 22, 2021. Available from Oak Ridge National Laboratory.

Rose, A.; Weber, E.; McKee, J.; Urban, M.; Lunga, D.; Yang, L.; Moehl, J.; Laverdiere, M.; Singh, N.; Tuttle, M.; Whitehead, M.; Huff, A.; Lakin, M.; Bhaduri, B. Mapping human dynamics. In Wright, D. J.; Harder, C. (Eds). *GIS for Science: Applying Mapping and Spatial Analytics.* Esri Press: Redlands, CA, 2020, pp. 70–83.

Rumora, L.; Miler, M.; Medak, D. Impact of various atmospheric corrections on Sentinel-2 land cover classification accuracy using machine learning classifiers. *ISPRS International Journal of Geo-Information* 2020, 9(4), 277. doi: 10.3390/ijgi9040277.

Stair, B. Computer modeling and homeland security. EurekAlert! December 31, 2001. https://www.eurekalert.org/news-releases/571936.

Stewart, R.; Urban, M.; Duchscherer, S; Kaufman, J.; Morton, A.; Thakur, G.; Piburn, J.; Moehl, J. A Bayesian machine learning model for estimating building occupancy from open source data. *Natural Hazards* 2016, *81*, 1929–1956. doi: 10.1007/s11069-016-2164-9.

Sun, W.; Chen, B.; Messinger, D. Nearest-neighbor diffusion-based pan-sharpening algorithm for spectral images. *Optical Engineering* 2014, *53*(1), 1–11. doi: 10.1117/1.OE.53.1.013107.

Swan, B.; Laverdiere, M.; Yang, H. L.; Rose, A. Iterative self-organizing SCene-LEvel Sampling (ISOSCELES) for large-scale building extraction. *GIScience & Remote Sensing* 2021. doi: 10.1080/15481603.2021.2006433.

Vanonckelen, S.; Lhermitte, S.; Van Rompaey, A. The effect of atmospheric and topographic correction methods on land cover classification accuracy. *International Journal of Applied Earth Observation and Geoinformation* 2013, *24*, 9–21. doi: 10.1016/j.jag.2013.02.003.

Yang, H. L.; Yuan, J.; Lunga, D.; Laverdiere, M.; Rose, A.; Bhaduri, B. Building extraction at scale using convolutional neural network: Mapping of the United States. *IEEE Journal of Selected Topics in Applied Earth Observations and Remote Sensing* 2018, *11*(8), 2600–2614. doi: 10.1109/JSTARS.2018.2835377.

Zhang, L.; Zhang, L.; Du, B. Deep learning for remote sensing data: A technical tutorial on the state of the art. *IEEE Geoscience and Remote Sensing Magazine* 2016, *4*(2), 22–40. doi: 10.1109/MGRS.2016.2540798.

Zhou, G.; Chen, W.; Kelmelis, J.A.; Zhang, D. A comprehensive study on urban true orthorectification. *IEEE Transactions on Geoscience and Remote Sensing* 2005, *43*(9), 2138–2147.

Zhu, X.X.; Tuia, D.; Mou, L.; Xia, G.; Zhang, L.; Xu, F.; Fraundorfer, F. Deep learning in remote sensing: A comprehensive review and list of resources. *IEEE Geoscience and Remote Sensing Magazine* 2017, *5*(4), 8–36. doi: 10.1109/MGRS.2017.2762307.

9 Distributed Deep Learning and Its Application in Geo-spatial Analytics

Nilkamal More and Jash Shah
Department of Information Technology,
Somaiya Vidyavihar University

V.B. Nikam
Department of Computer and Information Technology,
Veermata Jijabai Technological Institute

Biplab Banerjee
CSRE Department
Indian Institute of Technology

CONTENTS

9.1 INTRODUCTION

In today's era of the World Wide Web, much amount information is available. This information is available in images, audio, and video form. Geo-spatial data is a form of data that describes the events or location with coordinates [13]. Geo-spatial data also have temporal information associated with it. It includes explicitly vast sets of spatial data which is related to a location. It is available in any form and heterogeneous sources in various formats. It can include data about such as satellite optical or SAR imagery, sensor data, mobile data, drawings/ paintings, and data from social websites. Satellite images possess properties of big data. They are voluminous, generated at

DOI: 10.1201/9781003270928-12

high velocity [12], and available in various forms. In order to analyze this big data efficiently, we have to use high-performance computing technologies. To increase the efficiency of training deep learning models [1] that have billions of parameters that are free have been trained so as to learn features for massive amounts of data making use of GPU or a cluster of CPUs. Furthermore, what aids in training of deep learning models at a large scale are various approaches such as model parallelism and data parallelism [10]. For example, parallel training massive amounts of geo-spatial data which is not possible to be loaded into memory is then divided into a number of blocks. In contrast, over many CPU cores, back-propagation is performed. As a result, some amount of success has been gained in case of geo-spatial deep learning models in addressing the problems of geographic data management. Nevertheless, the size of existing models of deep learning that are trained for big data is heavily reliant on advancing high-performance computing systems such as GPU or cluster of CPUs [13]. Sadly, advances in computer performance are trailing significantly behind the rate of increase in big data. As a result, with the continued increase of data, substantially a greater number of large-scale deep learning models will be required. Depending on the prevailing methodologies and computer capability, such enormous models that have big data might no longer be efficiently trained. As a result, one potential approach to address this problem will be to create computing infrastructures and new frameworks of learning, such as distributed learning utilizing various frameworks based on Apache Spark [11]. It works on the principle of divide and conquers.

9.2 HIGH-PERFORMANCE COMPUTING (HPC)

High-performance computing (HPC) solves computationally intensive problems by utilizing supercomputers and parallel processing techniques [16]. HPC technology creates algorithms that are used for parallel processing by combining management [19]. High-performance computing is often used to solve complex issues and conduct research using simulation, computer modeling, and analysis. Through the parallel utilization of computer resources, HPC systems may provide sustained performance. Therefore, high-performance computing and supercomputing are phrases that are frequently used interchangeably.

9.2.1 Need for High-performance Computing

With a variety of real-world applications in every sector, a lot of opportunities are available for researchers to work in the domains of Artificial Intelligence and Machine Learning. However, we must deal with it in conventional deep learning applications.

a. **Big Data:**The ImageNet dataset contains a few terabytes of data [13]. Facebook users, for example, upload 800 million photographs each day. Deep learning models analyze all of these images in order to know more about the users.
b. **Parameter Storage:** Models are built up of hundreds of layers, each of which represents a few hundred to 2Bn parameters. It is equivalent to 0.1–8 GiB of storage space for the model (which resides in memory when in use, not on the hard drive). It should be noted that the training phase of the model often consumes more memory, making it even more challenging to fit the model on a single standard Personal Computer [10].
c. **Computation Power:** Labels tend to grow in size as tasks get more complicated. Typical image recognition systems, for example, categorize an item into one of 10–22 k categories, whereas face identification requires one label per individual. It takes weeks to train models with this many categories [10].

9.2.2 Parallel Computing

Parallel computing is a type of computational problem that makes use of several computing resources (Figure 9.1). A large problem is divided into tiny portions and worked on concurrently. Instructions

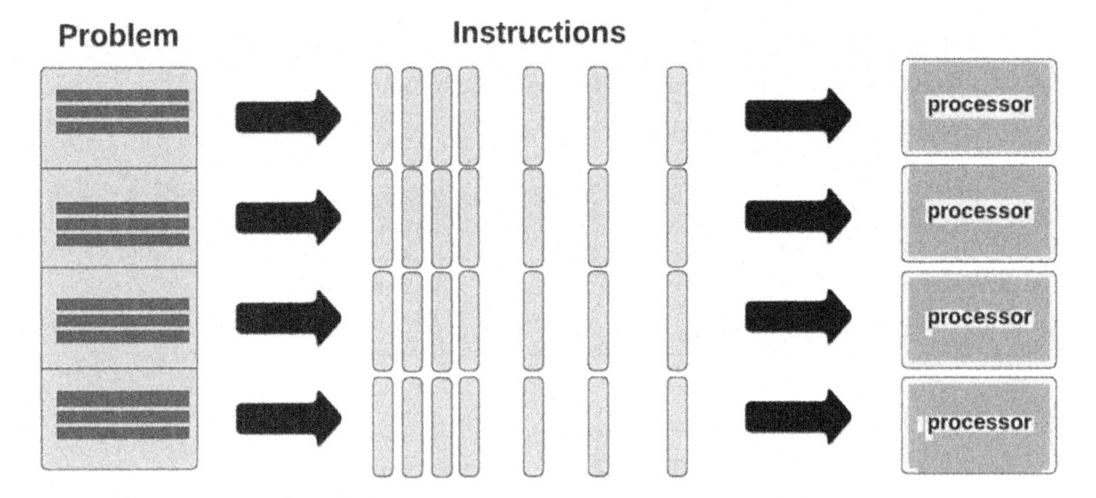

FIGURE 9.1 Parallel computing in which a large problem is divided into tinier portions.

further subdivide each small component, and each part is processed together on various processors. The speed-up must be ideally linear; if the number of processes is made twice then, it should reduce the duration to half. It is intrinsically difficult to parallelize some algorithms. For example, splitting a job and combining the results are activities that are linear and commonly carried out by a single process. The speed-up would be lower than linear if dividing and combining take up a large portion of the whole time of execution; at worst[4], the speed-up will be just marginally more rapid than if the job was completed serially [5,6]. There are various forms of parallelism, but Task Parallelism is most related to the task given in this thesis. A job is broken into sub-tasks, which are then assigned to specifically built hardware for execution [5]. Parallel computation gear varies; it can go from one processor with numerous cores to many independent computers connected to share tasks to massive racks, each led by specialized hardware and a network with great speed to link all of them [5]. Message Passing Interface (MPI) communication standard is used by parallel computing, which is now widely utilized in High-Performance Computing [15,16].

MPI represents a mechanism in distributed memory is a system which permits processes to interact with one-another despite the fact that they do not have immediate access to memory of one-another and direct communication is not possible. Processors may run in parallel and I/O can run concurrently on top of parallel systems, which is a feature of HPC hardware. According to the literature, this is essential for data-intensive applications like machine learning [23]; several purely parallel programming methods are available. Table 9.1 describes the most widely used pure parallel programming paradigms [3,17].

A multicore architecture is wherein the core logic of many processors is contained in one single physical processor. A single integrated circuit is used to package these processors. These single-chip integrated circuits are referred to as dying. The multicore architecture integrates numerous processing cores into one physical processor. The objective is to create a system that can execute more tasks simultaneously, hence boosting overall performance. This technique is most typically employed in multicore processors, which run two or more processing chips or cores simultaneously as a single system. Mobile devices, PCs, workstations, and services all employ multicore processors.

9.2.3 Distributed Computing

Distributed computing is the use of numerous computing resources linked by a network to address a single issue (Figure 9.2). Distributed Computing [17] is based on a loosely linked situation in which set of data and resources that belong to each processor are located remotely.

TABLE 9.1
Pure Parallel Programming Models Implementations

Implementation	Pthreads	OpenMP	MPI
Programming model	Threads	Shared memory	Message passing
System architecture	Shared memory	Shared memory	Distributed and shared memory
Communication model	Shared address	Shared address	Message passing for shared address
Granularity	Coarse or fine	Fine	Coarse or fine
Synchronization	Explicit	Implicit	Explicit or implicit
Implementation	Library	Computer	Library

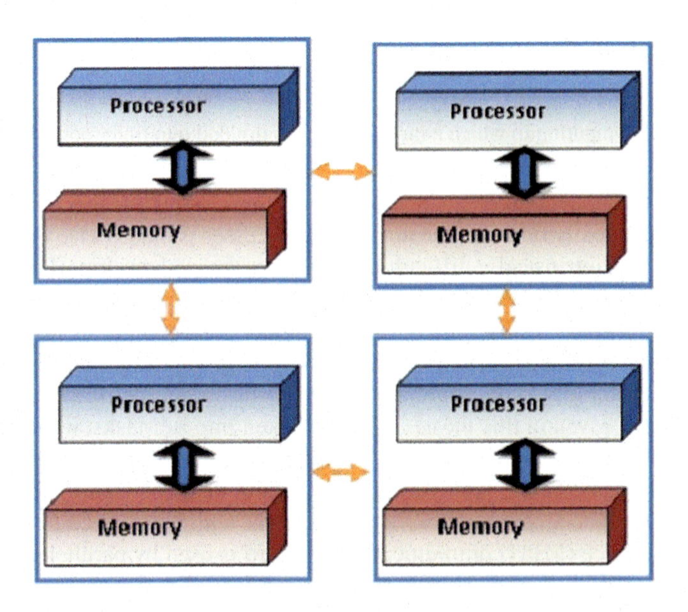

FIGURE 9.2 Distributed computing is illustrated.

This improves computational speed and efficiency. Distributed computing allows for the processing of large volumes of data in fewer time (Figure 9.3). Furthermore, distributed systems may interact and coordinate their actions with other network systems by sharing information and the state of their processes. Such distributed systems enable enterprises to operate a network of considerably smaller and less expensive computers instead of a single massive server with a larger capacity.

9.2.3.1 Distributed Computing for Geo-Spatial Data

Various geo-spatial big data platforms available for distributed computing are listed here. On Big data platforms, geo-spatial data is being progressively used [18]. As data-collecting techniques outstrip the data processing and computational capabilities of traditional geo-spatial platforms, it is clear that strategies for handling large geographic data have attracted and continue to garner significant attention. Many platforms have developed in the past like Spatial Hadoop [18], GeoSpark [21], Geomesa [20], and HIVE Spatial also implemented in Hadoop GIS [6]. A fast development has been seen in each of them, with several version releases of GeoSpark and Geomesa recorded in recent times. A comparison of these platforms based on performance on different jobs that are related to spatial data [22] resulted in benchmarking for available versions. Although most of the comparisons remain valid, the study looks out of date about a year after the GeoSpark Core was released.

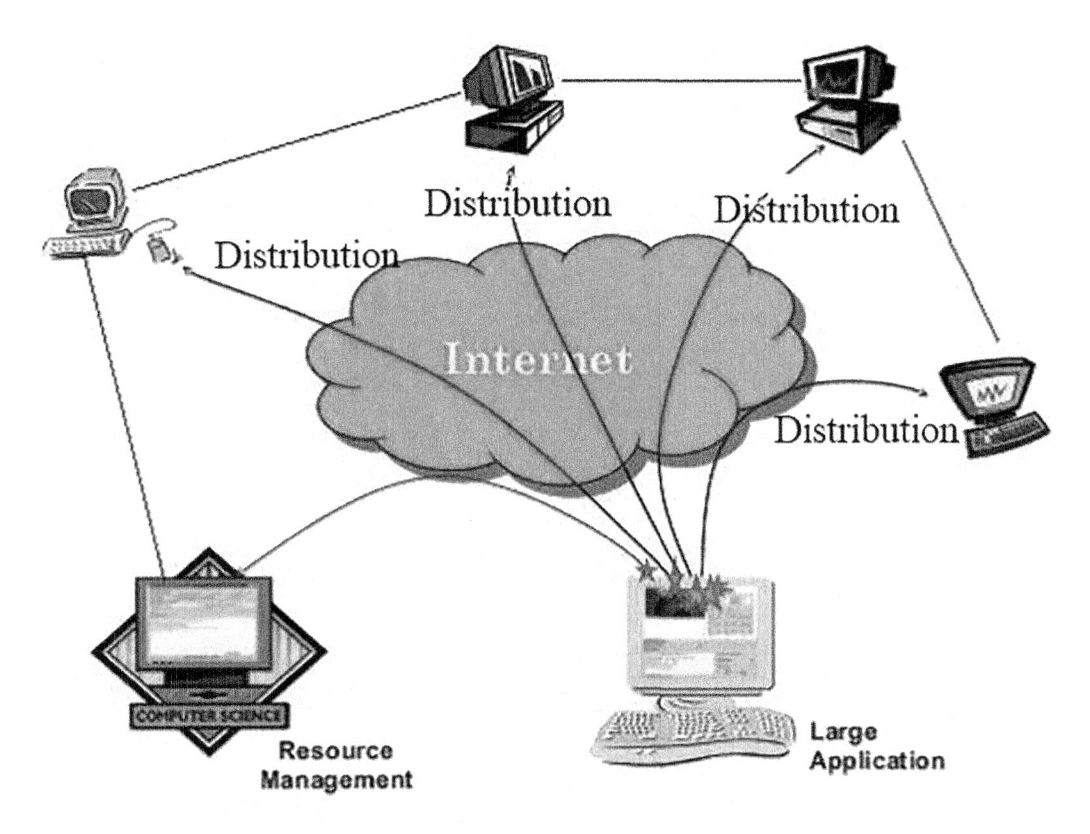

FIGURE 9.3 Distributed computing of large volumes of data.

Geomesa has also received considerable upgrades. According to the study, spatial join queries were hard to perform upon those platforms and frequently failed. The requirement for reindexing such systems was also shown when the data changes.

Table 9.2 shows differentiation among different geo-spatial distributed frameworks which are available as an open source.

Other parallels Haut et al. [10] evaluated feature and performance comparisons of massive spatial data frameworks and demonstrated that GeoSpark does not support specialized queries (such as ContainedBy and WithinDistance). Similarly, DATA Reply's performance review study [6] identified many flaws in the GeoSpark platform. As a result, the aforementioned research investigated several versions of spatial implementations on Spark platforms [8]. Similarly, there are a variety of ways to process geographic data on Hadoop. There is also oracle spatial and ESRI spatial frameworks, in addition to the open-source platform. However, GeoSpark and Spatial Hadoop remained the top two open-source platforms for our experimental assessment.

9.2.4 Challenges in High-performance Computing

Some of the challenges faced by researchers while adapting high-performance computing for handling geo-spatial data are as follows

a. **Adaptation of New Tools and Technologies:** Before advancing technologies, people used traditional CPU computing for analysis purposes [14]. So adapting to this new technology was a challenge for the researcher. As geo-spatial data is big data, handling this with high-performance computing requires special tools and techniques. So training needs to be given before introducing these technologies.

TABLE 9.2
Comparison of Geo-Spatial Distributed Frameworks

	Spatial Hadoop	Sedona (GeoSpark)	GeoMesa	HIVE Spatial
Machine Learning Support	With MAHOUT	Native SQL and MLlib library	Spark MLlib for Spatio-temporal fusion machine learning	Native SQL and Hivemall components
Framework Support	Hadoop framework	Spark framework	Spark framework	Hadoop framework
Supported Query Language	Pigeon-based query language	SQL-based query language	SQL-based query language	Hive Query Language (HQL) which is like SQL
Developed with Programming Language	Java language is used	Java and Scala languages are used	Scala language is used	Java language is used
Spatial Libraries Provided	Own implementation	JTS	JTS	Own implementation
Spatial Data Structures	Points and polygons	Polygons, points, lines	Polygons, points, lines	Polygons, points, lines
Spatial Operations Supported	Union and KNN queries	Union, spatial join intersection, difference, and KNN queries	Union, intersection, difference, and user defined functions	Union, spatial join, dynamic partition
Indexing Structures Used	R-tree index, R+-tree index, and grid file index	R-tree index and quad-tree index	Z-curve index and XZ-curve index	Bitmap indexing and compact indexing

b. **Massively Parallel Computers:** Programming complex massively parallel computing needs a different paradigm than traditional computing [15]. First, knowledge of parallel computing and tools used for processing needs to be researched. A variety of data exists, so data needs to standardize before applying high-performance computing techniques.
c. **Expensive and Time Consuming:** In high-performance computing often needs CUDA-like environment. CUDA [14] programming is different from other programming languages. Limited resources are available to learn CUDA-like environments. Therefore, it is often expensive and time-consuming.
d. **Special Infrastructure:** High-performance computing [14,15] can be achieved with GPU or FPGA-like additional hardware. In addition to that, the server machines should be accompanied by tools and programming environments compatible with hardware configurations.

9.3 DISTRIBUTED DEEP LEARNING FOR GEO-SPATIAL ANALYTICS

Different technologies in deep learning are evolving, which can be used for geo-spatial analytics. In this topic, we will see parallel and distributed deep learning [17].

9.3.1 Distributed Deep Learning

Data is distributed in distributed deep learning approach. Parallelization in distributed deep learning can be achieved using three possible ways [17]. Within distributed implementation of deep learning applications, one of these three approaches can be employed.

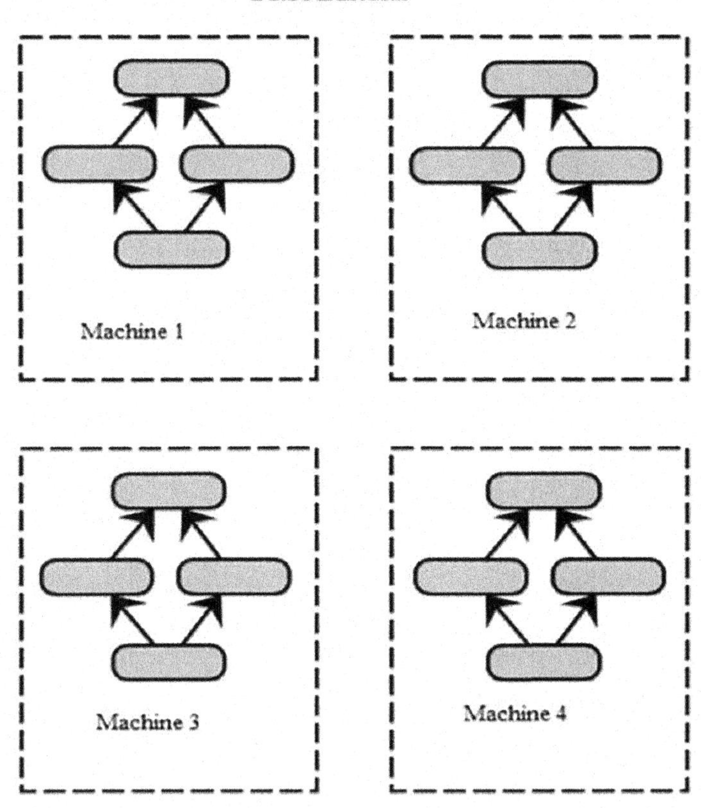

FIGURE 9.4 Data parallelism.

a. **Data Parallelism:** One of the parallelization techniques is data parallelism (Figure 9.4). It partitions the data into subparts. The number of partitions in data parallelism is equal to the number of machines. These machines are called as the worker machine [17]. In the next step, each worker has an independent partition, and it operates on that data. As multiple nodes are processing the data in parallel, more amounts of data can be processed as compared to a single node. Higher throughput can be achieved with distributed parallel computing [17]. The typical example of data parallelism is DistBelief [3]. It is a system developed by Google.
b. **Model Parallelism:** Model parallelism is a bit more complex than data parallelism. In model parallelism, data is not divided into multiple partitions (Figure 9.5). But model parallelism tries to divide the machine learning model. It distributes the process to numerous machines. These machines are called as computational workers [15]. A significant amount of speedup is achieved with the model parallelism.
c. **Hybrid Approach:** Data parallelism and model parallelism is combined in a hybrid approach of parallelism to get advantages of both these techniques. In this approach, data partitioning is used for some specific parts of a neural network, and model partitioning is used to achieve correctness in some parts. Most of the processing is done in convolutional layers in AlexNet. AlexNet neural network the parameters are observed in the fully connected layers. A speedup in the processing up to six times can be achieved for eight GPUs over one using the hybrid approach. Other hybrid implementations have had similar outcomes. In other cases, the hybrid technique resulted in a 3.1 times speedup for four GPUs. An asynchronous approach of training of neural network training on CPUs is done with

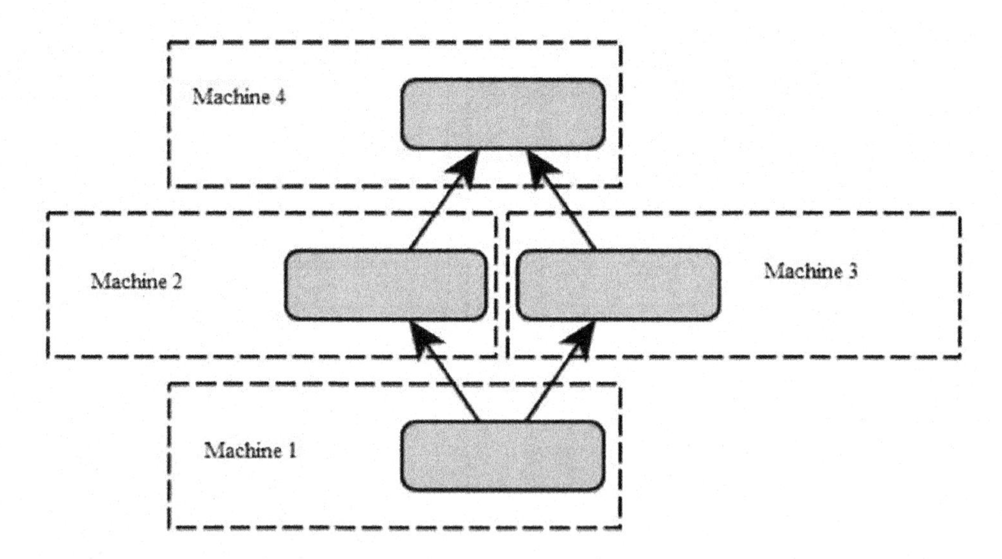

FIGURE 9.5 Model parallelism.

AMPNet. Fine-grained model parallelism is achieved with it. The recurrent, tree-based and graph-based neural networks are available with AMPNet. It shows very diverse features. It allows for changing sample lengths and a dynamic control flow. Data parallelism, model parallelism and hybrid approach are used in DistBelief [3]. It is a distributed deep learning system. The training process is carried out on multiple copies of the model in parallel. Every copy of the model is trained on different samples.

9.4 APACHE SPARK FOR DISTRIBUTED DEEP LEARNING

Apache Spark is among the most powerful engines for managing large amounts of data (Gif 6). It is also observed that it is faster than the Map-Reduce approach of distributed computing. The main reason for speedier processing is that Spark runs on random access memory, a primary memory. So, the processing is done much faster than on a hard disk (Figure 9.6).

Spark is used more popularly for multiple things, like implementing distributed SQL, creating pipelines of data, database ingesting, modeling machine learning algorithms, graph processing, data streams generation, and much more.

Resilient Distributed Dataset (RDD) is one of the important features of Apache Spark (Figure 9.7). The RDD can be executed in parallel and is a system of elements that can tolerate faults. RDDs can be generated by parallelizing the collection in the driver program. It can also be referenced as a dataset in the external system of storage of data, for example, a shared filesystem like Hadoop distributed file system, HBase, or a data source which offers a Hadoop Input Format.

In Apache Spark, all transformations are performed very slowly. Instead of generating results immediately, they memorize the modifications applied to the dataset. The transformation operations are performed when a result is sent to the main program. This is one of the essential features of Spark. Each transformed RDD can be calculated when action is run on it (Figure 9.8). However, RDD may be stored persistently in memory using persist method. For faster access, Apache Spark keeps the elements on the cluster. The support is available in Spark for persisting RDDs on disk. RDDs can also be replicated across multiple nodes.

Apache Spark uses a Data Frame (Figure 9.9). Dataset is organized into named columns in a data frame. Conceptually data frame is similar to a relation or table in a relational database. Spark's data

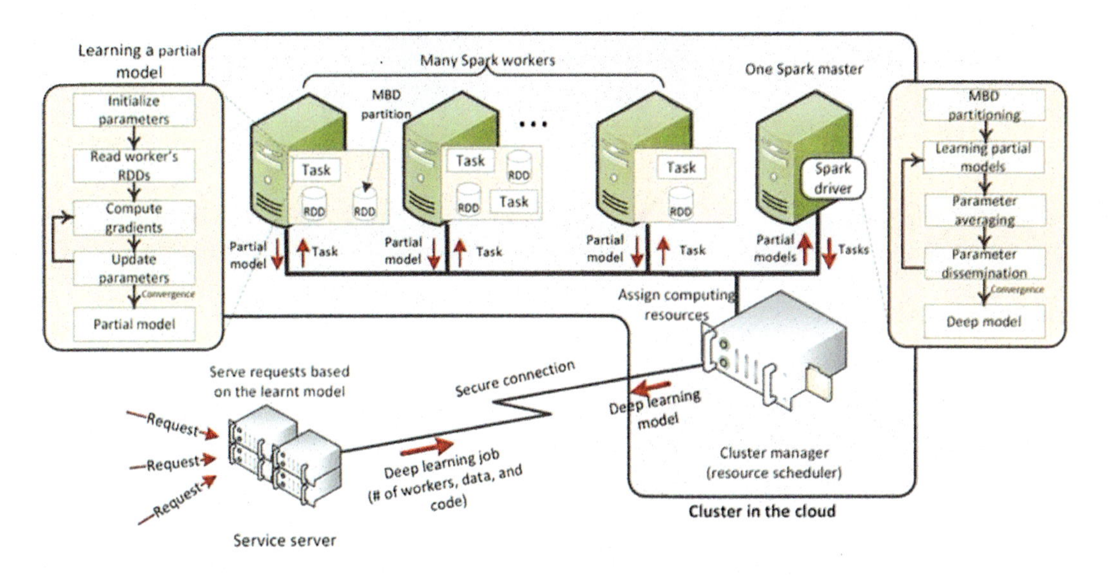

FIGURE 9.6 Apache Spark.

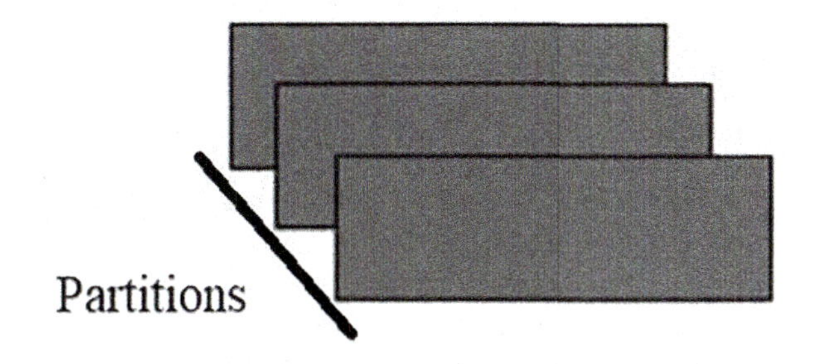

FIGURE 9.7 Resilient Distributed Dataset (RDD), an important feature of Apache Spark.

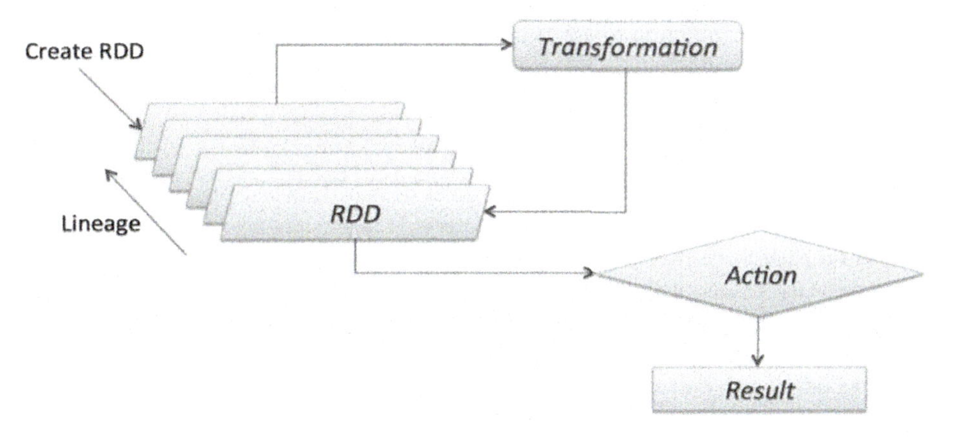

FIGURE 9.8 Illustrates how the transformed RDD can be calculated when action is on it.

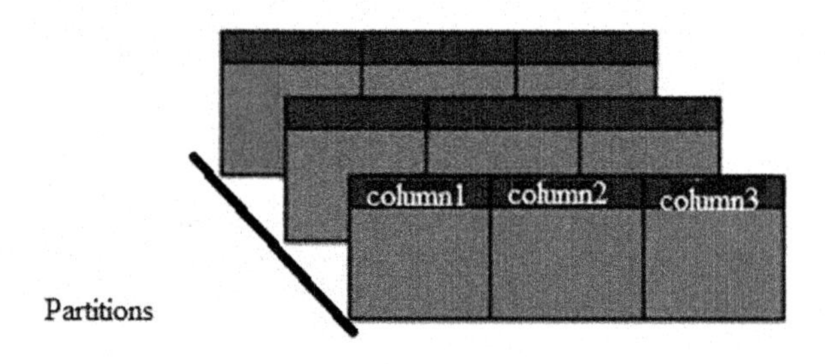

FIGURE 9.9 Data frames.

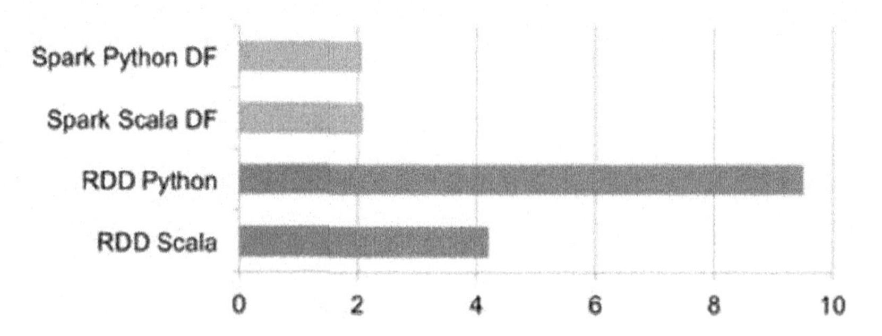

FIGURE 9.10 Optimization of data frames in Spark.

frame is the same as a data frame in R or Python. But data frames in Spark are richly optimized. Data Frames are generated from various sources such as structured files, Hive tables, databases, or previously created RDDs. Data frame API in Spark to make it easy to work with Data. But these data frames have various other advantages. Data frames can be spread across a cluster. They are optimized and work with data in a variety of forms (Figure 9.10).

Optimization in data frames is achieved with the help of Catalyst. A Catalyst is like a wizard. It accepts queries and generates an optimized plan to support distributed computing. Figure 9.10 explains the optimization of data frames in Spark.

The process of optimization in Spark is complex. But it is transparent to the programmer but helps in a distributed processing.

9.4.1 Distributed Hyper-Parameter Optimization

Optimization of hyper-parameters is one of the critical steps in deep learning technologies. Elephas carry out hyper-parameter optimization using hyperas [13]. Results are collected in Spark using several trails on each Spark worker, and the model with the best results is returned. hyperopt in Distributed mode uses MongoDB. It isn't easy to configure as it is error-prone at the implementation time. Random search is the only available optimization algorithm. Table 9.3 gives information about the Apache spark based distributed deep learning frameworks.

Data bricks have created deep learning pipelines. It is an open-source library. High-level APIs for executing scalable deep learning using Apache Spark on Python are used to achieve scalability.

a. Easy-to-use APIs are provided that allow deep learning in very less lines of code which is similar to Spark and MLlib.
b. It emphasizes ease of access and integration over performance.
c. The Apache Spark's inventors developed it; therefore; it is more likely than others to be merged as an official API.

TABLE 9.3
Apache Spark-Based Distributed Deep Learning Frameworks

Name of Framework	Support	Features
Elephas [3]	It is an extension of Keras. It supports deep learning models with distributed computing. Data-parallelism is supported for implementing deep learning. It also works with Spark	Applications supported by Elephas are: i. Distributed training with ensemble models. ii. Training the models with Data-parallel approach.
CERN distributed Keras [7]	It works with spark as well as Keras. Data-parallelism is supported	Applications supported by Distributed Keras are: i. Innovative distributed optimizer ii. Applications which require reduced training time.
Qubole [6]	The works with Keras and Dist-Keras. This framework is used to achieve data parallel model approach on the Apache Spark	Applications supported by Qubole are: i. Cross-validating deep neural networks on Spark ML Pipeline. ii. It uses an application-specific parameter grid.
Yahoo! Inc.: TensorFlow on Spark [7]	It works on Spark. Distributed deep learning is achieved on a cluster of GPUs and CPU servers. Model and data parallelism are possible. But it is not supported by the Windows operating system	It is useful for applications which require direct server-to-server communication
Intel corporation: BigDL [2]	Data parallelism is supported but not Model parallelism	It is used to develop applications on big data platforms in industries
Deepspark [3]	It accepts input from Caffe and it is also compatible with network structures and designs	It is suitable for applications which require faster training and communication
ChainerMN [5]	It supports only Data parallel approach	It is useful for applications i. which can speed up the deep learning process such as ResNet-50 model to the ImageNet dataset. ii. It supports 128 GPUs with the 90% of parallel efficiency.

d. It is developed in Python to work with all of its well-known libraries. It currently employs the deep learning capabilities of TensorFlow and Keras, the two most popular libraries at the time.

9.4.2 Deep Learning Pipelines

The deep learning pipeline is a significant feature in distributed deep learning. So here, we will use this library for distributed deep learning [17]. The diagram below shows MLIB Pipeline, which takes images as input. Then the preprocessing is carried out to remove noise. A deep learning image featurizer is used to extract features, and then the logistic regression model is used for the prediction. Figure 9.11 explains the MLIB pipeline.

a. **Less LOC**: Deep learning pipeline provides easy-to-use APIs. So, it allows us to implement deep learning with fewer lines of code as compared to Spark and Spark MlLib.
b. Deep learning models can be implemented without affecting the performance.
c. The developers of Apache Spark have created it. So, it is also an official API.
d. Python is used to write the deep pipelines. It uses popular libraries of python.

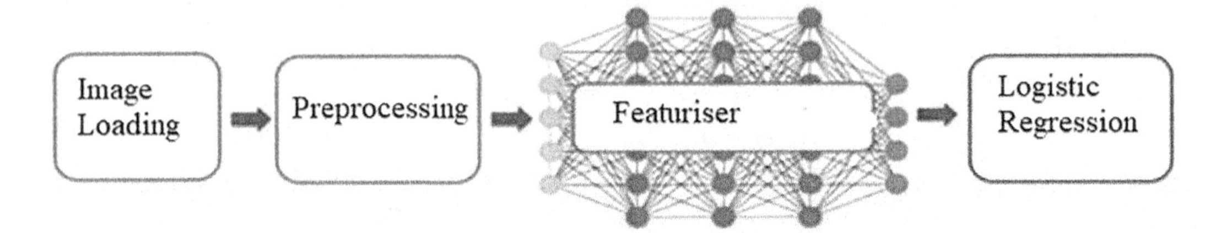

FIGURE 9.11 MLIB pipeline.

e. It is rich with power of TensorFlow and Keras. Deep Learning Pipelines are built on top of Apache Spark's ML library. The steps involved in implementing deep learning are as follows.

Step 1: Loading an image.

Step 2: Applying transfer learning where we use pre-trained models with eights as transformers.

Step 3: Scale the Deep Learning models.

Step 4: Tuning of Distributed hyper-parameter is used.

Step 5: Deploying models in Data Frames and SQL.

9.4.3 Apache Spark on Deep Learning Pipeline

Transfer learning on images can be done on with the help of deep learning pipelines. It provides utilities to perform it. One of the advantages is it can be done with fewer lines of code. Because of special APIs and utility, transfer learning is done faster. The last layers of a pre-trained neural network are taken out by the deep image featurizer. The output from former layers is used in Logistic Regression (LR). LR is a fast and straightforward method. So, this training of transfer learning integrates with significantly fewer images than building a whole new deep learning model. It enables running models that are pre-trained for transfer learning with a distributed approach on Spark. It is available in batch as well as streaming data processing. It contains top-rated models. It enables users to perform deep learning without much training a model. The prediction of the model is carried out with the functionality of Spark. Built-in models can be used with Keras plug-in and Graphs in Tensor Flow in a Spark prediction pipeline. So, a huge amount of data is handled by single-node in a distributed manner.

9.5 APPLICATIONS OF DISTRIBUTED DEEP LEARNING IN REAL WORLD

Distributed deep learning is used in the development of applications to address the real-world problems [24]. Some of the applications of developed using the distributed framework of Apache spark are Tree canopy Assessment [12], Land-use land cover classification [13], Spectral unmixing [9], and satellite image interpretation [10]. We are exploring building framework detection as a case study of distributed deep learning. Apache spark is used as a distributed framework for implementation purposes. Worldview 2 and Worldview 3 dataset is used for building detection. The hyper-parameter values used are learning rate of 0.002, a batch size of 6, and Adam as an optimizer [11]. Figure 9.12 shows the building detected from a satellite image.

This research is carried out on various regions to check scalability and efficiency of the algorithm developed using Apache spark.

The diagram shown in Figure 9.13 demonstrates the different steps involved in distributed deep learning using Apache Spark [13]. The first data is split into smaller chunks. It is given to each data node for feature extraction. So, the feature extraction and training process is carried out in parallel. Feature extraction and training are the most critical yet time-consuming operations in the

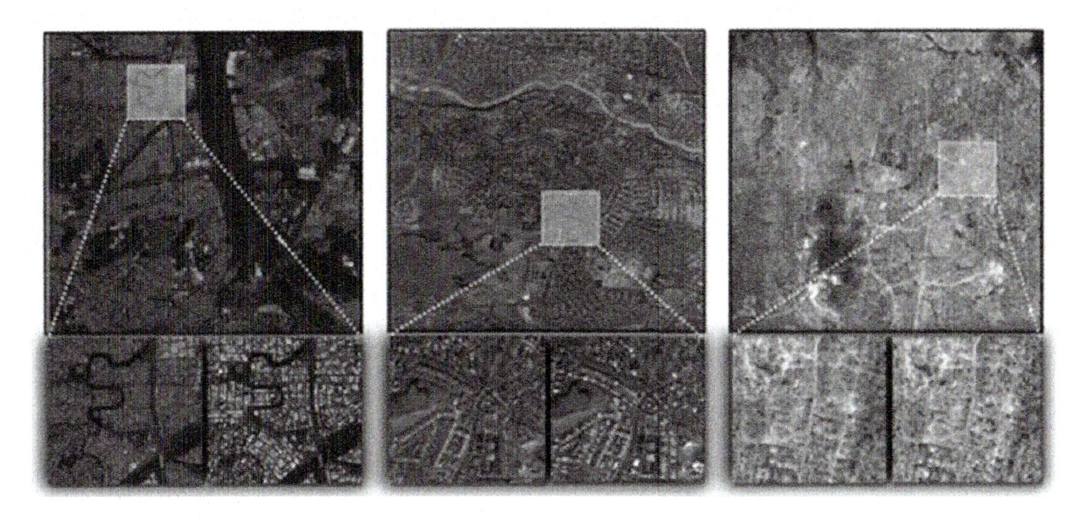

FIGURE 9.12 Building detected from a satellite image.

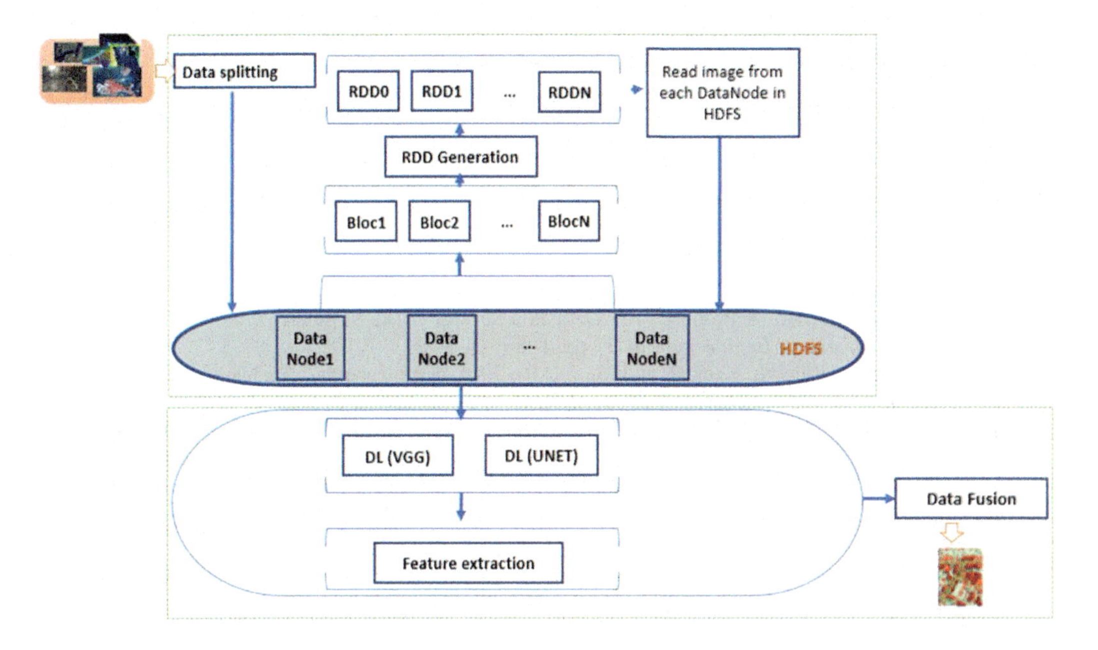

FIGURE 9.13 Steps involved in distributed deep learning using Apache Spark.

classification process. But, as it is done parallel in Apache Spark, it reduces the time required for the execution.

9.6 CONCLUSION SUMMARY AND PERSPECTIVES

Today, deep learning is perhaps the most widely used machine learning topic. Deep learning models have indeed been prominent in various domains in recent times, including computer vision, satellite image processing, recommendation systems, health-care systems, and processing of natural language. The vast generation of data provides a sufficient number of training objects, which aids deep learning performance. In addition, powerful devices of computation and architectures, like GPUs and clusters of distributed computing, facilitate the training of various deep learning models for learning of big data features. Deep learning models are now able to take care of a considerable number of parameters, generally millions, and many training objects. While extensive data provides a sufficient number of training items, it also poses a difficulty to deep learning. As a result, several models for considerable learning of data have been created in recent years. An overview of big data deep learning models and distributed deep learning has been presented. The volume, variety, veracity, and velocity these define big data. It denotes a large amount of data, diverse sorts of data, day today life data, and reduced data. As a result, we describe deep learning methods for learning of big data from four perspectives. We looked at large-scale deep learning models. In case of heterogeneous data, use of multi-modal deep learning models can be made. Deep computation models and incremental deep learning models are used for real-time data. Reliable deep learning models are used for data of lower quality. So, we can use distributed deep learning for large amounts of data in detail.

REFERENCES

1. Pouyanfar, S., Sadiq, S., Yan, Y., Tian, H., Tao, Y., Reyes, M.P., Shyu, M.L., Chen, S.C. and Iyengar, S.S., 2019. A survey on deep learning: Algorithms, techniques, and applications. *ACM Computing Surveys (CSUR)*, 51(5), p. 92.
2. Dai, J.J., Wang, Y., Qiu, X., Ding, D., Zhang, Y., Wang, Y., Jia, X., Zhang, C.L., Wan, Y., Li, Z. and Wang, J., 2019. Bigdl: A distributed deep learning framework for big data. In *Proceedings of the ACM Symposium on Cloud Computing,* Santa Cruz, CA (pp. 50–60). ACM.
3. Dean, J., Corrado, G., Monga, R., Chen, K., Devin, M., Mao, M., Ranzato, M.A., Senior, A., Tucker, P., Yang, K. and Le, Q.V., 2012. Large scale distributed deep networks. In S.A. Solla, K.-R. Müller and T.K. Leen (Eds.), *Advances in Neural Information Processing Systems* (pp. 1223–1231). MIT Press: Cambridge, MA.
4. Ben-Nun, T. and Hoefler, T., 2019. Demystifying parallel and distributed deep learning: An in-depth concurrency analysis. *ACM Computing Surveys (CSUR)*, 52(4), p. 65.
5. Akiba, T., Fukuda, K. and Suzuki, S., 2017. ChainerMN: Scalable distributed deep learning framework. arXiv preprint arXiv:1710.11351.
6. Al-Hakeem, M.S., 2016. A Proposed Big Data as a Service (BDaaS) model. *International Journal of Computer Sciences and Engineering*, 4(11), pp. 1–6.
7. Johnsirani Venkatesan, N., Nam, C. and Shin, D.R., 2019. Deep learning frameworks on apache spark: A review. *IETE Technical Review*, 36(2), pp. 164–177.
8. Zaharia, M., Xin, R.S., Wendell, P., Das, T., Armbrust, M., Dave, A., Meng, X., Rosen, J., Venkataraman, S., Franklin, M.J. and Ghodsi, A., 2016. Apache spark: A unified engine for big data processing. *Communications of the ACM* 59(11), pp. 56–65.
9. Plaza, A., Plaza, J., Paz, A. and Sanchez, S., 2011. Parallel hyperspectral image and signal processing applications corner. *IEEE Signal Processing Magazine* 28(3), pp. 119–126.
10. Haut, J.M., Paoletti, M.E., Moreno-Álvarez, S., Plaza, J., Rico-Gallego, J.-A. and Plaza, A. 2021. Distributed deep learning for remote sensing data interpretation. *Proceedings of the IEEE* 109(8), pp. 1320–1349.
11. Lunga, D., Gerrand, J., Yang, L., Layton, C. and Stewart, R. 2020. Apache spark accelerated deep learning inference for large scale satellite image analytics. *IEEE Journal of Selected Topics in Applied Earth Observations and Remote Sensing* 13, pp. 271–283.
12. Yang, H.L., Yuan, J., Lunga, D., Laverdiere, M., Rose, A. and Bhaduri, B. 2018. Building extraction at scale using convolutional neural network: Mapping of the United States. *IEEE Journal of Selected Topics in Applied Earth Observations and Remote Sensing* 11(8), pp. 2600–2614.
13. Chebbi, I., Mellouli, N., Farah, I.R. and Lamolle, M. 2021. Big remote sensing image classification based on deep learning extraction features and distributed spark frameworks. *Big Data and Cognitive Computing* 5(2), p. 21.
14. Lu, Y. 2020. Deep learning for remote sensing image processing. Doctoral dissertation, Old Dominion University.
15. Fan, Z., Qiu, F., Kaufman, A. and Yoakum-Stover, S. 2004. GPU cluster for high performance computing. In *SC'04: Proceedings of the 2004 ACM/IEEE Conference on Supercomputing*, Pittsburgh, PA (pp. 47–47). IEEE.

16. Chen, S., He, Z., Han, X., He, X., Li, R., Zhu, H., ... Niu, B. 2019. How big data and high-performance computing drive brain science. *Genomics, Proteomics & Bioinformatics*, 17(4), 381–392.
17. More, N., Galphade, M., Nikam, V. B. and Banerjee, B. 2021. High-performance computing: A deep learning perspective. In A. Suresh and S. Paiva (Eds.), *Deep Learning and Edge Computing Solutions for High Performance Computing* (pp. 247–268). Springer: Cham.
18. More, N.P., Nikam, V.B. and Sen, S.S. 2018. Experimental survey of geo-spatial big data platforms. In *2018 IEEE 25th International Conference on High Performance Computing Workshops (HiPCW)*, Bengaluru, India (pp. 137–143). IEEE.
19. Riedel, M., Cavallaro, G. and Benediktsson, J.A. 2021. Practice and experience in using parallel and scalable machine learning in remote sensing from HPC over cloud to quantum computing. In *2021 IEEE International Geoscience and Remote Sensing Symposium IGARSS*, Brussels, Belgium (pp. 1571–1574). IEEE.
20. Hughes, J.N., Annex, A., Eichelberger, C.N., Fox, A., Hulbert, A. and Ronquest, M. 2015. Geomesa: A distributed architecture for spatio-temporal fusion. In M.F. Pellechia, et al. (Eds.), *Geo-Spatial Informatics, Fusion, and Motion Video Analytics V* (Vol. 9473, pp. 128–140). SPIE: Bellingham, WA.
21. Yu, J., Zhang, Z. and Sarwat, M. 2019. Spatial data management in apache spark: The geospark perspective and beyond. *GeoInformatica*, 23(1), pp. 37–78.
22. De Oliveira, S.S., Rodrigues, V.J. and Martins, W.S. 2021. SmarT: Machine learning approach for efficient filtering and retrieval of spatial and temporal data in big data. *Journal of Information and Data Management*, 12(3), pp. 273–289.
23. Sabek, I. and Mokbel, M.F. 2021. Machine learning meets big spatial data. In *2021 22nd IEEE International Conference on Mobile Data Management (MDM)*, Toronto, ON (pp. 5–8). IEEE.
24. Goodman, S.M. 2021. Filling in the gaps: Applications of deep learning, satellite imagery, and high performance computing for the estimation and distribution of geo-spatial data. Dissertations, theses, and masters projects, The College of William and Mary.

10 High-Performance Computing for Processing Big Geospatial Disaster Data

Pratyush V. Talreja, Surya S Durbha, Rajat C. Shinde, and Abhishek V. Potnis
Centre of Studies in Resources Engineering, Indian Institute of Technology Bombay, India

CONTENTS

10.1 INTRODUCTION

With the boom in the sensor industry, data acquired for disasters is also growing at a rapid pace. There are different kinds of sensors that can be used to collect disaster data. The most used sensors are the imaging sensor and LiDAR sensor. The data acquired by imaging sensors is improving in quality considering the advancements in the sensor technology and thus the size of the data obtained from the imaging sensors is huge. LiDAR is known for its volume as it contains depth information. The exponential increase of data also gives rise to the increasing complexity of disaster data, like

DOI: 10.1201/9781003270928-13

the higher dimensionality characteristics of the data. For current scenarios, real-time disaster data has a lot more to it than it appears at first, and extracting the usable information in an efficient manner leads to a huge computing difficulty, such as analyzing, aggregating, and storing data collected remotely (Rathore et al., 2015). With such a large amount of data comes the issues of computational capability and efficiency to handle the huge data. CPU gives very less efficiency for huge data resulting in the worst time complexity. The solution to this problem is Graphics Processing Units (GPUs). Computing innovations based on massively parallel General Purpose Graphics Processing Unit (GPGPU) have given personal computers incredible computing power. The use of a graphics processing unit (GPU) in conjunction with a CPU to speed scientific, analytical, engineering, consumer, and corporate applications is known as GPU-accelerated computing (Parker, 2017). GPU accelerators are being used in government labs, colleges, companies, and small and medium organizations all over the world to power energy-efficient data centers. GPUs are used to accelerate applications on a variety of platforms, including vehicles, mobile phones and tablets, drones, and robotics. With the introduction of GPU technology, it is now feasible to perform computationally heavy (i.e., time-consuming) operations in real or near real-time.

Various types of data collected from a range of sensors are important in a number of areas. Wireless sensing networks (WSN), which can comprise a range of agro-meteorological sensors, can provide data in the agriculture domain. In situ (spectroradiometers, chlorophyll, LAI, fluorescence, sap, and other measuring devices) as well as airborne and spaceborne sensors could be used to make remote-sensing-based measurements (Talreja et al., 2018). We may also get various types of data from web sources. Because of the high dimensionality of data, the total quantity of data created is enormous, and processing such a big volume of data is computationally costly. It is desirable to distribute information quickly in agriculture for several decision-making reasons. However, due to a lack of high-performance computing (HPC) infrastructure on a mobile platform, this capability is currently limited.

The volume of disaster data has expanded tremendously with the introduction of satellite remote sensing, global navigation satellite systems (GNSS), and aerial surveys utilizing photographic cameras, digital cameras, sensor networks, radar, and LiDAR. The main issue that needs to be addressed is the high computational complexity. On desktop computers, general-purpose computing on graphics processing units is a relatively low-cost way to achieve high computational throughput (Talreja et al., 2018). Traditional algorithms cannot be run on a GPU due to the design of GPUs being fundamentally different from that of CPUs. Current advances in computing technology and paradigm shifts have discredited long-held assumptions regarding algorithm performance.

10.2 RECENT ADVANCES IN HIGH-PERFORMANCE COMPUTING FOR GEOSPATIAL ANALYSIS

With the introduction of NVIDIA CUDA, a C programming language extension that allows GPUs to be programmed in a general-purpose fashion (GPGPU), GPUs can now be used in a wide range of science and engineering applications. The growing GPU capabilities have attracted many scientists and engineers, especially scientists in remote sensing sectors, to employ it as a cost-effective high-performance computing platform. Clusters, distributed networks, and specialized hardware devices are examples of high-performance computing (HPC) infrastructure that can speed up computations linked to information extraction in remote sensing (Lee et al., 2011).

Geospatial big data is a subset of big data that includes geographic information. Location information is essential in the big data era since the majority of data today is essentially geographical and is collected via widely used location-aware devices like satellites, GPS, and environmental observations. Climate science, disaster management, public health, precision agriculture, and smart cities are just a few of the sectors where geospatial big data might help advance scientific discoveries (Li, 2020). In fact, the usage of HPC systems in spatial big data processing, such as picture categorization, land use change analysis, and urban expansion modeling, has grown in popularity in recent years. When processing multi-temporal land cover data with greater resolution using more advanced processes, a very significant computing burden is often unavoidable when doing land-use change analysis over massive geographical data on a national scale. A theoretical technique to capture the computing loads can be utilized to drive the two necessary steps: geographic breakdown and task scheduling, in particular, to tackle this geospatial challenge (Kang et al., 2018).

10.3 DAMAGE ASSESSMENT AND SOURCES OF DISASTER DATA

Assessment of damages caused due to extreme events can be done using approaches such as change detection, crowd-sourcing, and Machine Learning techniques. Table 10.1 shows the classification of damage because of the earthquake using European Macro-seismic Scale (EMS-98). There are different types of damages that occur due to earthquakes which include damages to buildings or other built-up structures, damages to infrastructure such as roadways and water lines, and secondary damages caused due to events triggered by an earthquake.

TABLE 10.1
Classification of damage because of Earthquake

EMS-98 Intensity	Felt	Magnitude	Building Damage
I	Not felt	2	
II-III	Weak	3	
IV	Light	4	
V	Moderate		
VI	Strong	5	
VII	Very strong	6	
VIII	Severe		
IX	Violent	7	
X+	Extreme		

Source: Adapted from Christina Corbane (2011).

The aim of collecting disaster data is to utilize different types of sensors mounted on Aerial platforms, ground platforms, satellites, etc., to acquire the characteristics of the object using the variations of electromagnetic radiation (EMR) reflected by the targeted object which forms the basis of remote sensing. There are different sources from which we can acquire disaster data. The following are the major sources of disaster data.

10.3.1 Images

Remotely sensed images are considered to be one of the most vital sources of collecting disaster data. Almost all remote sensing applications make use of images to perform analysis. Images can be obtained from a variety of sources. The two common sources of acquiring remotely sensed images are Airborne and Satellite.

10.3.1.1 Airborne

Aerial photography has two applications: (1) collecting detailed measurements from aerial images in order to prepare maps; and (2) determining land use, environmental conditions, and geologic information (outcrop locations, lineation, etc.) (Baskaran & Tamilenthi, 1990). It's important to remember that aerial images aren't maps. Maps are correct directionally and geometrically because they are orthogonal representations of the earth's surface.

Aerial photographs often have a lot of radial distortion that needs to be fixed. Aerial photography is divided into three categories: black and white, color, and infrared.

10.3.1.2 Satellite

A variety of national and international government and business agencies acquire satellite imagery. The majority of this data is copyright controlled, therefore individual users or institutions must negotiate access. Through collaboration with NASA and NASA-funded universities, free access is possible. The instructor must examine the value of this data in the instructional context because data products cover the useable electromagnetic spectrum in a variety of resolutions (Baskaran & Tamilenthi, 1990). Free or low-cost data frequently has a resolution cell size of more than 100 m/pixel, therefore this form of imagery often provides a broad regional view, but may not provide comprehensive insight into geologic features on the ground. Passively obtained data and actively gathered data are the two basic categories of satellite-based remotely sensed data. Figure 10.1 shows the Haiti earthquake 2010 GeoEye -1 satellite imagery obtained from DigitalGlobe.

10.3.2 LiDAR

LiDAR stands for Light Detection and Ranging and is one of the remote sensing methods which is used to generate 3D point clouds that contains information about latitude, longitude, and height (or depth). LiDAR is an altimeter system that is known for calculating distance by time of travel of light pulse. LiDAR uses light in the form of a pulsed laser to measure the depth information and when this light hits a surface or any area on the ground, it is reflected back to the receiver sensor which records the range (Favorskaya & Jain, 2017). When this range of information is clubbed with the location information, it generates detailed information which is termed as a point cloud.

Light energy is a collection of photons so as the photon moves toward the surface of the earth, they hit some surfaces such as buildings and trees, and reflects back to the sensor. Some light moves toward the ground considering the small shape of the object or some gaps in the object so multiple reflections are possible from one pulse of light. Figure 10.2 shows the LiDAR dataset related to the Haiti earthquake which was obtained from OpenTopography.

A point cloud is a collection of points $C = \{p_1, p_2, \ldots, p_n\}$ and these points represent the structure of the 3D object. At the minimum, each point contains longitude, latitude, and elevation, i.e., x, y, and z information. There can be other information such as scan angle, normal, classification,

FIGURE 10.1 Haiti earthquake satellite imagery.

FIGURE 10.2 LiDAR point cloud - Haiti earthquake.

intensity, R, G, B represented by a single point in a point cloud. The main reason for using point cloud data is because the elevation information obtained from the point cloud can be used to study the change detection in a 3D environment and thus help in detecting damaged buildings. There are different types of systems where LiDAR sensors can be used based on the application. Airborne LiDAR is a LiDAR unit that is installed in a plane or UAV or any other flying object. Airborne LiDAR can be used for urban applications, disaster damage mapping, and assessment, forestry. Terrestrial LiDAR, as the name suggests is a Ground-based system. It can be mounted on a stationary object or a moving vehicle. Autonomous driving, road surveys, and city surveys are some of the applications where Terrestrial LiDAR comes into play. Bathymetric LiDAR is used in conjunction with airborne LiDAR or terrestrial LiDAR with the addition of a green pulse that has the ability to pass through the water and back to the receiver. It is used for coastal engineering majorly.

10.4 KEY COMPONENTS OF HIGH-PERFORMANCE COMPUTING

10.4.1 Domain Decomposition

Domain decomposition in HPC uses the divide-and-conquer technique and breaks a big problem into a set of small concurrent problems and uses multiple cores to process the data in parallel. Domain decomposition considers the data as a problem and divides it into small chunks to be processed. For example, consider a geospatial problem of creating a heat map of COVID-19 spread around the world, the domain decomposition problem will divide entire data into small chunks and process each chunk in parallel to obtain a heat map of the world in an efficient way. Processing of data using parallel implementation accelerates geospatial analysis by an order of magnitude.

Usually, domain decomposition is used to break down distinct dimensions of data based on the fact that geospatial data has a five-dimensional (5D) tuple (*X*, *Y*, *Z*, *T*, and *V*) (Li, 2020). One of the most vital factors that need to be considered while performing domain decomposition is the dependence among subdomains. For example, while performing IDW interpolation in parallel, neighboring decomposed chunks depends on each other so they need to be considered while processing the IDW interpolation in parallel. In a similar way, while rasterizing a LiDAR point cloud, there are no dependencies among the subdomains and thus it is much faster as the operations are performed in parallel without depending on the data from the subdomain.

10.4.2 Spatial Indexing

As the data is divided into 'n' chunks for processing in parallel, the multiple cores must concurrently retrieve these chunks of data. For this task, we need spatial indexing. Spatial indexing is used to find and access the data quickly (Li, 2020). The efficiency of the concurrent spatial data traversing is dependent on the performance of the spatial indexing. Thus, spatial indexing directly affects the performance of parallel data processing.

There are different data structures that are used for spatial indexing. The most famous ones are quad-tree, R-tree, and their variants. Quad-tree recursively decomposes a two-dimensional space into four quadrants which is based on the maximum capacity of the leaf cell. R-tree is used for efficient nearest-neighbor searches and it handles both, the point data as well as geometric bounding boxes.

10.4.3 Task Scheduling

The distribution of subtasks to concurrent computing units is task scheduling. Task scheduling is a vital component of HPC as the time of finishing a sub-task has a direct effect on the performance of the parallelization. Task scheduling depends upon programming paradigms and the platform, the problem to be parallelized, and the underlying computing resources.

When executing task scheduling, two factors must be considered: data locality and load balance. When doing a data processing activity, load balancing makes sure that each computing unit is given roughly the same (if not exactly the same) amount of subtasks, allowing them to complete at the same time. This is significant since the job's completion time is decided by the last completed task in parallel computing (Li, 2020). As a result, the number of subdomains and their workloads, as well as the number of concurrent computing units available for load balancing, should be evaluated. The proximity of data to its processing sites is referred to as data locality; a lower distance indicates better data locality. Because good data proximity necessitates less data travel during parallel data processing, it improves performance.

10.4.4 Evaluation Metrics

The performance of any parallel computing work is measured using speedup, which is a fractional measure of one processor's improvement over another or, in the case of parallel computing, an improvement over the total number of processors (Armstrong, 2020). The speedup is given by

$$\text{Speedup} = t_1/t_n$$

where t_1 is the time taken by a sequential processor and t_n is the time taken by parallel processors. Speedup is standardized using efficiency which is given as

$$\text{Efficiency}_n = \text{Speedup}_n/n$$

10.5 HARDWARE AND ITS PROGRAMMING MODEL

10.5.1 Graphics Processing Unit

GPU computing is the use of a GPU in conjunction with a CPU to accelerate general-purpose and logical applications. Due to the enormous dimensionality of data, hyperspectral image processing is a promising candidate for GPU-based computing. When compared to CPU-based processing, GPU computing typically produces faster results due to its parallel computational capabilities. The GPU's many-core design allows for a large number of threads to be used for parallel processing, whereas the CPU has a set number of cores dedicated to serial processing (Zhang et al., 2010). The reason for the difference in floating-point capacity between the CPU and the GPU is that the GPU is optimized for compute-intensive, highly parallel calculations, which is what graphics rendering is all about. As a result, more transistors are dedicated to data processing rather than data caching and stream control (Figure 10.3).

10.5.2 General Architecture of GPU

At its core, the NVIDIA GPU is made up of a number of multiprocessors, each of which has a number of parallel thread processors. GPU works with the CPU using its own set of rules and has its own device memory. The thread processors run synchronously using the Single Instruction Multiple Thread (SIMT) technique, which means that each processor runs the same program in parallel on different data (Wolfe, 2009).

The progressive memory design is implemented by the GPU. The GPU has its own memory, known as gadget memory, which may now be as large as 4GB (Wolfe, 2009). The memory handling approach of the GPU is responsible for the way it outperforms the CPU in terms of execution. The memory comes in the way of the CPU's handling most of the time, so the CPU uses cache memory to speed up the retrieval of information. However, this strategy is ineffective for GPUs due to the fact that GPUs require streaming access to large datasets that cannot be accommodated in the cache. The GPU solves this problem by employing a multi-threading strategy. When the thread requires memory, it is slowed until the value is retrieved from memory. In a multithreading

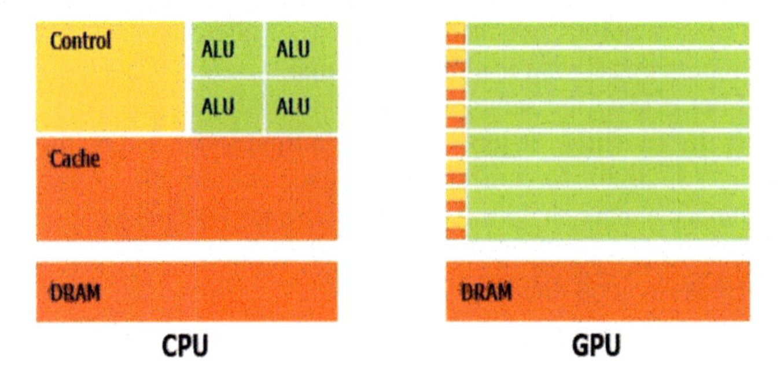

FIGURE 10.3 CPU vs GPU (Velasco-Forero & Manian, 2009).

worldview, the GPU runs another thread while the previous thread performs memory operations, resulting in a high amount of parallelism.

The multiprocessor SIMT unit creates, supervises, schedules, and executes threads in warps, which are groups of 32 parallel threads. Warps are the simplest scheduling unit in the GPU. The main distinction between the GPU and CPU is that the GPU, unlike the CPU, does not rely on large stores for execution change; rather, it uses multithreading to increase execution. Because a similar program is executed for each information component in SIMT design, there is a lower requirement for refined flow control, and because it is executed on multiple information components and has high number arithmetic power, the memory access latency can be covered up with computations rather than huge information caches (Wolfe, 2009) (Figure 10.4).

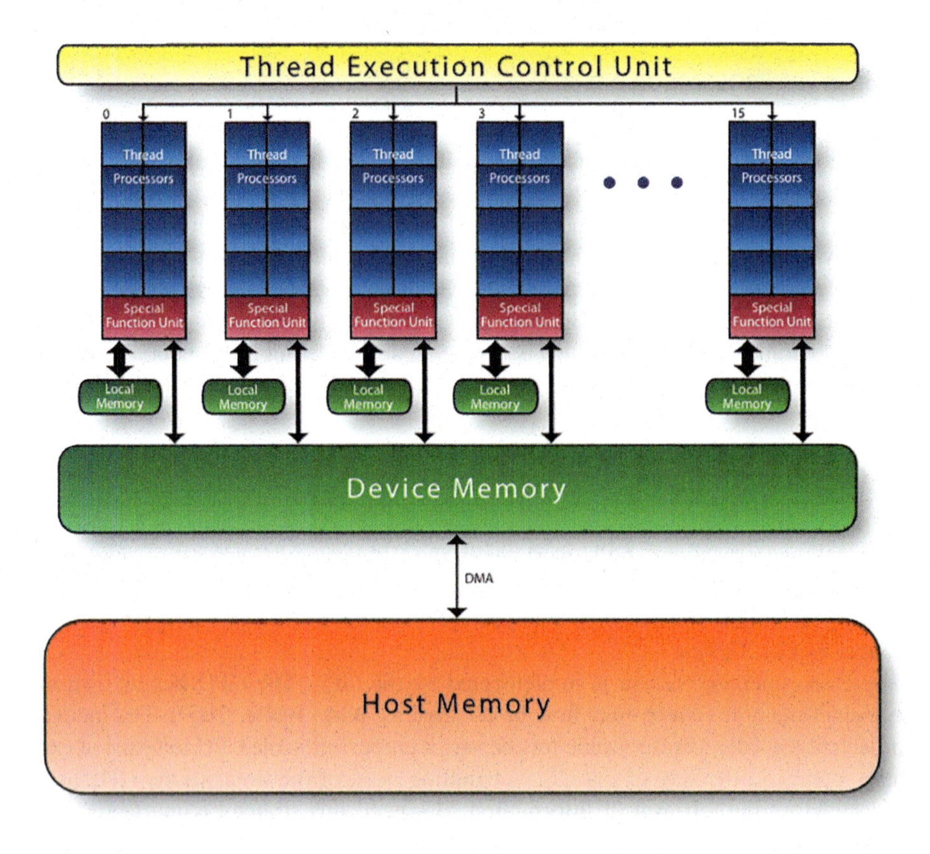

FIGURE 10.4 Architecture of GPU (Wolfe, 2009.)

FIGURE 10.5 NVIDIA's Jetson Nano. (https://developer.nvidia.com/embedded/jetson-nano-developer-kit.)

10.5.3 Jetson Nano—Embedded HPC

High-performance computing (HPC) refers to the use of large, computationally intensive programs to solve complex numerical calculations that are commonly used in areas such as image processing and simulation. Climate monitoring, flood detection and monitoring, forest fire identification, and hyperspectral image categorization are examples of HPC-based Geospatial applications. If you want a computer to simulate something in the actual world, you'll need a massive amount of computational power. In general, a supercomputer would be located in a large office with lines and racks of flickering lights, massive aeration and cooling, and possibly a water-based cooling system, if you were to really see one.

Hydrological systems, transit planning and watershed investigations, ecological modeling and surveillance, crisis response, and military activities all rely on geospatial data. As the availability of geospatial data has grown, so has its volume, creating a slew of challenges and complications that render current frameworks incapable of providing the necessary processing capability. For such applications, high-performance computing (HPC) enters the picture (Figure 10.5).

The Jetson Nano is equipped with a 128-core Maxwell architecture GPU from NVIDIA. The Jetson Nano includes two CSI camera slots, allowing it to be used for image acquisition and processing on the fly. For practical workloads, the Jetson Nano uses between 5 and 10 W of electricity, and it has the potential to be employed as an onboard computing platform. With 4 GB of RAM, the board measures 100 mm broad by 80 mm long by 29 mm thick (Talreja et al., 2021).

10.6 HPC FOR BUILDING DAMAGE DETECTION FOR EARTHQUAKE-AFFECTED AREA

High-performance computing can be applied to various applications for geospatial analysis, and one of the major applications is damaged building detection during earthquakes. For this case study, the input dataset is pre-earthquake LiDAR data and post-earthquake LiDAR data. The acquisition of the pre-earthquake data was done on 15th April 2016 and that of post-earthquake was done on 23rd April 2016. The dataset was collected by Air Survey Co., Ltd. Japan. Figure 10.6 shows the proposed method used to detect damaged buildings.

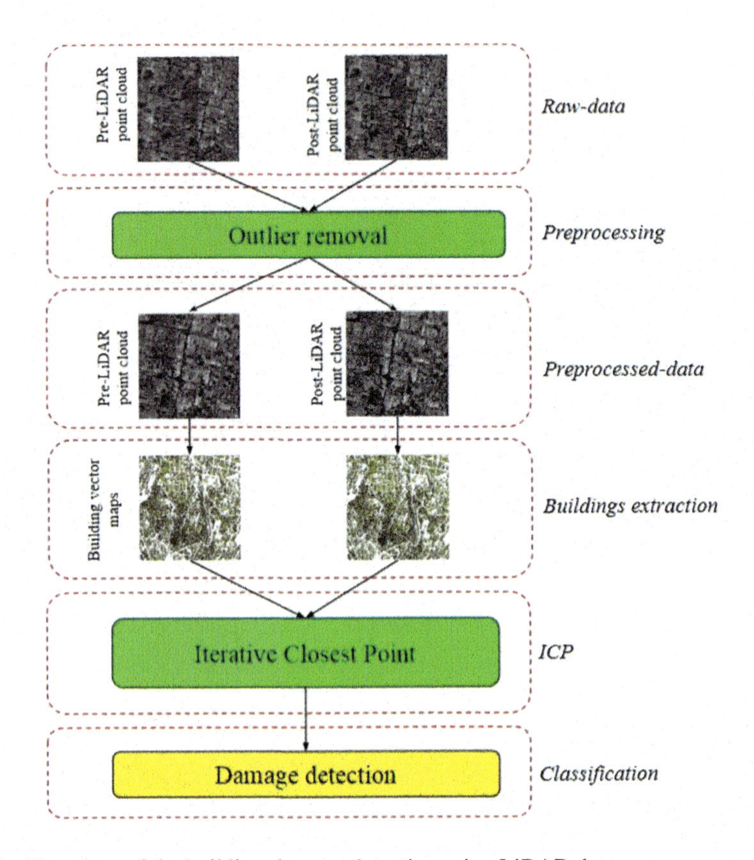

FIGURE 10.6 Flowchart of the building damage detection using LiDAR data.

10.6.1 Point Cloud Outlier Removal

For our work, we have made use of the k-nearest neighbor search (Rusu et al., 2008) for removing the outlier points from the raw LiDAR scan. One of the advantages of removing outliers is improving the performance of the point cloud by making use of only required information and eliminating the unwanted points (outliers). Also, the time complexity of processing the point cloud is improved as the number of points will be less as compared to the unprocessed raw point cloud.

The k-nearest neighbor algorithm is implemented using Point Data Abstraction Library (PDAL) which is a C++ library for working with point cloud data. The time and space complexity of the algorithm implemented is $O(N)$ where N is the number of points in the point cloud. The algorithm is implemented in two passes. The first pass is required to calculate the threshold value based on global statistics and the second pass is used to figure out the outliers in the point cloud. The threshold (T) is calculated in the first pass after calculating the global mean and standard deviation. The equations (Rusu et al., 2008) are given as

$$\mu = \frac{1}{N}\sum_{i=1}^{N}\mu i \tag{10.1}$$

$$\sigma = \sqrt{\frac{1}{N-1}\sum_{i=1}^{N}(\mu i - \mu)^2} \tag{10.2}$$

$$T = \mu + m\sigma \tag{10.3}$$

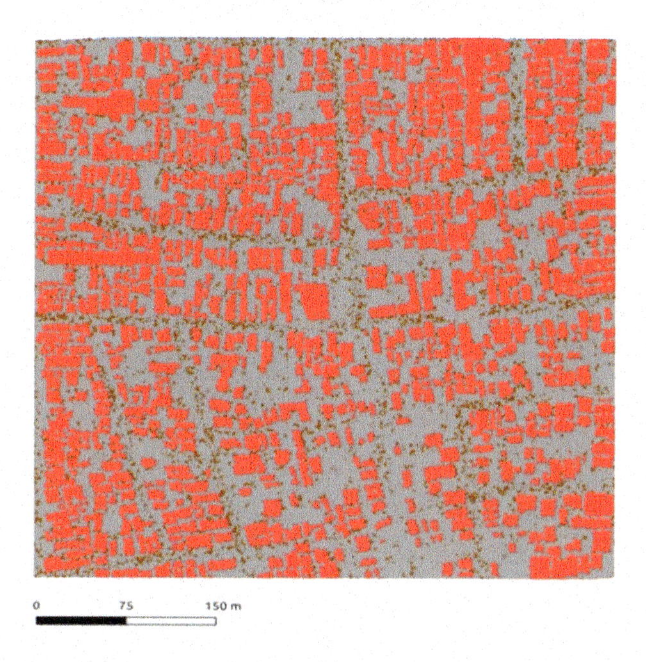

FIGURE 10.7 Point cloud of the earthquake affected area.

where *M* is the user-defined multiplier which is fixed to three after performing multiple experiments on our data. In the second pass of the algorithm, the pre-computed mean distances given by μi are compared with the threshold value calculated in equation (10.3). If μi is less than the threshold value then it is considered to be a point that is not an outlier. All other points are marked as outliers and are not taken into account for processing (Figure 10.7).

10.6.2 Buildings Extraction

After point cloud outlier removal in the pre-processing step, all the buildings are extracted from the pre-event LiDAR data as well as the post-event LiDAR point cloud using a vector map of the study area. The major advantage of separating the building class from the non-building class is the increase in the accuracy of the classification algorithm since only the building points are considered. Figure 10.8 shows the point cloud of the earthquake-affected region. Buildings are represented by dark orange color.

The buildings extracted are shown in Figure 10.9. The building extraction is done using PDAL (Point Data Abstraction Library) pipeline.

Following are the steps employed to extract the buildings from the point cloud:

1. Using QGIS/ArcGIS, split the polygons (shown in Figure 10.9) using the primary key.
2. Save the polygons in geojson format.
3. Using the geojson library in python, convert the polygon from geojson to WKT (Well-known text).
4. Create a PDAL pipeline and save it in JSON format. Following is one of the PDAL pipelines to extract a single building from the point cloud:

```
1. [
2. "/home/username/post.laz",
3. {
4. "type": "filters.crop",
```

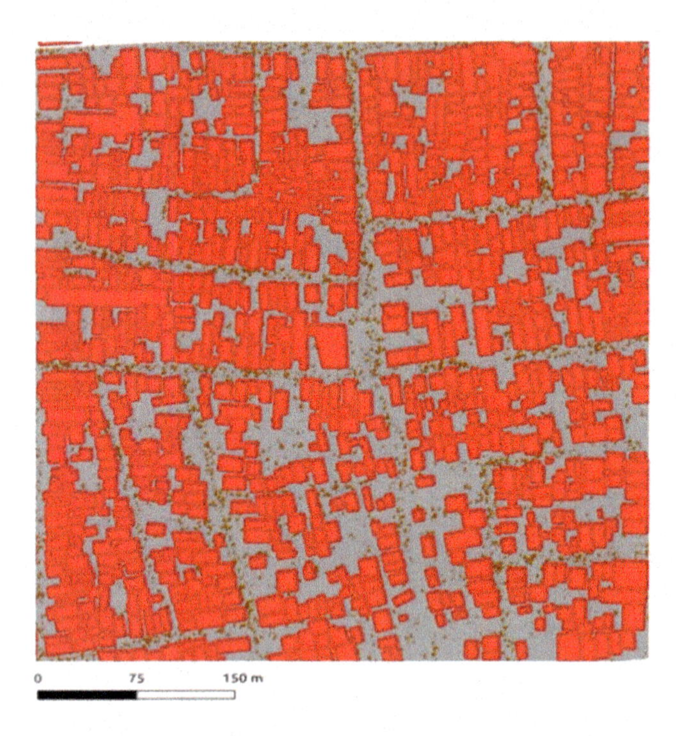

FIGURE 10.8 Pre-event buildings vector map overlayed on the point cloud.

FIGURE 10.9 Extracted buildings from the point cloud.

```
 5. "polygon": "POLYGON ((130.783362 32.780771, 130.783386 32.780775,
    130.78338 32.780802, 130.783518 32.780823, 130.783543 32.780703,
    130.783381 32.780679, 130.783362 32.780771))"
 6. },
 7. {
 8. "type": "writers.las",
 9. "filename": "/home/username/pointcloud/post_9.laz"
10. }
11. ]
```

Line No. 2 in the code above mentions the input point cloud (point cloud of the pre-event or post-event for our case study). Line No. 4 mentions the crop filter which is used to crop the point cloud. Line No. 5 provides the polygon geometry in WKT (created in Step No. 3) to crop from the input point cloud. Line No. 8 mentions the LAS writer and Line No. 9 mentions the output file containing the cropped polygon (point cloud file).

5. Execute the PDAL pipeline using the command: *pdal pipeline <name_of_the_pipeline>.json.*
6. Repeat the steps for all the polygons (buildings) of the area in the pre-event as well as post-event point cloud file.

10.6.3 Iterative Closest Point

In the iterative closest point (ICP) algorithm, one of the two-point clouds is kept as a reference and the other is transformed to best match the reference. The advantage of using ICP is high accuracy. ICP algorithm (Figure 10.10) works by iterating the following steps:

a. Correspondence estimation (finding the correspondence set represented using equation (10.4) includes point selection, point matching, and point rejection

$$X = \{(p, q)\} \tag{10.4}$$

b. Updating the transformation T by minimizing the error metric $E(T)$ for point-to-point correspondence represented using equation (10.5)

$$E(T) = \sum_{\{(p,q)\}\in X} p - Tq^2 \tag{10.5}$$

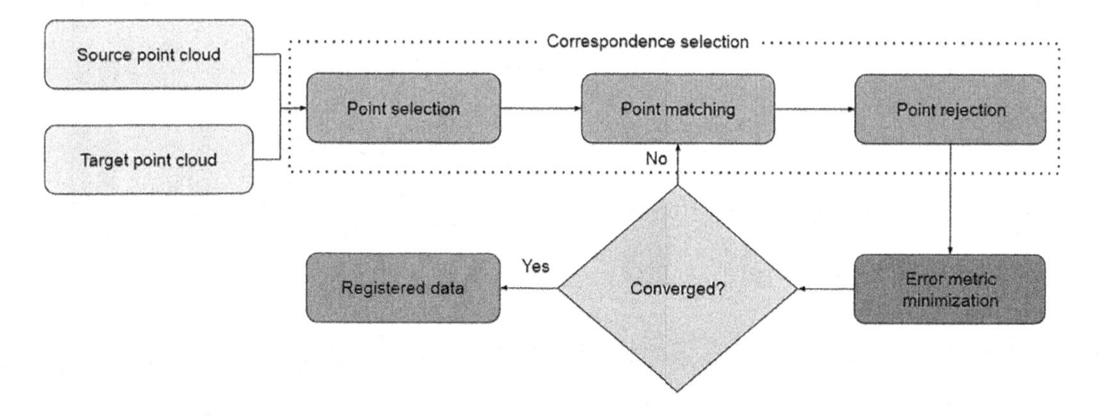

FIGURE 10.10 ICP algorithm. (Adapted from Li et al. (2021).)

For this case study, we will be using CUDA-PCL which is a Point Cloud Library by NVIDIA that can perform ICP on GPU as well as CPU. In this way, we will be able to compare the results of ICP on both the platforms and we can prove that HPC is indeed a solution when it comes to processing computation-intensive tasks. The ICP algorithm that we will use gives a fitness score which is a measure of ICP transform (Lei, 2021). The less the fitness score, the better transformation it provides. For our case study of detecting damaged buildings using the pre-event and post-event building point cloud, if the fitness score is more than 0.7, the building will be classified as damaged.

10.6.4 Classification Results

For our case study, we have extracted 100 buildings from the pre-event as well as post-event LiDAR dataset and performed damage detection using both the GPU and CPU. For our study, we have used NVIDIA Jetson Nano to perform the operation for damage detection. Figure 10.11 shows the sample output of ICP performed on a single building using both GPU as well as CPU.

Figure 10.12 shows the time required to perform ICP on 100 buildings using GPU as well as CPU. The experiment was performed on 10 buildings at a time gradually increasing the number of buildings to 100. The time to run the algorithm on CPU as well as GPU is displayed in the graph.

```
------------checking CUDA ICP(GPU)---------------
CUDA ICP by Time: 29.2893 ms.
CUDA ICP fitness_score: 0.267341
matrix_icp calculated Matrix by Class ICP
Rotation matrix :
    | 0.999915 0.013059 0.000856 |
R = | -0.013059 0.999915 0.000153 |
    | -0.000854 -0.000164 1.000001 |
Translation vector :
t = < 316.171875, -278.404297, -22.235579 >
```

```
------------checking PCL ICP(CPU)---------------
PCL icp.align Time: 161.629 ms.
has converged: 1 score: 0.274065
CUDA ICP fitness_score: 0.274065
transformation_matrix:
     0.999951   0.00991934  0.000757636     240.404
  -0.00991671     0.999953 -0.000642321    -211.081
 -0.000763863  0.000634737     0.999999   -0.867834
            0            0            0           1
```

FIGURE 10.11 Sample output (single building) on GPU and CPU.

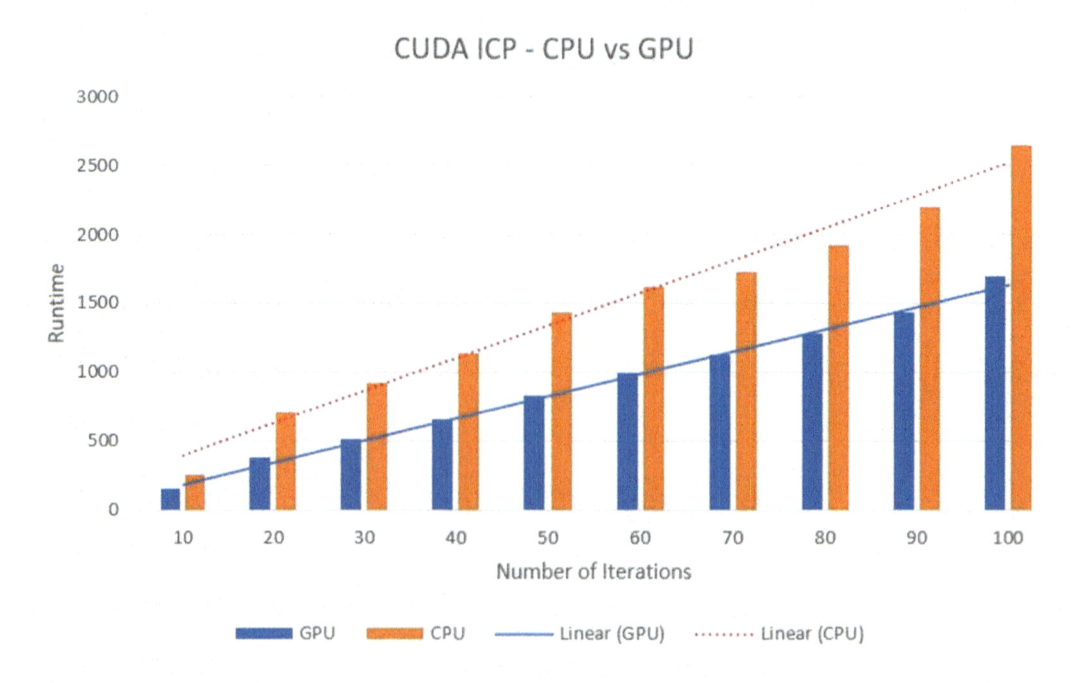

FIGURE 10.12 Running time of ICP on GPU and CPU.

After evaluating the classification results with the reports generated by NILIM, BRI (2016), we have found the accuracy of our proposed method to be 84%.

10.7 SUMMARY AND FUTURE WORK

Due to the high dimensionality of data and the quick response nature of most of the geospatial applications, High-Performance Computing proves to be useful in order to process and analyze data in real or near-real-time. We have proven the fact that embedded HPC devices such as Jetson Nano come in handy when it comes to real-time processing of the data obtained from the sensors. Considering the voluminous nature of the geospatial data, we have seen that decomposing the data can help reduce the volume which can be used for batch processing in parallel.

In this chapter, we have seen different categories of sensors that are used to collect big disaster data. We have also briefed about the key components of High-Performance Computing and how domain decomposition, spatial indexing, and task scheduling help process big disaster data in real or near-real-time. We have also described in brief the architecture of GPU, NVIDIA's programming model – CUDA, and the embedded GPU platform – NVIDIA Jetson Nano.

We have also explained a case study with the aim of detecting damaged buildings due to earthquakes which is one of the vital geospatial applications. The case study mentions how to use pre-earthquake and post-earthquake LiDAR data to perform building damage detection. From our results which performed damage detection on 100 buildings, GPU outperformed CPU by a factor of 1.56. As the number of buildings increases the time factor between GPU and CPU will also increase.

To make the process of detecting damaged buildings automated, a deep learning framework can be developed considering the future work for this case study. Also, data fusion using geospatial data from multiple sensors can help increase the accuracy of the future deep learning model.

REFERENCES

Armstrong, M. P. (2020). *High-Performance Computing for Geospatial Applications*. Springer: Berlin/Heidelberg, Germany.

Baskaran, R., & Tamilenthi, S. (1990). Using geospatial technology article. International Transaction Journal of Engineering *2*(2). http://TuEngr.com/V02/183-195.pdf.

Favorskaya, M. N., & Jain, L. C. (2017). *Handbook on Advances in Remote Sensing and Geographic Information Systems*. http://www.kesinternational.org/organisation.php.

Kang, X., Liu, J., Dong, C., & Xu, S. (2018). Using high-performance computing to address the challenge of land use/land cover change analysis on spatial big data. *ISPRS International Journal of Geo-Information*, *7*(7). doi: 10.3390/ijgi7070273.

Lee, C. A., Gasster, S. D., Plaza, A., Chang, C. I., Chang, C. I., & Huang, B. (2011). Recent developments in high performance computing for remote sensing: A review. *IEEE Journal of Selected Topics in Applied Earth Observations and Remote Sensing*, *4*(3), 508–527. doi: 10.1109/JSTARS.2011.2162643.

Li, Z. (2020). Geospatial big data handling with high performance computing: Current approaches and future directions. In: W. Tang & S. Wang (Eds.), *High Performance Computing for Geospatial Applications* (pp. 53–76). Springer: Cham. doi: 10.1007/978-3-030-47998-5_4.

Li, L., Wang, R., & Zhang, X. (2021). A tutorial review on point cloud registrations: Principle, classification, comparison, and technology challenges. Mathematical Problems in Engineering *2021*. doi: 10.1155/2021/9953910.

Parker, M. (2017). *Digital Signal Processing 101*. Newnes: Boston, MA.

Rathore, M. M. U., Paul, A., Ahmad, A., Chen, B. W., Huang, B., & Ji, W. (2015). Real-time big data analytical architecture for remote sensing application. *IEEE Journal of Selected Topics in Applied Earth Observations and Remote Sensing*, *8*(10), 4610–4621. doi: 10.1109/JSTARS.2015.2424683.

Rusu, R. B., Marton, Z. C., Blodow, N., Dolha, M., & Beetz, M. (2008). Towards 3D point cloud based object maps for household environments. *Robotics and Autonomous Systems*, *56*(11), 927–941. doi: 10.1016/j.robot.2008.08.005.

Talreja, P. V., Durbha, S. S., & Potnis, A. V. (2018). On-board biophysical parameters estimation using high performance computing. *IGARSS 2018-2018 IEEE International Geoscience and Remote Sensing Symposium*, Valencia, Spain.

Talreja, P., Durbha, S. S., Shinde, R. C., & Potnis, A. V. (2021). Real-time embedded HPC based earthquake damage mapping using 3D LiDAR point clouds. *In 2021 IEEE International Geoscience and Remote Sensing Symposium IGARSS*, Brussels, Belgium, pp. 8241–8244. doi: 10.1109/igarss47720.2021.9554481.

Velasco-Forero, S., & Manian, V. (2009). Accelerating hyperspectral manifold learning using graphical processing units. *Algorithms and Technologies for Multispectral, Hyperspectral, and Ultraspectral Imagery XV, 7334*, 73341R. doi: 10.1117/12.820176.

Wolfe, M. (2009). The PGI accelerator programming model on NVIDIA GPUs part 1 PGI accelerator programming model for NVIDIA GPUs part 1. http://www.pgroup.com/lit/articles/insider/v1n1a1.htm.

Zhang, J., You, S., & Gruenwald, L. (2010). Indexing large-scale raster geospatial data using massively parallel GPGPU computing. *In GIS'10: 18th SIGSPATIAL International Conference on Advances in Geographic Information Systems*, San Jose, CA.

Section IV

Geovisualization

Innovative Approaches for Geovisualization and Geovisual Analytics for Big Geospatial Data

11 Dashboard for Earth Observation

Manil Maskey, Rahul Ramachandran, Brian Freitag, and Aaron Kaulfus
Earth Science Branch, Marshall Space Flight Center
National Aeronautics and Space Administration

Aimee Barciauskas, Olaf Veerman, and Leo Thomas
Development Team
Development SEED

Iksha Gurung and Muthukumaran Ramasubramanian
Earth System Science Center
University of Alabama

CONTENTS

DOI: 10.1201/9781003270928-15

11.1 INTRODUCTION

The term *dashboard* is widely used, yet its exact meaning can vary from one instance to another. Put simply, dashboards have multiple definitions and descriptions. A dashboard can be a visual display of the most important information needed to achieve one or more objectives where that information is consolidated and arranged on a single screen to be monitored at a glance [1]. In another implementation, a dashboard can be a "faceted analytical display" of interactive charts (often graphs and tables) that simultaneously reside on a single screen, each of which presents a somewhat different analytical view of a common dataset. Wexler et al. [2] define a dashboard as "a visual display of data used to monitor conditions and/or facilitate understanding," which can include graphical elements and interactive visualizations. In essence, then, a dashboard is a type of display or style of presentation, not a specific type of information or technology.

While there may be many different definitions, all dashboards are built to support data-driven insights and decision-making. To fulfill this fundamental purpose, a dashboard must provide certain functionalities. It must display data and information in a concise, clear, and intuitive manner. It must provide high-level summaries, including exceptions, to communicate key information at a glance. A dashboard needs to quickly convey what's happening (like the gauges, meters, and indicator lights on a car) but also needs to serve as the starting point for digging deeper into why it is happening, letting users drill down further detail to perform an analysis.

Dashboards should provide the ability to be customized such that the data and the information on a dashboard can be tailored specifically to the requirements of a given person, group, or function; otherwise, it would fail to serve its purpose. For example, dashboards had been commonly used in a diverse set of domains including economics [3] and public health [4], and then COVID-19 pandemic introduced new needs for such dashboards, especially to track cases, deaths, and recovery over time for geographic areas [5]. The use of dashboards has expanded beyond a certain set of domains and has evolved beyond just providing a snapshot visualization and single-screen infographics. Now the expectation is that dashboards provide an intuitive, easily understandable tool to the intended user. This requires dashboard designers to be mindful of the stakeholders, appropriate visual attributes, and contextual framing.

This chapter focuses on science dashboards specifically using Earth observation (EO) data and the specific challenges associated with designing EO dashboards. These challenges are:

- **Understanding the Overall Objective**: The objectives of the dashboard could be operational, analytical, or strategic. An operational dashboard focuses on short time frames and operational processes such as data production and distribution. A strategic dashboard is focused on programmatic scientific strategies and displays key performance indicators that are measured to monitor program success. An analytical dashboard allows vast amounts of data to be readily utilized, letting users drill down into further detail to perform analyses on large volumes as well as a wide variety of data.
- **Understanding the Scientific Context**: The designers of the dashboard must understand the scientific objective, the curated data and information that will be used, community-accepted display plots, and color maps as well as other analysis functionality that are commonly used by the different stakeholders. Only a thorough understanding of the science context will allow designers to select the correct data elements, display the data and information to convey the scientific results and provide guided narratives.
- **Making the Appropriate Technology Choices to Big Data**: EO data poses a big data problem in terms of volume, variety as well as velocity. The current volume and complexity of Earth observation (EO) far exceed the memory, storage, and processing capabilities of individual users' machines. New technological solutions are needed to enable users to fully

utilize these large archives of EO data. A rapid shift in server-side technology toward cloud computing is underway. The use of cloud-based spatial data infrastructures (SDI) is needed to support the new EO dashboards. SDIs are platforms to facilitate the access, integration, and analysis of multi-source spatial data [6]. Traditional SDIs have been built based on legacy technologies that represent and store geospatial data in files and databases as well as serve spatial data, metadata, and processes. These legacy technologies include file formats that focus on data preservation rather than accessibility or ease of use. While these legacy technologies allow server-side data processing, they have limitations in scaling. These limitations are based on the available underlying on-premise hardware. With the advent and large-scale adoption of cloud computing, the next generation of SDIs has been developed that allow users to access, visualize, process, and analyze large amounts of geospatial data

The chapter also details the different functional dashboard design requirements that must be considered. These design requirements include the web interface, the access and views for different types of users (both exploratory and explanatory), and the right information refresh intervals. Dashboard designers must determine the manner in which heterogeneous information can be cohesively displayed, whether through a graphical, tabular, map, or similar visualizations. The design must provide both overview visualizations as well as functionality to drill into details on demand. In addition, other functional requirements such as interoperability with legacy systems, being mobile-ready, and providing sharing capabilities to promote collaboration will be discussed along with the important components of the underlying data systems, metadata management using a catalog, providing the right application programming interfaces (APIs) to discover and serve data, and data pipelines that produce science products for the dashboard.

Furthermore, EO dashboards need to align with open science principles to accelerate science process and knowledge dissemination. These principles add additional requirements to make sure that the dashboard is reusable and open, enables collaboration and knowledge sharing, and provides a reusable platform to support future data science challenges, training, and workshops.

This chapter tackles these challenges to designing reusable EO dashboards for a wide spectrum of Earth science use cases. The chapter will provide details of the VEDA concept (Visualization, Exploration, and Data Analysis: Scalable and Interactive System for Science Data) as a response to these challenges and requirements. The design and development process utilized to materialize VEDA from a concept to actual platform will also be described. Assessment of the current technologies that support the VEDA ecosystem and the design choices related to their selection will be covered, as well as current limitations of design.

11.2 CANONICAL USE CASES AND HIGH-LEVEL REQUIREMENTS (SCIENCE)

Several dashboard use cases are described in this section. These use cases form the basis for the development of the VEDA concept and drive a focus on developing a common platform and an ecosystem of tools to address the different requirements.

11.2.1 COVID-19 Dashboard

The COVID-19 pandemic caused authorities to limit or lock down cities which resulted in changes in human behaviors that impacted the Earth system. Studying such impacts on the Earth system requires an integrated study of relevant parameters using remotely sensed and socio-economic data. The COVID-19 dashboard [7] was developed to bring together EO datasets along with socio-economic datasets to visualize, explore, and communicate the environmental effects of human behavior due to COVID-19.

To enable the COVID-19 dashboard to support understanding of how the Earth system was affected and communicate those findings, researchers, software developers, and data engineers collaborated to combine relevant EO data with spatial data layers, maps, charts, colors, graphics, and stories, in an intuitive and easy to navigate the system. The remote sensing datasets were used to derive indicators—measurable parameters that can be easily visualized and explored within a dashboard.

Various visualization techniques and attributes were used within the dashboard. On the map interface, users can compare the indicator information against a normal baseline, step through different times, look at before and after images, dynamically change the color scales, and draw bounding boxes to focus on certain areas. Other visualization techniques such as small multiples are also used to demonstrate temporal changes over a fixed amount of time in a single display. A guided narrative section under the "Discoveries" menu guides the general public in understanding integrated findings using an explanatory storytelling approach.

Additionally, the COVID-19 dashboard was also used to showcase results from artificial intelligence (AI)-based economic indicators. Specifically, AI algorithms were used to detect shipping and air traffic activities on high-resolution satellite imagery. The dashboard contextualizes these AI detections on maps, showing quantitative information as dynamic charts and a guided narrative as a story. Figure 11.1 shows the main interface of the COVID-19 dashboard.

11.2.2 MAAP Dashboard

The Multi-Mission Algorithm and Analysis Platform (MAAP) [8] is a cloud-based platform development collaboration between NASA and the European Space Agency (ESA) that supports the scientific needs of the aboveground biomass research community. The MAAP is designed to address the data and information sharing needs of the biomass community and enable users to easily process, visualize, and analyze NASA and ESA mission data. The MAAP science research and development team identified that supporting data and information publication to an Earth observation dashboard would satisfy two needs: (1) rapid visualization and internal sharing of science products produced on the MAAP for product review within the MAAP science community, and (2) engaging policymakers and the broader community with biomass estimation and the uncertainties of biomass

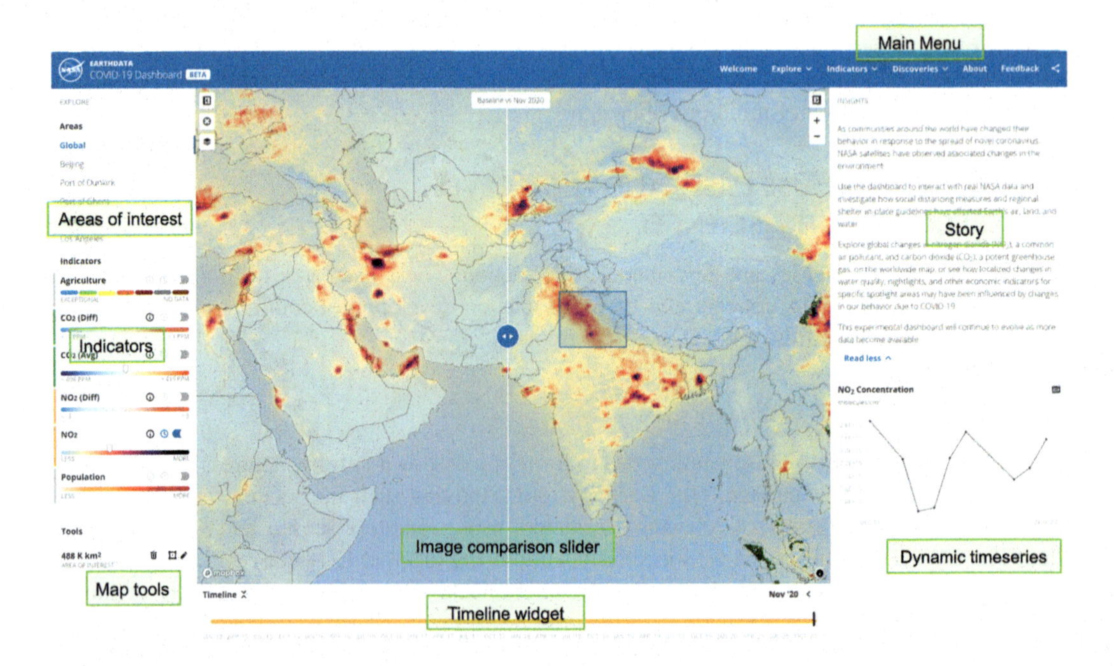

FIGURE 11.1 COVID-19 dashboard.

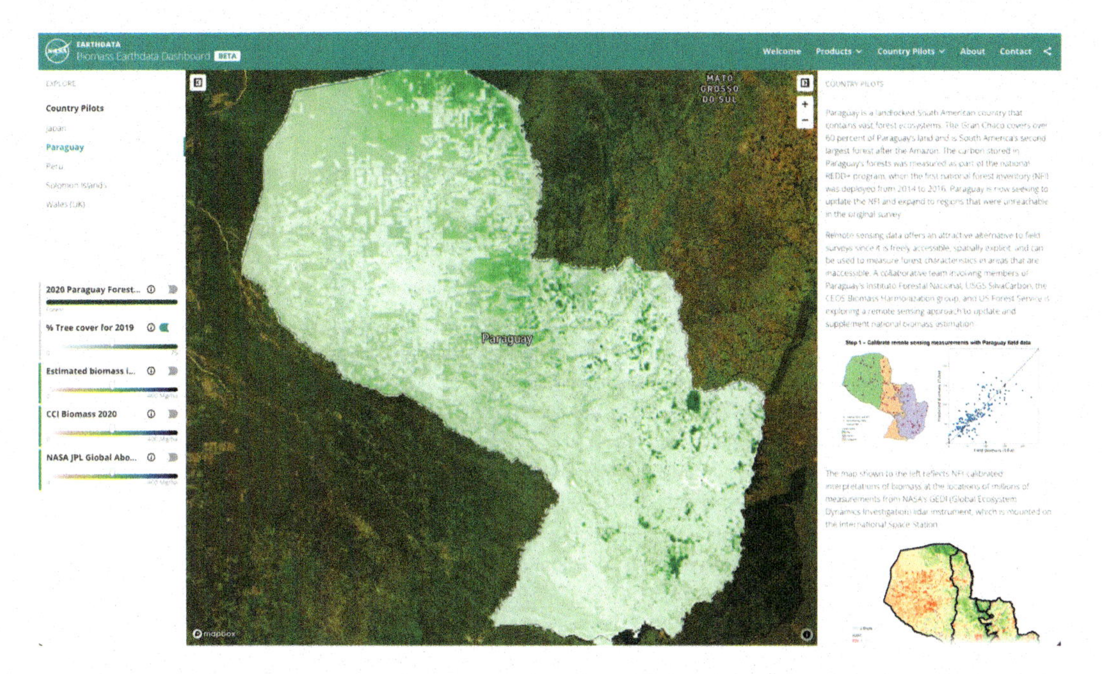

FIGURE 11.2 MAAP biomass dashboard.

estimation products. Despite being distinctly different audiences, both dashboard needs required combining relevant data products with spatial data layers, metadata, and stories in an intuitive system that easily communicates the science.

The MAAP dashboard was developed and deployed alongside the overall MAAP platform in Amazon Web Services (AWS) cloud [9]. This dashboard benefitted from the availability of the COVID-19 dashboard open-source code and its supporting data management system in that the overall web application design, data, and information publication workflow, as well as data pre-processing workflows, were adapted for the MAAP use cases. A major advantage offered by this architecture is the ability of the MAAP science teams to focus on the generation and delivery of science products and supporting information while the MAAP developers provide and monitor the automated publication and visualization generation workflows. Furthermore, the storytelling approach of the dashboard supports the visualization of global products—providing large-scale inferences on the change and biomass from the perspective of multiple algorithms—and regionally scaled visualizations with a focus on areas with poor in-situ inventories or biomass estimations or where biomass is rapidly changing. MAAP biomass dashboard is illustrated in Figure 11.2.

11.2.3 Community Workshops, Tutorials, and Hackathons

Citizen science is an integral part of open science. Community workshops, hackathons, and tutorials are mechanisms of citizen science that maximize the use of data, develop solutions to outstanding problems, enhance existing solutions, and educate communities. The COVID-19 dashboard hackathon [10] is one successful example, demonstrated by the growing interest within the scientific community in using the dashboard and its associated data in the cloud. Participants demonstrated how open data, open code, open platform, and open API can advance scientific research. Application tutorials support community involvement as a key pillar of the open science initiative. In tutorials, users can utilize a catalog of datasets, application programming interfaces, stories, and visualization widgets. They engage in hands-on activities that allow them to utilize data and platforms to tell novel stories about the changing Earth and present those stories in a dashboard environment.

TABLE 11.1
High-Level Use Case Requirements

Data system	• Incorporate various types of data: raster, vector, tabular • Integrate heterogeneous datasets from different satellites/sensors with different spatial resolution, temporal resolution • Handle data gaps, i.e., when data is unavailable for certain locations or the inability to capture useful data in certain atmospheric conditions • Address inconsistent metadata and different metadata profiles
Platform and system	• Provide a uniform data ingestion pipeline for data transformation • Utilize cloud computing platform for collaboration across teams • Provide a light-weight metadata catalog • Provide an API for discovery and access • Provide a mapping interface • Facilitate auto refresh once new data is ingested that updates catalog and visualizations • Provide a pipeline for AI-driven insights to be ingested and properly displayed
Usability	• Support different personas in both exploratory and explanatory modes • Provide the appropriate visualizations based on data type and narratives including customizable widgets for personalization • Provide overview information and detail on demand • Ability to localize information
General	• Modular design to enable replacement of components as technologies evolve • The platform and the system should be performant to support interactivity with data at scale
Open science	• Open and reusable software (both system and ecosystem) • Community buy-in is part of the development process to engage a larger pool of both developers and users • Design choices (APIs, formats, etc.) align with existing as well as evolving community standards or best practices

Based on these use cases, high-level requirements were developed along five different dimensions, and these are listed in Table 11.1.

11.3 TECHNOLOGY LANDSCAPE ANALYSIS

Data systems form the foundational component in an overall dashboard system. Much of the legacy EO data systems were designed to support discovery and accessibility only. With the migration to cloud computing and storage, modes of access are focused on analysis rather than file download. This shift minimizes the need to move or transfer data and enables users to leverage on-demand scalable cloud computing on data in the same cloud data center as the data archives.

A geospatial dashboard system can be viewed as four conceptual layers:

- *Data stores,*
- *Data processing,*
- *Data services,*
- *Visualization front end.*

In this section, we review different new technology components in each of these layers to address issues related to big data. We describe and evaluate potential components based on their technology maturity, adoption, and extensibility. Furthermore, we list known issues and concerns that need to be factored in if a specific solution is incorporated into the implementation.

11.3.1 Data Stores

Cloud-based storage technologies are provided by different cloud providers such as AWS, Google Cloud Platform [11], or Azure [12]. These cloud providers support optimizing for cost via different storage temperatures. For example, AWS provides Simple Storage Service (S3) [13] as standard storage, which is the most expensive but provides the lowest latency (fastest access). AWS Infrequent Access (IA) [14] costs less per GB of storage but more to access. Similarly, S3 Glacier [14] costs the least for storage but can take more time to retrieve. S3 Glacier can be configured for different needs, from instant retrieval to "Deep Archive" in terms of how fast data is delivered when requested. Microsoft's Azure Blob Storage [14] and Google's Cloud Storage [15] offer similar variations.

Most cloud-based archives for geospatial data use object storage, such as AWS Simple S3 for storing files in their native formats such as NetCDF [16]. These object stores provide flexible and reliable storage, and all users can access the same files. However, access is through S3 or HTTPS network protocols. This is the main limitation of object storage: it requires transferring data over a network and thus can incur costs and latency greater than files accessed from a locally mounted disk.

In addition to using cloud storage services, various new cloud-optimized data formats need to be considered. Legacy data formats (e.g., BUFR, HDF) were designed to store data in defined structures and enable reader libraries to utilize the structure to read the contents. These formats focused on long-term preservation (i.e., use of the data) and optimizing the packing as much data as possible. The data in these file formats are mostly read sequentially. With the shift toward analytics, new cloud-optimized data formats have emerged focused on improving data read performance via metadata and subsetted range access. Data within and across files are chunked to optimal sizes to suit analysis workflows. The files contain indexes that allow read routines to efficiently access the chunks needed. Cloud-optimized GeoTIFFs (COGs) [17] and Zarr [18] are two popular cloud-optimized formats for Earth data.

Figure 11.3 shows how Development Seed COGs [19] provide structured metadata within each file. This means the file itself is slightly larger than it might be in a non-cloud-optimized version, but the file itself isn't intended to be transferred in its entirety. Each COG includes an image file directory (IFD) to inform clients about the TileOffsets and TileByteCount. This information enables data reader libraries to know where all the data is stored by reading only the first few bytes. The IFD supports subsetted access to COGs in cloud object stores via HTTP GET range requests. COGs have become the default format for cloud-optimization of raster data. Internal overviews (i.e., reduced resolutions of the data) make dynamic visualization possible without storing numerous copies of the data (i.e., storing image tiles for every zoom level, also referred to as pyramids).

Zarr is another popular cloud-optimized format for n-dimensional gridded data. Zarr is both an open-source python library and a format for the storage of chunked, compressed, n-dimensional arrays. The metadata is stored external to the data files themselves. The data itself is often reorganized and compressed into files that can be accessed according to which chunks the user is interested in. For example, from 3-dimensional data where a single variable is stored in x (latitude), y (longitude), and z (time) dimensions, a user might access the 0.0.0 chunk of data which itself is a file representing the first chunk along the x, y and z dimensions within the larger Zarr store. Figure 11.4 shows how Zarr format stores how it can store metadata and chunked data.

Since most EO data are multi-dimensional arrays, Table 11.2 is a comparison of COG and Zarr that can be used to make an informed decision as to which of the two formats to store data as best suits the use case.

While COG and Zarr have become popular for gridded datasets, cloud optimizations for other types of data are still evolving. Some that are currently gaining interest support the category of point and vector data include:

- *GitHub* - open geospatial/geo parquet [21],
- *COPC* - Cloud Optimized Point Cloud [22]; and
- *FlatGeoBuf* [23].

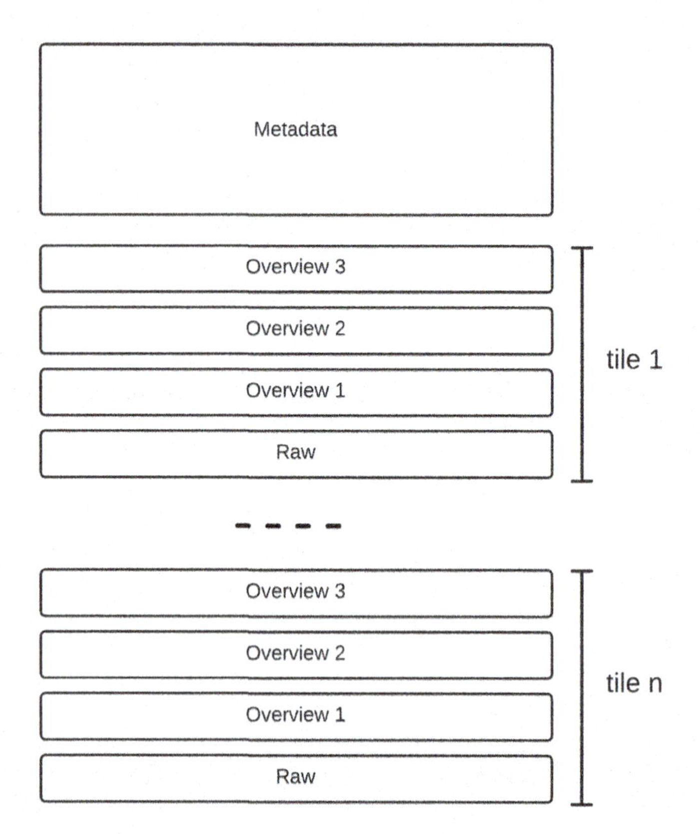

FIGURE 11.3 Metadata and tile representation in a mosaic of Cloud-Optimized GeoTIFF. (Image adapted from Vincent Sarago [19].)

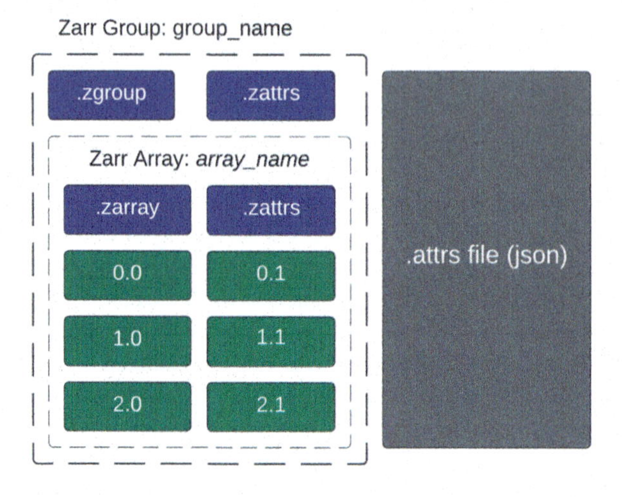

FIGURE 11.4 Zarr format specification with metadata and chunked data. (Image Adapted from Ryan Abernathy [20].)

TABLE 11.2
COG/Zarr Comparison

	COG	Zarr
Maturity of the code base	Mature. Open source and adopted by most geospatial data providers	Mature. Open source and adopted by most geospatial data providers
Extensible for use/adoption	Robust community adoption across Earth science domains	Robust community adoption in geospatial, however more popular in certain communities. Drivers beyond python are relatively new
Cost profile	More expensive to store COGs than GeoTIFFs, since they are larger, but cheaper to access. COGs are also compressed typically which often offsets the added space of overviews	Zarr stores are often compressed and store more limited numbers of variables than archival files
Known issues	May not include all data and metadata described in the archival format	May not include all data and metadata described in the archival format
		While xarray and Dask support lazy loading of Zarr datasets and very powerful on-demand n-dimensional analytics (i.e., time series analysis) development of libraries to support serving Zarr to common visualization API specifications (XYZ, WMTS, WMS) is ongoing

11.3.2 Data Processing

Geospatial data processing requires increasingly large-scale computational capability which is suited for cloud computing. Challenges in processing large-scale geospatial data include cost, efficiency, security of computing platforms, and managing data life cycle during pre, during, and post-processing.

Platforms like Google Earth Engine [24] and Microsoft Planetary Computer [25] have used respective cloud computing platforms with the intention of data-proximate processing. They achieve this by making a copy of the most often used geospatial datasets within a computing platform. These datasets are associated with a lightweight catalog (e.g., STAC [26]). Online interfaces such as Google Colab [27] and Jupyter [28] provide a processing entry point that uses the catalog and computing backend scaled in the cloud. Although this provides a great deal of simplicity for scientists to process large amounts of geospatial data efficiently, there are several limitations:

- A limited set of functions available and adding new functions is not easy;
- Only certain types of datasets are supported;
- Adding new, large-scale data to the processing platform requires a community consensus;
- Writing code requires knowledge of a specific programming language; and
- There is not an existing solution to verify the authenticity of the data.

Containerization technologies (e.g., Docker [29]) have simplified some of these hurdles involved in executing processing codes in cloud computing. MAAP and Harmonized Landsat Sentinel (HLS) processing [30] are examples of containerized-based processing. MAAP and HLS processing provide flexibility in containerizing existing legacy code without needing to rewrite it. Both are secure platforms, work with authoritative datasets, and allow the management of datasets pre, during, and post-processing. Each provides collaborative algorithm development and scaling using task-dependent processing options.

11.3.3 Data Services

11.3.3.1 Discovery Services

Providing discovery services, specifically, data is essential for any dashboard implementation. Instead of implementing and operating a traditional database to store metadata, simpler solutions are needed that are easy to implement and have low operating cost while addressing all the high-level requirements. One such approach that is now being widely adopted is the SpatioTemporal Asset Catalog (STAC).

STAC is designed to make it as easy as possible for data providers to expose their data to the world. Most geospatial catalogs require the provider of data to maintain servers and databases to enable search. This can be a considerable challenge with huge amounts of data. STAC uses the analogy of creating web pages. A HTML page is very simple to create, and searching those pages can be done by anyone, with experts in search emerging over time. Geospatial search engines as well as traditional search engines can index and support discovery of data via spatial and temporal queries.

STAC has a simple schema for catalogs, collections, items, and assets that is intuitive to understand and flexible to use. The STAC spec itself is encoded in JSON format to describe various metadata elements. STAC supports more complex discovery needs via filterable properties (see stac-spec/item-spec.md at master) and extensions. STAC serves as a foundation for leveraging popular tools that optimize access to cloud-based geospatial archives.

11.3.3.2 Data Access Services

Cloud-optimized client libraries support discovery and "lazily loading" data so it's only actually read when computations are required. STAC metadata is used to store references to files in memory but clients can only read data when final computation is requested. An optimized workflow [31] may include using *Sat-search* [32] library for discovering the data using spatial and temporal parameters and then Intake-stac and dask [33] libraries for lazily loading and streaming data into a Xarray.

11.3.3.3 Mapping Services

Mapping services based on Open Geospatial Consortium (OGC) [34] APIs are used for interoperability with common tools, such as desktop geographic information systems (GIS) and mapping clients like leaflet [35]. Commonly used APIs for geospatial data are the Web Coverage Service (WCS) [36] for subsetted access and Web Map Tile Service (WMTS) [37] or XYZ for visualization. Most WTMS use XYZ specification to request tiles, where X and Y correspond to row and column numbers in a Web-Mercator (e.g., EPSG:3857/WGS 84) projection and Z corresponds to the zoom level, where 0 is typically a zoom level where the entire globe fits within a single tile. The tiles themselves are typically either 256×256 pixels or 512×512 pixels.

Web Map Service (WMS) [37] is a standard protocol developed by the Open Geospatial Consortium in 1999 for serving georeferenced map images. A limitation of WMS is that it must fetch the entire image for a whole viewport before allowing the user to pan the map. In contrast, when the user pans the map, WMTS seamlessly joins individually requested image (or vector data) files and fetches new tiles as needed. This improves the user experience compared to WMS. WMTS also allows individual tiles to be pre-computed (a task easily parallelized).

WMTS is further improved by the use of raster overviews, where tiles corresponding to various zoom levels are pre-computed. An analogy to raster overviews would be a map that returns only state boundaries and major cities at the lowest zoom level, highways and state parks when zooming in to the state level and then neighborhoods, parks and landmarks when zooming into the city level. Pre-computing these maps allows the WMTS to serve tiles at any zoom level without increasing the response delay.

While OGC's API-based services such as WMTS and Open Layers XYZ API's deliver image tiles (PNGs and JPEGs) to client libraries such as Mapbox [38], Folium [39], and Leaflet and OpenLayers [40]. These services require pyramids of these image tiles to be stored, requiring pre-generation and storage. A recent open-source library TiTiler [41] has gained wide adoption for dynamic tiling, generating these PNG and JPEG image tiles at the time of request. These image tiles can be cached so visualizations are still relatively performant.

A common need is to visualize many COGs at once, because COGs are grouped to visualize a larger region, or multiple COGs over time are required to deliver a cloud-free image. A new framework—described in eoAPI [42]—uses a Postgres database to store STAC metadata with the pgSTAC [43] schema. The use of pgSTAC supports storing STAC queries which TiTiler can now use via [44] to create "mosaics" from multiple source COGs at once. Figure 11.5 shows the schematic of mosaicking using dynamic tiler.

11.3.4 Visualization Front End

11.3.4.1 Visualization Libraries

Mapping technology on the web has come a long way in the past decades and developers have a large number of visualization libraries to choose from. These libraries have a lot in common and map raster and vector data in different coordinate systems and projections. There are also key differences that may make a library more appropriate for a particular use case. The three libraries we will take a closer look at are Leaflet, Mapbox GL JS, and OpenLayers. In Table 11.3, we show the three main visualization libraries and their relevant information.

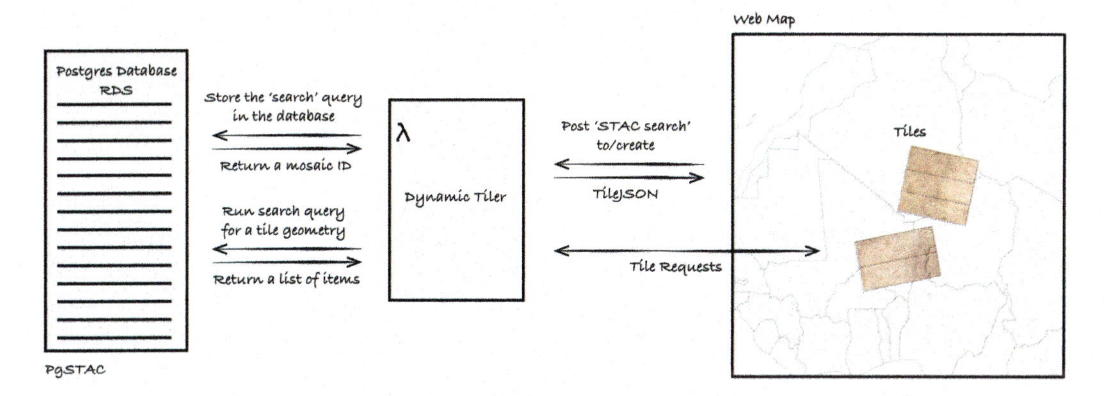

FIGURE 11.5 Schematic of mosaicking using dynamic tiling.

TABLE 11.3
Visualization Libraries

	Leaflet	Mapbox GL JS v2	OpenLayers
Website	https://leafletjs.com/	https://docs.mapbox.com/mapbox-gl-js	https://openlayers.org/
Codebase	https://github.com/Leaflet/Leaflet	https://github.com/mapbox/mapbox-gl-js/	https://github.com/openlayers/openlayers
Github Stars	33.5k	8.5k	8.9k
License	BSD-2	Proprietary From v2.0 on, Mapbox uses a proprietary license. The codebase is publicly available and can be partially modified. Active usage of the client requires a Mapbox account	BSD-2

11.3.4.2 Technical Considerations

There are a number of technical considerations when choosing a visualization client. This section outlines some of the key considerations and compares the three alternatives: Leaflet, Openlayers, and Mapbox GL. These considerations focus primarily on raster data.

Data server protocol and how data is being served, may determine which visualization library is a better fit for a given use case. WMS is a protocol for serving maps, developed by the OGC. It is a feature rich protocol that supports more traditional GIS operations. OpenLayers has the broadest support for data served through a WMS and implements the full specification. Working with a WMS is more challenging in Leaflet and Mapbox GL. They don't support features like *GetCapabilities*, and data can only be requested for a small set of coordinate systems.

Tiled web maps are a popular approach to serving map data. These services are easier to use and more performant than a WMS. A number of conventions and standards exist for tiled maps, but they all use a similar approach. A client makes a request for tiles at a particular location and zoom level; the server returns a number of smaller tiles (e.g., 256×256 pixels), which get stitched together in the browser. Other popular protocols are Slippy Maps used by OpenStreetMap, the Tile Map Service (TMS) developed by OSGeo, and the Web Map Tile Service developed by OGC (Table 11.4).

As a rule of thumb, OpenLayers supports a broader set of services and schemas. If compliance with OGC-type services is important, OpenLayers provides the most flexibility out of the box. If the use case data sources follow schemas optimized for the web, it is likely that Leaflet and Mapbox GL will provide everything you need.

11.3.4.3 Dynamic Tilers

TiTiler dynamically serves map tiles from the COG raster data file format. In this case, dynamic means that the data served by the tiling service doesn't have to be pre-computed. By reading the COG file's headers, the tiler is able to make use of the HTTP range request header to read into memory *only the bytes corresponding to the requested tile*, without having to read the entire source file into memory.

An extension of the TiTiler library, called Titiler.pgstac can be used to dynamically generate mosaics. A mosaic is a technique to serve data from multiple COGs at the same time as if it were a single COG. For example, one might have separate map tiles covering Paris and London. A mosaic is a JSON file that contains information about the different tiles and the total area they cover, such that we can issue an API query covering an area over the English Channel against this mosaic and retrieve a map tile that pulls data from both data files.

11.3.4.4 User Interactivity and Engaging End User Experience

Mapping frameworks allow you to build rich experiences in which the map is animated or reacts to the underlying data or user input. All libraries will allow for basic operations like zooming and panning to a place on the map. Beyond that, the degree of interactivity varies.

Over the past years, there have been a lot of advances in how people can interact with vector data in the browser. Users can explore thousands of features and manipulate them without any

TABLE 11.4
Visualization Libraries Support Comparison

	OpenLayers	Leaflet	Mapbox GL
WMS	Supported	Limited support	Limited support
XYZ	Supported	Supported	Supported
TMS	Supported	Supported	Supported
WMTS	Supported	Supported	Not supported
Coordinate Reference System (CRS)	Anything supported by Proj4s	Anything supported by Proj4s, through a plugin	Only EPSG:3857

performance issues. Mapbox GL in particular has a powerful expression language that dynamically styles map data based on its properties or user input. Similar development on the client side for raster data has been limited. Performing calculations like generating the mean land surface temperature for a user-defined area or dynamically re-scaling a layer with monthly CO2 values still has to be done server-side.

When it comes to interactivity and creating a dynamic experience, Mapbox GL is the library with the most complete feature set. Beyond data-driven styling, it allows one to add camera animations that create a fly-over experience or overlay a raster on top of a digital elevation model creating a 3D terrain effect.

11.3.4.5 Map Projections

All two-dimensional maps employ a projection to represent the globe on a flat surface. While there are different kinds of projections (i.e., different ways to translate three-dimensional coordinates into two dimensions) all projections must distort some aspect of the map, whether shape, size, or distance.

The Web Mercator (WGS84/EPSG:3857) has been the de-facto standard for web cartography since Google adopted it as the base map for their navigation products in 2005. Web Mercator is a natural choice for street-level navigation since it maintains north straight up at all points of the map, preserves angles between lines, as well as relative sizes at higher zoom levels. The disadvantage of Web Mercator is that at lower or even global zoom levels the relative sizes of territories further from the equator are grossly exaggerated.

All visualization libraries support Web Mercator (used by EPSG:3857 CRS) out of the box and provide support for alternative projections.

- Mapbox GL v2 and higher supports six alternative projections, in addition to Web Mercator. A future release has support for a globe projection, allowing users to seamlessly switch between a 2D and 3D experience.
- OpenLayers has built-in support for any CRS that is supported by Proj4js [45], both for the source raster data and the visualization layer.
- Leaflet comes out of the box with support for three different projections. Through the use of the Proj4js plugin, support for other projections can be added as well. It does expect vector data to be in Web Mercator.

TiTiler is also capable of dynamically re-projecting raster and vector data. However, since the front end is built using Mapbox v2.6, which doesn't accept input data in projections other than W Web Mercator, this functionality isn't needed. It could be used, in theory, to enable serving tiles from source COG's in projections other than Web Mercator. This workflow would look like this: (1) a non-Web Mercator source file is opened and read by the dynamic tiler API; (2) the API re-projects the requested tile to Web Mercator and serves it to the map client; and (3) the map client dynamically projects the tile back to the original projection. In practice, this workflow is a bit cumbersome, and non-Web Mercator source files are not an important enough use case to warrant it. It is easier to simply pre-process the source data and store it using the Web Mercator projection.

The key takeaway is that there are no easy choices as to the optimal mapping library to use. The leaflet is an excellent choice for simple use cases in which one wants to show a 2D map on a page. The core library is lightweight but can be extended using a rich ecosystem of plugins. OpenLayers is a package that comes with batteries included. OpenLayers is best suited if the use case requirement is an environment that relies heavily on OGC services or if the use case has to support a large number of different data sources. Of the three libraries, Mapbox GL allows the best interactive experience. Some of its features make it feel like a mix of a mapping library and a 3D rendering engine. If the lack of an open-source license is not a dealbreaker, Mapbox GL is the best option for creating a rich and engaging experience.

11.3.4.6 Analysis Clients

Data visualization is often a data user's introduction to a dataset of interest. In addition to the visualization clients and services listed above, data analysis clients can be used for the interactive exploration of geospatial data. Effective data analysis clients address needs across the user proficiency spectrum from beginners to experts, support large multivariate datasets, support cloud-native data and data formats, and provide a collaborative workspace. Such clients accelerate scientific discovery and the adoption of new datasets into existing science application workflows. Examples of popular data analysis clients include:

- Jupyter notebooks/notebook hubs;
- Tableau [46]; and
- Apache Superset [47]

While beneficial for statistical analysis of geospatial datasets, Jupyter notebooks are most used for collaborative, interactive data analysis. Notebook environments allow for executable code and markdown text in a single document. The versatility of notebook environments allows for the exploration of new datasets by a single user and also allows for collaborative development of derived higher-level data products from lower-level source products. The true value of future data analysis notebook clients lies in between these two use cases where the collaborative analysis and visualization of multivariate geospatial datasets can be leveraged to accelerate human understanding of the Earth system.

Tableau and Apache Superset (among others) eliminate the programming elements of notebook clients and integrate them as part of a visual programming interface. One major advantage of visual programming interface data analysis clients is the elimination of scientific programming. Where notebook hubs require scientific code for inspection of input data at any level, visual programming interfaces load a dataset and allow users to drag and drop variables for dataset analysis and exploration. Visual programming interfaces, however, are typically more rigid in terms of the data format that can be loaded into the client. For many of the visual programming interfaces surveyed and their documentation, input data is required to have a simpler structure such as a TXT, CSV, JSON, or database file. In their current state, visual programming interfaces do not support more complex datasets such as TIF, NetCDF, HDF, or the integration of additional data layers for more detailed analysis.

With rapid increases in free and open data, the use and development of data analysis clients are paramount to data exploration and information discovery. As the number of analysis clients increases, a new data user must determine the analysis client appropriate for their dataset. While visual programming interfaces are typically the quickest way to explore a dataset and derive meaning from it, only a select number of data formats are supported. Notebook hubs provide more powerful capabilities but require some experience with scientific programming to extract metadata, identify the most meaningful components, and analyze/visualize the data.

In addition, AI inference results are usually demonstrated using notebooks. While this is good for deeper interactions, debugging the model, and verification of outputs, it is not scalable. For scalability, efficient interactions, and quicker analysis the inference results can be stored in a cloud-optimized format accompanied by some metadata of the results. These results can be served using the API, and the dashboard can be leveraged to visualize, interact, and perform an analysis on the results.

In summary, some solutions listed in this section are built upon new technologies driven by the cloud platform to deliver scalable and optimized access. Some of these tools are still working toward operational maturity. The opportunity of these libraries is in being open source, so adopters are able to understand and contribute to them. Limitations include new tools that have been mostly designed around a few formats (COG and Zarr) and STAC, so many existing data collections in legacy format are still not accessible via these methods. Initiatives like pangeo-forge and Microsoft's

Planetary Computer are trying to make more massive shifts in the creation and adoption of these cloud-optimized tools and workflows for data systems.

11.4 VEDA

11.4.1 Overview

Visualization, Exploration, and Data Analysis (VEDA) is being designed as a scalable and interactive system for science data in response to the challenges and requirements discussed earlier. In particular, VEDA proposes to build a data system with an ecosystem of mostly existing services and analysis clients. The ecosystem of tools and services will be modular and interoperable, hence, any new data systems needs can be addressed by selecting a set of components from the ecosystem rather than VEDA as a whole. VEDA is designed with open science principles in mind: openness and reusability are pillars of VEDA development and mostly community-backed software stacks are selected for implementation. It addresses the needs of a wide range of users in both exploratory and explanatory modes.

11.4.2 Implementation

VEDA is being implemented in AWS cloud primarily due to NASA's decision to archive its data in AWS. VEDA's goal is to provide data proximate computing to minimize data movement and limit cost. The main components of VEDA implementation include (1) federated data stores, (2) data processing to address extract, transform, and load, (3) data services including APIs, and (4) visualization, exploration, and analysis clients; all deployed in AWS. High-level architecture for VEDA is shown in Figure 11.6.

11.4.2.1 Federated Data Stores

VEDA proposes to generate analysis-ready cloud-optimized (ARCO) data stores. Two main types of ARCO will be supported: COGs and Zarr. COGs will support most of the raster data needs for mapping clients while Zarr will support analysis needs. The idea behind federated data stores is to minimize the duplication of ARCO datasets. If there are already existing authoritative ARCO datasets then VEDA will publish the metadata to the VEDA catalog.

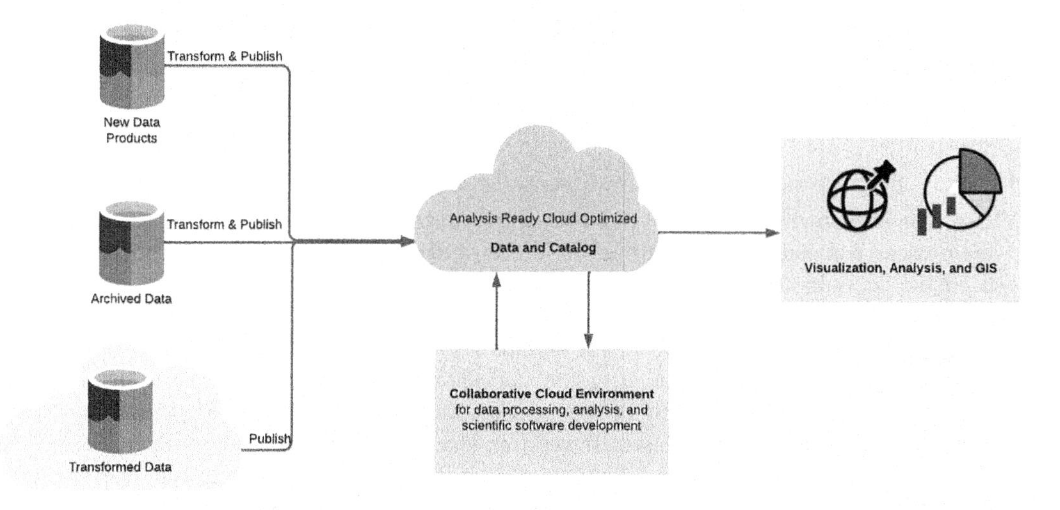

FIGURE 11.6 VEDA high-level view.

11.4.2.2 Data Processing (Extract, Transform, and Load)

VEDA proposes to develop standard workflow and pipelines for the transformation of science data from ARCO archives. The transformed data will be uploaded into AWS S3 and registered into a catalog. VEDA will provide validation tools and services which include a COGs viewer and sample notebooks for comprehensive validation of data transformations. Additionally, VEDA will include a collaborative cloud computing environment which provides beginner and advanced science users a computing environment with data analysis and processing capabilities, along with the ability to directly generate ARCO datasets that are easily integrated into the dashboard. Initial data analysis use cases will be supported by the Pangeo stack which will be deployed close to the ARCO data store. Large-scale data processing will leverage the Hybrid Cloud Science Data System (HySDS) for data processing (a component of MAAP) [48].

11.4.2.3 Data Services

VEDA will provide a set of data services which includes catalog and mapping services. VEDA is developing a lightweight STAC-based metadata catalog to support search, discovery, and access of ARCO data. The metadata catalog will include lineage to source data for ARCO. The VEDA will use TiTiler to dynamically serve map tiles. By reading the COG file's headers, the tiler is able to make use of the HTTP range request header to read into memory only the bytes corresponding to the requested tile, without having to read the entire source file into memory. VEDA implementation will use the Titiler.pgstac extension of the TiTiler library to dynamically generate mosaics as described in Section 11.3.

11.4.2.4 APIs

VEDA will support four main types of APIs: Publish API, Search API, Visualization API, and Access API.

- Publish API is used to add new metadata records to STAC. Credentials are required to ingest metadata into STAC catalog.
- Search API provides search interface to data in ARCO.
- Visualization API provides standards-based data visualization requests to data in ARCO. These queries allow requests for specific subset of the data in the ARCO based on spatial bounds, temporal bounds, and science variables. The response includes raster or vector information that is easily visualized.
- Access API returns data in ARCO that can be directly integrated into scientific analysis workflow.

VEDA will also provide a set of Process APIs to run analysis and processing jobs.

11.4.2.5 Data Visualization, Exploration, and Analysis Clients

VEDA will develop visualization, exploration, and analysis clients and also support GIS interfaces. A main visual exploration client will be the VEDA dashboard. User groups that include the general public, decision-makers, and data scientists interact with VEDA using the dashboard. The dashboard will support direct integration of data in ARCO using visualization APIs that support dynamic tiling and OGC standard services for data visualization: WMS and WMTS.

Dashboard-provided capabilities include a storytelling platform, map interface, and a dataset exploration interface. The storytelling platform provides a configurable interface to develop stories based on the ARCO-published datasets. Interactive, story-telling visualization components include configurable slider widgets, comparison widgets, and dynamic time series. A map interface provides map-based access to data published in ARCO. The VEDA dashboard makes use of TiTiler to dynamically serve map tiles.

VEDA dashboard stories will be linked to interactive notebooks for advanced users to perform an independent analysis of their own. Jupyter notebook hubs will be implemented such that analysis can be performed at larger scales directly from the clients. Common GIS tools such as QGIS and ESRI ArcGIS will directly integrate with ARCO datasets using visualization and access APIs.

11.5 SUMMARY

Dashboards are visual displays of data used to monitor conditions and/or facilitate understanding. They are built to support data-driven insights and decision-making. The use of dashboards has expanded beyond a certain set of domains and needs to provide an intuitive, easily understandable tool to the intended user.

Science dashboards specifically platforming EO data need to address specific challenges associated with using EO data. EO dashboards need to enable big data analysis, letting users drill down into further detail. The designers of an EO dashboard must understand the scientific objective, identify curated data and information that will be used, and display plots and color maps as well as other analytic functionality commonly used by the stakeholders. The current volume, velocity, and complexity of EO data far exceed the capabilities of individual users' machines and servers. Cloud-based spatial data infrastructures are a potential path forward to support the new EO dashboards.

Detailed use cases are important for developing requirements for a generalized platform and an ecosystem of tools for EO dashboard. Three use cases with different objectives were utilized to drive the requirements presented here.

The COVID-19 dashboard was developed to communicate the environmental effects of human behavior due to COVID-19. Its goal is to further understand how the Earth system was affected during the pandemic and communicate those findings to researchers. The COVID-19 dashboard required existing datasets to be transformed to derive different indicators and display temporal changes in a single display. The MAAP is designed to address the biomass community's needs for rapid visualization and internal sharing of science products and to engage policymakers and the broader community. Community workshops, hackathons, and tutorials are a strategic pillar of the open science initiative with the objective of maximizing the use of EO data and broadening community involvement. Based on these use cases, high-level requirements were developed along the dimensions for the data system, platform, usability, open science, and general functionality.

A review of different technology components was conducted for different components to address issues related to big data. This review focused on existing and emerging new technologies driven by the cloud platform to deliver scalable and optimized access. Some of the tools reviewed are still working toward operational maturity.

The Visualization, Exploration, and Data Analysis (VEDA) design is presented as a solution for generalized scalable and interactive system for EO data to meet the listed requirements

One of the key lessons learned during the design of VEDA is that there is no one-size-fits-all solution, whether for architecture, data transformation, visualization, or user targeting. Therefore, VEDA is designed as an ecosystem of services and analysis clients that are modular and interoperable such that they can address changes in future requirements as well as the availability of new technologies. Initially, the Earth Information System (EIS) will be the primary focus of VEDA. VEDA implementation supports EIS activity that translates scientific results into accessible, actionable information. EIS synthesizes NASA's Earth science observations and models to produce new science and support decision-making. EIS users will configure the dashboard with new data products to disseminate science results to a broader community. There are four EIS topics: fire, sea-level change, freshwater, and greenhouse gas center. Each one of these will use VEDA for its data systems. Additionally, the public-facing visualization capabilities of VEDA supported by the dashboard will cover the needs of a wide range of stakeholders for new initiatives including environmental justice and an Earth action center.

REFERENCES

1. Few, S., 2005. Intelligent dashboard design. *Information Management, 15*(9), p. 12.
2. Wexler, S., Jeffrey, S. and Cotgreave, A., 2017. *The Big Book of Dashboards: Visualizing Your Data Using Real-World Business Scenarios.* John Wiley & Sons: Hoboken, NJ.
3. Cohen, A.M., 2011. Envisioning a global economic dashboard. *The Futurist, 45*(3), pp. 11–12.
4. Lechner, B. and Fruhling, A., 2014. Towards public health dashboard design guidelines. *HCI in Business - Proceedings of First International Conference, HCIB 2014, Held as Part of HCI International 2014,* CityHeraklion, Crete, pp. 49–59.
5. Dong, E., Du, H. and Gardner, L., 2020. An interactive web-based dashboard to track COVID-19 in real time. *The Lancet Infectious Diseases, 20*(5), pp. 533–534.
6. Abbas, R. and Williamson, I., 2001. Spatial data infrastructures: concept, SDI hierarchy and future directions. *In Proceedings of GEOMATICS'80 Conference*, Tehran, Iran, vol. 10.
7. Maskey, M., Falkowski, M., Murphy, K., Veerman, O., Mestre, R., Gurung, I., Ramasubramanian, M., Thomas, L., Yi, Z., Bollinger, D., Seadler, A. and Ivey, Y., 2021. Visualizing, exploring, and communicating environmental effects of COVID-19 using earth observation dashboard. *2021 IEEE International Geoscience and Remote Sensing Symposium IGARSS*, pp. 1370–1373. doi: 10.1109/IGARSS47720.2021.9553461.
8. Albinet, C., Whitehurst, A., Jewell, L., Bugbee, K., Laur, H., Murphy, K. and Frommknecht, B., 2019. A joint ESA-NASA multi-mission algorithm and analysis platform (MAAP) for biomass, NISAR, and GEDI. *Surveys in Geophysics, 40*(4), pp. 1017–1027.
9. Wittig, M. and Wittig, A., 2018. *Amazon Web Services in Action.* Simon and Schuster: New York.
10. Zhongming, Z., Linong, L., Xiaona, Y., Wangqiang, Z. and Wei, L., 2021. Join ESA, NASA and JAXA for the Earth observation COVID-19.
11. Bisong, E., 2019. An overview of google cloud platform services. In: Bisong, E. (Ed.), *Building Machine Learning and Deep Learning Models on Google Cloud Platform.* Apress: New York, pp. 7–10.
12. Copeland, M., Soh, J., Puca, A., Manning, M. and Gollob, D., 2015. *Microsoft Azure.* Apress: New York, pp. 3–26.
13. Palankar, M.R., Iamnitchi, A., Ripeanu, M. and Garfinkel, S., 2008, June. Amazon S3 for science grids: A viable solution? *In Proceedings of the 2008 International Workshop on Data-Aware Distributed Computing*, New York, pp. 55–64.
14. Daher, Z. and Hajjdiab, H., 2018. Cloud storage comparative analysis amazon simple storage vs. Microsoft azure blob storage. *International Journal of Machine Learning and Computing, 8*(1), pp. 85–9.
15. Daher, Z. and Hajjdiab, H., 2018. Cloud storage comparative analysis amazon simple storage vs. Microsoft azure blob storage. *International Journal of Machine Learning and Computing, 8*(1), pp. 85–9.
16. Kivachuk Burdá, V. and Zamo, M., 2020, May. NetCDF: Performance and storage optimization of meteorological data. In *EGU General Assembly Conference Abstracts*, p. 21549, Vienna, Austria.
17. Durbin, C., Quinn, P. and Shum, D., 2020. Task 51-cloud-optimized format study (No. GSFC-E-DAA-TN77973).
18. Gowan, T.A., Horel, J.D., Jacques, A.A. and Kovac, A., 2022. Using cloud computing to analyze model output archived in Zarr format. *Journal of Atmospheric and Oceanic Technology, 39*(4), pp. 449–462.
19. Sarago, V., 2022. COG talk: Part 1 What's new? https://developmentseed.org/blog/2019-05-03-cog-talk-part-1-whats-new, Accessed June 1, 2022.
20. Abernathy, R., 2022. Cloud native climate data with Zarr and XArray. https://speakerdeck.com/rabernat/cloud-native-climate-data-with-zarr-and-xarray?slide=14, Accessed June 1, 2022.
21. GeoParquet, 2022. https://github.com/opengeospatial/geoparquet, Accessed May 25, 2022.
22. Cloud Optimized Point Cloud, 2022. https://copc.io/, Accessed May 25, 2022.
23. FlatGeoBuf, 2022. https://github.com/flatgeobuf/flatgeobuf, Accessed May 30, 2022.
24. Gorelick, N., Hancher, M., Dixon, M., Ilyushchenko, S., Thau, D. and Moore, R., 2017. Google Earth Engine: Planetary-scale geospatial analysis for everyone. *Remote Sensing of Environment, 202*, pp. 18–27.
25. Microsoft, 2022. A planetary computer for a sustainable future. https://planetarycomputer.microsoft.com/, Accessed May 30, 2022.
26. Hanson, M., 2019, December. The open-source software ecosystem for leveraging public datasets in Spatio-Temporal Asset Catalogs (STAC). In *AGU Fall Meeting Abstracts*, Vol. 2019, pp. IN23B–07, San Francisco, USA.

27. Kuroki, M., 2021. Using Python and Google Colab to teach undergraduate microeconomic theory. *International Review of Economics Education, 38*, p. 100225.
28. Granger, B. and Pérez, F., 2021. Jupyter: Thinking and storytelling with code and data. *Computing in Science & Engineering, 23*(2), pp. 7–14.
29. Reis, D., Piedade, B., Correia, F.F., Dias, J.P. and Aguiar, A., 2021. Developing Docker and Docker-compose specifications: A developers' survey. *IEEE Access, 10*, pp. 2318–2329.
30. Freitag, B., Ju, J., Harkins, S. and Masek, J., 2021, December. Update on the production of global harmonized Landsat/Sentinel-2 data products. *In American Geophysical Union Fall Meeting,* New Orleans, USA.
31. Henderson, S., 2022. Skip the download! Stream NASA data directly into Python objects. https://medium.com/pangeo/intake-stac-nasa-4cd78d6246b7, Accessed June 1, 2022.
32. Sat-search, 2022. https://github.com/sat-utils/sat-search, Accessed June 1, 2022.
33. Rocklin, M., 2015, July. Dask: Parallel computation with blocked algorithms and task scheduling. *In Proceedings of the 14th Python in Science Conference*, Austin, TX, SciPy, Vol. 130, p. 136.
34. Open Geospatial Consortium, 2021. OGC standards. http://www.opengeospatial,org/stand-ards/is.
35. Horbiński, T. and Lorek, D., 2022. The use of Leaflet and GeoJSON files for creating the interactive web map of the preindustrial state of the natural environment. *Journal of Spatial Science, 67*(1), pp. 61–77.
36. Baumann, P., 2021, November. The OGC/ISO coverage API standards: Heavy-lifting APIs for massive multi-dimensional data. *In Proceedings of the 3rd ACM SIGSPATIAL International Workshop on APIs and Libraries for Geospatial Data Science*, pp. 1–2, Beijing, China.
37. Masó, J., 2022. Geospatial web services. In: Kresse, W. and Danko, D.M. (Eds.), *Springer Handbook of Geographic Information*. Springer: Cham, pp. 493–530.
38. Miller, M., 2020. Mapbox. js: An engaging open-source web mapping tool for teaching data visualization theory. *Bulletin-Association of Canadian Map Libraries and Archives (ACMLA), 165*, pp. 32–37.
39. Folium, 2022. https://python-visualization.github.io/folium/, Accessed June 2, 2022.
40. Yongguang, L.U., 2018. A web publishing solution of geospatial information based on MapServer and OpenLayers. *Bulletin of Surveying and Mapping, 6*, p. 130.
41. TiTiler, 2022. https://developmentseed.org/titiler/, Accessed June 2, 2022.
42. eoAPI, 2022. https://github.com/developmentseed/eoAPI/, Accessed June 2, 2022.
43. pgSTAC, 2022. https://github.com/stac-utils/pgstac, Accessed June 2, 2022.
44. TiTiler pgSTAC, 2022. https://github.com/stac-utils/titiler-pgstac, Accessed June 2, 2022.
45. PROJ4JS, 2022. JavaScript library to transform coordinates from one coordinate system to another, including datum transformations. http://proj4js.org/, Accessed June 4, 2022.
46. Battle, L. and Heer, J., 2019, June. Characterizing exploratory visual analysis: A literature review and evaluation of analytic provenance in tableau. In: Gleicher, M., Leitte, H. and Viola, I. (Eds.), *Computer Graphics Forum*. John Wiley & Sons Ltd: Hoboken, NJ, Vol. 38, No. 3, pp. 145–159.
47. Superset, 2022. Apache Superset is a modern data exploration and visualization platform. https://superset.apache.org/, Accessed June 4, 2022.
48. Malarout, N., Hua, H. and Manipon, G., 2019, December. Science data processing for NISAR using hybrid cloud computing. *In AGU Fall Meeting Abstracts,* Vol. 2019, pp. IN13C–0722, San Francisco, CA.

12 Visual Exploration of LiDAR Point Clouds

Satendra Singh and Jaya Sreevalsan-Nair
Graphics-Visualization-Computing Lab,
International Institute of Information Technology Bangalore

CONTENTS

12.1 INTRODUCTION

A point cloud is a geometric dataset that represents three-dimensional (3D) shapes or objects in space using a set of 3D or 4D position coordinates, in the space-time continuum. In this chapter, we discuss static point clouds, i.e., those that do not change with time. Each point position is represented using the Cartesian coordinate system, i.e., x, y, and z values. 3D point clouds must be seen through the lens of point samples of surfaces of relevant objects. Depending on the data acquisition mode and purpose, the point clouds relate to different objects. Most of the point clouds are directly acquired using the laser scanning method or stereographic measurements, or re-sampled from other data acquisition methods, such as images. Here, we focus on the laser-scanned datasets. The objects of interest in computer graphics applications of rendering archaeological objects, creating busts of

DOI: 10.1201/9781003270928-16

the head of state, etc. involve scanning a single object, with complex surface geometry of folds and holes in the manifold. At the same time, the objects of interest in geospatial applications belong to larger regional capture, where the resolution of data capture per object is not of as high fidelity as is needed for digital geometric reconstruction of the object in its entirety.

In geospatial applications, Light Detection and Ranging (LiDAR) technology is widely used, which gives point clouds. The points in these datasets belong to semantic/object classes in the scanned region such as road, vegetation, low vegetation, buildings, poles, transmission lines, vehicles (cars, trucks). Visualizing the point clouds as-is gives us a rough impression of the objects in the region/scene and is the first step in *sense-making* (Klein, Moon, and Hoffman 2006). However, some classes with similar physical characteristics would be indistinguishable, e.g., asphalt road and dirt road, a building and a dense tree. Hence, performing *semantic classification/segmentation* becomes an essential step, where a distinct color mapped to a semantic class used for rendering gives a better semantic understanding of the scene (Figure 12.1). The visualizations of the point cloud with the same color or height-based color scheme are based on observed variables, i.e., the position of the points. Different from the observed data, the class information is either given in the annotated dataset or is computed using machine or deep learning methods (Weinmann, Jutzi, Hinz, and Mallet 2015). Hence, the visualizations related to semantic classification provide insight into a more complete sense-making of the dataset.

In conventional visualization of these point clouds, the points are either used as-is in their raw unstructured format (Schütz 2016) or are converted into a structured format, i.e., triangle models or images (Sreevalsan-Nair, Mohapatra, and Singh 2021). These different data formats lead to rendering using different visualization methods. However, many tools that provide the 3D rendering of the raw point clouds, even with observed attribute-based colors, are limited to summary visualizations. They do not usually visualize relevant intermediate results which may be global or local aspects

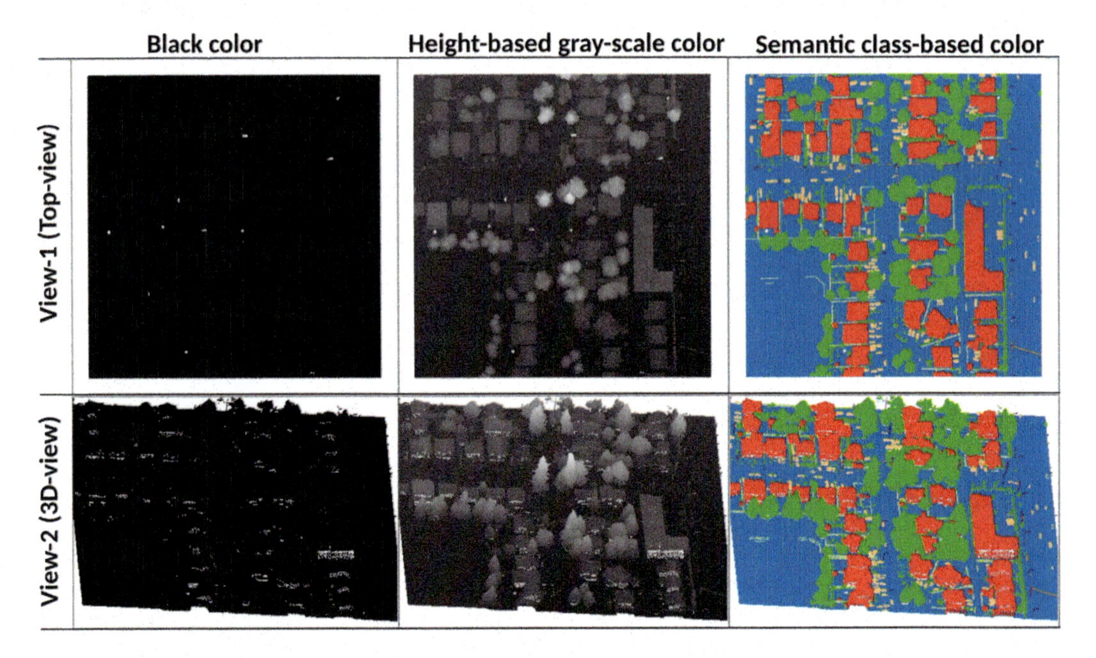

FIGURE 12.1 The visualization shows how adding color based on height and semantic classes improves the semantic understanding of the airborne LiDAR point cloud than rendering the points black in color. View-2 shows that with a certain orientation of the point cloud, one can infer the region the best using a semantic class-based color scheme, e.g., trees and buildings are better rendered from left to right. Semantic color scheme: ground (blue), vegetation (dark green), power lines (light green), poles (orange), buildings (red), fences (light blue), trucks(yellow), cars (pink), and unknown (dark blue).

of point clouds, e.g., the overall class distribution of the point cloud, the features used in semantic classification using machine learning or deep learning. Such a requirement is fulfilled by a *visual analytics* system that explores the 3D point cloud data for interactive selection and data analysis. *Visual analytics* is formally defined as data science workflows that provide a feedback loop from the human-in-the-loop (Keim, Andrienko, Fekete, Görg, Kohlhammer, and Melançon 2008), in the parlance of data visualization.

Such a visualization system supports not just point cloud visualization in its raw format, but also the intermediate and outcomes of semantic classification, such as extracted features. To decouple the visualization from the analysis to the greatest extent, for the sake of modularity, an interactive visualization system can be designed to use services to and from a cloud-based distributed system for data analysis. For the latter, Apache Spark-Cassandra integration has been used in the state-of-the-art (Singh and Sreevalsan-Nair 2020).

The traditional approach to visualize these point clouds requires transferring data to stand-alone devices and installing a visualization tool. Additionally, this requires specific configurations to be present on the device, such as CPU, RAM, or GPU. As the data size increases, e.g., in the case of airborne LiDAR, loading and transferring the data from one device to another becomes challenging on a single system. For instance, interactive visualization and analysis reports by a researcher team require sharing the data and loading it on the target device. When performing semantic classification of the point clouds, there arises the need to inspect the feature vectors used for supervised learning, such as the random forest classifier (Breiman 2001). Thus, we focus on the visual exploration of raw data and hand-crafted feature vectors for the semantic classification of the point clouds. This entails point-wise analysis and graphical rendering. However, the conventional method of data management, and storage as well as computation, is not scalable for large-scale point clouds. Thus, we use a cloud-based distributed system designed for handling multiscale feature extraction and semantic classification of large-scale LiDAR point clouds. Here, we discuss in detail an interactive and effective browser-based visualization tool that is also integrable with the cloud-based distributed system.

The Open Graphics Library (OpenGL) is the de-facto accepted standard specification used for 3D rendering in computer graphics (Shreiner 2004). With the advances in OpenGL extensions to web browsers, namely, WebGL (Matsuda and Lea 2013), both 3D graphical rendering and real-time user interactivity have become available on standard browsers. These are natively supported by all browser engines, and even on mobile devices. This has facilitated the distribution of 3D content on the browser seamlessly without the need for a specific device. Also, the advancement in web technologies to load data asynchronously and process using the worker threads allow the user to keep the web browser application interactive while doing all the heavy lifting in the background in parallel. These issues are addressed by a thin-client-based visualization system, discussed here.

For smaller datasets that can fit in the device memory where the browser is running, the data loading and visualization are straightforward, without any special memory management. But, as the data size increases, the data may not fit in the memory, and the time to download may range from minutes to hours, thus affecting user interactivity. Thus, data management becomes a critical process in the case of large-scale point cloud datasets on browser-based visualization tools. The use of a big data framework addresses the challenge of managing data with limited browser cache, and yet maintaining the user interactivity of the tool.

12.2 VISUALIZATION SYSTEMS FOR AIRBORNE LIDAR POINT CLOUDS

For rendering large-scale LiDAR point clouds with sufficient interactivity for querying, stylized rendering, etc., out-of-core visualization is done using the geometry shader and hierarchical spatial data structures such as the kd-tree (Richter, Kyprianidis, and Döllner 2013). Such a visualization system is capable of implementing focus+context rendering of point clouds. This has been implemented on server-client architecture, where thick and thin clients run WebGL for point rendering (Discher, Richter, and Döllner 2018). The rendering engine in such an architecture includes a level-of-detail selector, memory manager, and image compositor for interactive user capabilities of highlighting, annotation, querying, etc., of point clouds. Image compositing is used for rendering on thin clients and rendering on cloud resources is used in the case of thick clients.

Visualization tools are often designed to comply with open standards for the portability of data representation across tools. One such rigorous standard applicable for point clouds is *3D Tiles* (OGC International Community Standard 2019) by OGC (Open Geospatial Consortium), which was originally published by Cesium. OGC 3D Tiles recommends the use of a hierarchical organization of tiles which refers to the content that can be rendered. This standard specifies the file format for storing the readable content. For point clouds, there are special rendering options as a batch-wise feature applied to a subset of points, or a point-wise feature. OGC Indexed 3D Scene Layer (I3S) (OGC International Community Standard 2017) is an alternative to 3D Tiles and was originally published by ESRI. While I3S is similar to 3D Tiles in using the hierarchical organization of data, I3S uses regions, referred to as "nodes," instead of tiles. I3S is designed for rapidly streaming and distributing large datasets across enterprise applications to mobile, web, and desktop clients. Thus, I3S is used in applications that use server-client models. I3S manages nodes in the form of scene layers. The overarching standard for visualization of GIS data that is used in conjunction with I3S and 3D Tiles is the OGC 3D Portrayal Service (OGC Implementation Standard 2015) (3DPS) (Koukofikis and Coors 2018). The 3DPS is a service implementation standard for ensuring seamless and interoperable 3D portrayals in servers and clients. It defines how to view 3D data integrated from heterogeneous geoinformation sources.

Similar to the distributed system-based visualization tool described in this chapter, there are several other browser-based visualization tools for LiDAR point clouds. iTowns (IGN and LASTIG, France 2016; Picavet, Brédif, Konini, and Devaux 2016) is one such visualization tool that complies with the 3D Tiles standard. Built using Three.js (Danchilla 2012; Dirksen 2013), iTowns facilitates the rendering of geometric primitives using WebGL in the form of layers. It is used for visualizing 3D geospatial as well as precise measurement data. It provides options to render PointCloudLayer and PotreeLayer for point clouds. Potree (Schütz 2016) is a web-based point rendering system that implements point cloud rendering on limited-resource devices. This is done by supporting visualization on standard web browsers and implementing point cloud editing using the modifiable nested octree (MNO) structure. Potree allows users to select regions of point clouds by using a volumetric brush. Different from Potree, a distributed system can be used as a backbone to provide real-time visualization of analytic processes (Singh 2021), e.g., feature extraction and semantic classification. When visualization processes are executed as services to the distributed system (Singh 2021), one can create interactive applications that perform real-time operations, analytics, and rendering of a large-scale point cloud simultaneously. Potree achieves multi-resolution rendering of the point cloud using hierarchical data structures, namely octree.

A similar tool for tiling, hosting, and managing 3D spatial data is Cesium (2011) which also complies with the 3D Tiles standard. Unlike iTowns which is developed in lines of geospatial datasets, Cesium is developed in the space of high-fidelity graphics rendering for geospatial data, including LiDAR point clouds. Cesium has extensions to work seamlessly with graphics engines such as Unreal and O3DE (Open 3D Engine) to expand their capabilities to render geospatial datasets. Cesium is also a Javascript browser-based tool. CesiumJS library is available to be used as API calls in other Javascript projects.

There are several other alternatives to Potree among the free and open-source software (FOSS) for web-based point cloud visualization systems. This includes Plasio with support for LAS file format that is exclusively used for LiDAR point clouds, Point Data Abstraction Library (PDAL) for abstraction layer with point management operations, Entwine as an indexing library built using PDAL, Greyhound as a RESTful HTTP server that works with the Entwine library, etc., (Martinez-Rubi, Verhoeven, Van Meersbergen, Van Oosterom, Gonçalves, Tijssen, et al. 2015).

The aforementioned tools and libraries are specifically available for point cloud data management and rendering. However, there is a gap in support for data science workflows and tools for large-scale point clouds. The system discussed in this chapter describes an architecture and its implementation to seamlessly integrate machine learning and visualization.

Using alternative coordinate systems for visualization, the LiDAR point clouds can be projected to barycentric space to visualize the probabilistic geometric features in a triangle referred to as the dimensionality density diagram (Brodu and Lague 2012). Adding semantic class information using color encoding has enabled making the image output of the visualization a global descriptor of such point clouds for comparison (Sreevalsan-Nair et al. 2021).

12.3 DISTRIBUTED SYSTEM ARCHITECTURE FOR VISUAL ANALYTICS TOOL

In this chapter, we describe a system architecture (Figure 12.2) that integrates data analytics as well as interactive visualization (Singh 2021). This distributed system has three components, namely, (1) the browser-based visualization tool, (2) the Apache Spark-Cassandra cluster for processing and managing the data, and (3) a set of services between the tool and the cluster, referred to as *backend services*. These services include scheduling jobs, loading point clouds, and performing data analytics. The browser-based visualization tool has in-built features to allow interaction and dispatch requests to these services to perform various user-defined tasks.

12.3.1 Browser-based Visualization Tool

There are specific design choices to enable the visualization of a large-scale point cloud on the web browser. The visualization tool is decoupled from the distributed system for semantic classification, but at the same time, they communicate with each other through the backend services. The visualization tool has three main components, namely, (i) BufferGeometry for the canvas rendering, (ii) Web workers for asynchronous and parallel processing needed for the rendering at interactive frame rate, and (iii) a Service interface to communicate with the backend services.

12.3.1.1 Canvas Rendering Using BufferGeometry

The canvas is a region of the web browser which exclusively supports an *OpenGL context* for 3D rendering. This context involves setting up and running the *OpenGL state machine* for its implementation. The context is implemented using WebGL, which is written using the Javascript programming

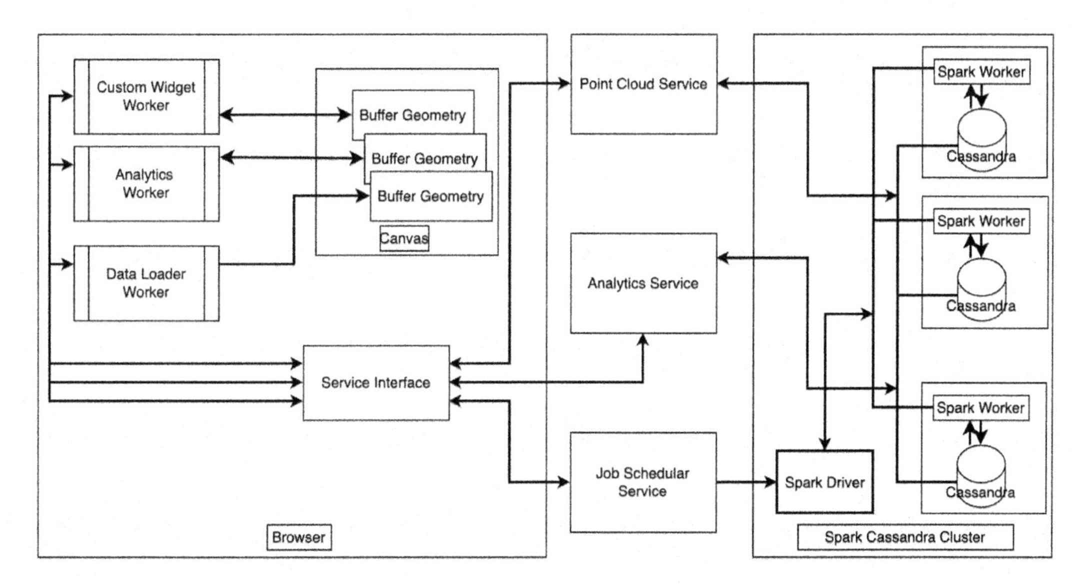

FIGURE 12.2 Distributed system architecture for an interactive visualization system that is decoupled from, but peripheral to an Apache Spark-Cassandra integrated system for semantic classification (Singh 2021).

language. Higher-level language abstractions of WebGL in Javascript, namely, Three.js, p5.js, D3.js, etc., are relevant alternatives to WebGL. Three.js (Danchilla 2012; Dirksen 2013) v0.142.0[1] has been used for a proof-of-concept (POC) implementation (Singh 2021) owing to its implementation of high-quality 3D graphics rendering of point primitives on a web browser. Alternatives to Three.js are applicable here for high-fidelity rendering.

Here, Three.js is used for rendering the point clouds, for which BufferGeometry provided in the library is used. BufferGeometry is a representation of 3D geometric primitives, such as mesh, line, or point geometry. BufferGeometry refers to a collection of buffers, where each buffer is dedicated to different attributes, namely, vertex positions, face indices, normals, colors, UVs, and custom attributes. Since BufferGeometry stores all the relevant data for rendering, it reduces the communication cost of passing data to GPU for rendering. These memory buffers act like fixed buffers and are reused to update the attribute values selectively. We use BufferGeomtery to update selected attributes interactively for specific visualizations. Thus, BufferAttributes in Three.js are used to create these buffers. The entire point cloud is stored in a single instance of BufferGeometry. In this system, three different BufferGeometry are used for: (1) point cloud rendering, (2) analytics plots/charts, and (3) widgets on the GUI.

When using BufferGeometry for point rendering, the user has the additional flexibility to decide the number of points that are finally rendered on the screen, which is referred to as the point budget. This size determines efficient and interactive rendering by the browser. Thus, a critical step in using Three.js for rendering point primitives here is in determining an appropriate point budget. For instance, a fixed value has been used in the POC implementation (Singh 2021). Also, the BufferGeometry itself cannot be initialized to different sizes for different file inputs. This is because the visualization and the distributed system are initialized before the actual loading of the file.

Hence, a maximum point budget, say 10 million points, is used which is also used to determine the BufferGeometry size. The rationale is that a large enough point budget value helps contain the entire input point cloud in the BufferGeometry instance. This option is essential to address the computational complexity of repetitive configuration of the BufferGeometry.

12.3.1.2 Asynchronous and Parallel Processing

Javascript is a single-threaded program, and multiple scripts cannot run at the same time. Hence, to accommodate the requirement of Javascript program implementation for browser visualization, we design handling of two different types of tasks, namely, the IO- and compute-intensive tasks. If any highly compute-intensive task is run, the thread tends to get blocked leading the graphical user interface (GUI) to become unresponsive. Then, the user has to wait until the task gets completed to become responsive. Hence, we use a *non-blocking strategy* to improve the interactivity of the GUI. Callback functions are used to run compute-intensive tasks without blocking the program running on the same thread. This strategy allows access to the frame buffer in the GPU as soon as it has completed the *write* operation.

This system has *web workers* which are deployed on the browser to run parallel tasks using background threads. *Asynchronous methods* are additionally used for many of the web API interactions and for dispatching the tasks to the web workers. Thus, a parallel approach is strategically used to efficiently generate analytics (e.g., class distribution), data processing, or performing any mathematical operations on the data. For instance, in the POC implementation (Singh 2021), web workers are used for transforming the data into objects required by the graph API for plotting the analytics in the visualization. We have three different types of web workers in our system, namely, the *Custom Widget* worker, *Analytics* worker, and *Data Loader* worker.

12.3.1.3 Service Interface

A specific interface is provided between the backend services to the distributed system and the web workers referred to as the service interface. The service interface enables the scheduling of the backend services. This interface resides within the visualization system.

12.3.2 Distributed System for Semantic Classification

The goal for the distributed system is two-fold, namely (1) storing and managing large datasets of millions of points in a point cloud, and (2) performing computationally intensive tasks on large datasets. Compute-intensive tasks include the hand-crafted feature extraction of point clouds, as well as running supervised learning algorithms for classification, such as random forest classifier (Breiman 2001). The hand-crafted features for semantic classification are extracted at each point using three steps—(1) computation of local geometric descriptors, i.e., covariance matrix of position coordinates of local neighbors, at the point, (2) the subsequent eigenvalue decomposition of the matrices, and (3) computation of height- and eigen-based features (Weinmann, Jutzi, and Mallet 2013). Given that LiDAR point clouds are environmental datasets with sufficient inherent uncertainty, these features are computed at multiple spatial scales, i.e., the size of local neighborhoods to improve the accuracy of the computed features (Demantké, Mallet, David, and Vallet 2011). These computations cannot be done in real time even for less than a million points and require parallel computation (Kumari, Ashe, and Sreevalsan-Nair 2014). For instance, for the DALES airborne LiDAR dataset (Varney, Asari, and Graehling 2020), for raw data of 8 and 3 GB of training and testing, respectively, the corresponding extracted features require 70 and 26 GB, for three scales. Thus, for a large-scale point cloud, a parallel computing environment is required to also support distributed storage, thus alleviating the dependency on a single system.

As a compute or a unified analytics engine, Apache Spark uses a sequence of data transformations on a Resilient Distributed Dataset (RDD), which is a partition of fault-tolerant collection of objects (Zaharia et al. 2016). Spark incorporates libraries with composable Application Programming Interface (API) for machine learning (MLlib) (Meng et al. 2016), which can be used for supervised learning for classification. Spark has already been used for tree crown extraction from large-scale point clouds (Boehm, Liu, and Alis 2016). The requirement for a distributed storage system is to perform two-way communication with the classification module, each submodule of the multiscale feature extraction, and the visualization tool/engine. The distributed storage is population using one-way communication from the normalization module which handles the reading of the input data. Thus, distributed storage is truly the central component of our visual analytics system (Figure 12.3).

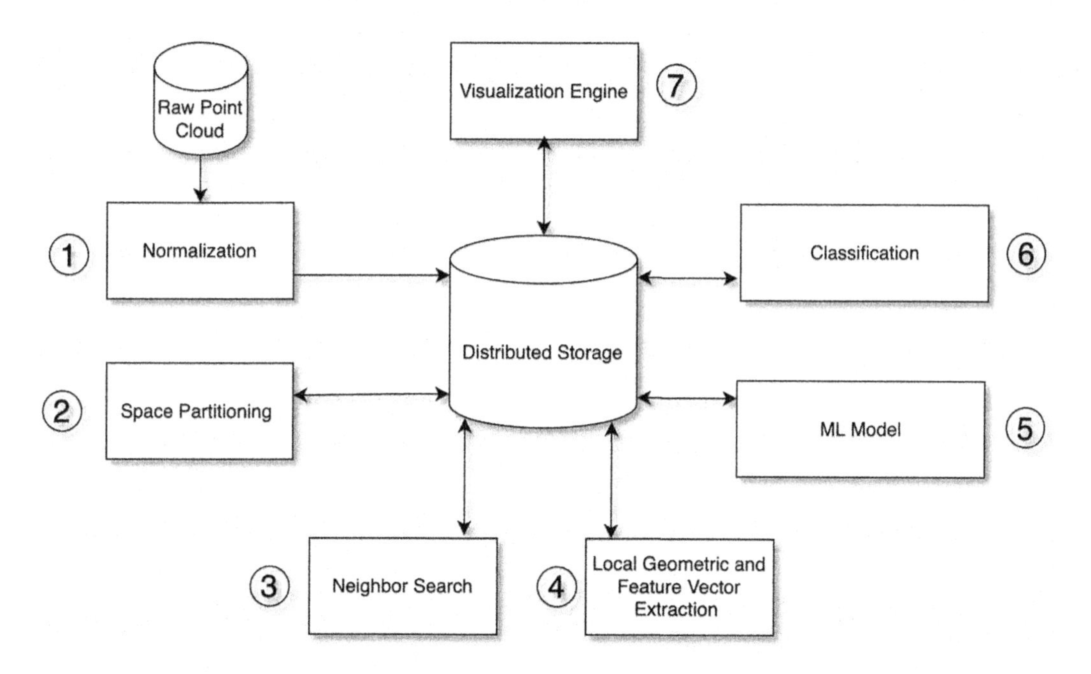

FIGURE 12.3 Requirements for a distributed storage system to support both visualization and semantic classification.

Apache Cassandra (Lakshman and Malik 2010) is an open-source, distributed, NoSQL database, that uses a partitioned wide-column storage model with eventually consistent semantics.

Here, the system requirements are met by using an integration of widely used big data frameworks, namely Spark and Cassandra (Singh and Sreevalsan-Nair 2020) for computing and storage needs. The distributed system integrates the horizontally scalable compute engine and storage framework, namely, Spark and Cassandra (Singh and Sreevalsan-Nair 2020, 2021). The choice of technologies in the distributed system fulfills the requirements of processing the large-scale point cloud dataset and performing semantic classification. This system has been utilized to implement adaptive multiscale feature extraction (Singh and Sreevalsan-Nair 2022), and perform supervised learning, such as random forest classifier (RFC) (Breiman 2001) and gradient boosted tree (GBT) classifier (Friedman 2001). It must be noted that the system is limited to using the classification models available on Spark MLlib.

Specifically, in the case of semantic segmentation (or classification) of airborne LiDAR point clouds, RFC continues to be widely used (Liao et al. 2022; Weinmann et al. 2015; Westfechtel et al. 2021). This is because RFC is highly optimized in comparison with other supervised learning algorithms and performs comparably to deep learning methods. The use of deep learning for airborne point cloud classification is still not established owing to the requirement of a large amount of labeled data required for training and the non-generalizability of a pre-trained model to new geographical regions (Liao et al. 2022). For airborne LiDAR point clouds, manual annotation continues to be widely used, which gets labor- and time-intensive for big data point clouds. Thus, the visualization tool with the facility of performing RFC remains relevant for airborne LiDAR point cloud classification.

The overall architecture of the distributed system has four components (Figure 12.4), namely, a cloud computing environment, a Spark master node, a cluster manager, and a cluster of worker nodes. At one end, the raw point cloud is stored in a *cloud computing setup*, for which the Amazon Web Services (AWS) S3 is used. At the other end of the architecture, the *cluster of worker nodes* is deployed in such a way that each run both the Spark Executor and Cassandra (data) Node.

Spark loads the raw point cloud data points from cloud computing resources into the Spark RDDs in the worker nodes. The *cluster manager* manages and allocates resources required by Spark Executors in the cluster on which the Spark application runs. This distributes the data randomly across all the worker nodes using its default *hash partitioning* program, and the data is stored as RDDs in each worker node. Now, the worker nodes run the logic of the custom partitioning algorithm (Section 12.4.4) on the data in the RDD to assign the partition key to each point locally. The partition key is the ID of the physical data partition to which the point belongs and is used to re-distribute the point cloud to the worker nodes. The data partitions finally reside in the local Cassandra Nodes in the worker nodes. In the POC implementation (Singh 2021), this custom algorithm is a data-driven spatial partitioning one (Section 12.4.4) that partitions the LiDAR point cloud data into multiple contiguous regions. Thus, each Cassandra Node stores one or more data partitions, i.e., regions, based on load balancing. The parameters for the partition algorithm are either given as user-defined inputs as job parameters or can be determined from the data. For the latter, the local parameters are identified in the Spark Executors from the initial partitions using the default hash partitioning. These local parameters are then *map-reduced* in the Spark Master and persisted back to the Cassandra Nodes in the worker nodes.

The data is stored in Cassandra, in its nodes in the worker nodes using a primary key for each data item, i.e., a point in the point cloud. In our case, the primary key used in Cassandra is (`RegionID, X, Y, Z`) where `RegionID` is the partition key, i.e., the partition to which the point belongs, and `X, Y, Z`, i.e., the position coordinates of the point, is its clustering key. While each node can hold multiple data partitions for load balancing, the cluster manager ensures that a data partition itself is not split across nodes. By design, Cassandra uses *consistent hashing* to distribute and manage the data in the cluster. The partition keys are required for managing consistent hashing partitions. The partition keys are stored in the primary keys. The clustering keys are needed within the partition.

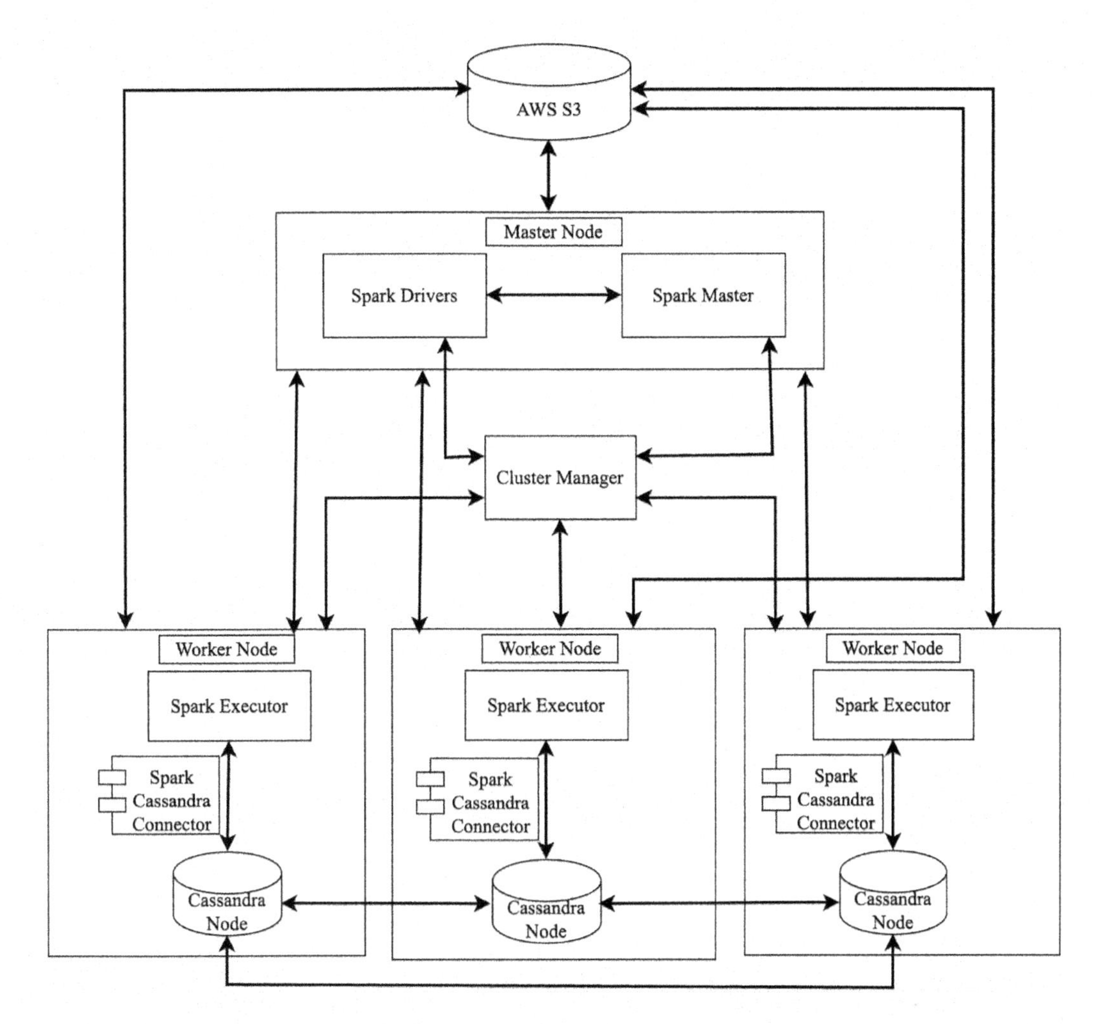

FIGURE 12.4 Architecture of the cloud-based Apache Spark-Cassandra integrated system, which interfaces with the visualization system using the backend services (Singh 2021).

Thus, in each *worker node*, the Spark Executor program assigns the `RegionID` to each point in the point cloud using a custom partitioning algorithm (Section 12.4.4). The resultant RDD is now stored in Cassandra across worker nodes using `RegionID` as the partition key. This ensures the data for each region in the point cloud resides entirely in the local node for processing. Any intermediate result, that requires to be persisted, is also stored on locally available Cassandra Node. Each worker node consists of the Spark Executor, the Spark Cassandra Connector (DataStax 2014), and the Cassandra Node (Figure 12.4). The role of the connector is to load the local data and save the intermediate and outcomes from the data analytics in Spark to the Cassandra Node.

The data analytics is performed in-memory on the RDD in the Spark Executor using MLlib for semantic classification, clustering, regression models, etc. The training for the models that are supported in MLlib, e.g., Random Forest Classifier (RFC), naïve Bayes, generalized linear models, etc., are executed in a distributed manner through the use of distributed file system. The implementation of MLlib on Spark enables data communications to arrive at a single model. The self-contained worker node avoids any I/O (input/output) latency, thus improving the overall performance. This architecture is designed to be horizontally scalable, thus allowing us to process any data size by just adding extra worker nodes (Singh 2021).

Any supervised learning classifier in MLlib that works with hand-crafted feature vectors can be used for airborne LiDAR point clouds. However, probabilistic learning and ensemble methods with bagging are strongly preferred for semantic segmentation of airborne LiDAR point clouds (Weinmann et al. 2015).

12.3.3 Backend Services

These services are provided to manage the data in Cassandra nodes, for data mobility to the visualization system, and for performing user-defined selections made on the GUI.

- **Point Cloud Service**: A dedicated service is needed to expose the set of APIs to manage the point cloud data and its intermediate processed data stored in the Cassandra cluster. It directly interacts with the Cassandra cluster to retrieve data.
- **Analytics Service**: There is a requirement to inform the user of the visualization system of the "status" of the current operations. This entails building an aggregated report or exposing the analytics information generated on a particular analysis done on the point cloud directly or on the processed data. This is achieved using the Analytics service. This service is also responsible for computing lightweight analytics reports, such as class distribution reports on data requested to generate the view-ready report.
- **Job Scheduler Service**: The Spark-Cassandra cluster can accommodate a limited number of requests only, at any given time, depending on the cluster configuration. This requires us to build a service that can schedule the submitted task for processing and manage the status of the task.

12.4 SYSTEM IMPLEMENTATION

In this section, we discuss the design decisions and considerations made for the implementation of the system described in Section 12.3. For the visualization tool, the predefined visualization tasks and the supporting design for the GUI are explained. For efficient rendering, we use a subsampling strategy. In the distributed system for classification, we describe the custom partitioning algorithm used in Spark for distributing data across different worker nodes. We discuss in detail the distributed data model needed for Cassandra integration for efficient storage, query, and retrieval of data. The system specifications in the POC implementation (Singh 2021) are summarized here.

12.4.1 Visualization Tasks

For enabling visual analytics through this tool, the GUI is specifically designed to implement the following visualization tasks:

- Compare the outcomes of different classifiers on the entire point cloud or a selected region of interest (ROI), or between an ROI and the rest of the point cloud,
- Visualize the statistical distributions of the semantic class and features in the point cloud or in a specific class, to understand its influence on the choice of a classifier.

To accomplish these tasks, the GUI supports the following user interactions:

- 3D exploration of the point cloud, as a complete instance, through the navigation controls on the canvas,
- Progressive visualization of real-time classification outcomes to monitor the accuracy and performance of the classifier.
- Update of the analytics report charts whenever the dataset is changed or ROI is selected,

- Selection of an ROI to exclusively perform 3D point cloud visualization, and classification using different classifiers,

12.4.2 Graphical User Interface Design

Figure 12.5 shows the dashboard of the GUI designed to accomplish the visualization tasks using the distributed system (Section 12.3). The GUI is divided into three panels, namely, *left, middle,* and *right* panels. They correspond to the *control widgets*, the canvas for point cloud rendering, and the *data analytics charts*, respectively. These three panels correspond to the Web workers in the visualization tool (Section 12.3.1). The right panel also has a secondary canvas region to render the ROI, referred to as *ROI visualization* box. The mouse and the widget-based GUI controls are described further here.

12.4.2.1 Navigation Controls on the Canvas

A single navigation control mechanism is not sufficient to fulfill all the desired interactivity, i.e., rotation, panning, and zooming of the point cloud. Hence, *OrbitControls* are used in the Three.js library as the navigation tool. As the name suggests, it orbits around a target or pivot. This is customized to the requirement of visualization of the LiDAR point clouds. The rotation is constrained along the Y-axis by locking it in the positive up direction. It prevents any tilting of the axis and allows the user to orbit, zoom and pan using the mouse buttons.

12.4.2.2 Classification Tool

The *classification tool* in the control widgets provides the list of pre-built classifiers widely used for point clouds, e.g., random forest classifier, gradient boosted tree classifier, etc., available on Spark MLlib. Clicking on the tool activates the classifier and the classification job gets submitted to the job scheduler service. This backend service internally schedules the classification job on the Spark-Cassandra cluster. As the classification of points in the point cloud is complete, the corresponding results are stored in the Cassandra cluster in real time. The background worker can now pull these results progressively and update the buffer for *progressive visualization* of the classification results in real time.

12.4.2.3 Analytics Widgets

The visualization tool, by design, exposes a set of methods to create widgets using the message-passing by the web workers. Each widget receives the bounding box information of the selection box message in the worker method. Then, the web worker uses this information to get details for an analytics report from the backend services, i.e., either the analytics service or the point cloud service. The analytics report or data is then built by the worker method. Examples of these widgets include the class distribution chart, probability distribution of a feature, etc. These widgets are seen as data analytics charts in the rightmost panel in the dashboard (Figure 12.5).

12.4.2.4 Selection and Exploration of Regions of Interest

This visualization tool has the feature of interactively selecting a rectangular ROI in X-Y plane in the point cloud. With a focus on ROI, the user can perform various tasks, such as real-time classification using pre-built machine learning (ML) models available on Spark MLlib. The user interactively selects the ROI in the main canvas of the middle panel using the magenta box, the ROI selector. The updated rendering of the ROI is displayed in the secondary canvas in the right panel.

12.4.3 Subsampling for Efficient Rendering

The original point cloud data can be very dense and large scale, in our application. Loading and visualizing these points in the browser for a larger area will require a large memory footprint, and

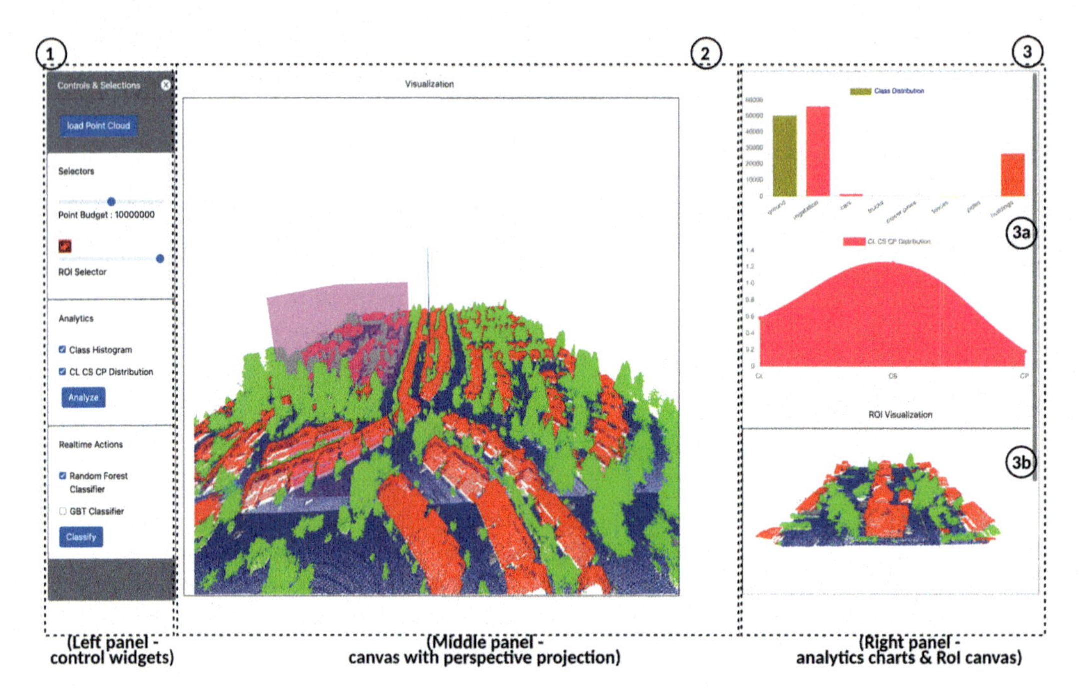

FIGURE 12.5 The graphical user interface (GUI) of the visualization system (Singh 2021) with the (1) left, (2) middle, and (3) right panels. In the right panel, the analytics and exploration of the point cloud are seen, for (3a) analytics charts, and (3b) region of interest (ROI) visualization. Here, the ROI selector is shown in the canvas in (2).

transferring the data incurs network latency. In addition, this leads to rendering and frame refresh on a browser becoming slow. Thus, rendering large point clouds impacts the interactivity of the visualization application. Here, the widely-used solution of optimal sampling of the original point cloud data is used. The selective sampling preserves the underlying (geometric) surface model. The design requirement of the tool also includes that the number of points in the sampling must be limited by the maximum that can be rendered on the targeted device browser.

Voxel-grid subsampling approximates the set of points inside a voxel with the centroid of these points (Figure 12.6). While approximation using the voxel center is faster than using the centroid, the irregularities inherent in the point cloud are best represented using the centroid. Thus, the centroid of points in a voxel approximates the set of points in the voxel and its underlying surface more accurately. We choose the voxel-grid-based sampling instead of the Poisson disk sampling used in Potree (Schütz 2016) to reduce the number of computations. The size of the voxel determines the sample size and the amount of information that is retained from the original data. The smaller the voxels, the more information is retained, but the number of samples itself increases. The visualization tool in the distributed system also implements intensive computations and communication for user interactivity and analytics. The voxel-grid sampling strategy is an optimal trade-off between computational complexity and accuracy (Singh 2021).

12.4.4 Custom Partitioning in Spark

For partitioning the data across multiple Cassandra Nodes and Spark Executors, the partitioning strategy has to be such that inter-worker-node communication is minimized. Hence, for a geospatial application, an appropriate *spatial partitioning strategy* can be used for the data partitioning algorithm in the distributed system.

Figure 12.7 demonstrates one way of contiguous partitioning of the airborne LiDAR data. As an example (Singh and Sreevalsan-Nair 2020), for airborne LiDAR point clouds, we contiguously

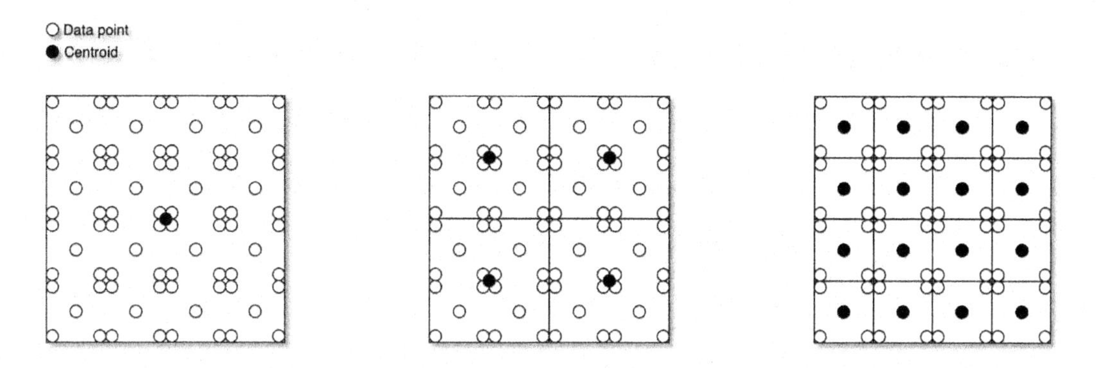

FIGURE 12.6 Example demonstrating the change in voxel grid sub-sampling with change in voxel sizes.

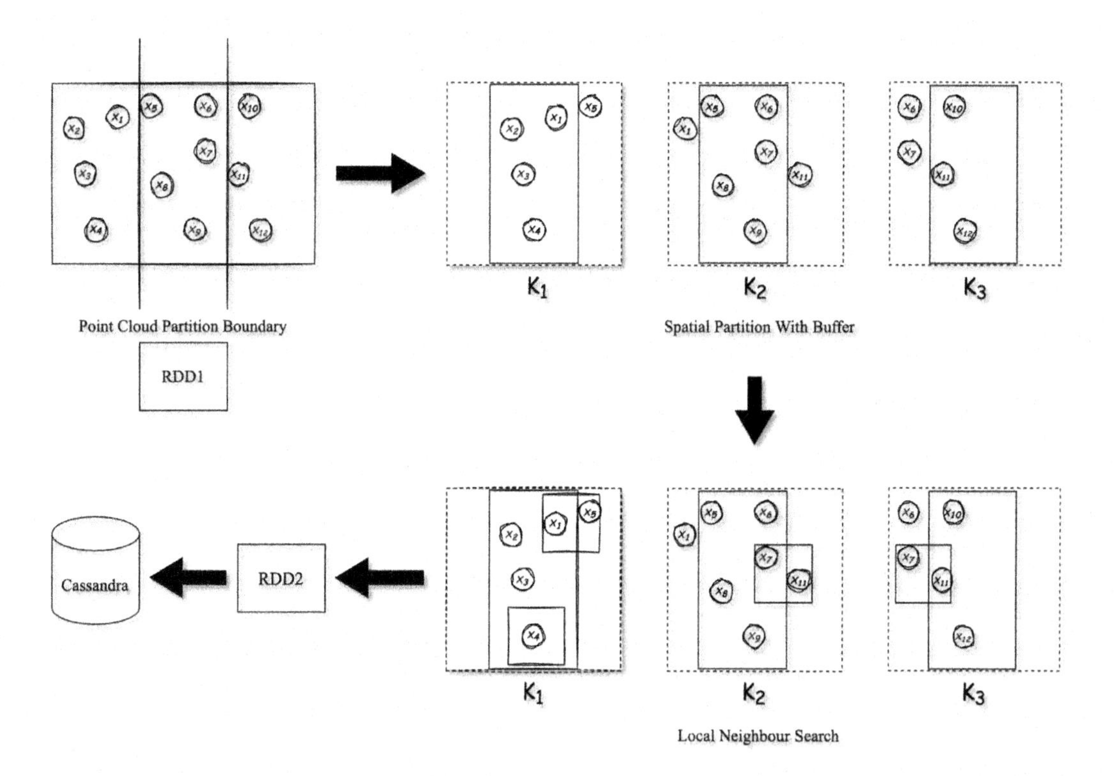

FIGURE 12.7 Spatial partitioning algorithm for airborne LiDAR point clouds serves as the custom partitioning program in Spark in the POC implementation (r).

partition along the X-Y plane, and the partitioning is along the axis with a longer range, referred to as the *principal axis*. We ensure that the number of partitions matches the number of worker nodes. Thus, for partitioning along a single axis, say X-axis, we get partitions of regions within uniform intervals of X-value, but across all Y and Z values. In addition to the actual partition, we include buffer regions on either side of the partition along the principal axis. This is to ensure that the points in the actual partition have all their local neighbors available in the local worker node. Thus, by design, the points in the buffer region are processed for feature extraction in the nodes where that region is part of the actual partition for the worker node. To reduce the number of computations, we use a cuboidal local neighborhood, as opposed to *spherical*, *cylindrical*, or *k-nearest* neighborhoods (Weinmann et al. 2015).

The already partitioned point cloud is now loaded into multiple regions in Spark as RDD, where each region also contains the left and right buffer points as a neighbor. Then, the points in the partition are subsampled using the voxel-grid method. The resultant RDD with the subsampled data then is persisted into the Cassandra for visualization. Additionally, the different voxel sizes across partitions are chosen to adaptively subsample at different resolutions, and then persist the sub-sampled points for visualization on demand. Further, the subsampled data are stored in each region in the cache available in the Point Cloud Service using the cache-aside strategy. The use of cache allows efficient loading of the subsampled data on the browser over the internet/network, thus, reducing the redundant read calls to Cassandra.

In our case study of the airborne LiDAR point cloud, we have used a voxel size of (0.5, 0.5, 0.5 m) for the partitions. Experimenting with different voxel sizes may be considered for future work.

12.4.5 Distributed Data Model in Cassandra

As discussed in Section 12.4.4, there is a specific way by which the raw point cloud is partitioned and stored across different Cassandra nodes. Also, any intermediate data like local geometric descriptors, feature vectors, and classification results get stored using the unique identifier, `RegionID`, based on this spatial partitioning. This method of storing data allows querying Cassandra most optimally to retrieve the data inside a voxel bounded by specific (`X, Y`) values.

Different from other NoSQL databases, the data model should be defined based on the query requirements in Cassandra. The objective of data modeling in Cassandra is to write and store data in such a way that can improve the performance of the read query. The governing rules of the data model are:

1. **To Spread Data Evenly around the Cluster**: the data should be evenly distributed across the cluster. By saying that the choice of partition key should be such that can balance the data and distribute it evenly.
2. **To Minimize the Number of Partitions Read**: the objective here is to allow the query to read from one partition and should not be spread across multiple partitions. Even when the partitions lie on the same node, it is complex to read from multiple partitions.

However, both these rules contradict each other. As per Rule (2), one should ideally have only one partition for the whole data to reduce the queries across partitions. But this violates Rule (1). That also means that if Rule (1) is satisfied, then Rule (2) will be violated. Overall, a trade-off between the two rules is needed, which is achieved using a data model in Cassandra that optimizes the performance.

Cassandra also has several restrictions on the operators that can be used, including:

- The partition key columns support only two operators: = and IN.
- Clustering columns support the =, IN, >, ≥, ≤, <, CONTAINS and CONTAINS KEY operators in single-column restrictions and the =, IN, >, ≥, ≤, <, operators in multicolumn restrictions.

As per the requirements in the distributed system, there are additional constraints to filter the points from one region at a time and to always execute a query on a single partition.

Thus, to implement an optimal data model, one executes the query to load the points in batches on multiple pages to maximize the throughput and minimize the network latency. To define multiple-page searches, the principal axis in the region is now partitioned into multiple sub-intervals, and for each sub-interval, the range search loads one page. An example query:

```
SELECT * FROM varname.PointCloud
WHERE RegionID = 1 AND (X, Y) >= (1.5, 3.0) AND (X, Y) <= (3.0, 4)
LIMIT 100000;
```

Here, the above query is used for implementing the distributed data model in Cassandra. Improving the quality of 3D graphical rendering for visualization and adding GUI features already available in Potree for data exploration can be done efficiently using Cassandra for persistent storage as an improvement to the POC implementation (Singh 2021).

12.4.6 System Specifications

In the POC implementation (Singh 2021), Apache Spark 2.4 and Apache Spark ML, integrated with Cassandra 3.0., have been used with one master node and five executor nodes on Apache Spark. Each of the six nodes uses Intel i7 processor @2.80 GHz, 4 cores, 8 logical processors, 8GB RAM, and runs Ubuntu 18.04.1 LTS (GNU Linux 5.4.0–1049-aws) x86_64 operating system. AWS (Amazon Web Services) S3 bucket has been additionally used for storing the classifier model and raw point cloud. The total volume of data and the number of objects one can store are unlimited. Each Amazon S3 object ranges from 0 to 5 TB in size.

An EC2 image has been created, which is installed with Apache Spark, Cassandra, and Java support. The instances have been created from the image, which has been spawned in the same private network or a virtual private network, i.e., a virtual private cloud (VPC) in AWS, to reduce the network latency. The S3 bucket has also been created in the same region. To run the applications in the Spark cluster, they have been deployed using the stand-alone mode in the POC implementation. The stand-alone cluster manager mode has also been adopted on top of it. Owing to the horizontally scalable design, whenever the size of the data increases, new instances from the EC2 image have to be spawned, and then added to the cluster configuration. The GitHub repository for the Spark setup is `https://github.com/GVCL/Pclspark` and for the visualization tool is `https://github.com/GVCL/PointCloudVizBoard`.

Thus, this overall architecture is scalable and can be deployed to process unlimited data. The usability of this system is demonstrated through our case study 5. Its performance testing has been under a specific request load but the tool can be extensively tested further for higher data volumes.

12.5 CASE STUDY: VISUAL ANALYTICS OF AIRBORNE LIDAR POINT CLOUDS

Here, we demonstrate the use of the tool with the distributed system on a tile of the Dayton Annotated Laser Earth Scanning (DALES) dataset (Varney et al. 2020). The dataset is a large-scale airborne LiDAR point cloud available as a benchmark dataset. It is a high-resolution annotated point cloud dataset with 0.5 billion points, covering 10 km^2 of area. It contains 40 tiles of dense, labeled point data, including urban regions, of which 29 are training and the remaining, testing files. Each tile is of size 0.5 km^2, with a point density of 50 ppm (points per square meter) as average, and 20 ppm as a minimum. The eight semantic classes available in DALES are ground, vegetation, cars, trucks, poles, power lines, fences, and buildings. Points that are to be discarded in any learning activity are labeled as 0. We demonstrate the application of the distributed system for visual analytics in an application of semantic classification and visualization.

Figure 12.5 shows the dashboard with an active ROI selection. The ground truth class labels are used for coloring the point cloud, and the ROI is progressively rendered using the labels computed from the random forest classifier, which is selected on the control panel. Here, we also visualize the ROI in juxtaposed view with the entire point cloud, thus, enabling visualization of the differences in the computed results and the ground truth. Figure 12.8 shows the classification implemented in the ROI using the random forest classifier. Currently, we have options for random forest classifier (RFC) and gradient boosted tree classifier (GBT), as available in the classification tool in the control widgets.

For the complete dataset, we get 0.817 and 0.792 overall accuracies, with ~51 million points training on five tiles, and ~12 million points testing from tile `5080 _ 54470` using an 11-dimensional feature vector (Singh and Sreevalsan-Nair 2021) and roughly 80/20 split for training and testing.

TABLE 12.1
Representative Set of Performance Measurements of the Case Study for One Tile of LiDAR Point Clouds on the Visual Analytics Tool Using Apache Spark-Cassandra Integration

Action	Number of Points (in millions)	Run Time
Spatial partition of a tile	12	~7 minutes
Feature extraction in a tile	67	~10 hours
Random forest classifier: training	51	~4.4 hours
Random forest classifier: testing	12	~45 minutes
Loading points on browser from Cassandra cluster	2.6	~2 minutes

Here, ten uniformly distributed scales are used with the cubical neighborhoods of each point for feature extraction. The neighborhood sizes start with $l = 1$ m at the minimum scale, and at increments of $\Delta l = 1$ m. An 11-dimensional feature vector is computed per point using eigenvalue decomposition of the covariance matrix. The features are aggregated over different scales by averaging the feature vectors. The representative performance of the system is given in Table 12.1. Since this chapter discusses the visualization tool, the readers are redirected to previous work with implementation details of the classification methods (Singh 2021; Singh and Sreevalsan-Nair 2020, 2021).

As per the design, Apache Spark is used for feature extraction and ML models, however, Cassandra persists the features that need to be used for both analytics as well as visualizations. This allows progressive rendering, as shown in Figure 12.8, where the points in the ROI are *progressively* labeled using the RFC.

12.6 CONCLUSIONS

In this chapter, the system architecture, design, and prototype (or POC) implementation of a browser-based visualization tool for LiDAR point clouds are discussed in detail. In particular, the chapter provides the detailed design of a visualization tool that is a part of a larger visual analytics system for the semantic classification of point clouds. The data analytics engine in this system is driven by the supervised learning classifiers implemented using a distributed system. This distributed system is an Apache Spark-Cassandra integrated framework to support storing, managing, and processing large-scale point clouds both in its raw format, intermediate processed outcomes, and classifier outcomes. While Spark MLlib is usually not a preferred framework for machine learning owing to the data communication bottlenecks (Zhang, Jiang, Wu, Zhang, Yu, and Cui 2019), the system architecture discussed here demonstrates how a horizontally scalable system seamlessly supports visualization and data analytics.

The POC further demonstrates the specialized user interactive features such as the region of interest (ROI) selection. The ROI selection completes the feedback loop in the visual analytics workflow. Such a system can be improved further with respect to usability through juxtaposed views and higher-dimensional modeling of feature space, which can be explored further using analytics charts provided in the tool. This system can be further improvised to reformat the input point cloud with the OGC 3D Tiles standards based on the exploratory analysis using the visual analytics tool.

ACKNOWLEDGMENT

The authors are thankful to the members of GVCL and IIITB for their constant support. This article would not have been complete without the help of anonymous reviewers.

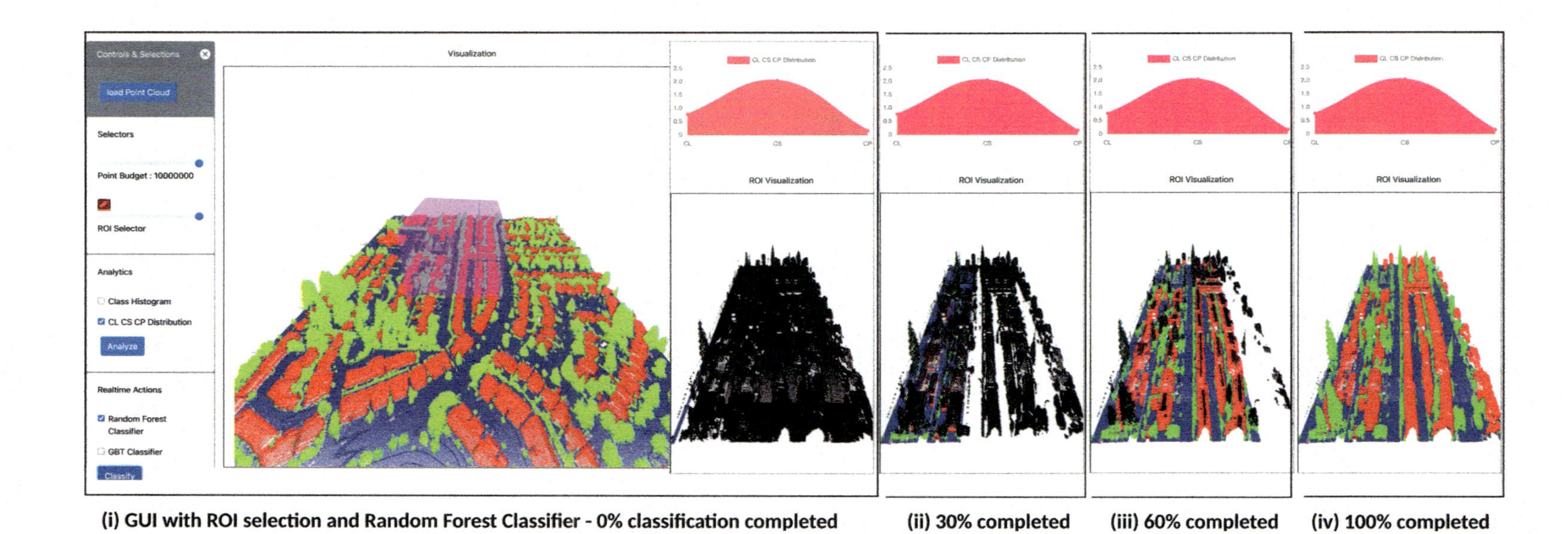

FIGURE 12.8 Semantic classification of the ROI in the DALES dataset implemented using the random forest classifier, selected in the left panel in the control widgets, and the updated classes shown as progressive rendering in ROI visualization in the right panel of the data analytics charts. The black points are unlabeled in the ROI visualization.

NOTE

1 https://github.com/mrdoob/three.js/; originally developed by Ricardo Cabello.

REFERENCES

Boehm, J., K. Liu, and C. Alis (2016). Sideloading-ingestion of large point clouds into the apache spark big data engine. *International Archives of the Photogrammetry, Remote Sensing and Spatial Information Sciences-ISPRS Archives*, 41, 343–348. doi: 10.5194/isprsarchives-xli-b2-343-2016.

Breiman, L. (2001). Random forests. *Machine Learning* 45(1), 5–32.

Brodu, N. and D. Lague (2012). 3D terrestrial LiDAR data classification of complex natural scenes using a multi-scale dimensionality criterion: Applications in geomorphology. *ISPRS Journal of Photogrammetry and Remote Sensing* 68, 121–134.

Cesium (2011). Cesium: The platform for 3D geospatial. https://cesium.com/, Last accessed on June 30, 2022.

Danchilla, B. (2012). Three.js framework. In: *Beginning WebGL for HTML5*, pp. 173–203. Apress: New York.

DataStax, A. (2014). Spark Cassandra connector. https://github.com/datastax/spark-cassandra-connector, Accessed on June 29, 2020.

Demantké, J., C. Mallet, N. David, and B. Vallet (2011). Dimensionality based scale selection in 3D LiDAR point clouds. *The International Archives of the Photogrammetry, Remote Sensing and Spatial Information Sciences* 38(Part 5), W12. doi: 10.5194/isprsarchives-xxxviii-5-w12-97-2011.

Dirksen, J. (2013). *Learning Three. js: the JavaScript 3D Library for WebGL.* Packt Publishing Ltd: Birmingham.

Discher, S., R. Richter, and J. Döllner (2018). A scalable webGL-based approach for visualizing massive 3D point clouds using semantics-dependent rendering techniques. *In Proceedings of the 23rd International ACM Conference on 3D Web Technology*, Poznań, Poland, pp. 1–9.

Friedman, J.H. (2001). Greedy function approximation: a gradient boosting machine. *Annals of Statistics* 29(5), 1189–1232.

IGN and LASTIG, France (2016). iTowns. http://www.itowns-project.org/itowns/, Last accessed on June 30, 2022.

Keim, D., G. Andrienko, J.-D. Fekete, C. Görg, J. Kohlhammer, and G. Melançon (2008). Visual analytics: Definition, process, and challenges. In: Kerren, A. et al. (Eds.), *Information Visualization*, pp. 154–175. Springer: Berlin, Heidelberg.

Klein, G., B. Moon, and R.R. Hoffman (2006). Making sense of sensemaking 1: Alternative perspectives. *IEEE intelligent systems* 21(4), 70–73.

Koukofikis, A. and V. Coors (2018). *Employing OGC's 3D Portrayal Service to Interoperate Hierarchical Data Structures: A Case Study on Visualizing I3S in Cesium. , DGPF.*

Kumari, B., A. Ashe, and J. Sreevalsan-Nair (2014). Remote interactive visualization of parallel implementation of structural feature extraction of three-dimensional LiDAR point cloud. In Srinivasa, S. and Mehta, S. (Eds.), *Big Data Analytics*, pp. 129–132. Springer: Berlin, Heidelberg.

Lakshman, A. and P. Malik (2010). Cassandra: A decentralized structured storage system. *ACM SIGOPS Operating Systems Review* 44(2), 35–40.

Liao, L., S. Tang, J. Liao, X. Li, W. Wang, Y. Li, and R. Guo (2022). A supervoxel-based random forest method for robust and effective airborne LiDAR point cloud classification. *Remote Sensing* 14(6), 1516.

Martinez-Rubi, O., S. Verhoeven, M. Van Meersbergen, P. Van Oosterom, R. Gonçalves, T. Tijssen, et al. (2015). Taming the beast: Free and open-source massive point cloud web visualization. *In Capturing Reality Forum 2015*, 23–25 November 2015, Salzburg, Austria. The Servey Association.

Matsuda, K. and R. Lea (2013). *WebGL Programming Guide: Interactive 3D Graphics Programming with WebGL.* Addison-Wesley: Boston, MA.

Meng, X., J. Bradley, B. Yavuz, E. Sparks, S. Venkataraman, D. Liu, J. Freeman, D. Tsai, M. Amde, S. Owen, et al. (2016). MLlib: Machine learning in Apache Spark. *The Journal of Machine Learning Research* 17(1), 1235–1241.

OGC Implementation Standard (2015). *OGC R 3D Portrayal Service 1.0*, Hagedorn, B., S. Thum, T. Reitz, V. Coors, and R. Gutbell (Eds.), https://www.ogc.org/standards/3dp, Last accessed on June 30, 2022.

OGC International Community Standard (2017). *OGC Indexed 3d Scene Layer (I3S) and Scene Layer Package (*.slpk) Format Community Standard Version 1.2*, Reed, C. and Belayneh, T. (Eds.), https://www.ogc.org/standards/i3s, Last accessed on June 30, 2022.

OGC International Community Standard (2019). *3D Tiles Specification 1.0*, Cozzi, P., S. Lilley, and G. Getz (Eds.), https://www.ogc.org/standards/3DTiles, Last accessed on June 30, 2022.

Picavet, V., M. Brédif, M. Konini, and A. Devaux (2016). iTowns, framework web pour la donnée géographique 3D. *Revue XYZ* 147, 49–52.

Richter, R., J.E. Kyprianidis, and J. Döllner (2013). Out-of-core GPU-based change detection in massive 3D point clouds. *Transactions in GIS* 17(5), 724–741.

Schütz, M. (2016). Potree: Rendering large point clouds in web browsers. Technische Universität Wien, Wiedeń. url, https://www.cg.tuwien.ac.at/research/publications/2016/SCHUETZ-2016-POT/.

Shreiner, D. (2004). *OpenGL (R) 1.4 Reference Manual.* Addison Wesley Longman Publishing Co., Inc.: Boston, MA.

Singh, S. (2021). A distributed system for multiscale analysis and visualization of large-scale airborne LiDAR point clouds, Master's Thesis, IIITB, https://www.iiitb.ac.in/gvcl/pubs/MS2017004_Thesis_2021-11-29.pdf.Singh, S. and J. Sreevalsan-Nair (2020). A distributed system for multiscale feature extraction and semantic classification of large-scale LiDAR point clouds. *In 2020 IEEE India Geoscience and Remote Sensing Symposium (InGARSS)*, Ahmedabad, India, pp. 74–77, IEEE.

Singh, S. and J. Sreevalsan-Nair (2021). A distributed system for optimal scale feature extraction and semantic classification of large-scale airborne LiDAR point clouds. *In 17th International Conference on Distributed Computing and Internet Technology (ICDCIT), Lecture Notes in Computer Science*, Bhubaneswar, India, pp. 280–288, Springer International Publishing.

Singh, S. and J. Sreevalsan-Nair (2022). Adaptive multiscale feature extraction in a distributed system for semantic classification of airborne LiDAR point clouds. *IEEE Geoscience and Remote Sensing Letters* 19, 1–5.

Sreevalsan-Nair, J., P. Mohapatra, and S. Singh (2021). IMGD: Image-based multiscale global descriptors of airborne LIDAR point clouds used for comparative analysis. In: Frosini, P., D. Giorgi, S. Melzi, and E. Rodolá (Eds.), *Smart Tools and Apps for Graphics (STAG 2021): Eurographics Italian Chapter Conference*, pp. 61–72. The Eurographics Association. doi: 10.2312/stag.20211475.

Varney, N., V.K. Asari, and Q. Graehling (2020). DALES: A large-scale aerial LiDAR data set for semantic segmentation. *In Proceedings of the IEEE/CVF Conference on Computer Vision and Pat Tern Recognition Workshops*, pp. 186–187. doi: 10.1109/CVPRW50498.2020.00101.

Weinmann, M., B. Jutzi, S. Hinz, and C. Mallet (2015). Semantic point cloud interpretation based on optimal neighborhoods, relevant features and efficient classifiers. *ISPRS Journal of Photogrammetry and Remote Sensing* 105, 286–304. doi: 10.1016/j.isprsjprs.2015.01.016.

Weinmann, M., B. Jutzi, and C. Mallet (2013). Feature relevance assessment for the semantic interpretation of 3d point cloud data. *ISPRS Annals of the Photogrammetry, Remote Sensing and Spatial Information Sciences* 5, W2. doi: 10.5194/isprsannals-II-5-W2-313-2013.

Westfechtel, T., K. Ohno, T. Akegawa, K. Yamada, R.P.B. Neto, S. Kojima, T. Suzuki, T. Komatsu, Y. Shibata, K. Asano, et al. (2021). Semantic mapping of construction site from multiple daily airborne LiDAR data. *IEEE Robotics and Automation Letters* 6(2), 3073–3080.

Zaharia, M., R.S. Xin, P. Wendell, T. Das, M. Armbrust, A. Dave, X. Meng, J. Rosen, S. Venkataraman, M.J. Franklin, et al. (2016). Apache spark: A unified engine for big data processing. *Communications of the ACM* 59(11), 56–65.

Zhang, Z., J. Jiang, W. Wu, C. Zhang, L. Yu, and B. Cui (2019). MLlib*: Fast training of GLMs using spark MLlib. *In 2019 IEEE 35th International Conference on Data Engineering (ICDE)*, Macao, pp. 1778–1789. IEEE Computer Society.

Section V

Other Advances in Geospatial Domain

13 Toward a Smart Metaverse City

Immersive Realism and 3D Visualization of Digital Twin Cities

Haowen Xu and Andy Berres
Computing and Computational Sciences Directorate,
Oak Ridge National Laboratory

Yunli Shao, Chieh (Ross) Wang, and Joshua R. New
Energy Science and Technology Directorate,
Oak Ridge National Laboratory

Olufemi A. Omitaomu
Computing and Computational Sciences Directorate,
Oak Ridge National Laboratory

CONTENTS

13.1 INTRODUCTION

Recently, the Metaverse has become a popular topic among major Information Technology (IT) and Artificial Intelligence (AI) companies, capturing the leading industry's imagination and

DOI: 10.1201/9781003270928-18

investments [1]. The concept and its mixed reality technologies have been predicted as a significant trend of the next-generation Internet. They also enable unique opportunities and innovative approaches to facilitate the ongoing initiative of developing a smart and sustainable city [2] and visualizing 3D urban data [3]. The major strengths of the Metaverse are its capability for immersive and realistic geovisualization and two-way interaction between users and the virtual environment [4]. These strengths can be effectively synchronized with a digital twin that connects the virtual city with its physical form to allow researchers and urban residents to directly interact with the real world through the immersive reality that is powered by the ubiquity of access and immersive geovisualization. The proposed synergy aims to enable science-guided solutions and bring the human factor and socio-technical dimension into smart urban management to improve the real-world city for better security, livability, and economic growth. In this book chapter, we propose a new concept, debuted as the "Smart Metaverse City", to integrate functional Metaverse technologies into digital twin city applications to enable immersive and realistic geovisualization for generating new opportunities and research agendas toward human-centered, participatory, and decentralized urban development and management [5,6]. To support our vision, we prototyped a technical framework for creating a realistic 3D smart Metaverse city environment and realistic visualization in a systematic manner. We also discuss relevant 3D geospatial datasets and open-source web technologies to implement the prototype, as well as to scale the virtual environment to larger geographic areas. Our prototyping framework primarily focuses on the Geographic Information System (GIS) perspective of collecting, assembling, and processing 3D urban environment data and constructing 3D virtual environments using game engines, which can deliver interactive 3D visualization of an urban environment through web browsers and mixed reality devices. We demonstrate the feasibility and potential of our vision through a real-world digital twin city application developed at Oak Ridge National Laboratory (ORNL). The application debuts as the "ORNL-DT", it constructs a 3D digital twin city to replicate the ORNL campus (e.g., building, infrastructure, terrain, and parking facilities) to enhance participatory, smart, and sustainable campus management, as well as support spatial decisions. We apply our proposed concept on top of the existing application to demonstrate the potential use cases and benefits of the proposed "Metaverse City" concept. Then, we discuss a suite of state-of-art GIS and virtual reality technologies, and geospatial datasets for creating a scalable "Metaverse City" framework.

13.2 METAVERSE FOR DIGITAL TWIN CITIES

In this section, we first provide necessary definitions of the existing digital twin and Metaverse concept to provide some context about developing a Metaverse city. Then, we introduce the vision to synergize the two concepts into a "Smart Metaverse City" that builds two-way connectivity between urban residents and the physical city through immersive digital twin cities to promote geovisualization of urban data and decentralized human-centered smart city governance.

13.2.1 Digital Twin Cities

A digital twin city is the application of the Digital Twin (DT) concept and technologies for creating a cyber replica of the physical city in a digital environment [7]. The replica is synchronized with the real world through a network of sensors to mirror the dynamics in different urban systems [8]. With the advent of the 5G network, edge-computing, and AI, the use and potential of the concept of digital twin cities as a significant driver for smart and sustainable cities have become increasingly promising [9,10]. Compared with the traditional smart city platforms and cyberinfrastructure relying on digital models and digital shadows [11], a digital twin city employs the concept of a Cyber-Physical System (CPS) to create two-way connections between the digital city representation and the physical city [9]. The digital-physical connectivity enables bidirectional changes that allow the physical world to alter the digital representation and vice versa [12].

In smart city applications, digital-physical connectivity can enable a low-cost solution to replicate an urban system and simulate its behavior and responses to hypothetical scenarios and conditions. Examples of these scenarios and conditions include extreme weather events [13], alternative urban policies, governance, and planning strategy [14], emerging mobility technologies and traffic management solutions [15,16], and modification of urban infrastructure [17]. The early efforts of digital twin cities started as New York's strategy for building "the world's most digital city" [18,19] and were promoted by the smart city plans of 15 major cities worldwide [20].

13.2.2 Metaverse

We believe the recent emerging "Metaverse" concept could be incorporated into digital twin cities to expand the digital-physical connectivity by placing humans into the loop to promote effective public participation and engagement in citizen science and human-centered smart city governance. The term "Metaverse" is a combination of "meta" (meaning beyond) and the stem "verse" from "universe", it originally denotes a hypothetical synthetic environment linked to the physical world, where users represent themselves virtually through avatars [1]. It is considered the next stage of the "Reality-Virtuality Continuum" [21], covering a wide spectrum of technologies (e.g., virtual reality, augmented reality, extended reality, artificial intelligence, and blockchain), digital hardware (e.g., IoT and robotics, and computing edge and cloud), and application areas (e.g., content creation, immersive entertainment, and virtual economy). Past studies have provided comprehensive discussion and reviews of the "Metaverse" concept and its historical evolution, past applications, and future research agenda [1,22]. A few studies have discussed visions that apply the Metaverse and its related technologies to benefit urban planning applications [23,24]. Different from many existing virtual reality applications, the "Metaverse" has five unique features, including a 3D graphical interface and integrated audio, multi-user remote access and interactivity, persistence and operability without particular users, immersive nature, and user-generated activities and objects [25]. The combination of these features allows a Metaverse to render a compelling and accessible 3D virtual environment for human socio-cultural and socio-technical interactions [26,27]. The environment allows humans to remotely modify the existing digital copies (i.e., digital twin cities) and make new creations in the virtual environment [1]. When applied to GIS and smart city applications, these Metaverse features can potentially create two-way connectivity between humans and virtual cities through 3D geovisualizations that are immersive, realistic, collaborative, and interactive. The human-virtual connectivity can complement a digital twin city's virtual-physical connectivity, shedding light on a "human-virtual-physical" continuum that synchronizes human interactions, virtual creations, and collaboration into a Machine-to-Machine (M2M) workflow that is data-driven, simulation-informed, and CPS-enabled. The synchronization can further promote public engagement, citizen science, decentralized governance, and collaborative urban management.

13.2.3 Smart Metaverse City

Based on the proposed "human-virtual-physical" continuum, we envision integrating the recent emerging Metaverse concept into digital twin cities to overcome challenges related to the social dimension of human-centered urban science and management. This proposed integration is coined as the "Smart Metaverse City" and aims to create a synthetic environment that is fully immersive, collaborative, and visually explicit. The participatory nature of the environment allows it to be seamlessly synchronized with many well-established urban research and management practices, such as shared vision planning, social learning, serious gaming, and neighborhood games. It allows different urban management participants (e.g., citizens, urban planners, policy-makers, and researchers) to freely explore and interact with complex urban system dynamics and concepts which

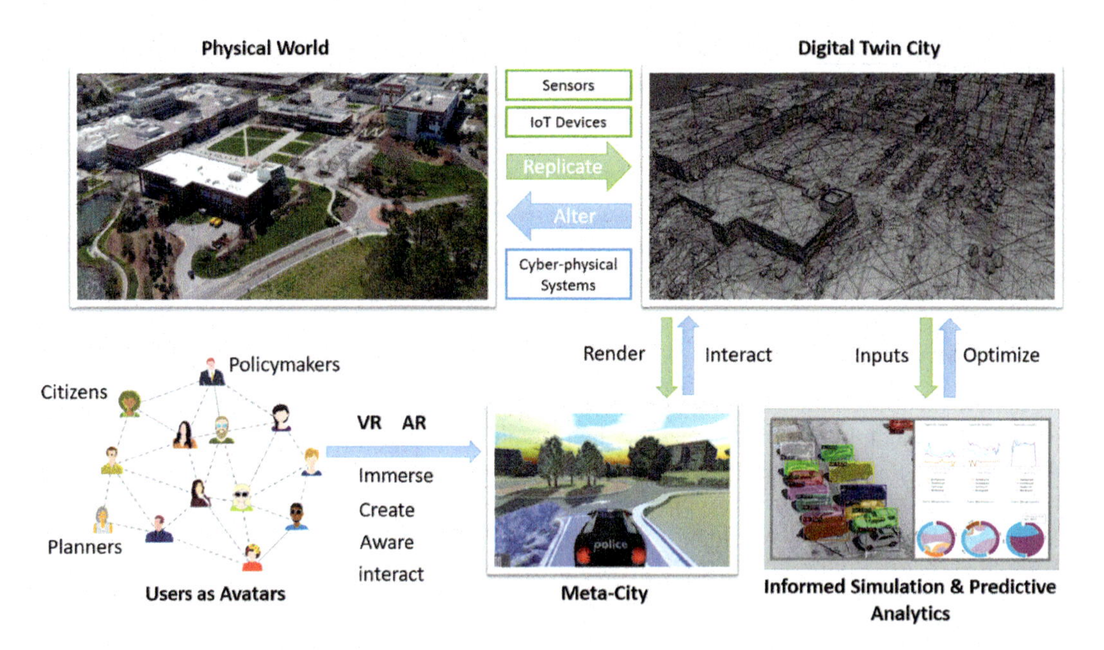

FIGURE 13.1 Reality-virtuality continuum: From digital twin cities to Metaverse cities using the ORNL-DT as an example.

may affect their future life through interactive and intuitive geovisualizations. Figure 13.1 illustrates the rationale of our proposed concept and vision. Our "Smart Metaverse City" concept is different from existing 3D web-based GIS platforms and 3D geospatial viewers, such as Cesium, as our concept focuses on building a photo-realistic and gamified digital twin city, allowing citizens to freely interact with the DT and create a new virtual object through both non-VR and extended reality devices. A Metaverse city should be a synergy between the 3D digital twin city, Metaverse immersive realism, and the interactive design of serious gaming (scientific gamification) using real-world geospatial data. Our proposed concept also differs from many existing Metaverse applications and open-world video games, such as the Grand Theft Auto series and Minecraft. A Metaverse city takes the form of an open-world game but is built using real-world GIS data to mirror the physical world, while video games are created mostly using imaginary, artistic, and unrealistic scenes. More importantly, our proposed concept will connect the Metaverse-enabled virtual 3D scenes with capabilities enabled through a digital twin city, which includes real-time situational awareness through IoT sensors and predictive analytics enabled through scientific models.

13.2.3.1 Immersive Realism

It is important to ensure that non-technical participants can understand these domain-specific dynamics and concepts. Thus, the virtual environment should utilize the Metaverse's nature of "immersive realism" to create intuitive 3D scientific visualizations. These techniques engage participants in complex urban planning and problem-solving processes. Within the virtual environment, different participants and their underlying communities should be able to represent themselves and interact with each other through avatars, no matter where they are physically located. Through the virtual environment, citizens can view models of the proposed construction of new buildings, roads, or parks. In addition to existing virtual walking tour applications, a Metaverse city would allow them to interact and modify the 3D environments and create new virtual entities based on their interests and needs. Powered with the DT's predictive analytics, a Metaverse city could also present citizens with science-based predictions on the consequences of their virtual creation.

13.2.3.2 Scientific Virtual Object Creation

Many past and existing object creations in virtual spaces are used for entertainment and art [28], such as the application of the video game SimCity for college education and urban collaborative planning [29]. In addition to artistic and entertaining creations, we believe the virtual environment should also guide citizens to develop science-based creations. This engagement process can effectively promote public education, engagement, and citizen science, bringing complex urban science concepts into citizens' daily lives in the form of immersive realism and serious games. For instance, citizens are provided with the flexibility to create and deploy green infrastructure and green roofs in their communities. Their virtual roof creation is converted as inputs into the digital twin city's predictive analytics powered by a series of urban and environmental models. Examples of these models include the Watershed Analysis Tool (HEC-WAT) for soil-water conservation analysis [30] and EnergyPlus building simulations [31]. Through predictive analytics, citizens could speculate realistic effects of their creation to reduce surface runoff flood risks and optimize building energy usage. By connecting the immersive environment with more domain science models, the citizens could remotely access and see visualizations of invisible variables, such as air quality, wind, or weather predictions.

13.2.3.3 Public Engagement through Avatars

Avatars are virtual representations of individuals used to promote citizen-government or citizen-business interactions [32] for decentralized smart city governance. The use of avatars can create opportunities for the urban science communities to study and evaluate the satisfaction and effectiveness of a specific management practice or plan. Domain scientists and software developers can monitor how citizen users' avatars interact with smart city services and respond to hypothetical scenarios simulated through digital twin cities. These avatars' reactions and interactions naturally reflect citizens' opinions regarding the policy and their underlying urban management needs (e.g., plans for new roads and traffic rules, demands of water distribution systems, and new designs of street landscape). Through avatars, citizens could also directly interact with urban planners, policymakers, and urban scientists, delivering their feedback and visions for improving the design of the physical city, its operations, and the smart city software itself. Different from previous e-governance web platforms, citizens can use avatars as surrogates to co-develop and co-create urban management plans with other urban management participants. They can view and interact with different road layouts or building designs, as well as virtually create entities (e.g., infrastructure, floor plans, and decorations) and integrate them into the digital twin city. Their avatars can send their feedback and their virtual creation to the city. Our proposed paradigm aims to create a dynamic and transparent bridge of trust, positive relationships, and communication between end-users and smart city builders (e.g., urban planners, policymakers, scientists, app developers). In this setting, the city builders could invite citizens to co-develop and evaluate features of urban policy and plans, guiding the scope of a smart city to maximize the interest and satisfaction of its citizens. The paradigm can also improve the acceptance and trust of smart city services, software, and urban science models among citizens.

13.2.4 Potential Applications

We outline a few urban applications that can benefit from the proposed Metaverse city paradigm. The demonstration of these applications will be detailed in Section 13.4 using the ORNL-DT as examples. First, the 3D Metaverse city can be applied to generate hypothetical scenarios of different environmental, traffic, and building energy simulations, enabling exploratory visual analytics on urban data to generate new insights and research hypotheses. The immersive and realistic visualization of these hypothetical scenarios can facilitate public engagements and support spatial decisions.

In the vehicle, traffic, and energy simulation disciplines, this type of application is categorized as "scenario generation" [33,34]. Figure 13.2 presents a hypothetical scenario when a deer is crossing the road within the ORNL campus. Such a scenario could be used to raise driver awareness to

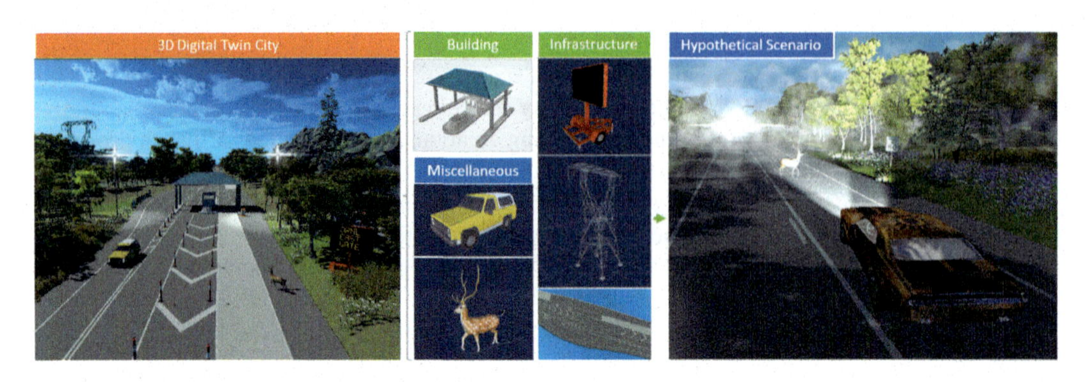

FIGURE 13.2 ORNL-DT examples: East entrance of the national lab at night with a deer crossing the road, and an example of scenario generation: Driving on Bethel Valley Road on a foggy overcast day to evade a deer on the road. The photo-realistic scene simulates potential traffic safety risks (deer on the road) under bad weather and low visibility.

stay alert on the road and to train autonomous driving functions for autonomous vehicles. It can also be used to train new drivers on how to navigate the road under uncertainty of the road conditions such as animal crossing, snowfall accumulations, and reckless driving. It can also inform how the public interacts with autonomous vehicles in different urban settings and environmental conditions.

Second, the immersive photo-realistic visualization of a city can be utilized to automate label generation to facilitate scenario generation for city planners and citizens. The approach aims to extend the capability of existing Computer Vision (CV) applications to detect complex urban objects and spatial features (e.g., roads covered with snow) from different perspectives and under different weather and lighting conditions. A similar vision that uses computer-generated graphics to train visual AIs has been proposed by Ref. [35]. Third, we believe that Metaverse cities align with the visions proposed by Refs. [36,37], which employ advanced interaction and immersion techniques to support the visualization and exploration of complex scientific processes and dynamics, and spatial co-simulation [38] and collaborative decision support [39]. Furthermore, with advances in cognitive science, Metaverse cities can achieve some elements of cognition. Thus, Metaverse cities will exploit implicit knowledge drawn from the experience of existing smart cities and enable the transfer of higher performance decisions and control and improve performance at scale.

13.3 GEOSPATIAL FRAMEWORK FOR METAVERSE CITIES

This section outlines a geospatial framework that connects state-of-the-art GIS technologies to help researchers generate 3D Metaverse Cities using mostly open geographic data and information. We refer to the framework as the "Meta-City framework." The purpose of the framework is to help researchers lower the technical barriers of generating realistic digital twin cities over any geographic area using open-access geospatial datasets and convert them into Metaverse cities to promote public engagement and education, citizen science, and decentralized urban research and management. In this regard, a Metaverse city should entail 3D realistic data of digital terrain, buildings, vegetation, and infrastructure. Its digital representation should have the following technical capabilities:

1. Terrain data and 3D models should be stored in common 3D object formats and can be readily imported into contemporary game engines and visualization platforms.
2. A digital twin city should support both pre-rendered and rendered views to enable digital realism (e.g., realistic lighting effects, photo-realistic textures, and high-quality geometry), and interactively rendered views that are optimized for performance.

3. 3D objects should be classified based on their entity type (e.g., traffic signs, roads, buildings, and trees), and they should support assignments of physical properties and 3D animations.

13.3.1 Overall Framework Design

Given these requirements, a Meta-City framework should enable the integration of available 3D terrain, building, and infrastructure data. It should be deployed on cyberinfrastructure with adequate computing power for spatial data mining and analytics. Our framework consists of three main components:

- **Geospatial Data Acquisition**: This component is responsible for the discovery and acquisition of appropriate 3D urban data. Possible data sources include public data, spatially mined data, purchasable data, and derived data.
- **Digital Twin City Construction**: The second component is responsible for constructing a detailed digital twin city (e.g., buildings, terrains, in-door environment, and urban infrastructure) using an automated data processing pipeline.
- **Immersive 3D Geovisualization**: The third component imports the 3D digital twin city generated from the previous component into 3D game engines to construct a 3D immersive visualization environment that is compatible with multiple platforms (e.g., desktop, web browsers, smartphones and tablets, and VR/AR devices).

Along with the framework development, a semi-automated data-processing pipeline is developed to connect the different components, enabling a system-based workflow to discover available 3D data from distributed resources, clean and process 3D data to construct 3D digital twin cities, and create an interactive and immersive virtual environment to visualize the 3D digital twin cities.

13.3.2 Geospatial Data Acquisition

Based on their utility nature, we categorize 3D digital city entities either as static or dynamic, as depicted in Figure 13.3. Static entities are the 3D objects that are less likely to be modified in a virtual environment by users or computer programs. These objects are created to replicate their counterparts in the physical world. Examples of statics entities include 3D digital terrain, buildings, and some large urban infrastructure (e.g., bridges, culverts, and dams). In other smart city applications

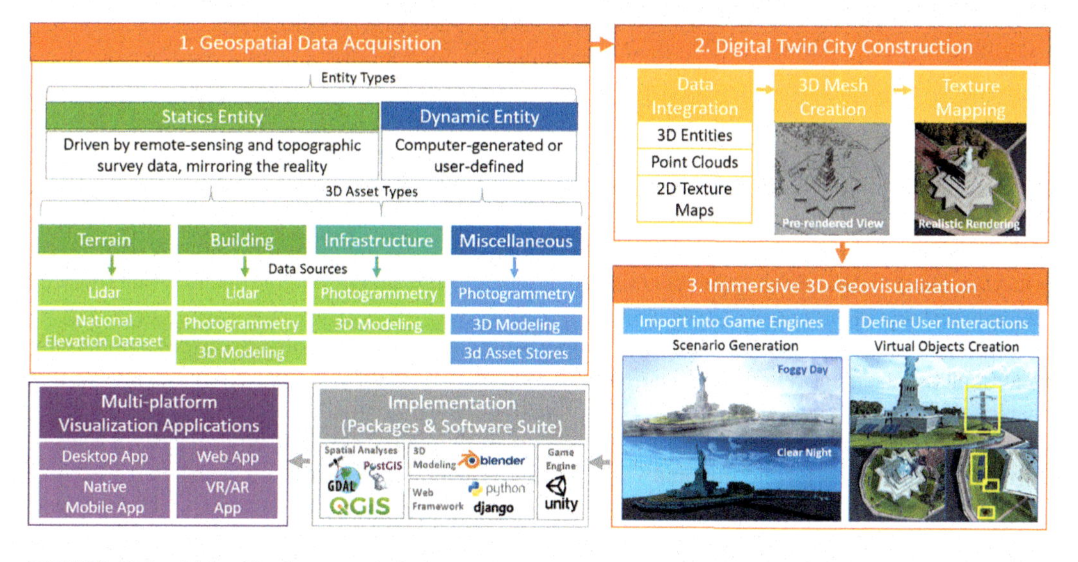

FIGURE 13.3 Meta-City framework design and data sources.

that focus on collaborative urban design, buildings and bridges are subjected to changes and are defined as dynamic entities. On the contrary, some entities are to be created or modified in the virtual environment by users or computer programs (e.g., real-time simulation outputs). These entities are categorized as dynamic and are not required to mirror the exact real-world behavior. Examples of these entities include miscellaneous objects, such as moving vehicles, humans, avatars, and animals. Some infrastructure, such as traffic signals, temporary construction zones, and Dynamic Message Signs (DMS), should also be classified as dynamic entities. The dynamic message system depicted in Figure 13.2 is a typical example as their messages, configurations, and/or locations could be changed during different campus management practices. At the urban scale, there are many useful data sources for 3D digital city entities. Some of these data sources are free and open to public access, and others are published by private corporations. Some data are not available publicly but can be derived from satellite imagery.

13.3.3 Digital Twin City Construction

A Metaverse city integrates digital representations of digital city entities, like terrain, roads, bridges, and buildings, into a virtual environment to create a realistic representation that mirrors the real world. At the technical level, the creation of such a scene would require the construction and integration of 3D data (e.g., point clouds, meshes, and textures) into a 3D scene. The 3D representations of these entities need to be cleaned, processed, mapped with photo-realistic textures or shaders, and converted to common 3D file formats that are compatible with major 3D modeling software and game engines. Examples of these formats include FBX and OBJ. 3D entities that are acquired using LiDAR point clouds need to be converted into objects to allow the assignment of physical properties for creating potential user interactions with these entities. For lightweight displays in virtual applications, these 3D entities should be simplified and tessellated into a simpler geometric model.

The placement of 3D models of buildings, vegetations, and infrastructure should be guided by the real-world locations of their physical counterparts. This process can be conducted manually in 3D modeling software with geospatial capabilities (e.g., BlenderGIS) or automated using 2D GIS data (e.g., building footprints and landcover data) combined with customized scripting in game engines. As an example, the derivation and placement of 3D road geometries can be automated using multiple GIS data and methods. Examples of these methods include using road centerlines tagged with lane information [40], lane detection in satellite imagery [41], as well as lane detection in street-level imagery [42,43].

13.3.4 Immersive 3D Geovisualization

The immersive 3D geovisualization of a digital city and user interactions are enabled through the rendering and scripting capabilities in a 3D game engine, such as Unity and Unreal. Dynamic objects, such as some infrastructure and trees, are assigned with physical properties to define potential user interactions and are added to the static scene. Many contemporary game engines can export the virtual environment as cross-platform applications (e.g., desktop games, web-based games, and VR/AR games) to deliver the immersive visualization of urban data to support divers urban research and management efforts.

Visualizing a photorealistic Metaverse world is important for supporting various public education and engagement-related applications. The photo realism of a virtual environment is often enabled through high-quality 3D models (e.g., fine details) with realistic texture mapping and shaders and advanced 3D rendering software [44]. Many of these concepts pertain to the disciplines of 3D modeling, digital arts, and film and video game production, with which many urban researchers and GIS specialists are not familiar. To develop a photorealistic visualization of a 3D city, these concepts and their underlying techniques need to be integrated with the GIS framework to create 3D urban environments and objects.

13.4 USE CASES

This section presents a digital twin city developed at ORNL as an example to demonstrate different use cases enabled through the Metaverse city vision.

13.4.1 Public Engagement, Education, and Training

A Metaverse city enables a participatory environment that brings citizen users into the management of an urban area. The ORNL-DT allows different users to be immersed in its digital campus through different types of avatars, such as humans, vehicles, and animals. Users can conduct a virtual campus tour to explore various buildings, infrastructure, and vehicles. Figure 13.4a depicts a scenario where different user avatars are interacting near ORNL's visitor center. As a mixture between virtuality and reality, the ORNL-DT was developed using real-world GIS data, and it allows the incorporation of unlikely entities, such as a koala, into the scene. In addition, a prototyping feature, which is yet to be implemented in the ORNL-DT, would be the capability to allow users to create virtual objects. Figure 13.4b illustrates the proposed rationale that a Metaverse city should allow users to alter the digital realism through customized virtual creation in an emulated geographic space. An avatar navigated to a location and created a water fountain based on the user's interests. In an immersive, photorealistic environment, the avatar could visualize the visual effect of their creation and share it with other avatars that represent different types of participants in campus management, creating an alternative virtual scenario with realistic visual effects to support spatial decisions.

13.4.2 Training Computer Vision Applications

A Metaverse city can automatically render photorealistic graphics to visually simulate hypothetical environments that are not common in the physical world. These computer-generated graphics often carry labels to highlight different virtual objects, which can automate the training of computer vision algorithms for detecting various objects under special conditions (e.g., lighting conditions, low visibility, and shadow interference) in different geographic regions. The automated labeling feature enabled through the game engine could efficiently reduce labor-intensive efforts to gather training data and create labels. Figure 13.4c demonstrates how our ORNL-DT can simulate different lighting and weather conditions at a real-world parking lot to help train the IoT-connected cameras and computer vision applications to detect vehicles in parking spaces. These graphics can automatically train real-world computer vision applications to help cameras detect vehicles, humans, and animals [38].

13.4.3 Spatial Co-simulation

In applications such as transportation studies, researchers often apply digital twins in simulators using photorealistic game engines. These simulators are considered serious games and can be used to solve real-world geospatial decision problems. Here, we use the traffic co-simulation as an example.

Within a Metaverse city, technologies such as perception for automated vehicles (using cameras, LiDAR, radar, etc.) can be thoroughly evaluated and tested. Realistic sensor streams of these vehicles can be emulated in such a virtual environment and used as inputs to test various automated driving algorithms. There are many existing simulators or serious games software such as CARLA, AirSim, and LGSVL Simulator. Users can customize control scripts for a target vehicle, often referred to as the "ego" vehicle using different algorithms. Behaviors of background traffic objects can be defined within these serious games. This approach is typically used to study interactions centered around the ego vehicle. In the interest of studying or understanding mobility at a larger scale, i.e., along a corridor or in a city, serious games are connected to traffic simulators such as SUMO, VISSIM, or Aimsun for co-simulation [45]. In such co-simulation settings, only the ego vehicle(s) will be controlled by users within the Metaverse city, and traffic simulators simulate the motions and dynamics of all background traffic objects.

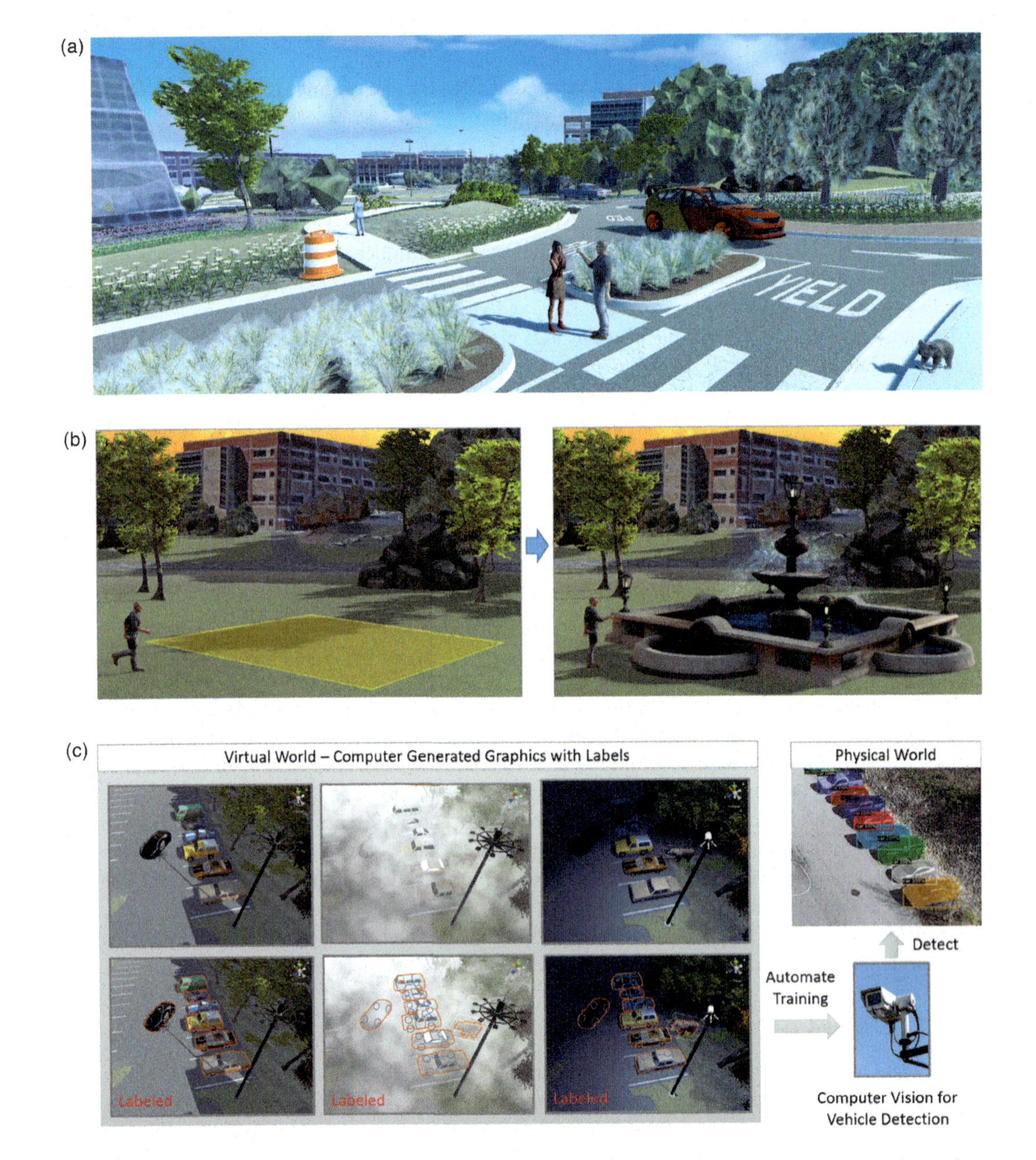

FIGURE 13.4 Use case of the ORNL-DT's immersive visualization. (a) Users can represent themselves using customized avatars, such as humans and animals. (b) Users can conduct virtual campus tours with virtual object creation capabilities. The virtual creation can be a direct import from smartphone-based photogrammetry, which scans the user's real-world surrounding environment. (c) A computer-generated hypothetical scenario with automated labels can help train computer vision applications for object detection in various GIS applications.

13.5 FUTURE OPPORTUNITIES

The book chapter presents a vision that synergizes digital twin cities with a Metaverse virtual world to create an innovative "Smart Metaverse city" concept for supporting decentralized human-centered urban planning, management, and governance. The synergy aims to combine DT's merits (real-time situational awareness, science-based predictive analytics, and cyber-physical systems) with the unique advantage of a Metaverse world (immersive interaction and realism). Along with the concept, we prototyped a technical framework through the integration of a variety of GIS, 3D modeling, Geo-AI, visualization, and extended reality techniques to help researchers and urban planners create functional Metaverse cities using mostly publicly available data and web technologies. The technical framework was presented as a vision to expand the current technical capability for acquiring, integrating, processing, and rendering 3D spatial data to enable a photorealistic and interactive urban environment in a virtual world. We believe our proposed framework can generate several future opportunities once our proposed technical framework is fully implemented. When the full-scale framework is deployed on big data cyberinfrastructure, it will automate the technical workflow for mining, cleaning, and integrating available web-accessible 3D data to generate Metaverse cities at any user-defined regions and areas that have adequate 3D data.

DISCLAIMER

This manuscript has been authored by UT-Battelle, LLC, under contract DE-AC05-00OR22725 with the US Department of Energy (DOE). The US government retains and the publisher, by accepting the article for publication, acknowledges that the US government retains a nonexclusive, paid-up, irrevocable, worldwide license to publish or reproduce the published form of this manuscript, or allow others to do so, for US government purposes. DOE will provide public access to these results of federally sponsored research in accordance with the DOE Public Access Plan (https://www.energy.gov/downloads/doe-public-access-plan).

ACKNOWLEDGEMENT

We thank the United States Geological Survey (USGS) and National Oceanic and Atmospheric Administration (NOAA) for providing digital elevation models and lidar data that allowed us to create our digital twins. Satellite images from the Google Maps Platform are used as a visual guide to assist the development of the digital twin. Road networks are created using Open Street Map information and MathWorks RoadRunner software. Purchased 3D assets from the Unity Asset Store are used in our 3D virtual environment to support the proof-of-concept for research and educational purposes. We thank the Unity Asset Store and third-party asset vendors (JUNNICHI SUKO, WDALLGRAPHICS, STUDIO NEW PUNCH, ISTVAN SZALAI, MANNEKO, MAKSIM BUGRIMOV, RIVERMILL STUDIOS, ALEX LENK, KOBRA GAME STUDIOS, PATRICK CHENOWETH, PXLTIGER, MEHDI RABIEE, SABRI AYES, FEROCIOUS INDUSTRIES, REM STORMS, NOBIAX / YUGHUES, ALP8310, SHAPES, MEHDI RABIEE, FORST, PIXEL GAMES, MONO CREATIVE, VIS GAMES, RENDER KNIGHT) for providing these assets.

REFERENCES

1. L.-H. Lee, T. Braud, P. Zhou, L. Wang, D. Xu, Z. Lin, A. Kumar, C. Bermejo, and P. Hui, "All one needs to know about metaverse: A complete survey on technological singularity, virtual ecosystem, and research agenda", arXiv preprintarXiv:2110.05352, 2021.
2. H. Duan, J. Li, S. Fan, Z. Lin, X. Wu, and W. Cai, "Metaverse for social good: A university campus prototype", *In Proceedings of the 29th ACM International Conference on Multimedia*, China, 2021, pp. 153–161.

3. A. Coltekin, A. L. Griffin, A. Slingsby, A. C. Robinson, S. Christophe, V. Rauten-bach, M. Chen, C. Pettit, and A. Klippel, "Geospatial information visualization and extended reality displays", In: H. Guo, M. F. Goodchild, and A. Annoni (Eds.), *Manual of Digital Earth*. Springer, Singapore, 2020, pp. 229–277.
4. S. Mystakidis, "Metaverse", *Encyclopedia*, vol. 2, no. 1, pp. 486–497, 2022.
5. A. Simonofski, E. Hertoghe, M. Steegmans, M. Snoeck, and Y. Wautelet, "Engaging citizens in the smart city through participation platforms: A framework for public servants and developers", *Computers in Human Behavior*, vol. 124, p. 106901, 2021.
6. G. White, A. Zink, L. Codeca, and S. Clarke, "A digital twin smart city for citizen feedback", *Cities*, vol. 110, p. 103064, 2021.
7. H. Lehner and L. Dorffner, "Digital geotwin vienna: Towards a digital twin city as geodata hub", *PFG – Journal of Photogrammetry Remote Sensing and Geoinformation Science*, vol. 88, no. W1, p. 63, 2020.
8. M. Farsi, A. Daneshkhah, A. Hosseinian-Far, and H. Jahankhani, *Digital Twin Technologies and Smart Cities*. Springer: Berlin/Heidelberg, Germany, 2020.
9. A. Fuller, Z. Fan, C. Day, and C. Barlow, "Digital twin: Enabling technologies, challenges and open research" *IEEE Access*, vol. 8, pp. 108 952–108 971, 2020.
10. T. H. Luan, R. Liu, L. Gao, R. Li, and H. Zhou, "The paradigm of digital twin communications" arXiv preprint arXiv:2105.07182, 2021.
11. W. Kritzinger, M. Karner, G. Traar, J. Henjes, and W. Sihn, "Digital twin in manufacturing: A categorical literature review and classification", *IFAC-Papers OnLine*, vol. 51, no. 11, pp. 1016–1022, 2018.
12. Q. Qi, F. Tao, Y. Zuo, and D. Zhao, "Digital twin service towards smart manufacturing", *Procedia Cirp*, vol. 72, pp. 237–242, 2018.
13. D. N. Ford and C. M. Wolf, "Smart cities with digital twin systems for disaster management", *Journal of Management in Engineering*, vol. 36, no. 4, p. 04020027, 2020.
14. L. Wan, T. Nochta, and J. Schooling, "Developing a city-level digital twin–propositions and a case study", In: M. J. DeJong, J. M. Schooling, and G. M. B. Viggiani (Eds.), *International Conference on Smart Infrastructure and Construction 2019 (ICSIC) Driving Data-Informed Decision-Making*. ICE Publishing: London, 2019, pp. 187–194.
15. H. Wang, M. Zhu, W. Hong, C. Wang, G. Tao, and Y. Wang, "Optimizing signal timing control for large urban traffic networks using an adaptive linear quadratic regulator control strategy", *IEEE Transactions on Intelligent Transportation Systems*, vol. 23, no. 1, pp. 333–343, 2022.
16. W. Hong, G. Tao, H. Wang, and C. Wang, "Traffic signal control with adaptive online-learning scheme using multiple-model neural networks", *IEEE Transactions on Neural Networks and Learning Systems and Learning Systems*, pp. 1–13, 2022.
17. F. Jiang, L. Ma, T. Broyd, W. Chen, and H. Luo, "Digital twin enabled sustainable urban road planning", *Sustainable Cities and Society*, vol. 78, p. 103645, 2022.
18. City of New York, "City of New York, New York City's digital leadership (New York: 2013)", 2021a, https://www.slideshare.net/etsarkov/new-york-citys-digital-leadership, accessed: 2021-11-15.
19. City of New York, "Roadmap for the digital city; achieving New York city's digital future", 2021b, www.nyc.gov/html/media/media/PDF/90dayreport.pdf, accessed: 2021-11-15.
20. M. Angelidou, "The role of smart city characteristics in the plans of fifteen cities", *Journal of Urban Technology*, vol. 24, no. 4, pp. 3–28, 2017.
21. P. Milgram and F. Kishino, "A taxonomy of mixed reality visual displays", *IEICETRANSACTIONS on Information and Systems*, vol. 77, no. 12, pp. 1321–1329,1994.
22. S.-N. Suzuki, H. Kanematsu, D. M. Barry, N. Ogawa, K. Yajima, K. T. Nakahira, T. Shirai, M. Kawaguchi, T. Kobayashi, and M. Yoshitake, "Virtual experiments in metaverse and their applications to collaborative projects: The framework and its significance", *Procedia Computer Science*, vol. 176, pp. 2125–2132, 2020.
23. Q. Feng and R. Cai, "data hegemony: Reflections for the application and development direction of metaverse technology in urban design based on digital", *Journal of World Architecture*, vol. 5, no. 6, pp. 52–61, 2021.
24. J. Kemp and D. Livingstone, "Putting a second life metaverse skin on learning management systems", *In Proceedings of the Second Life Education Workshop at the Second Life Community Convention*, vol. 20. The University of Paisley CA, San Francisco, 2006.
25. R. Gilbert, "The prose project: A program of in-world behavioral research on the metaverse", *Journal of Virtual Worlds Research*, vol. 4, no. 1, pp. 3–18, 2011.
26. J. D. N. Dionisio, W. G. B. III, and R. Gilbert, "3d virtual worlds and the metaverse: Current status and future possibilities", *ACM Computing Surveys(CSUR)*, vol. 45, no. 3, pp. 1–38, 2013

27. J. Sanchez, "Second life: An interactive qualitative analysis", *In Society for Information Technology & Teacher Education International Conference*, San Antonio, TX, Association for the Advancement of Computing in Education (AACE), 2007, pp. 1240–1243.
28. L.-H. Lee, Z. Lin, R. Hu, Z. Gong, A. Kumar, T. Li, S. Li, and P. Hui, "When creators meet the metaverse: A survey on computational arts", arXiv preprintarXiv:2111.13486, 2021.
29. J. Minnery and G. Searle, "Toying with the city? Using the computer game simcity4 in planning education", *Planning Practice and Research*, vol. 29, no. 1, pp. 41–55, 2014.
30. C. N. Dunn and P. R. Baker, "A watershed modeling tool, HEC-WAT", *In Watershed Management 2010: Innovations in Watershed Management under Land Use and Climate Change*, Madison, Wisconsin, 2010, pp. 1101–1112.
31. D. J. Sailor, "A green roof model for building energy simulation programs", *Energy and Buildings*, vol. 40, no. 8, pp. 1466–1478, 2008.
32. O. Gil, M. E. Cortes-Cediel, and I. Cantador, "Citizen participation and the rise of digital media platforms in smart governance and smart cities", *International Journal of E-Planning Research (IJEPR)*, vol. 8, no. 1, pp. 19–34, 2019.
33. W. Ding, C. Xu, M. Arief, H. Lin, B. Li, and D. Zhao, "A survey on safety-critical driving scenario generation: A methodological perspective", pp. 1–17, 2022. Available: http://arxiv.org/abs/2202.02215.
34. B. Yue, S. Shi, S. Wang, and N. Lin, "Low-cost urban test scenario generation using microscopic traffic simulation", *IEEE Access*, vol. 8, pp. 123 398–123 407, 2020.
35. B. Alvey, D. T. Anderson, A. Buck, M. Deardorff, G. Scott, and J. M. Keller, "Simulated photorealistic deep learning framework and workflows to accelerate computer vision and unmanned aerial vehicle research", *In Proceedings of the IEEE/CVF International Conference on Computer Vision*, Montreal, 2021, pp. 3889–3898.
36. S. Grainger, F. Mao, and W. Buytaert, "Environmental data visualization for non-scientific contexts: Literature review and design framework", *Environmental Modelling & Software*, vol. 85, pp. 299–318, 2016.
37. C. Tague and J. Frew, "Visualization and ecohydrologic models: Opening the box", *Hydrological Processes*, vol. 35, no. 1, p. e13991, 2021.
38. Y. Shao, C. Wang, A. Berres, J. Yoshioka, A. Cook, and H. Xu, "Computer vision enabled smart traffic monitoring for sustainable transportation management", *In Proceedings of ASCE International Conference on Transportation & Development (ASCE-ICTD 2022)*, Seattle, WA, 2022 (pp. 34–45).
39. Y. Sermet and I. Demir, "An immersive decision support system for disaster response", *In 26th ACM Symposium on Virtual Reality Software and Technology*, Ottawa, 2020, pp. 1–3.
40. H. Xu, A. Berres, S. A. Tennille, S. K. Ravulaparthy, C. Wang, and J. Sanyal, "Continuous emulation and multiscale visualization of traffic flow using stationary roadside sensor data", *IEEE Transactions on Intelligent Transportation Systems*, 23(8), 10530–10541, 2021.
41. A. Zang, R. Xu, Z. Li, and D. Doria, "Lane boundary extraction from satellite imagery", *In Proceedings of the 1st ACM SIGSPATIAL Workshop on High-Precision Maps and Intelligent Applications for Autonomous Vehicles*, New York, 2017, pp. 1–8.
42. J. Tang, S. Li, and P. Liu, "A review of lane detection methods based on deep learning", *Pattern Recognition,* vol. 111, p. 107623, 2021. Available: https://www.sciencedirect.com/science/article/pii/S003132032030426X.
43. Q. Wang, J. Gao, and Y. Yuan, "Embedding structured contour and location prior in siamesed fully convolutional networks for road detection", *IEEE Transactions on Intelligent Transportation Systems*, vol. 19, no. 1, pp. 230–241, 2018.
44. J. Birn, *Digital Lighting & Rendering.* Pearson Education: London, 2014.
45. Y. Shao, D. Deter, A. Cook, C. Wang, B. Thompson, and N. Perry, "Real-sim interface: Enabling multi-resolution simulation and x-in-the-loop development for connected and automated vehicles", *SAE International Journal of Connected and Automated Vehicles*, vol. 5(12-05-04-0026), pp. 327–339, 2022.

14 Current UAS Capabilities for Geospatial Spectral Solutions

David L. Cotten, Andrew Duncan, Andrew Harter, Matt Larson, and Brad Stinson
Geospatial Science and Human Security Division,
Oak Ridge National Laboratory

CONTENTS

DOI: 10.1201/9781003270928-19

14.1 HISTORY

In 1859, French inventor Gaspard-Félix Tournachon captured the first remotely acquired aerial image when he lofted a tethered balloon equipped with a camera over Paris. His idea of taking photos from the air became popular, and people around the world began experimenting with new ways to capture images from this new perspective. By World War I, cameras were attached to birds for reconnaissance of troop movements (Tiner et al., 2015). And once the value of remotely sensed data was realized in battle, it became a staple for both military and science applications. Through World War II and beyond, significant investments were made into the design of specially equipped aircraft that could push the limits of remote sensing. In the late 1950s, the first meteorological satellite was launched by the United States. Known as Explorer 6, it took the first images of Earth from a satellite on August 14, 1959, and, while crude, the images pushed remotely sensed data collection into outer space (Colomina & Molina, 2014).

These early attempts at remote sensing almost always included a degree of danger. Early aircraft were dangerous to fly, and early rockets blew up during launch attempts. In the 1970s and 1980s, the National Aeronautics and Space Administration (NASA) unsuccessfully launched unmanned aircraft for high-altitude atmospheric sampling (Watts et al., 2010). The idea of acquiring more data with less risk (and cost) drove an interest in developing ways to acquire data without putting humans at risk.

Remotely piloted vehicles (RPVs) existed well before satellites were orbiting the Earth, but they were not widely accepted as a tool for robust scientific measurement. While various studies were carried out using RPVs, it was not until 2004 that a significant number of studies involving unmanned aerial systems (UAS) were published at one time. Only four submissions to the 2004 International Society for Photogrammetry and Remote Sensing conference addressed UAS for remote sensing tasks (Colomina & Molina, 2014). In 2005, a joint project between NASA and the US Forest Service demonstrated that UAS could be used to monitor wildfires (Watts et al., 2010), highlighting the early usefulness of the technology.

Since the early 2000s, the technology surrounding UAS has advanced dramatically, and UAS are now widely used by academics, government agencies, and industry. This growth, in conjunction with the popularity of UAS as a recreational activity, fueled the creation of a new market, with the number of vendors increasing every year. As of February 2022, the Federal Aviation Administration (FAA) reported a total of 860,983 registered drones with 328,670 classed as commercial and the other 528,725 classed as recreational. In addition, the FAA reported 261,952 certified remote pilots. The total number is likely much higher considering nonregistered users, hobbyist operators, and the number of UAS operated outside the United States (Godlewski, 2022).

A demand for UAS technology has prompted the entry of many companies into this market. While it is true that competition drives innovation, it must be noted that the race to bring new UAS and accompanying sensors to market has resulted in the appearance of many lower-performing units. The landscape of companies and offerings change at such a rapid pace that UAS offerings should be reviewed at least every 2 years, if not yearly.

The majority of the small UAS developed in the early 2000s and into the 2010s were built by hobbyists, designed to carry payloads consisting of basic RGB cameras focused on producing high-quality photos, and usually operated with traditional remote control technology. Most of these systems were not mounted on a gimbal and lacked radiometric calibration and geolocation data, so their scientific applications were limited.

While autonomous control technology was advancing, computing power began to reach a point where photogrammetry could be run on modern desktops. Both of these technological developments were driven by additional concurrent technological advancements: the miniaturization of electronics and the increase of processing power per watt. These advancements, in combination with UAS-acquired aerial imagery, meant that images could be collected and processed by a single user. Collecting aerial imagery became less reliant on large-scale aircraft and complex processing techniques that could take weeks to implement and execute from planning to results.

In 2013, a relatively small electronics company known as Shenzhen Da-Jiang Innovations Sciences and Technologies Ltd. (DJI) released their first ready-to-fly, commercially available, global positioning system (GPS)–stabilized drone, named the Phantom 1. By today's standards, the Phantom had rudimentary capabilities. Users had to purchase their own GoPro camera, the operational range of the UAS was limited to about 500 m, and it had a maximum flight time of 10 minutes. It was, however, one of the first low-cost drones that could be easily purchased anywhere in the world and flown with relatively little training.

Scientific or remote sensing projects using these early UAS were difficult because flights had to be conducted manually, data had to be downloaded after the flight, and georeferencing required complicated postprocessing techniques. Furthermore, before the Phantom 1, a self-stabilizing unmanned aerial vehicle (UAV) was nearly nonexistent; therefore, flying a UAV required the most skilled pilots. Although fixed-wing RPVs have been available for decades, they were primarily applicable for a small market of hobbyists and featured unreliable, noisy, and vibration-prone internal combustion engines.

14.2 CURRENT STATE OF THE ART

14.2.1 Types of Platforms

Today's unmanned aircraft manufacturers offer a wide array of designs and configurations with numerous options for various remote sensing applications. Investments into open-source flight control technologies have led to the development of many unique airframe designs capable of carrying a vast array of payloads. In this section, we will explore some of the major types of airframes and the latest developments in propulsion systems.

The FAA currently classifies small UAS as any unmanned aircraft weighing <55 lb, including payload. For the purposes of this study, we chose to subdivide these into three categories based on their mass. Small platforms have a wingspan under 1 m and fit into a small backpack (see Figure 14.1 for an example). Medium platforms have wingspans that vary from 1 to 2 m and fit into a medium-sized rugged case. Large systems have a wingspan >2 m. For the sake of brevity and the fact that most remote sensing publications address smaller UAS, this chapter will primarily focus on varieties of small to medium-sized platforms.

14.2.1.1 Multirotor

Multirotor aircraft have become extremely popular due to their low cost, ease of use, and ability to operate in confined spaces. Although they are limited in range compared to fixed-wing systems, advances in propulsion systems have improved efficiency and increased range to the point that multirotor aircraft have become powerful remote sensing platforms.

The most popular configuration is the quadrotor, which, as the name implies, consists of four propulsion motors and propellers. This configuration is an efficient design that maximizes flight times and lowers costs. The drawback to this design, however, is that the aircraft cannot continue flying if any of the four propulsion systems fail. For this reason, it is not uncommon to find multirotor platforms with 6, 8, 10, or even 12 rotors. These aircraft are typically used to carry larger and more expensive payloads because the aircraft can safely recover from one or multiple propulsion system failures. Examples of these systems include the DJI Matrice M600 hexacopter (Figure 14.2) and the Freefly Alta 8 octocopter.

FIGURE 14.1 Skydio X2D, a typical small quadcopter designed for intelligence, surveillance, and reconnaissance.

FIGURE 14.2 DJI Matrice M600 with LiDAR.

14.2.1.2 Fixed-Wing

Long-duration missions or large-area surveys are the ideal use case for fixed-wing UAS. These platforms are inherently more efficient than multirotor ones and therefore can fly much longer, further, and higher. They are also usually more resistant to poor weather conditions such as wind and rain. There are, however, downsides to using fixed-wing aircraft. If they are light enough, they can be hand-launched (Figure 14.3), but this presents some risk of physical injury to the operator if not properly trained. Heavier aircraft require runways or specialized launchers and an understanding of aircraft approach and departure procedures.

14.2.1.3 Hybrid Airframes

The latest configuration of UAS to enter the market combines the vertical takeoff and landing (VTOL) capabilities of a multirotor with the range of a fixed-wing platform. Hybrid airframes vary in motor configuration. Some look like typical fixed-wing systems with vertical lift motors added while others utilize tilting motor mounts to keep the fuselage horizontal throughout the flight (see Figure 14.4 for an example). Other systems tilt their entire fuselage during the transition from fixed-wing to VTOL flight. Regardless of the configuration, all of these systems ascend vertically to a safe altitude before transitioning to horizontal flight, negating the need for runways or specialized launch/recovery equipment and enhancing the overall safety of flight operations. Until recently, these systems were not possible due to the high power requirements required for lifting these typically heavy units (Berie & Burud, 2018).

14.2.2 Propulsion Systems

Lithium-battery electric propulsion systems are currently the most prolific systems in use by UAS manufacturers and are the primary propulsion systems for multirotor aircraft (Figure 14.5). They offer high energy density and hundreds of battery cycles and are low cost and easy to charge and maintain. Lithium-polymer (LiPo) and lithium-ion battery chemistries are the two most popular in use today. However, these systems present a significant drawback in that they are prone to combustion if damaged or punctured. It is important that aircraft operators take care to properly store, charge, and maintain lithium batteries.

FIGURE 14.3 Hand-launching an RQ-20 Puma.

FIGURE 14.4 Wingtra One, a hybrid VTOL fixed-wing survey aircraft.

FIGURE 14.5 Assortment of modern Lithium-Polymer batteries for UAVs.

FIGURE 14.6 Hybrid Project Supervolo, a hybrid VTOL quad-plane aircraft utilizing a hybrid gas-electric engine.

Many large, fixed-wing aircraft are available with internal combustion engines while smaller systems typically rely on electric propulsion. Due to the inherently higher energy density of fuels, internal combustion engines can provide longer flight times than similarly weighted electric propulsion systems.

Because electric VTOL propulsion systems are so ubiquitous, several companies have developed hybrid gas-electric propulsion systems (see Figure 14.6 for an example). These systems use a gasoline engine to power an onboard generator that replenishes batteries used for flight. In the case of some VTOL fixed-wing aircraft, the gas engine is also used to provide forward thrust when the platform is in horizontal flight. These systems often result in platforms that can continuously operate in excess of 8–12 hours.

Hydrogen fuel cell technology has recently become available as a fuel source on UAS. These systems can provide extremely high endurance but suffer from the low availability of sufficiently pure sources of hydrogen fuels compared to other fuels.

Figure 14.7 provides a comparison of the typical flight time of various UAS platforms compared to their platform style.

14.2.3 Sensor Payloads

Today, most commercially available small UAS are equipped with RGB camera payloads. Live video from these payloads is often streamed wirelessly to an operator in real time and allows for monitoring of the collected data. Over the last 10 years, significant improvements have been made to these sensors. Improvements such as automated geotagging of images turn today's standard camera

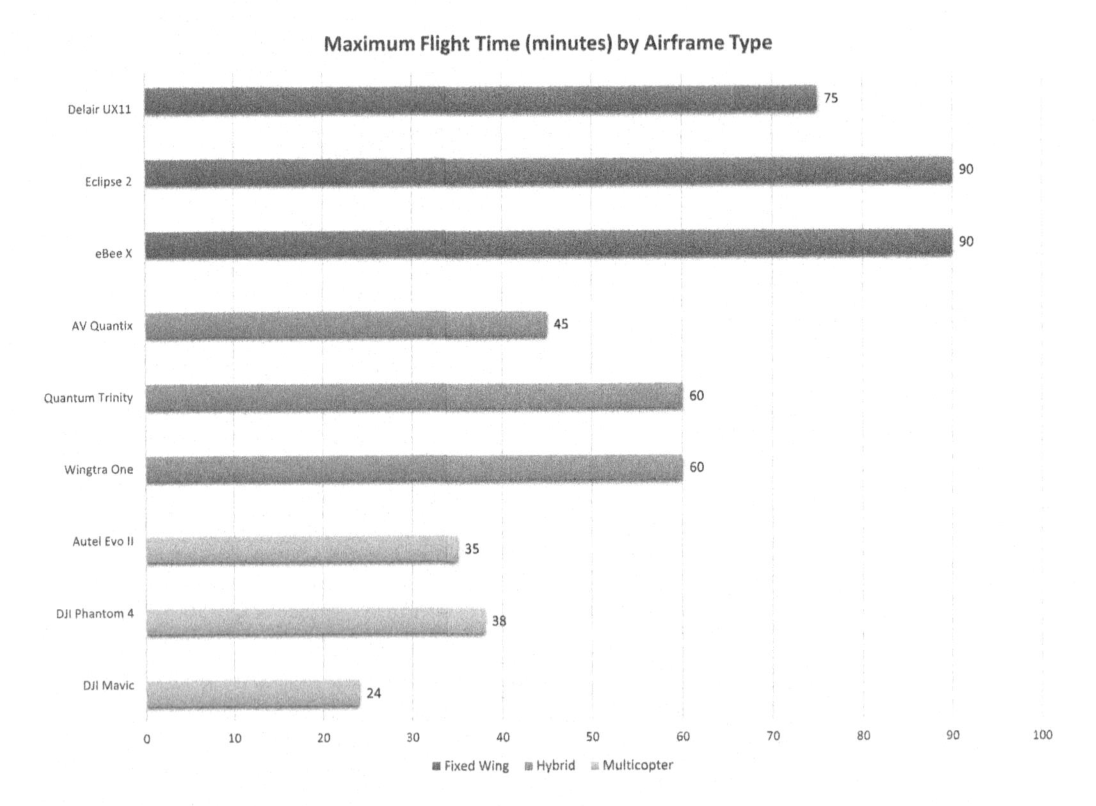

FIGURE 14.7 Relative flight times based on aircraft platform style from manufacturer specifications.

payloads into useful scientific tools. Many other platforms offer a wide range of additional payload options or can be fitted with multiple sensors. At the same time, multispectral sensors, such as the MicaSense, are becoming more common while light detection and ranging (LiDAR) systems are getting smaller, lighter, and more efficient.

The following sections represent a subset of the units available in January 2022. They are meant to inform the reader about the range of capabilities, performance, and cost points of some of the available UAS spectral-based sensor platforms. The units mentioned are based on the experience of the authors; inclusion in this chapter is not meant to serve as an endorsement for or against any units but rather to illustrate the wide range of platforms available at this time.

14.2.3.1 RGB

The most common sensor on nearly all platforms are basic visible/RGB imagers (see Figure 14.8 for an example of a UAS outfitted with an RGB camera). These sensors typically exist in the 400–900 nm range and can even consist of off-the-shelf (OTS) point-and-shoot cameras. They are ideal for projects that require true color imagery and are most easily integrated into photogrammetric software. When considering these sensors for geospatial research, it is important to understand their method of storing metadata. Some sensors automatically write the data to the image file, while other systems require postprocessing.

Sensors currently available on the market have a wide range of prices and capabilities. For example, the GoPro Hero 7 Black costs under $400 for a 12 MP camera, includes a 4K 60p video option, and is powered by its own battery. On the other end of the spectrum, the Phase One Industrial iMX–100 costs two orders of magnitude more (~$40,000) but has a resolution of 100 MP. This equates to a ground sample distance of 1.3 cm when flown at 400 ft, but the sensor does not take video and requires 14 W of external power.

FIGURE 14.8 Parrot Anafi UAS with integrated RGB camera.

14.2.3.2 Multispectral

When there is a need to obtain more information about specific processes or to quantify features seen at specific bandwidths, multispectral or hyperspectral imaging payloads can provide a useful solution. Multispectral sensors on multirotor systems have been used for leaf area index (LAI) detection of chestnut trees (Padua et al., 2020); carbon estimation from tree identification (Effiom et al., 2019); and suspended sediment monitoring in rivers (Larson et al., 2018). Multispectral sensors are also extremely popular in the agricultural industry, which conducts normalized difference vegetation index (NDVI) analyses on crops (Gómez-Sapiens et al., 2021).

Numerous companies offer multispectral solutions. MicaSense is considered a leader in the industry with UAS multispectral cameras. MicaSense offers a five-band solution: blue = 475 nm, green = 560 nm, red = 668 nm, red edge = 717 nm, and near infrared (NIR) = 840 nm with an 8 cm ground sample distance (GSD) at 400 ft. MicaSense also offers a six-band solution that has an additional thermal band of 8–14 μm. The six-band option, the MicaSense Altum, has 5.2 cm GSD for the five bands and an 81 cm GSD for the thermal band, both at 400 ft. Mapir Kernal cameras offer a custom solution that has 17 different options of bands that fall between 400 and 950 nm with a 4 W power draw and 4.31 cm GSD at 400 ft. A more expensive option (>$15,000), MAIA offers nine-band solutions that have the same spectral bands as the Worldview or Sentinel-2 space-based sensors, and they yield a 4.7 cm GSD.

14.2.3.3 Hyperspectral

Some studies require more precise spectral data than can be obtained using a multispectral sensor. Hyperspectral sensors are ideal to identify specific wavelengths with bandwidths smaller than a few nm. A hyperspectral payload may be needed, such as the Resonon Pika L sensor seen in Figure 14.9. These sensors are heavier and more expensive, typically costing more than $15,000. As a result, they are generally installed on heavy-lift aircraft.

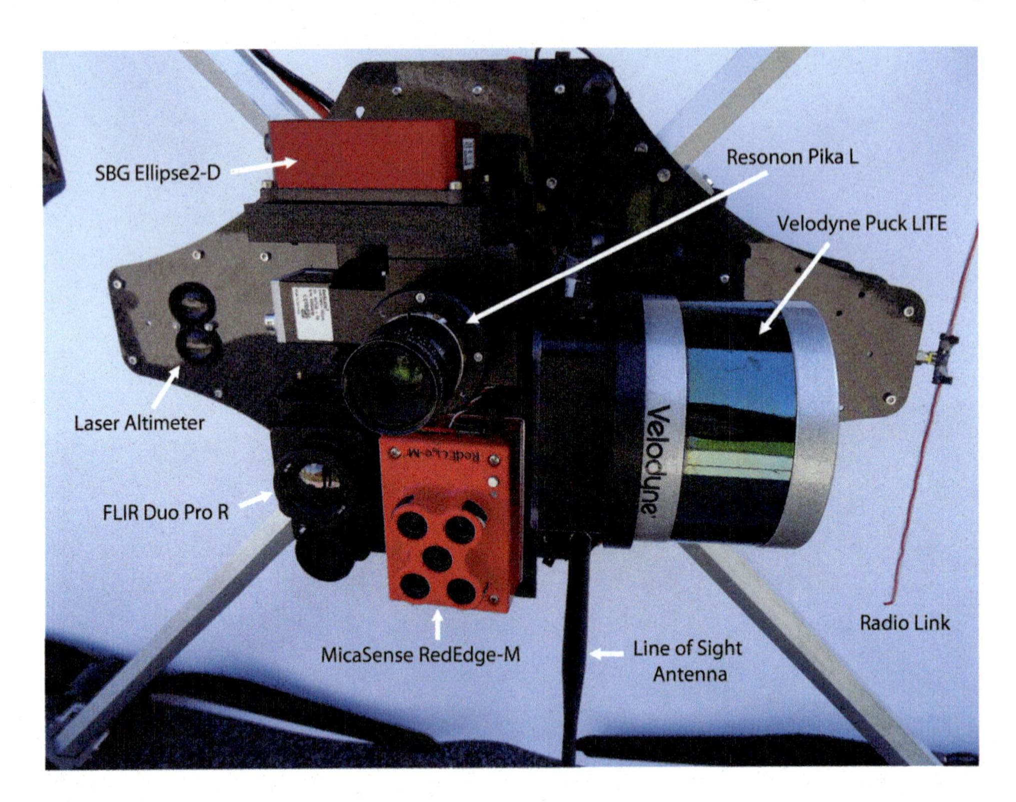

FIGURE 14.9 ORNL ARTEMIS payload with hyperspectral sensor.

The payload solutions for hyperspectral sensors can vary greatly in both their spectral range and the number of bands. Headwall is one of the leading manufacturers of hyperspectral push broom imagers, and they offer a variety of imagers that cover different wavelength regions in the electromagnetic spectrum. When compared to other companies, Headwall's options may be slightly larger in size. Headwall's most sophisticated, the Headwall Co-Aligned VNIR-SWIR, offers a spectral range of 400–2500 nm and up to 270 bands. This unit is fairly large at 27.2×20.8×16.5 cm^3 and weighs 2.83 kg. Headwall offers smaller units with 270 bands and a range of 400–1,000 nm that weighs 0.5 kg. Other options include the BaySpec OCI series, which can achieve a roughly 2 cm GSD from 400 ft and have options from 450 to 1,000 nm and total bands from 20 to 150, depending on the model. All of their units are 8×6×6 cm and weigh 180 g, which is very manageable on a UAS. The two BaySpec imagers designed for use onboard a UAV include both a push broom and snapshot sensor option, and while a snapshot hyperspectral imager is ideal due to stability concerns of the UAS, its resolution and number of spectral bands are limited when compared to the typical push broom sensor. Corning also offers solutions in the 400–2,500 nm for sensors that are meant for UAS and even space-based applications.

14.2.3.4 Thermal/LWIR

Thermal or long wave infrared (LWIR) sensors are typically used for studies that involve monitoring objects emitting rather than reflecting, such as visible range sensors. One study focused on particularly monitoring objects that were warmer than their surroundings, such as animals, particularly at night (Bushaw et al., 2019). Thermal bands are typically found between 7 and 15 μm (Figure 14.10). While Teledyne/FLIR is considered the leading thermal camera manufacturer for UAVs, MicaSense now offers a thermal band option for their multispectral imagers. When considering a thermal sensor for scientific applications, it is important to select a radiometric sensor. Many low-cost UAS are equipped with nonradiometric sensors designed to be used in emergency response or military intelligence, surveillance, and reconnaissance applications.

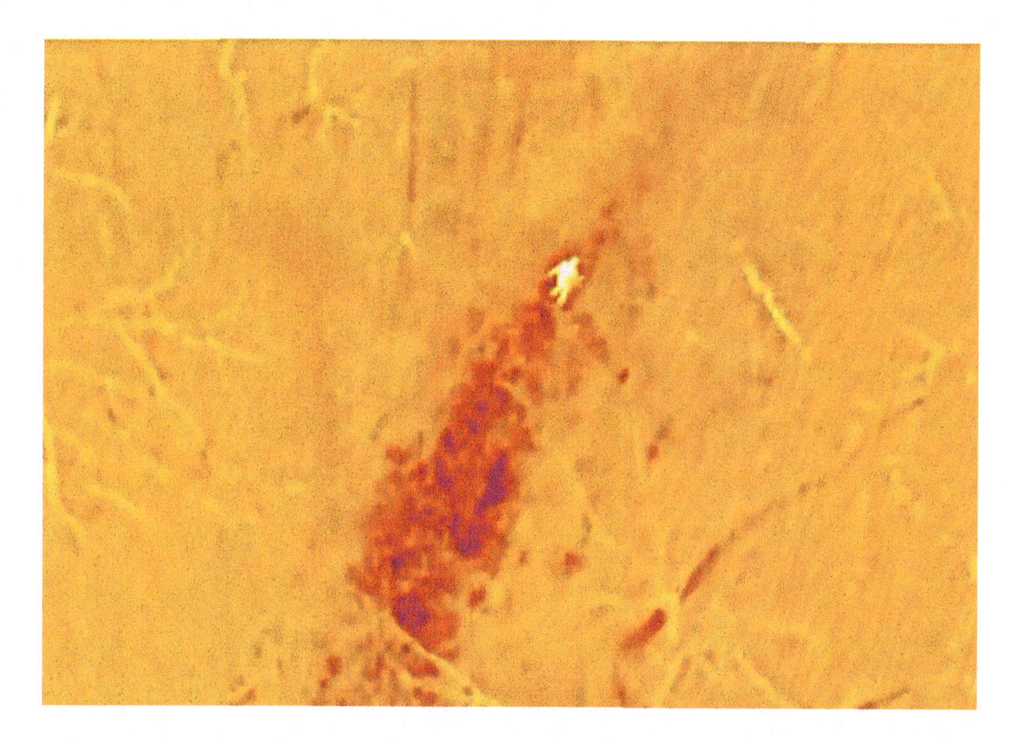

FIGURE 14.10 Thermal image of test subjects in a wooded area.

FLIR offers the FLIR Duo Pro R, which has an attached visible sensor, and the FLIR Vue Pro R that is thermal only. The Duo costs >$5,000 but has a spectral range of 7.5–13.5 μm and an array size of 336×256 and is 8.5×8.1×6.8 cm^3. The Vue has the same spectral range and array size and is smaller at 5.7×4.5 cm^2. The MicaSense Altum, discussed above has a lower-resolution thermal sensor (160×120 pixels) when compared to FLIR options.

14.2.3.5 LiDAR

LiDAR sensors have been available for decades for terrestrial use and on manned aircraft. These systems have been miniaturized and are now capable of being flown on a small UAS. A few groups were found to have utilized Rigel LiDAR units on custom built UAS (Peng et al., 2021b; Zhang et al., 2019a).

Velodyne is considered one of the leaders in LiDAR development and manufacturing. Their current customers include automotive manufacturers such as Ford, Volvo, and Mercedes-Benz as well as the UAS manufacturer DJI. The Velodyne Puck LITE costs <$5,000, has a range of 100 m, an accuracy ±3 cm, emits at a rate of 300–600 points per second (pps). It has a horizontal field of view (HFOV) of 360° and a vertical field of view (VFOV) of ±15°. Its size is manageable at 9×7 cm, weighs 590 g, and requires 8 W of power but has 16 channels and a 5–20 Hz rotation rate.

LeddarTech makes a variety of LiDAR systems for automotive purposes, UAV, traffic systems, and commercial vehicles. The LeddarVu was specifically designed for use with UAV and can cost <$1,000 but has limitations compared to other systems. Its range is up to 185 m, but it has an accuracy of±5cm and limited field of view (HFOV: 20°, 48°, 100°; VFOV: 0.3° or 3°). However, the LeddarVu weighs ~100 g and uses only 2 W. Ouster has developed one of the smallest and lightest LiDAR systems on the market, which is available in a 16- or 64-channel scanner. Their LiDAR systems have a sampling rate of over 1.3 million pps, ranges >100 m, HFOV: 360°, VFOV: ±11.25° or ±16.6°, and they weigh around 380 g.

Currently, all LiDAR units require their own internal measurement unit (IMU) and are controlled independently from the UAS. Preassembled and preconfigured LiDAR sensors with integrated

FIGURE 14.11 Flight crew performs a preflight check on a LiDAR payload from Phoenix LiDAR.

IMUs and co-bore sighted cameras are available from companies such as Phoenix LiDAR (see Figure 14.11 for an example LiDAR payload), Riegl, and YellowScan.

14.2.3.6 SAR

Synthetic aperture radar (SAR) is one of the newest sensor suites to be incorporated onto UAS and is just beginning to be used by research institutions. At this stage, mainly viability studies are being conducted. Manufacturers such as SAR Aero make small SAR systems for use on UAS platforms. These systems are not affected by cloud cover but require extremely accurate positional accuracy beyond what typical UAS IMUs can provide. For instance (Huang et al. 2021a), who utilized an SAR sensor for data collection, had to develop their own algorithms to account for the required high-precision inertial navigation systems, which are not typical of UAS.

IMSAR is one of the leaders in providing small, lightweight, low-power, and affordable systems. Many of their SAR systems are built for use with UAS. All IMSAR systems can use their proprietary software suite for real-time image processing, change detection, sensor control, and flight planning. Depending on the SAR system, they may include features such as coherent change detection, ground moving target indicator, maritime search and detection, foliage penetration, ground penetration, magnitude change detection, and moving target indication. All SAR systems can output complex NITF, TIFF, JPG, PNG, BMP, and KML file formats.

14.2.3.7 Sensor Precautions

One major consideration when using UAS for remote sensing research involves the integration between the sensor and UAS. For instance, many systems automatically georeference images in standard formats so that once they are downloaded, they can easily be used to create an orthorectified image or a 3D model. The location data for this georeferencing may be derived from the UAS or from stand-alone global navigation satellite system (GNSS) receivers. In some cases, data can be later combined with flight logs to manually georeference it, but this is not always a simple task. For example, placing a LiDAR system on a UAS and obtaining accurate georeferenced returns is extremely difficult to achieve in the postprocessing phase due to the relative inaccuracy of typical UAS IMUs. Even though UAS IMUs are accurate enough for most autonomous flights, payloads such as LiDAR and SAR require precision location knowledge to calibrate and process the high return rates required for these systems; thus, it is suggested that preintegrated solutions should be obtained wherever possible.

Another issue with many imagers is that some are not radiometrically calibrated and therefore could not be used to obtain accurate reflectance values (Suomalainen et al., 2018). All electro-optical sensors require at least some basic understanding of image saturation and balance level because many cameras automatically adjust various levels throughout the flight. While these auto-adjustments work well for stand-alone images, they can cause difficulty for image matching algorithms, orthomosaics, and/or 3D models because images acquired on nearby passes may have used different camera parameters, thus creating false differences in otherwise uniform vegetation or terrain. However, precise radiometric measurements are possible, and many groups have used a mix of custom and OTS units with various sensors to obtain accurate radiometric data (Wierzbicki et al., 2018; Kedzierski et al., 2019; Sekrecka et al., 2020).

One important caveat about these sensors is that they require frequent calibration. It should be noted that many multispectral and hyperspectral sensors come with a calibration panel that must be used before and after each flight to understand how the reflectance values have changed throughout the duration of the flight. Furthermore, active sensors such as LiDAR and SAR may require very specific precollection and postcollection flight paths to accurately calibrate the sensor. Many of these payloads require flights at specific heights, speeds, and angles relative to the ground to calibrate the rate of return from the active sensors. Failure to accurately follow the recommended calibration flight paths could result in data product accuracies an order of magnitude more, or worse, than the system can achieve due to the complexity and precision of the payload.

14.2.4 Use Cases/Literature Review

UAS have found a place in nearly all aspects of remote sensing, with a truly wide range of scientific applications. A large portion of use cases involve leveraging photogrammetric techniques to answer questions previously unobtainable using traditional collection methods. Additionally, great value has been found in combining UAS-acquired data with ground-based measurements, large-scale aircraft data, and/or satellite data.

14.2.4.1 Statistics

A brief survey conducted in November 2021 found 2,267 publications listed in Web of Science that mention UAVs and remote sensing. Roughly 300 entries were reviewed from that search, in which 205 of the publications listed the specific platform used in the study. These manuscripts focused on agriculture, forestry, coastal studies, animal movements, wetlands, land surface temperature, archeology, mining, and infrastructure, to name a few, and are summarized in the following subsections (see Table 14.1). One of the takeaways from the publications surveyed is that a UAS platform, and its sensors, can be used to answer almost any geospatial question. However, the precision of the answer is a function of the cost of the sensor and/or platform. It is not to say that cheaper platforms cannot provide very useful solutions but rather the high-precision sensors and platforms generally cost more and require more skilled operators.

Table 14.1 summarizes the range of platforms, sensors, and photogrammetry software used in the studies featured in the Web of Science search. While the majority of studies performed structure from motion (SfM) to some extent, not all of the studies did. Additionally, some studies used multiple platforms and sensors for their research. While a large number of studies used the standard available sensors, an RGB imager and multispectral imagers were used for roughly a quarter of the

TABLE 14.1
Applicant Area, Main Platforms Used, Sensors Used, and SfM Software Used

Subject Area		Platforms		Sensors		SfM	
Agriculture	52	DJI	139	Standard	79	Pix4D	70
Forestry	27	*DJI Phantom*	*74*	Multispectral	56	Agisoft Photoscan	63
Riparian	19	*DJI Phantom 4*	*56*	*Parrot Sequoia*	*21*	N/A	58
Coastal	12	*DJI M600*	*22*	*MicaSense*	*21*	other	16
Infrastructure	12	*DJI Inspire*	*19*	*Tetracam*	*6*		
Animals	12	*DJI Phantom 3*	*14*	IR	30		
Land cover	9	*DJI Phantom 3*	*14*	*FLIR*	*7*		
Mining	7	*DJI M100*	*7*	LiDAR	7		
Archaeology	7	*DJI Mavic Pro*	*6*	Hyperspectral	7		
Mangroves	5	*DJI S1000*	*3*				
Atmospheric dynamics	5	*DJI Phantom 2*	*2*				
Land surface	5	SenseFly	19				
Overview	5	3DR	11				
Optimization	4	Custom	15				
Cryosphere	4	Other	35				
Soil moisture	3						

studies. The following subsections briefly discuss the various sensor types used to answer scientific questions from a range of disciplines.

14.2.4.2 Agricultural

UAS have played a pivotal role in numerous scientific studies regarding the agriculture sector; note that, for the brief survey conducted for this work, nearly one-quarter of the entries focused on the agriculture sector. More specifically, these platforms lend themselves perfectly to the precision agriculture movement with their ability to quickly and efficiently reduce the workload of a farmer while saving on vital inputs such as water and fertilizers and helping to increase yields (Mogili & Deepak, 2018).

One of the most significant use cases for OTS UAS and RGB imagers is biomass measurements, which can be derived from photogrammetric techniques. Rice biomass was measured during its growing season (Peprah et al., 2021); crop heights and biomass of corn were determined with UAS (Fathipoor et al., 2019); and wheat height and vigor have also been monitored (Khan et al., 2018). Using OTS systems with basic RGB sensors was effective for the detection and monitoring of spice plants (Zhu et al., 2021); and creating vegetation indices using only visible wavelength sensors (Fu et al., 2020; Zhang et al., 2019b).

Studies using NIR cameras (i.e., modified Canon cameras) have been used to determine wheat nitrogen levels (Jiang et al., 2019) and optimal crop separation (Böhler et al., 2019). Besides yields, protein content using neural networks (NN) and various vegetation indices have been derived from multispectral UAS (Kang et al., 2021). Studies done on determining the amount of potassium in rice showed that NIR, acquired from a modified Canon PowerShot, worked well for achieving moderately accurate potassium values, but using multispectral data produced excellent results (Lu et al., 2021). Their work directly illustrated that cheaper systems can answer unique scientific questions, but more sophisticated sensors can provide a more precise answer.

Many other groups have utilized the wide range of solutions that can be achieved only with multispectral imagery. A few groups used these sensors to monitor cotton, specially monitoring root rot (Wang et al., 2020) and water stress (Bian et al., 2019). Multispectral data was heavily utilized when monitoring corn/maize with techniques ranging from chlorophyll measurements (Singhal et al., 2019); nitrogen level measurements (Thompson & Puntel, 2020); and NDVI measurements to determine early crop losses (Han et al., 2018; Sun et al., 2019).

LAI and biomass measurements of assorted crops have been made using multispectral sensors (Hu et al., 2021). Analysis of crop characteristics utilizing UAS with multispectral sensors have also been involved in genotype association studies with barley (Herzig et al., 2021); studies of the effects of multiangle views on LAI calculations of potatoes (Roosjen et al., 2018); and studies of weed detection in sugar beets (Khoshboresh-Masouleh & Akhoondzadeh, 2021). Furthermore, new studies are combining multispectral UAS data with chamber-based net ecosystem exchange (NEE) data to create daytime NEE variation estimation models that are scaled to the landscape level (Peng et al., 2021a). Only recently has multispectral UAS–acquired data been attempted using eddy covariance flux towers combined with satellite data. Some recent studies have been able to mount hyperspectral imagers on UAS to predict soybean yields (Hassanzadeh et al., 2021) and the effectiveness of irrigation on olive trees (Santos-Rufo et al., 2020). UAS-mounted LiDAR has even been used with multispectral sensors to monitor apple orchards (Hadas et al., 2019).

Another emerging trend in UAS use for agriculture is the use of convolutional neural networks (CNN) to help identify specific spatial features. The CNNs were used with multispectral imagery to monitor individual citrus trees for growth, fruit production, pests, and disease occurrence instead of the traditional manual delineation, which is more labor intensive (Csillik et al., 2018). Beyond citrus, any feature that can be identified through an image can utilize NNs for classification such as potato blight using RGB and NIR imagery (Duarte-Carvajalino et al., 2018), locating cheatgrass in range lands because of its effects on fire dynamics (Horning et al., 2020), and even for detection of compost piles over large areas (Song & Park, 2021).

14.2.4.3 Forestry

In the forestry sector, UAS have been used to measure biomass in both aboveground canopies (Hernández-Cole et al., 2021; Kameyama & Sugiura, 2020; Moukomla et al., 2018) and chopped-down stands to determine the size of wood chip piles. The traditional method of estimating the volume of wood piles uses GNSS devices, but using UAS and photogrammetry techniques produced fairly accurate measurements faster (Mokros et al., 2016). Traditional RGB imagers were also used to calculate the cover fractions of various plat species (Alexander et al., 2018; Fankhauser et al., 2018; Kattenborn et al., 2020; Yancho et al., 2019) and to detect various tree species (Brovkina et al., 2018; Huang et al., 2021b; Selim et al., 2019).

Delineation of trees in complex forests and crown estimations using UAS were shown to be very promising with an ~85% accuracy (Fraser & Congalton, 2021). Various types of machine learning have been used for species identification in the Amazon with over 90% accuracy (Moura et al., 2021), assessment of fir tree damage from beetles with accuracies ranging from 80%–94% (Safonova et al., 2019), and infected pines with 97% accuracy (Xia et al., 2021). Multispectral sensors helped classify Aspen trees (Kuzmin et al., 2021) and assess photosynthetic pigment on Norway spruce (Kopačková-Strnadová et al., 2021). Even UAS-based LiDAR has been used recently for Eucalyptus tree height measurement and detection (Picos et al., 2020), as well as tree plantation height measurement, which showed that SfM techniques agreed closely with UAV-based LiDAR (Li et al., 2021).

One of the largest risks facing forests in modern times is the threat of wildfire, which can devastate human lives but also greatly affects forest health, ecosystems, climate, and the economy (Akhloufi et al., 2021). Some researchers have focused on postprescribed burn evaluation using multispectral imagers to quantify soil and vegetation burn severity (Perez-Rodriguez et al., 2020). On the other end of the spectrum, authors have focused on early wildfire detection by proposing UAS frequently fly specific routes in a forest to collect data, via Bluetooth, from an array of early warning sensors deployed on the forest surface (Li, 2019).

14.2.4.4 Riparian Zones

The emerging technologies surrounding UAS have also greatly benefited the study of river dynamics and riparian zone studies, from the detection of floating river debris (Kim et al., 2017) to monitoring plastic trash in river systems (Geraeds et al., 2019; Jakovljevic et al., 2020). A more recent study used images instead of 3D models to estimate river flow velocity to track how large-scale particles move in the river (Liu et al., 2021). River discharge in ungauged rivers could be reliably estimated over long time scales using UAS (Lou et al., 2020). Other groups coupled UAS-based platforms with unmanned submersible vehicles to monitoring how particles released in a lake dispersed over time (Powers et al., 2018), and to create 3D lake surveys by combining UAS-based 3D mapping with sonar (Kapetanović et al., 2020).

Riparian habitats have been assessed using OTS RGB sensors (Gómez-Sapiens et al., 2021; Hentz et al., 2018) including the monitoring of land cover change near and around a river using RGB and NIR sensors (Morgan et al., 2021). Ancient river depositions have been mapped using ground-penetrating radar and the 3D models created from OTS UAS (Zhang et al., 2021).

14.2.4.5 Wetlands and Coastal Systems

Marsh dynamics (Marcaccio et al., 2016) and wetland restoration (Dale et al., 2020) efforts have benefited from UAS monitoring programs because these regions are difficult to access. This work includes monitoring underwater marine habits in shallow water (Papakonstantinou et al., 2020), intertidal topographic data derived from a video-mounted sensor on a UAV (Angnuureng et al., 2020), and island vegetation mapping using CNNs (Hamylton et al., 2020). Even soil salinity (Zhang & Zhao, 2019) and red tide events were characterized using OTS multispectral sensors (Kim et al., 2020).

Studies using RGB and multispectral combined datasets were found to be the best at monitoring macroalgal biodiversity, but neither sensor could see below 3 m due to turbidity at the study sites (Tait et al., 2019). UAS have even been used to quantify environmental damage from land-based oil spills (Mandianpari et al., 2018) and monitor beach litter (Merlino et al., 2020). Various studies have also been done on dunes using UAS (Marzialetti et al., 2021) and have been combined with historical imagery (Grottoli et al., 2021). Additionally, authors have participated in mangrove mapping utilizing OTS RGB imagers (Khakhim et al., 2019; Navarro et al., 2020) and hyperspectral imagers (Cao et al., 2018; Domingo et al., 2018).

14.2.4.6 Land Surface Temperature

As thermal-based imagers are made more available for use on UAS, new application areas are opening up, such as the monitoring of land surface temperatures and soil moisture. Land surface temperatures are ideal for understanding soil moisture levels whether in a wheat field (Wang et al., 2018) or meadow (Wigmore et al., 2019) or when quantifying biocrust layers using object-based image analysis to classify areas around various vegetation (Havrilla et al., 2020). Diurnal land surface temperatures in desert and grasslands have been measured (Malbéteau et al., 2018), as well as urban land surface temperatures for microthermal monitoring (Feng et al., 2020). Thermal imagers have also been used to compare UAV- and satellite-based acquisition of urban land temperature differences (Kim et al., 2021).

14.2.4.7 Animals

UAS have found a significant role in monitoring wild to domesticated animals as a way to monitor species from a distance. Bird habitats have been quantified using an array of microphones coupled with digital terrain models created using UAS (Wilson et al., 2021) and beach bird nesting habits using thermal imagers (Mapes et al., 2020). Other studies have highlighted the advantages and disadvantages of UAS versus large-scale aircraft in monitoring near-shore, large marine life in Australia (Kelaher et al., 2020) and Alaska (Angliss et al., 2018).

On the terrestrial side, domesticated pairing of cow and calf have been modeled (Mufford et al., 2019). A group of researchers, with >92% accuracy, were able to calculate the volume of manure piles to confirm they were not polluting local water sources (Park et al., 2021). A unique study tracked coyotes at night, using UAS to calculate populations instead of the traditional trapping method. They showed that UAS with thermal imagers are a viable tool for monitoring these nocturnal animals at scales up to 30 ha (Bushaw et al., 2019).

14.2.4.8 Archeology

Once UAS became more common place, they began to shape the field of archeology because of their ability to quickly and accurately map cultural resources using only basic RGB imagers (Jordan et al., 2016). Very successful comparisons between terrestrial laser scanning and UAS of cultural sites have been reported (Barba et al., 2019). UAS are being used to create 3D models of historic places, not just for preservation but also to create augmented reality experiences that ensure a larger audience (Carvajal-Ramírez et al., 2019). Even at sites where no surface features exist, UAS with multispectral and thermal imagers have enabled the interpretation of buried archaeological artifacts at two separate sites (Brooke & Clutterbuck, 2020).

14.2.4.9 Atmospheric Dynamics

Spectral analysis of turbulence using a UAS found that the general turbulence spectra matched for low frequencies but diverged at higher frequencies, specifically for horizontal components (Shelekhov et al., 2020). Planetary boundary layer monitoring of meteorological elements using the vertical structure of aerosols was performed using a 2 m fixed-wing UAS with an iMet-XQ2 sensor (Wang et al., 2021a). UAS platforms are being combined with eddy covariance towers to expand energy balance models beyond tower footprints (Simpson et al., 2021), while dropsondes released from UAS for monitoring wind speed and direction between flight height and the surface were shown to be effective for atmospheric monitoring, thus not limiting the practice to large-scale aircraft only (Prior et al., 2021).

14.2.4.10 Optimization

As UAS and their sensors have become more precise and sophisticated, optimizing data has become more important than ever. The optimal acquisition height, overlap, and camera parameters for accurate data collection have been quantified (Garcia & de Oliveira, 2021), as well as the importance of ground control points, for 0.07 m precision, where the number of image points was found to be the most important (Santos Santana et al., 2021). Other groups have focused on multispectral image registration (Meng et al., 2021) and super-resolution enhancement of images after photogrammetric calculations (Burdziakowski, 2020).

14.2.4.11 Automation

UAS have also been growing in the area of automation and security. Automated landing has been explored by several groups (Lee et al., 2018; Loureiro et al., 2021), as well as collision avoidance (D'Amato et al., 2018; Lim et al., 2019) and development of wireless mesh networks for field access (Shinkuma & Mandayam, 2020).

14.3 COMMUNICATIONS

Another vital requirement of a UAV is its ability to effectively communicate with its ground control station (GCS) via telemetry over a wireless connection. This capability varies greatly among platforms based on the manufacturer, type of platform, and payload. Typically, UAS are controlled and operated via a ground station, usually from some ground-based computer, tablet, or phone that communicates with the platform.

14.3.1 Ground Control

UAS platforms can be controlled by a variety of GCS types, including a hand controller, computer, or cell phone/tablet. These GCSs can provide a first-person view of what the aircraft currently "sees," or even display telemetry (e.g., battery life, altitude) in real time. But, most importantly, they can be used to plan the flight or control the UAV midflight. The choice of GCS software is dependent on the platform being used, preferred interface (e.g., tablet, phone), and on flight planning.

The GCS software, which is a computer program or Android/IOS application, is usually proprietary to the UAS manufacturer and custom-tailored to the specific UAS brand. Examples include eMotion3 from SenseFly, Freeflight 6 from Parrot, Skydio Enterprise Application from Skydio, and DJI GO for DJI UAS. Platforms that use PX4 or ArduPilot flight control software (e.g., Alta, DeltaQuad, WingtraOne, Yuneec) all communicate over a standard communication protocol called MAVLink. Due to this common communication protocol, they can be controlled with the same GCS, which is usually QGroundControl or MissionPlanner. Because QGroundControl is an open-source GCS, many manufacturers have built their own custom versions, with QGroundControl having the ability to work on both computer and phone/tablet hardware options.

14.3.2 Networked UAS

Beyond typical radio-frequency (RF) communications, there are options to extend the range and capabilities of UAS platforms based on the communication options available to the systems. Networked communication is critical when exploring the option of beyond visible line of sight (BVLOS) operation while maintaining positive control of the vehicle. Typically, networked UAS depend on cellular connectivity, although more advanced systems may also utilize satellite networks or other forms of bonded communications. For cellular-connected systems, two types of architectures are available: centralized server hosting and vehicle-based hosting.

A centralized server architecture includes a server component accessible via the open Internet. Individual unmanned vehicles send and receive data to the server, and GCSs communicate directly with the centralized server. Such an architecture has the advantage of being highly scalable because the number of users interacting with the vehicle via the GCS software is limited only by the resources available on the server, which are typically easily scaled to support thousands of simultaneous users.

The Multi-Modal Autonomous Vehicle Network (MAVNet) is an Oak Ridge National Laboratory–developed example of a centralized server architecture for unmanned systems. MAVNet is also unique in that it allows a system to operate as both a networked UAS and traditional RF-connected UAS depending on the available network connectivity. For example, if cellular or satellite connectivity is available, MAVNet provides secure networking to a centralized server and allows remote users to interact with the system using a web-based GCS from any location with Internet access. In parallel with the networked connectivity, users can employ a traditional RF-connected GCS to interact with the vehicle (Stinson et al., 2019). The MAVNet solution offers network diversity for UAS platforms, allowing them to communicate over Wi-Fi, cellular networks, satellite, or traditional RF links simultaneously, while avoiding command duplication and stale vehicle-state data.

Vehicle-based hosting provides an open port and/or web-based interface directly from the hardware onboard the vehicle to the open Internet, typically via a cellular radio. For security, such applications are often used in combination with a virtual private network. In this scenario, the GCS software connects directly to the vehicle via networking protocols and/or the user may interact with the vehicle through a web-based user interface accessed with the HTTP or HTTPS protocol in a web browser. Such an architecture is preferred in some cases due to the simplicity and low cost, primarily because a hosted server infrastructure is not needed. However, this solution does not scale to support multiple simultaneous users because of the limited resources and bandwidth available from the compute and communication resources onboard the vehicle. For example, video streaming from a vehicle-based host over cellular is limited to only a few simultaneous receivers. In contrast, a centralized server architecture allows for nearly unlimited viewers of the video feed.

14.4 PROCESSING TECHNIQUES

There are different aspects of data collection techniques and processing that can affect the outcome of the data. It was recently quantified that SfM accuracy is primarily dependent on sensor resolution, the height of data acquisition, the amount of image overlap, and the placement and number of ground control points (Deliry & Avdan, 2021). Other studies have provided overviews regarding the different end-use cases of datasets, and there are overview papers for agricultural methods (Eskandari et al., 2020), ecology and conservation (Dujon, 2020), forestry (Guimarães et al., 2020), and remote sensing applications (Hardin et al., 2019; Yao et al., 2019).

14.4.1 Onboard Processing

The advent of high-performance computing onboard has led to various groups utilizing GPUs for real-time decision-making. A Parrot Bebop 2 connected to a Legion laptop with an NVIDIA 2060 GPU was able to avoid moving objects thrown at it in real time. The algorithm was tested on a Jetson Nano that could run the pipeline in 0.18 seconds. This is promising for onboard processing because some of the newest UAVs on the market (e.g., Skydio 2, HEIFU) have similar GPUs onboard (Pedro et al., 2021). Other authors have designed a system that used an NVIDIA Xavier NX for vehicle speed detection capable of being run at 29 fps (Balamuralidhar et al., 2021), demonstrating one of the first real-time traffic-monitoring systems to be run on a UAV under real-world conditions. Using the same NVIDIA Xavier NX with the Pixhawk2 autopilot, a deep learning algorithm was created to help with automated trash and litter detection in real time on a UAV (Kraft et al., 2021). Other groups have experimented with the Jetson Nano to achieve similar onboard processing goal (Koay et al., 2021).

The Skydio aircraft makes use of the NVIDIA Jetson TX2 compute capabilities to perform real-time simultaneous localization and mapping, which it uses to avoid obstacles in all directions. Similarly, the Parrot Anafi AI has onboard computing resources for obstacle avoidance, data processing, and transferring data over a cellular network. However, at the time of writing, accessing this onboard compute power was limited to obstacle avoidance.

14.4.2 Postprocessing

One of the most common postprocessing techniques performed on UAS imagery is the creation of orthomosaics (i.e., 2D maps) and 3D models. This typically involves using a computer to stitch images together using photogrammetric software. Two of the most common software platforms available are Pix4D and Agisoft's Photoscan, as can be seen in the previous sections and in Table 14.1. The open-source community has also developed a UAS imagery processing software called Open Drone Map, which has gained traction over the past few years. All postprocessing software needs georeferenced imagery to work to its full capacity. Software like Pix4D can take in nongeoreferenced imagery, but the results cannot be overlaid on a base map.

New technologies allow for processing on a phone to remove the latency associated with having to physically take the data to a desktop computer. Reveal AI developed Farsight Mobile, which has the capability of processing small mapping missions on a phone. The imagery can either be streamed live, or an SD card can be inserted into the phone after flight.

For imagery from nonvisible payloads such as thermal or multispectral, these can also be stitched together into orthomosaics using Pix4D. However, when using a five-band multispectral camera, each band will have a separate orthomosaic, which requires other software solutions to create one, multiband orthomosaic.

Processing of LiDAR is more involved than processing imagery because it is an active sensor. Companies like Phoenix LiDAR Systems have developed an all-in-one solution which georeferences the point cloud onboard the system. During flight, a GPS base station is used to record position data for postprocessing the GNSS data. Additional LiDAR processing must be done after the flight to refine the accuracy of the points to achieve centimeter-level accuracy and to filter out any noise in the point cloud.

14.5 CURRENT ISSUES

Bottlenecks currently faced when using most UAS stem from issues ranging from basic operation of the platform and battery life to more sophisticated problems such as communication range, data protection, and effective use of the payload.

Flight Duration is one of the most significant limiting factors for small UAS today. While alternative propulsion systems exist, electric battery power is by far the most common due to convenience. Research is continuing into advanced battery technologies such as solid-state batteries and advanced chemistries that promise higher energy densities and safer operations. Today's flight duration for typical electric rotorcraft systems will be anywhere from 15 to 45 minutes based on size, weight, and configuration. The flight duration for fixed-wing platforms using batteries has a more acceptable operational duration from 30 to 120 minutes. New airframe designs, such as hybrid VTOL airframes, are able to achieve the extended flight time of a fixed-wing aircraft while retaining the convenience of rotorcraft.

Communication Range is another limiting factor for small UAS. As long-endurance propulsion technologies become more readily available, the unmanned system user community will want to take full advantage of the extended flight times. Future communication technologies and protocols must allow for long-range command and control of these systems. The marketplace is beginning to move in that direction with the implementation of commercial systems with embedded cellular radios; however, this alone only satisfies operations in range of cellular towers. The ultimate solution would allow for multiple radio bands that could transition between line-of-sight, cellular, and satellite-based radios as the unmanned system travels between coverage areas. One implementation of such a system is the MAVNet ecosystem, discussed above. Users must be aware of the communication distance capabilities of their system to ensure they can maintain proper control. As a reminder, this distance is greatly affected by the surrounding landscape; for instance, urban areas will have less communication distance than open fields.

Communication Bandwidth is an issue that is greatly impacting current geographical datasets. The ideal scenario for most unmanned system users would be for any data taken by the platform to be streamed in real time and at full resolution to the GCS for observation and processing. Even with high bandwidth line-of-sight radios, this can be quite difficult due to the size of each image/data, the amount of images/data, and the radio signal quality. While cellular radios can assist by providing another gateway to use for routing the data, the bandwidth is still quite limited, and cellular is not always available. One solution to this problem, which will be particularly impactful for long-duration missions, is to move the location of where the data processing occurs from the GCS to onboard the platform using powerful embedded computers. An example scenario would be for a fixed-wing platform to fly for 60 minutes out to its area of interest, collect imagery, and then fly 60 minutes back to its launch point. Onboard processing could begin when image collection starts and continue throughout the duration of the flight. There are plenty of single-board computers that have small enough size, weight, and power that can perform these types of operations, such as the NVIDIA Jetson line of systems. However, at the time of writing, these units need to be added after market as an independent system.

Real-Time Processing is one of the most common challenges facing UAS at the time of collection. For instance, when producing point clouds from SfM, the results cannot be seen until after data is returned to the office and processed. This makes it extremely difficult to know whether the data collected was useful or not, or whether new data must be collected. This is directly linked to the disconnect between the advancement of UAS technologies and the relatively slower growth of management technologies for large datasets. This means that not only is an understanding of the risks of flying a UAS required but so are the typical analysis techniques required by geospatial analysts because most data require skilled input from data scientists.

Real-time processing also illustrates a bottleneck very much related to the limited communication bandwidth issues. As these images and datasets become increasingly larger, which will assist with the extraction of better information, users will have to address how the datasets can be properly handled. To get the information to users as quickly as possible, it makes sense to employ real-time processing where practical. While this does include the stitching together of imagery, it also includes the analysis of each image to look for feature sets and then communicating over a limited bandwidth connection when a feature is located by sending its GPS coordinate and corresponding file name, for example. Machine learning and artificial intelligence algorithms are continually being improved to operate at the edge for these types of applications. An example would be for an algorithm running on the embedded computer of an unmanned system to determine and notify emergency responders of the location of damaged infrastructure after a natural disaster. One of the current issues with real-time processing is the requirement that high-resolution images must be divided into smaller images for some of the object detection software to work at the desired speeds and accuracies (Kraft et al., 2021).

Current Aviation Policy and Regulation limit the operational uses of the vast majority of UAS to visual line of sight. This limitation requires the pilot (or dedicated visual observer) to maintain visual contact with the aircraft at all times to ensure the UAS does not become a hazard to other aircraft. This is a concept known as "see and avoid." Due to this requirement, extended-endurance aircraft are often limited to using only a fraction of their entire range. Today, a number of protected airspace "corridors" are equipped with ground-based radar systems to enable safe research into BVLOS operations. Research efforts are underway to develop and improve "detect and avoid" capabilities that will enable UAS to autonomously sense aircraft and other hazards, enabling them to alter flight paths to avoid conflicts.

14.6 FUTURE DIRECTIONS

Publication outputs based on UAS remote sensing have seen exponential growth, with the United States and China leading the way. One caveat that the authors addressed involves how the United States is the most prolific in terms of international cooperation and how this cooperation will directly influence the technologies surrounding UAS remote sensing. With many technologies leaning toward artificial intelligence and machine learning, cooperation among institutions and countries will continue to advance the field at a faster pace (Wang et al., 2021b).

Many of the solutions coming on the market address a number of challenges including adopting and spearheading new methods and technologies of UAS operation; integrating data from a host of conventional and emerging sources and spatial analysis; improving uncertainty management, including dealing with denial and deception; dealing with data volume issues, especially the need to automate human interpretation tasks; and ubiquity of access, including web-based systems and the effective reuse of existing data.

14.6.1 Sensors

A large number of authors have found strength in fusing multiple sensors into one dataset because some wavelengths are ideal for very specific problems but only highlight a small portion of the solution. This multisensory approach illustrates the direction of the field; many solutions can be derived from RGB, NIR, multispectral, and LiDAR, alone and achieve very good, scientifically accurate results. However, depending on the level of precision required, adding on more sensors typically equates to more precise datasets (Lu et al., 2021; Peng et al., 2021b), but at the cost of time and money.

14.6.2 Processing

Machine learning, artificial intelligence, and deep learning will all play a significant role in the future of UAS, from communications to data processing. As processors get smaller, more GPUs will become standard on platforms. Skydio X2D already comes standard with a GPU, but it strictly functions to improve the collision avoidance system on the platform. The availability of GPUs on UAS will open the door to more real-time object-based analysis onboard the platform and will provide the ability to produce orthomosaics and possibly 3D models in near real time. Many new platforms will need to include multitask learning approaches as well as new computational power because, by the time processing power catches up to the current sensor suites, the sensors will have become even more data heavy.

Currently, very few applications can accurately register multispectral images. While the multispectral registration of images exists, but is limited, LWIR to visible imagery registration will be a new direction required for more effective scientific datasets (Meng et al., 2021).

With increasing numbers of researchers using AI for image analysis, there will be a greater need for pretrained datasets; because training a dataset is only as good as the data with which it is trained, large datasets are needed for training. Future UAS research will have to involve the distribution of data for training between platforms so training weights can be as robust as possible (Lin et al., 2021). This same level of data sharing will be influenced by the possibility of distributed computing technologies.

14.6.3 Communications

The availability of this processing power will greatly affect communications as well. As cognitive radio techniques become more prevalent and the available frequency spectrum becomes more crowded, the best solution is to use machine learning to solve the problem. Cognitive radios are systems that utilize software-defined radios to learn the spectral environment so it can dynamically adapt and efficiently utilize the frequency spectrum for communication (Santana et al., 2018). However, the processing ability and power consumption available now are not sufficient to make this a standard feature. However, once the processing power is available, these platforms will be able to communicate more securely and effectively than current systems.

14.7 CONCLUSION

UAS technology has been growing at an increasingly rapid rate over the past two decades and will continue to grow and directly influence geospatial science. As platforms are becoming more advanced, a larger divide is developing between the user-friendly OTS systems and ones that require experienced pilots and scientists. However, this divide is allowing more inexperienced operators to become familiar with the flexibility and robustness of datasets that can be collected by UAS while also providing platforms and sensors that can answer some of the most difficult geospatial science questions facing our world today. UAS will continue to shape the geospatial landscape, and as new processing abilities become available on platforms, latency will be reduced between data collection and scientific analysis, thus propelling the discovery of new scientific solutions.

REFERENCES

Akhloufi, M. A., Couturier, A., & Castro, N. A. (2021). Unmanned Aerial Vehicles for Wildland Fires: Sensing, Perception, Cooperation and Assistance. *Drones*, *5*(1), 15. doi: 10.3390/drones5010015.

Alexander, C., Korstjens, A. H., Hankinson, E., Usher, G., Harrison, N., Nowak, M. G., Abdullah, A., Wich, S. A., & Hill, R. A. (2018). Locating Emergent Trees in a Tropical Rainforest Using Data from an Unmanned Aerial Vehicle (UAV). *International Journal of Applied Earth Observation and Geoinformation, 72*, 86–90. doi: 10.1016/j.jag.2018.05.024.

Angliss, R. P., Ferguson, M. C., Hall, P., Helker, V., Kennedy, A., & Sformo, T. (2018). Comparing Manned to Unmanned Aerial Surveys for Cetacean Monitoring in the Arctic: Methods and Operational Results. *Journal of Unmanned Vehicle Systems, 6*(3), 109–127. doi: 10.1139/juvs-2018-0001.

Angnuureng, D. B., Jayson-Quashigah, P.-N., Almar, R., Stieglitz, T. C., Anthony, E. J., Aheto, D. W., & Appeaning Addo, K. (2020). Application of Shore-Based Video and Unmanned Aerial Vehicles (Drones): Complementary Tools for Beach Studies. *Remote Sensing, 12*(3), 394. doi: 10.3390/rs12030394.

Balamuralidhar, N., Tilon, S., & Nex, F. (2021). MultEYE: Monitoring System for Real-Time Vehicle Detection, Tracking and Speed Estimation from UAV Imagery on Edge-Computing Platforms. *Remote Sensing, 13*(4), 573. doi: 10.3390/rs13040573.

Barba, S., Barbarella, M., Di Benedetto, A., Fiani, M., Gujski, L., & Limongiello, M. (2019). Accuracy Assessment of 3D Photogrammetric Models from an Unmanned Aerial Vehicle. *Drones, 3*(4), 79. doi: 10.3390/drones3040079.

Berie, H. T., & Burud, I. (2018). Application of Unmanned Aerial Vehicles in Earth Resources Monitoring: Focus on Evaluating Potentials for Forest Monitoring in Ethiopia. *European Journal of Remote Sensing, 51*(1), 326–335. doi: 10.1080/22797254.2018.1432993.

Bian, J., Zhang, Z., Chen, J., Chen, H., Cui, C., Li, X., Chen, S., & Fu, Q. (2019). Simplified Evaluation of Cotton Water Stress Using High Resolution Unmanned Aerial Vehicle Thermal Imagery. *Remote Sensing, 11*(3), 267. doi: 10.3390/rs11030267.

Böhler, J. E., Schaepman, M. E., & Kneubühler, M. (2019). Optimal Timing Assessment for Crop Separation Using Multispectral Unmanned Aerial Vehicle (UAV) Data and Textural Features. *Remote Sensing, 11*(-15), 1780. doi: 10.3390/rs11151780.

Brooke, C., & Clutterbuck, B. (2020). Mapping Heterogeneous Buried Archaeological Features Using Multisensor Data from Unmanned Aerial Vehicles. *Remote Sensing, 12*(1), 41. doi: 10.3390/rs12010041.

Brovkina, O., Cienciala, E., Surový, P., & Janata, P. (2018). Unmanned Aerial Vehicles (UAV) for Assessment of Qualitative Classification of Norway Spruce in Temperate Forest Stands. *Geo-Spatial Information Science, 21*(1), 12–20. doi: 10.1080/10095020.2017.1416994.

Burdziakowski, P. (2020). Increasing the Geometrical and Interpretation Quality of Unmanned Aerial Vehicle Photogrammetry Products Using Super-Resolution Algorithms. *Remote Sensing, 12*(5), 810. doi: 10.3390/rs12050810.

Bushaw, J. D., Ringelman, K. M., & Rohwer, F. C. (2019). Applications of Unmanned Aerial Vehicles to Survey Mesocarnivores. *Drones, 3*(1), 28. doi: 10.3390/drones3010028.

Cao, J., Leng, W., Liu, K., Liu, L., He, Z., & Zhu, Y. (2018). Object-Based Mangrove Species Classification Using Unmanned Aerial Vehicle Hyperspectral Images and Digital Surface Models. *Remote Sensing, 10*(1), 89. doi: 10.3390/rs10010089.

Carvajal-Ramírez, F., Navarro-Ortega, A. D., Agüera-Vega, F., & Martínez-Carricondo, P. (2019). Unmanned Aerial Vehicle Photogrammetry and 3D Modeling Applied to Virtual Reconstruction of an Archaeological Site in the Bronze Age. *The International Archives of the Photogrammetry, Remote Sensing and Spatial Information Sciences, XLII-2-W15*, 279–284. doi: 10.5194/isprs-archives-XLII-2-W15-279-2019.

Colomina, I., & Molina, P. (2014). Unmanned Aerial Systems for Photogrammetry and Remote Sensing: A Review. *ISPRS Journal of Photogrammetry and Remote Sensing, 92*, 79–97. doi: 10.1016/j.isprsjprs.2014.02.013.

Csillik, O., Cherbini, J., Johnson, R., Lyons, A., & Kelly, M. (2018). Identification of Citrus Trees from Unmanned Aerial Vehicle Imagery Using Convolutional Neural Networks. *Drones, 2*(4), 39. doi: 10.3390/drones2040039.

D'Amato, E., Notaro, I., & Mattei, M. (2018). Distributed Collision Avoidance for Unmanned Aerial Vehicles Integration in the Civil Airspace. *2018 International Conference on Unmanned Aircraft Systems (ICUAS)*, pp. 94–102. https://www.webofscience.com/wos/woscc/full-record/WOS:000454860700013.

Dale, J., Burnside, N. G., Hill-Butler, C., Berg, M. J., Strong, C. J., & Burgess, H. M. (2020). The Use of Unmanned Aerial Vehicles to Determine Differences in Vegetation Cover: A Tool for Monitoring Coastal Wetland Restoration Schemes. *Remote Sensing, 12*(24), 4022. doi: 10.3390/rs12244022.

Deliry, S. I., & Avdan, U. (2021). Accuracy of Unmanned Aerial Systems Photogrammetry and Structure from Motion in Surveying and Mapping: A Review. *Journal of the Indian Society of Remote Sensing, 49*(8), 1997–2017. doi: 10.1007/s12524-021-01366-x.

Domingo, G. A., Claridades, A. R. C., & Tupas, M. E. A. (2018). Unmanned Aerial Vehicle (UAV) Survey-Assisted 3d Mangrove Tree Modeling. In: A. A. Rahman & H. Karim (Eds.), *International Conference on Geomatics & Geospatial Technology (GGT 2018): Geospatial and Disaster Risk Management*, Vols. 42–4, Issue W9, pp. 123–127. Copernicus Gesellschaft Mbh. doi: 10.5194/isprs-archives-XLII-4-W9-123-2018.

Duarte-Carvajalino, J. M., Alzate, D. F., Ramirez, A. A., Santa-Sepulveda, J. D., Fajardo-Rojas, A. E., & Soto-Suárez, M. (2018). Evaluating Late Blight Severity in Potato Crops Using Unmanned Aerial Vehicles and Machine Learning Algorithms. *Remote Sensing*, *10*(10), 1513. doi: 10.3390/rs10101513.

Dujon, A. M. (2020). Citation Patterns of Publications Using Unmanned Aerial Vehicles in Ecology and Conservation. *Journal of Unmanned Vehicle Systems*, *8*(1), 1–10. doi: 10.1139/juvs-2019-0013.

Effiom, A. E., Leeuwen, L. M. van, Nyktas, P., Okojie, J. A., & Erdbrügger, J. (2019). Combining Unmanned Aerial Vehicles and Multispectral Pleiades Data for Tree Species Identification, a Prerequisite for Accurate Carbon Estimation. *Journal of Applied Remote Sensing*, *13*(3), 034530. doi: 10.1117/1.JRS.13.034530.

Eskandari, R., Mahdianpari, M., Mohammadimanesh, F., Salehi, B., Brisco, B., & Homayouni, S. (2020). Meta-Analysis of Unmanned Aerial Vehicle (UAV) Imagery for Agro-Environmental Monitoring Using Machine Learning and Statistical Models. *Remote Sensing*, *12*(21), 3511. doi: 10.3390/rs12213511.

Godlewski, M. (2022, February 7). FAA Drone Pilot Numbers Hit Milestone. FLYING Magazine. https://www.flyingmag.com/faa-drone-pilot-numbers-hit-milestone/Fankhauser, K. E., Strigul, N. S., & Gatziolis, D. (2018). Augmentation of Traditional Forest Inventory and Airborne Laser Scanning with Unmanned Aerial Systems and Photogrammetry for Forest Monitoring. *Remote Sensing*, *10*(10), 1562. doi: 10.3390/rs10101562.

Fathipoor, H., Arefi, H., Shah-Hosseini, R., & Moghadam, H. (2019). Corn Forage Yield Prediction Using Unmanned Aerial Vehicles Images at Mid-Season Growth Stage. *Journal of Applied Remote Sensing*, *13*(3), 034503. doi: 10.1117/1.JRS.13.034503.

Feng, L., Tian, H., Qiao, Z., Zhao, M., & Liu, Y. (2020). Detailed Variations in Urban Surface Temperatures Exploration Based on Unmanned Aerial Vehicle Thermography. *IEEE Journal of Selected Topics in Applied Earth Observations and Remote Sensing*, *13*, 204–216. doi: 10.1109/JSTARS.2019.2954852.

Fraser, B. T., & Congalton, R. G. (2021). Estimating Primary Forest Attributes and Rare Community Characteristics Using Unmanned Aerial Systems (UAS): An Enrichment of Conventional Forest Inventories. *Remote Sensing*, *13*(15), 2971. doi: 10.3390/rs13152971.

Fu, Z., Jiang, J., Gao, Y., Krienke, B., Wang, M., Zhong, K., Cao, Q., Tian, Y., Zhu, Y., Cao, W., & Liu, X. (2020). Wheat Growth Monitoring and Yield Estimation Based on Multi-Rotor Unmanned Aerial Vehicle. *Remote Sensing*, *12*(3), 508. doi: 10.3390/rs12030508.

Garcia, M. V. Y., & de Oliveira, H. C. (2021). The Influence of Flight Configuration, Camera Calibration, and Ground Control Points for Digital Terrain Model and Orthomosaic Generation Using Unmanned Aerial Vehicles Imagery. *Boletim de Ciências Geodésicas*, *27*. doi: 10.1590/s1982-21702021000200015.

Geraeds, M., van Emmerik, T., de Vries, R., & bin Ab Razak, M. S. (2019). Riverine Plastic Litter Monitoring Using Unmanned Aerial Vehicles (UAVs). *Remote Sensing*, *11*(17), 2045. doi: 10.3390/rs11172045.

Gómez-Sapiens, M., Schlatter, K. J., Meléndez, Á., Hernández-López, D., Salazar, H., Kendy, E., & Flessa, K. W. (2021). Improving the Efficiency and Accuracy of Evaluating Aridland Riparian Habitat Restoration Using Unmanned Aerial Vehicles. *Remote Sensing in Ecology and Conservation*, *7*(3), 488–503. doi: 10.1002/rse2.204.

Grottoli, E., Biausque, M., Rogers, D., Jackson, D. W. T., & Cooper, J. A. G. (2021). Structure-from-Motion-Derived Digital Surface Models from Historical Aerial Photographs: A New 3D Application for Coastal Dune Monitoring. *Remote Sensing*, *13*(1), 95. doi: 10.3390/rs13010095.

Guimarães, N., Pádua, L., Marques, P., Silva, N., Peres, E., & Sousa, J. J. (2020). Forestry Remote Sensing from Unmanned Aerial Vehicles: A Review Focusing on the Data, Processing and Potentialities. *Remote Sensing*, *12*(6), 1046. doi: 10.3390/rs12061046.

Hadas, E., Jozkow, G., Walicka, A., & Borkowski, A. (2019). Apple Orchard Inventory with a LiDAR Equipped Unmanned Aerial System. *International Journal of Applied Earth Observation and Geoinformation*, *82*, 101911. doi: 10.1016/j.jag.2019.101911.

Hamylton, S. M., Morris, R. H., Carvalho, R. C., Roder, N., Barlow, P., Mills, K., & Wang, L. (2020). Evaluating Techniques for Mapping Island Vegetation from Unmanned Aerial Vehicle (UAV) Images: Pixel Classification, Visual Interpretation and Machine Learning Approaches. *International Journal of Applied Earth Observation and Geoinformation*, *89*, 102085. doi: 10.1016/j.jag.2020.102085.

Han, L., Yang, G., Feng, H., Zhou, C., Yang, H., Xu, B., Li, Z., & Yang, X. (2018). Quantitative Identification of Maize Lodging-Causing Feature Factors Using Unmanned Aerial Vehicle Images and a Nomogram Computation. *Remote Sensing*, *10*(10), 1528. doi: 10.3390/rs10101528.

Hardin, P. J., Lulla, V., Jensen, R. R., & Jensen, J. R. (2019). Small Unmanned Aerial Systems (sUAS) for Environmental Remote Sensing: Challenges and Opportunities Revisited. *GIScience & Remote Sensing*, *56*(2), 309–322. doi: 10.1080/15481603.2018.1510088.

Hassanzadeh, A., Zhang, F., van Aardt, J., Murphy, S. P., & Pethybridge, S. J. (2021). Broadacre Crop Yield Estimation Using Imaging Spectroscopy from Unmanned Aerial Systems (UAS): A Field-Based Case Study with Snap Bean. *Remote Sensing*, *13*(16), 3241. doi: 10.3390/rs13163241.

Havrilla, C. A., Villarreal, M. L., DiBiase, J. L., Duniway, M. C., & Barger, N. N. (2020). Ultra-High-Resolution Mapping of Biocrusts with Unmanned Aerial Systems. *Remote Sensing in Ecology and Conservation*, *6*(4), 441–456. doi: 10.1002/rse2.180.

Hentz, Â. M. K., Kinder, P. J., Hubbart, J. A., & Kellner, E. (2018). Accuracy and Optimal Altitude for Physical Habitat Assessment (PHA) of Stream Environments Using Unmanned Aerial Vehicles (UAV). *Drones*, *2*(2), 20. doi: 10.3390/drones2020020.

Hernández-Cole, J., Ortiz-Malavassi, E., Moya, R., & Murillo, O. (2021). Evaluation of Unmanned Aerial Vehicles (UAV) as a Tool to Predict Biomass and Carbon of Tectona Grandis in Silvopastoral Systems (SPS) in Costa Rica. *Drones*, *5*(2), 47. doi: 10.3390/drones5020047.

Herzig, P., Borrmann, P., Knauer, U., Klück, H.-C., Kilias, D., Seiffert, U., Pillen, K., & Maurer, A. (2021). Evaluation of RGB and Multispectral Unmanned Aerial Vehicle (UAV) Imagery for High-Throughput Phenotyping and Yield Prediction in Barley Breeding. *Remote Sensing*, *13*(14), 2670. doi: 10.3390/rs13142670.

Horning, N., Fleishman, E., Ersts, P. J., Fogarty, F. A., & Wohlfeil Zillig, M. (2020). Mapping of Land Cover with Open-Source Software and Ultra-High-Resolution Imagery Acquired with Unmanned Aerial Vehicles. *Remote Sensing in Ecology and Conservation*, *6*(4), 487–497. doi: 10.1002/rse2.144.

Hu, P., Chapman, S. C., Jin, H., Guo, Y., & Zheng, B. (2021). Comparison of Modelling Strategies to Estimate Phenotypic Values from an Unmanned Aerial Vehicle with Spectral and Temporal Vegetation Indexes. *Remote Sensing*, *13*(14), 2827. doi: 10.3390/rs13142827.

Huang, Y., Liu, F., Chen, Z., Li, J., & Hong, W. (2021a). An Improved Map-Drift Algorithm for Unmanned Aerial Vehicle SAR Imaging. *IEEE Geoscience and Remote Sensing Letters*, *18*(11), 1966–1970. doi: 10.1109/LGRS.2020.3011973.

Huang, D., Zhou, Z., Zhang, Z., Zhu, M., Yin, L., Peng, R., Zhang, Y., & Zhang, W. (2021b). Recognition and Counting of Pitaya Trees in Karst Mountain Environment Based on Unmanned Aerial Vehicle RGB Images. *Journal of Applied Remote Sensing*, *15*(4), 042402. doi: 10.1117/1.JRS.15.042402.

Jakovljevic, G., Govedarica, M., & Alvarez-Taboada, F. (2020). A Deep Learning Model for Automatic Plastic Mapping Using Unmanned Aerial Vehicle (UAV) Data. *Remote Sensing*, *12*(9), 1515. doi: 10.3390/rs12091515.

Jiang, J., Cai, W., Zheng, H., Cheng, T., Tian, Y., Zhu, Y., Ehsani, R., Hu, Y., Niu, Q., Gui, L., & Yao, X. (2019). Using Digital Cameras on an Unmanned Aerial Vehicle to Derive Optimum Color Vegetation Indices for Leaf Nitrogen Concentration Monitoring in Winter Wheat. *Remote Sensing*, *11*(22), 2667. doi: 10.3390/rs11222667.

Jordan, T. R., Goetcheus, C. L., & Madden, M. (2016). Point Cloud Mapping Methods for Documenting Cultural Landscape Features at the Wormsloe State Historic Site, Savannah, Georgia, Usa. In L. Halounova, V. Safar, F. Remondino, J. Hodac, K. Pavelka, M. Shortis, F. Rinaudo, M. Scaioni, J. Boehm, & D. RiekeZapp (Eds.), *XXIII ISPRS Congress, Commission V*, Vol. 41, Issue B5, pp. 277–280. Copernicus Gesellschaft Mbh. doi: 10.5194/isprsarchives-XLI-B5-277-2016.

Kameyama, S., & Sugiura, K. (2020). Estimating Tree Height and Volume Using Unmanned Aerial Vehicle Photography and SfM Technology, with Verification of Result Accuracy. *Drones*, *4*(2), 19. doi: 10.3390/drones4020019.

Kang, Y., Nam, J., Kim, Y., Lee, S., Seong, D., Jang, S., & Ryu, C. (2021). Assessment of Regression Models for Predicting Rice Yield and Protein Content Using Unmanned Aerial Vehicle-Based Multispectral Imagery. *Remote Sensing*, *13*(8), 1508. doi: 10.3390/rs13081508.

Kapetanović, N., Kordić, B., Vasilijević, A., Nađ, Đ., & Mišković, N. (2020). Autonomous Vehicles Mapping Plitvice Lakes National Park, Croatia. *Remote Sensing*, *12*(22), 3683. doi: 10.3390/rs12223683.

Kattenborn, T., Eichel, J., Wiser, S., Burrows, L., Fassnacht, F. E., & Schmidtlein, S. (2020). Convolutional Neural Networks Accurately Predict Cover Fractions of Plant Species and Communities in Unmanned Aerial Vehicle Imagery. *Remote Sensing in Ecology and Conservation*, *6*(4), 472–486. doi: 10.1002/rse2.146.

Kedzierski, M., Wierzbicki, D., Sekrecka, A., Fryskowska, A., Walczykowski, P., & Siewert, J. (2019). Influence of Lower Atmosphere on the Radiometric Quality of Unmanned Aerial Vehicle Imagery. *Remote Sensing*, *11*(10), 1214. doi: 10.3390/rs11101214.

Kelaher, B. P., Peddemors, V. M., Hoade, B., Colefax, A. P., & Butcher, P. A. (2020). Comparison of Sampling Precision for Nearshore Marine Wildlife Using Unmanned and Manned Aerial Surveys. *Journal of Unmanned Vehicle Systems*, *8*(1), 30–44. doi: 10.1139/juvs-2018-0023.

Khakhim, N., Marfai, M. A., Wicaksono, A., Lazuardi, W., Isnaen, Z., & Walinono, T. (2019). Mangrove Ecosystem Data Inventory Using Unmanned Aerial Vehicles (UAVs) in Yogyakarta Coastal Area. *Sixth Geoinformation Science Symposium*, *11311*, 175–183. doi: 10.1117/12.2547326.

Khan, Z., Chopin, J., Cai, J., Eichi, V.-R., Haefele, S., & Miklavcic, S. J. (2018). Quantitative Estimation of Wheat Phenotyping Traits Using Ground and Aerial Imagery. *Remote Sensing*, *10*(6), 950. doi: 10.3390/rs10060950.

Khoshboresh-Masouleh, M., & Akhoondzadeh, M. (2021). Improving Weed Segmentation in Sugar Beet Fields Using Potentials of Multispectral Unmanned Aerial Vehicle Images and Lightweight Deep Learning. *Journal of Applied Remote Sensing*, *15*(3), 034510. doi: 10.1117/1.JRS.15.034510.

Kim, H.-M., Yoon, H., Jang, S., & Chung, Y. (2017). Detection Method of River Floating Debris Using Unmanned Aerial Vehicle and Multispectral Sensors. *Korean Journal of Remote Sensing*, *33*(5), 537–546. doi: 10.7780/kjrs.2017.33.5.1.7.

Kim, W., Jung, S., Kim, K., Ryu, J.-H., & Moon, Y. (2020). Mapping Red Tide Intensity Using Multispectral Camera on Unmanned Aerial Vehicle: A Case Study in Korean South Coast. *IGARSS 2020-2020 IEEE International Geoscience and Remote Sensing Symposium*, pp. 5612–5615. doi: 10.1109/IGARSS39084.2020.9323103.

Kim, D., Yu, J., Yoon, J., Jeon, S., & Son, S. (2021). Comparison of Accuracy of Surface Temperature Images from Unmanned Aerial Vehicle and Satellite for Precise Thermal Environment Monitoring of Urban Parks Using In Situ Data. *Remote Sensing*, *13*(10), 1977. doi: 10.3390/rs13101977.

Koay, H. V., Chuah, J. H., Chow, C.-O., Chang, Y.-L., & Yong, K. K. (2021). YOLO-RTUAV: Towards Real-Time Vehicle Detection through Aerial Images with Low-Cost Edge Devices. *Remote Sensing*, *13*(21), 4196. doi: 10.3390/rs13214196.

Kopačková-Strnadová, V., Koucká, L., Jelének, J., Lhotáková, Z., & Oulehle, F. (2021). Canopy Top, Height and Photosynthetic Pigment Estimation Using Parrot Sequoia Multispectral Imagery and the Unmanned Aerial Vehicle (UAV). *Remote Sensing*, *13*(4), 705. doi: 10.3390/rs13040705.

Kraft, M., Piechocki, M., Ptak, B., & Walas, K. (2021). Autonomous, Onboard Vision-Based Trash and Litter Detection in Low Altitude Aerial Images Collected by an Unmanned Aerial Vehicle. *Remote Sensing*, *13*(5), 965. doi: 10.3390/rs13050965.

Kuzmin, A., Korhonen, L., Kivinen, S., Hurskainen, P., Korpelainen, P., Tanhuanpää, T., Maltamo, M., Vihervaara, P., & Kumpula, T. (2021). Detection of European Aspen (*Populus tremula* L.) Based on an Unmanned Aerial Vehicle Approach in Boreal Forests. *Remote Sensing*, *13*(9), 1723. doi: 10.3390/rs13091723.

Larson, M. D., Simic Milas, A., Vincent, R. K., & Evans, J. E. (2018). Multi-Depth Suspended Sediment Estimation Using High-Resolution Remote-Sensing UAV in Maumee River, Ohio. *International Journal of Remote Sensing*, *39*(15–16), 5472–5489. doi: 10.1080/01431161.2018.1465616.

Lee, S., Shim, T., Kim, S., Park, J., Hong, K., & Bang, H. (2018). Vision-Based Autonomous Landing of a Multi-Copter Unmanned Aerial Vehicle using Reinforcement Learning. *2018 International Conference on Unmanned Aircraft Systems (ICUAS)*, pp. 108–114. https://www.webofscience.com/wos/woscc/full-record/WOS:000454860700015.

Li, S. (2019). Wildfire Early Warning System Based on Wireless Sensors and Unmanned Aerial Vehicle. *Journal of Unmanned Vehicle Systems*, *7*(1), 76–91. doi: 10.1139/juvs-2018-0022.

Li, M., Li, Z., Liu, Q., & Chen, E. (2021). Comparison of Coniferous Plantation Heights Using Unmanned Aerial Vehicle (UAV) Laser Scanning and Stereo Photogrammetry. *Remote Sensing*, *13*(15), 2885. doi: 10.3390/rs13152885.

Lim, C., Li, B., Ng, E. M., Liu, X., & Low, K. H. (2019). Three-Dimensional (3D) Dynamic Obstacle Perception in a Detect-and-Avoid Framework for Unmanned Aerial Vehicles. *2019 International Conference on Unmanned Aircraft Systems (ICUAS' 19)*, pp. 996–1004. https://www.webofscience.com/wos/woscc/full-record/WOS:000503382900124.

Lin, W., Adetomi, A., & Arslan, T. (2021). Low-Power Ultra-Small Edge AI Accelerators for Image Recognition with Convolution Neural Networks: Analysis and Future Directions. *Electronics*, *10*(17), 2048. doi: 10.3390/electronics10172048.

Liu, W.-C., Lu, C.-H., & Huang, W.-C. (2021). Large-Scale Particle Image Velocimetry to Measure Streamflow from Videos Recorded from Unmanned Aerial Vehicle and Fixed Imaging System. *Remote Sensing*, *13*(14), 2661. doi: 10.3390/rs13142661.

Lou, H., Wang, P., Yang, S., Hao, F., Ren, X., Wang, Y., Shi, L., Wang, J., & Gong, T. (2020). Combining and Comparing an Unmanned Aerial Vehicle and Multiple Remote Sensing Satellites to Calculate Long-Term River Discharge in an Ungauged Water Source Region on the Tibetan Plateau. *Remote Sensing, 12*(13), 2155. doi: 10.3390/rs12132155.

Loureiro, G., Dias, A., Martins, A., & Almeida, J. (2021). Emergency Landing Spot Detection Algorithm for Unmanned Aerial Vehicles. *Remote Sensing, 13*(10), 1930. doi: 10.3390/rs13101930.

Lu, J., Eitel, J. U. H., Engels, M., Zhu, J., Ma, Y., Liao, F., Zheng, H., Wang, X., Yao, X., Cheng, T., Zhu, Y., Cao, W., & Tian, Y. (2021). Improving Unmanned Aerial Vehicle (UAV) Remote Sensing of Rice Plant Potassium Accumulation by Fusing Spectral and Textural Information. *International Journal of Applied Earth Observation and Geoinformation, 104*, 102592. doi: 10.1016/j.jag.2021.102592.

Malbéteau, Y., Parkes, S., Aragon, B., Rosas, J., & McCabe, M. F. (2018). Capturing the Diurnal Cycle of Land Surface Temperature Using an Unmanned Aerial Vehicle. *Remote Sensing, 10*(9), 1407. doi: 10.3390/rs10091407.

Mandianpari, M., Salehi, B., Mohammadimanesh, F., Larsen, G., & Peddle, D. R. (2018). Mapping Land-Based Oil Spills Using High Spatial Resolution Unmanned Aerial Vehicle Imagery and Electromagnetic Induction Survey Data. *Journal of Applied Remote Sensing, 12*(3), 036015. doi: 10.1117/1.JRS.12.036015.

Mapes, K. L., Pricope, N. G., Baxley, J. B., Schaale, L. E., & Danner, R. M. (2020). Thermal Imaging of Beach-Nesting Bird Habitat with Unmanned Aerial Vehicles: Considerations for Reducing Disturbance and Enhanced Image Accuracy. *Drones, 4*(2), 12. doi: 10.3390/drones4020012.

Marcaccio, J., Markle, C. E., & Chow-Fraser, P. (2016). Use of Fixed-Wing and Multi-Rotor Unmanned Aerial Vehicles to Map Dynamic Changes in a Freshwater Marsh. *Journal of Unmanned Vehicle Systems, 4*(3), 193–202. doi: 10.1139/juvs-2015-0016.

Marzialetti, F., Frate, L., De Simone, W., Frattaroli, A. R., Acosta, A. T. R., & Carranza, M. L. (2021). Unmanned Aerial Vehicle (UAV)-Based Mapping of Acacia Saligna Invasion in the Mediterranean Coast. *Remote Sensing, 13*(17), 3361. doi: 10.3390/rs13173361.

Meng, L., Zhou, J., Liu, S., Ding, L., Zhang, J., Wang, S., & Lei, T. (2021). Investigation and Evaluation of Algorithms for Unmanned Aerial Vehicle Multispectral Image Registration. *International Journal of Applied Earth Observation and Geoinformation, 102*, 102403. doi: 10.1016/j.jag.2021.102403.

Merlino, S., Paterni, M., Berton, A., & Massetti, L. (2020). Unmanned Aerial Vehicles for Debris Survey in Coastal Areas: Long-Term Monitoring Programme to Study Spatial and Temporal Accumulation of the Dynamics of Beached Marine Litter. *Remote Sensing, 12*(8), 1260. doi: 10.3390/rs12081260.

Mogili, U. R., & Deepak, B. B. V. L. (2018). Review on Application of Drone Systems in Precision Agriculture. *Procedia Computer Science, 133*, 502–509. doi: 10.1016/j.procs.2018.07.063.

Mokros, M., Tabacak, M., Lieskovsky, M., & Fabrika, M. (2016). Unmanned Aerial Vehicle Use for Wood Chips Pile Volume Estimation. In L. Halounova, V. Safar, C. K. Toth, J. Karas, G. Huadong, N. Haala, A. Habib, P. Reinartz, X. Tang, J. Li, C. Armenakis, G. Grenzdorffer, P. LeRoux, S. Stylianidis, R. Blasi, M. Menard, H. Dufourmount, & Z. Li (Eds.), *XXIII ISPRS Congress, Commission I*, Vol. 41, Issue B1, pp. 953–956. Copernicus Gesellschaft Mbh. doi: 10.5194/isprsarchives-XLI-B1-953-2016.

Morgan, B. E., Chipman, J. W., Bolger, D. T., & Dietrich, J. T. (2021). Spatiotemporal Analysis of Vegetation Cover Change in a Large Ephemeral River: Multi-Sensor Fusion of Unmanned Aerial Vehicle (UAV) and Landsat Imagery. *Remote Sensing, 13*(1), 51. doi: 10.3390/rs13010051.

Moukomla, S., Srestasathiern, P., Siripon, S., Wasuhiranyrith, R., & Kooha, P. (2018). Estimating Above Ground Biomass for Eucalyptus Plantation Using Data from Unmanned Aerial Vehicle Imagery. *Remote Sensing for Agriculture, Ecosystems, and Hydrology XX, 10783*, 39–44. doi: 10.1117/12.2323963.

Moura, M. M., de Oliveira, L. E. S., Sanquetta, C. R., Bastos, A., Mohan, M., & Corte, A. P. D. (2021). Towards Amazon Forest Restoration: Automatic Detection of Species from UAV Imagery. *Remote Sensing, 13*(-13), 2627. doi: 10.3390/rs13132627.

Mufford, J. T., Hill, D. J., Flood, N. J., & Church, J. S. (2019). Use of Unmanned Aerial Vehicles (UAVs) and Photogrammetric Image Analysis to Quantify Spatial Proximity in Beef Cattle. *Journal of Unmanned Vehicle Systems, 7*(3), 194–206. doi: 10.1139/juvs-2018-0025.

Navarro, A., Young, M., Allan, B., Carnell, P., Macreadie, P., & Ierodiaconou, D. (2020). The Application of Unmanned Aerial Vehicles (UAVs) to Estimate above-Ground Biomass of Mangrove Ecosystems. *Remote Sensing of Environment, 242*, 111747. doi: 10.1016/j.rse.2020.111747.

Padua, L., Marques, P., Martins, L., Sousa, A., Peres, E., & Sousa, J. J. (2020). Estimation of Leaf Area Index in Chestnut Trees Using Multispectral Data from an Unmanned Aerial Vehicle. *IGARSS 2020-2020 IEEE International Geoscience and Remote Sensing Symposium*, 6503–6506. doi: 10.1109/IGARSS39084.2020.9324614.

Papakonstantinou, A., Stamati, C., & Topouzelis, K. (2020). Comparison of True-Color and Multispectral Unmanned Aerial Systems Imagery for Marine Habitat Mapping Using Object-Based Image Analysis. *Remote Sensing, 12*(3), 554. doi: 10.3390/rs12030554.

Park, G., Park, K., & Song, B. (2021). Spatio-Temporal Change Monitoring of Outside Manure Piles Using Unmanned Aerial Vehicle Images. *Drones, 5*(1), 1. doi: 10.3390/drones5010001.

Pedro, D., Matos-Carvalho, J. P., Fonseca, J. M., & Mora, A. (2021). Collision Avoidance on Unmanned Aerial Vehicles Using Neural Network Pipelines and Flow Clustering Techniques. *Remote Sensing, 13*(13), 2643. doi: 10.3390/rs13132643.

Peng, M., Han, W., Li, C., & Huang, S. (2021a). Improving the Spatial and Temporal Estimation of Maize Daytime Net Ecosystem Carbon Exchange Variation Based on Unmanned Aerial Vehicle Multispectral Remote Sensing. *IEEE Journal of Selected Topics in Applied Earth Observations and Remote Sensing, 14*, 10560–10570. doi: 10.1109/JSTARS.2021.3119908.

Peng, X., Liu, H., Chen, Y., Chen, Q., Wang, J., Li, H., & Zhao, A. (2021b). A Method to Identify Dacrydium pierrei Hickel Using Unmanned Aerial Vehicle Multi-source Remote Sensing Data in a Chinese Tropical Rainforest. *Journal of the Indian Society of Remote Sensing.* doi: 10.1007/s12524-021-01453-z.

Peprah, C. O., Yamashita, M., Yamaguchi, T., Sekino, R., Takano, K., & Katsura, K. (2021). Spatio-Temporal Estimation of Biomass Growth in Rice Using Canopy Surface Model from Unmanned Aerial Vehicle Images. *Remote Sensing, 13*(12), 2388. doi: 10.3390/rs13122388.

Perez-Rodriguez, L. A., Quintano, C., Marcos, E., Suarez-Seoane, S., Calvo, L., & Fernandez-Manso, A. (2020). Evaluation of Prescribed Fires from Unmanned Aerial Vehicles (UAVs) Imagery and Machine Learning Algorithms. *Remote Sensing, 12*(8), 1295. doi: 10.3390/rs12081295.

Picos, J., Bastos, G., Míguez, D., Alonso, L., & Armesto, J. (2020). Individual Tree Detection in a Eucalyptus Plantation Using Unmanned Aerial Vehicle (UAV)-LiDAR. *Remote Sensing, 12*(5), 885. doi: 10.3390/rs12050885.

Powers, C., Hanlon, R., & Schmale, D. G. (2018). Tracking of a Fluorescent Dye in a Freshwater Lake with an Unmanned Surface Vehicle and an Unmanned Aircraft System. *Remote Sensing, 10*(1), 81. doi: 10.3390/rs10010081.

Prior, E. M., Miller, G. R., & Brumbelow, K. (2021). Topographic and Landcover Influence on Lower Atmospheric Profiles Measured by Small Unoccupied Aerial Systems (sUAS). *Drones, 5*(3), 82. doi: 10.3390/drones5030082.

Roosjen, P. P. J., Brede, B., Suomalainen, J. M., Bartholomeus, H. M., Kooistra, L., & Clevers, J. G. P. W. (2018). Improved Estimation of Leaf Area Index and Leaf Chlorophyll Content of a Potato Crop Using Multi-Angle Spectral Data – Potential of Unmanned Aerial Vehicle Imagery. *International Journal of Applied Earth Observation and Geoinformation, 66*, 14–26. doi: 10.1016/j.jag.2017.10.012.

Safonova, A., Tabik, S., Alcaraz-Segura, D., Rubtsov, A., Maglinets, Y., & Herrera, F. (2019). Detection of Fir Trees (Abies sibirica) Damaged by the Bark Beetle in Unmanned Aerial Vehicle Images with Deep Learning. *Remote Sensing, 11*(6), 643. doi: 10.3390/rs11060643.

Santana, G. M. D., Cristo, R. S., Dezan, C., Diguet, J.-P., Osorio, D. P. M., & Branco, K. R. L. J. C. (2018). Cognitive Radio for UAV Communications: Opportunities and Future Challenges. *2018 International Conference on Unmanned Aircraft Systems (ICUAS)*, pp. 760–768. doi: 10.1109/ICUAS.2018.8453329.

Santos Santana, L., Araújo E Silva Ferraz, G., Bedin Marin, D., Dienevam Souza Barbosa, B., Mendes Dos Santos, L., Ferreira Ponciano Ferraz, P., Conti, L., Camiciottoli, S., & Rossi, G. (2021). Influence of Flight Altitude and Control Points in the Georeferencing of Images Obtained by Unmanned Aerial Vehicle. *European Journal of Remote Sensing, 54*(1), 59–71. doi: 10.1080/22797254.2020.1845104.

Santos-Rufo, A., Mesas-Carrascosa, F.-J., García-Ferrer, A., & Meroño-Larriva, J. E. (2020). Wavelength Selection Method Based on Partial Least Square from Hyperspectral Unmanned Aerial Vehicle Orthomosaic of Irrigated Olive Orchards. *Remote Sensing, 12*(20), 3426. doi: 10.3390/rs12203426.

Sekrecka, A., Wierzbicki, D., & Kedzierski, M. (2020). Influence of the Sun Position and Platform Orientation on the Quality of Imagery Obtained from Unmanned Aerial Vehicles. *Remote Sensing, 12*(6), 1040. doi: 10.3390/rs12061040.

Selim, S., Sonmez, N. K., Coslu, M., & Onur, I. (2019). Semi-Automatic Tree Detection from Images of Unmanned Aerial Vehicle Using Object-Based Image Analysis Method. *Journal of the Indian Society of Remote Sensing, 47*(2), 193–200. doi: 10.1007/s12524-018-0900-1.

Shelekhov, A. P., Afanasiev, A. L., Kobzev, A. A., & Shelekhova, E. A. (2020). Opportunities to Monitor an Urban Atmospheric Turbulence Using Unmanned Aerial System. In T. Erbertseder, N. Chrysoulakis, & Y. Zhang (Eds.), *Remote Sensing Technologies and Applications in Urban Environments V*, Vol. 11535, p. 1153506. SPIE, the Internatonal Society for Optical Engineering. doi: 10.1117/12.2573486.

Shinkuma, R., & Mandayam, N. B. (2020). Design of Ad Hoc Wireless Mesh Networks Formed by Unmanned Aerial Vehicles with Advanced Mechanical Automation. *16th Annual International Conference on Distributed Computing in Sensor Systems (Dcoss 2020)*, pp. 288–295. doi: 10.1109/DCOSS49796.2020.00053.

Simpson, J. E., Holman, F., Nieto, H., Voelksch, I., Mauder, M., Klatt, J., Fiener, P., & Kaplan, J. O. (2021). High Spatial and Temporal Resolution Energy Flux Mapping of Different Land Covers Using an Off-the-Shelf Unmanned Aerial System. *Remote Sensing*, *13*(7), 1286. doi: 10.3390/rs13071286.

Singhal, G., Bansod, B., Mathew, L., Goswami, J., Choudhury, B. U., & Raju, P. L. N. (2019). Chlorophyll Estimation Using Multi-Spectral Unmanned Aerial System Based on Machine Learning Techniques. *Remote Sensing Applications: Society and Environment*, *15*, 100235. doi: 10.1016/j.rsase.2019.100235.

Song, B., & Park, K. (2021). Comparison of Outdoor Compost Pile Detection Using Unmanned Aerial Vehicle Images and Various Machine Learning Techniques. *Drones*, *5*(2), 31. doi: 10.3390/drones5020031.

Stinson, B. J., Duncan, A. M., Vacaliuc, B., Harter, A., Roberts II, C., & Thompson, T. (2019). MAVNet: Design of a Reliable Beyond Visual Line of Sight Communication System for Unmanned Vehicles. Oak Ridge National Laboratory (ORNL), Oak Ridge, TN. https://www.osti.gov/biblio/1528701-mavnet-design-reliable-beyond-visual-line-sight-communication-system-unmanned-vehicles.

Sun, Q., Sun, L., Shu, M., Gu, X., Yang, G., & Zhou, L. (2019). Monitoring Maize Lodging Grades via Unmanned Aerial Vehicle Multispectral Image. *Plant Phenomics*, *2019*. doi: 10.34133/2019/5704154.

Suomalainen, J., Hakala, T., de Oliveira, R. A., Markelin, L., Viljanen, N., Nasi, R., & Honkavaara, E. (2018). A Novel Tilt Correction Technique for Irradiance Sensors and Spectrometers On-Board Unmanned Aerial Vehicles. *Remote Sensing*, *10*(12), 2068. doi: 10.3390/rs10122068.

Tait, L., Bind, J., Charan-Dixon, H., Hawes, I., Pirker, J., & Schiel, D. (2019). Unmanned Aerial Vehicles (UAVs) for Monitoring Macroalgal Biodiversity: Comparison of RGB and Multispectral Imaging Sensors for Biodiversity Assessments. *Remote Sensing*, *11*(19), 2332. doi: 10.3390/rs11192332.

Thompson, L. J., & Puntel, L. A. (2020). Transforming Unmanned Aerial Vehicle (UAV) and Multispectral Sensor into a Practical Decision Support System for Precision Nitrogen Management in Corn. *Remote Sensing*, *12*(10), 1597. doi: 10.3390/rs12101597.

Tiner, R. W., Lang, M. W., & Klemas, V. V. (2015). *Remote Sensing of Wetlands: Applications and Advances*. CRC Press: Boca Raton, FL.

Wang, W., Wang, X., Wang, L., Lu, Y., Li, Y., & Sun, X. (2018). Soil Moisture Estimation for Spring Wheat in a Semiarid Area Based on Low-Altitude Remote-Sensing Data Collected by Small-Sized Unmanned Aerial Vehicles. *Journal of Applied Remote Sensing*, *12*(2), 022207. doi: 10.1117/1.JRS.12.022207.

Wang, T., Thomasson, J. A., Yang, C., Isakeit, T., Nichols, R. L., Collett, R. M., Han, X., & Bagnall, C. (2020). Unmanned Aerial Vehicle Remote Sensing to Delineate Cotton Root Rot. *Journal of Applied Remote Sensing*, *14*(3), 034522. doi: 10.1117/1.JRS.14.034522.

Wang, H., Liu, A., Zhen, Z., Yin, Y., Li, B., Li, Y., Chen, K., & Xu, J. (2021a). Vertical Structures of Meteorological Elements and Black Carbon at Mt. Tianshan Using an Unmanned Aerial Vehicle System. *Remote Sensing*, *13*(7), 1267. doi: 10.3390/rs13071267.

Wang, J., Wang, S., Zou, D., Chen, H., Zhong, R., Li, H., Zhou, W., & Yan, K. (2021b). Social Network and Bibliometric Analysis of Unmanned Aerial Vehicle Remote Sensing Applications from 2010 to 2021. *Remote Sensing*, *13*(15), 2912. doi: 10.3390/rs13152912.

Watts, A. C., L. N. Kobziar, and H. F. Percival (2010). Unmanned aircraft systems for fire and natural, resource monitoring: technology overview and future trends, in Robertson, K. M., Galley, K. E. M., and Masters, R. E., Proceedings of the 24th Tall Timbers Fire Ecology Conference: the future of prescribed fire: public awareness, health, and safety. Tallahassee, FL. Tall Timbers Research, Inc.,Tallahassee, FL. 24, p. 86-89,Proceedings of the Tall Timbers Fire Ecology Conference.

Wierzbicki, D., Fryskowska, A., Kedzierski, M., Wojtkowska, M., & Delis, P. (2018). Method of Radiometric Quality Assessment of NIR Images Acquired with a Custom Sensor Mounted on an Unmanned Aerial Vehicle. *Journal of Applied Remote Sensing*, *12*, 015008. doi: 10.1117/1.JRS.12.015008.

Wigmore, O., Mark, B., McKenzie, J., Baraer, M., & Lautz, L. (2019). Sub-Metre Mapping of Surface Soil Moisture in Proglacial Valleys of the Tropical Andes Using a Multispectral Unmanned Aerial Vehicle. *Remote Sensing of Environment*, *222*, 104–118. doi: 10.1016/j.rse.2018.12.024.

Wilson, S. J., Hedley, R. W., Rahman, M. M., & Bayne, E. M. (2021). Use of an Unmanned Aerial Vehicle and Sound Localization to Determine Bird Microhabitat. *Journal of Unmanned Vehicle Systems*, *9*(1), 59–66. doi: 10.1139/juvs-2020-0021.

Xia, L., Zhang, R., Chen, L., Li, L., Yi, T., Wen, Y., Ding, C., & Xie, C. (2021). Evaluation of Deep Learning Segmentation Models for Detection of Pine Wilt Disease in Unmanned Aerial Vehicle Images. *Remote Sensing*, *13*(18), 3594. doi: 10.3390/rs13183594.

Yancho, J. M. M., Coops, N. C., Tompalski, P., Goodbody, T. R. H., & Plowright, A. (2019). Fine-Scale Spatial and Spectral Clustering of UAV-Acquired Digital Aerial Photogrammetric (DAP) Point Clouds for Individual Tree Crown Detection and Segmentation. *IEEE Journal of Selected Topics in Applied Earth Observations and Remote Sensing*, *12*(10), 4131–4148. doi: 10.1109/JSTARS.2019.2942811.

Yao, H., Qin, R., & Chen, X. (2019). Unmanned Aerial Vehicle for Remote Sensing Applications-A Review. *Remote Sensing*, *11*(12), 1443. doi: 10.3390/rs11121443.

Zhang, S., & Zhao, G. (2019). A Harmonious Satellite-Unmanned Aerial Vehicle-Ground Measurement Inversion Method for Monitoring Salinity in Coastal Saline Soil. *Remote Sensing*, *11*(14), 1700. doi: 10.3390/rs11141700.

Zhang, X., Gao, R., Sun, Q., & Cheng, J. (2019a). An Automated Rectification Method for Unmanned Aerial Vehicle LiDAR Point Cloud Data Based on Laser Intensity. *Remote Sensing*, *11*(7), 811. doi: 10.3390/rs11070811

Zhang, X., Zhang, F., Qi, Y., Deng, L., Wang, X., & Yang, S. (2019b). New Research Methods for Vegetation Information Extraction Based on Visible Light Remote Sensing Images from an Unmanned Aerial Vehicle (UAV). *International Journal of Applied Earth Observation and Geoinformation*, *78*, 215–226. doi: 10.1016/j.jag.2019.01.001.

Zhang, X., Lin, C., Zhang, T., Huang, D., Huang, D., & Liu, S. (2021). New Understanding of Bar Top Hollows in Dryland Sandy Braided Rivers from Outcrops with Unmanned Aerial Vehicle and Ground Penetrating Radar Surveys. *Remote Sensing*, *13*(4), 560. doi: 10.3390/rs13040560.

Zhu, M., Zhou, Z., Huang, D., Peng, R., Zhang, Y., Li, Y., & Zhang, W. (2021). Extraction Method for Single Zanthoxylum bungeanum in Karst Mountain Area Based on Unmanned Aerial Vehicle Visible-Light Images. *Journal of Applied Remote Sensing*, *15*(2), 026501. doi: 10.1117/1.JRS.15.026501.

15 Flood Mapping and Damage Assessment Using Sentinel – 1 & 2 in Google Earth Engine of Port Berge & Mampikony Districts, Sophia Region, Madagascar

Penchala Vineeth Kurapati and Ashish Babu
Ministry of Agriculture, Livestock and Fisheries

Kesava Rao Pyla, Prasad NSR, and Venkata Ravi Babu Mandla
National Institute of Rural Development and Panchayati Raj

CONTENTS

15.1 INTRODUCTION

15.1.1 BACKGROUND

A flood is characterized as the flow of water in abnormal conditions or above the standard water surface which inundated higher ground (Rahman 2006). It is a natural disaster and occurs frequently due to excessive rainfall and causes loss of human life, property, wide area of agriculture, and forest/vegetation, which in turn impacts the country economically, environmentally, and socially (Zhou and Zhang 2017). Therefore, for efficient management, it has created interest for the researchers, policymakers, and governmental organizations to assess the information related to floods accurately. In general, the management of a disaster can be divided into three

DOI: 10.1201/9781003270928-20

phases: the preparatory phase which is applied before the disaster by identifying threat zones; the mitigation phase in which activities such as emergency evacuation, tracking, and execution of contingency plans are conducted only prior to or during the disaster; ultimately, response phase which includes tasks such as damage estimation, recovery measures are conducted shortly after the damage (Jeyaseelan 2004).

Recent developments in space technology enable the researchers and agencies in utilizing the satellite images, it can also provide the flooding period and extent approximately (Peter et al., Matjaž, and Krištof 2013) and mapping of flooded areas is a key practice for understanding the affected land use/cover (D'Addabbo et al. 2018). An automatic threshold-based method is implemented to separate existing and non-existing water pixels. Different Application uses approaches like the split combination, otsu image threshold, split selection and image tiling, kittler and Illingworth's technique, quality index, and Global minimum method for automatic threshold-based techniques. From studies, methods like otsu image thresholds are widely used in flood mapping as it gives better results in Sentinel 1 SAR images (Martinis, Twele, and Voigt 2009; Papamarkos and Gatos 1994; Xue and Zhang 2012; Miasnikov, Rome, and Haralick 2004).

Most tropical climatic regions experience extreme rainfall which will result in floods. As per the study conducted by Ali, Modi, and Mishra (2019) Coupled Model Inter comparison Project (CMIP5), and the Noah-MP land surface hydrological model, shows continuous heavy rainfall causes extreme flooding at river basins in tropical areas which will further be increased in future years due to climate change. Studies associated with the hydrological model needed up-to-date DEM and real-time AWS (Automatic Weather Station) which is not available for underdeveloped countries like Madagascar. So, for countries like Madagascar, real-time water inundation studies are suitable (Maniruzzaman et al. 2020; Lal et al. 2020; Vishnu et al. 2019). Therefore, these types of studies will provide valuable information for decision-making and the government to take swift action in policy-making and risk assessment (Armenakis et al. 2017).

The use of optical data is very difficult due to cloud cover and accessibility of free Synthetic Aperture Radar (SAR) data by the European Space Agency (ESA), i.e., sentinel-1 created an advantage for monitoring the flood extent because the radar sensor does not depend on solar illumination and can penetrate through the clouds (Uddin, Matin, and Meyer 2019). Madagascar is a flood-prone zone, most of which occurred due to excessive rainfall, a report from National Office for Risk and Disaster Management (BNGRC), Madagascar stated that during January 2020 almost 116,675 people were affected from 7 regions, i.e., Alaotra Mangoro, Analamanga, Betsiboka, Boeny, Diana, Melaky, and Sofia. This study is primarily focused on floods associated with heavy rainfall. Damages associated with floods will have a higher impact on poorer people as it will affect crops, dwellings, infrastructure, and loss of life. This will cause financial strain to them. So, studies like these will help countries like Madagascar to prepare for better economic planning and management (Tiwari et al. 2020).

15.2 STUDY AREA

Port Berge and Mampikony districts lie between—15.5645° south, 47.6685° east, i.e., south-central, and –16.05105° south—47.606163° east, i.e., Southwestern, respectively, of the Sofia region as shown in Figure 15.1. These locales have a tropical maritime environment described by two particular seasons, dry from May to October and moist from November to April. The temperatures in the region are quite favorable for agriculture and vary according to the altitude, where the average annual temperature reaches in the Sofia region is 26°C. The annual average temperature recorded at Mampikony was 22.2°C. Port-Berge is a coastal zone that has a mean annual temperature above 25°C. Rainfall in this region is characterized by a high degree of irregularity, the wet season usually begins in December and is mainly concentrated over 4 months of a year (December to April). There can be heavy rainfall for a few hours during the day.

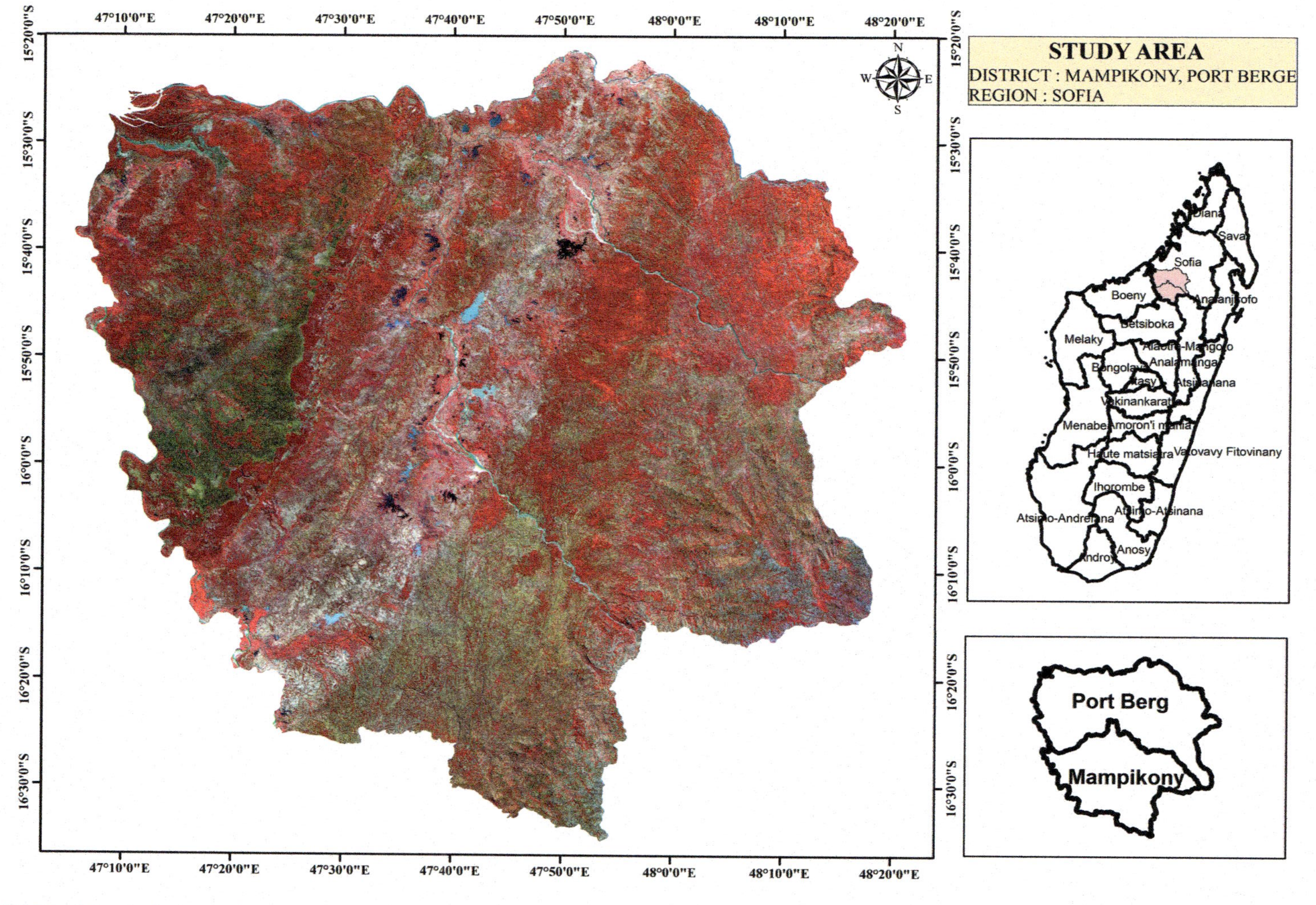

FIGURE 15.1 Map depicting study area.

15.3 DATA USED AND METHODOLOGY

15.3.1 Data Used

The complete details of each datasets used for this study are shown in Table 15.1. For estimating flooded areas, a mosaic of sentinel-1 level-1 Ground Range Detected (GRD) having Interferometric Wide Swath (IW) and Vertical-Horizontal (VH) polarized products were acquired for the dates of pre-flood events between 5 and 16 January 2020 and post-flood event between 17 and 27 January 2020. In addition to SAR data, for assessing damage a mosaic of pre-flood cloud-free sentinel-2 satellite data collected between 1 January 2019 and 14 November 2019 was utilized for Land use/Land cover (LULC) mapping. Google Earth Engine (Gorelick et al. 2017) is used for image processing, and analysis, and ArcMap 10.5 (ESRI 2016) is used for map generation.

15.3.2 Methodology

The detailed method utilized for the study is shown in Figure 15.2. Precisely, sentinel-1 data is used for identifying flooded areas with the initial processing of speckle filter, change detection, exclusion of permanent water bodies, and using sentinel-2 data Land Use and Land Cover (LULC) is created for estimating damage that occurred due to the flood.

Flood Mapping: Google Earth Engine provides Ground Range Detected (GRD) items comprising of SAR-centered information that has been identified, multi-looked, and projected to the ground range utilizing an Earth ellipsoid model (Canty and Nielsen 2017). Using the height of the terrain, the ellipsoid projection of GRD products was rectified. A speckle filter is applied to the GRD products for pre and post-flood events, generally, speckle noise occurs within one resolution cell due to irregular interference of several elementary reflectors and has a great effect on radiometric resolution (Moreira et al. 2013). Speckles were removed using either two techniques, multi-look processing or filtering techniques, for this analysis filtering technique is utilized and a mean filter was applied with a smoothing radius of 50 meters, a circle kernel was used for this purpose. Pre (*R*) and post (*F*) flood images were selected, for change detection techniques (Long, Fatoyinbo, and Policelli 2014) image ratio is applied to extract the flooded areas (CD).

$$\mathrm{CD} = F/R \tag{15.1}$$

After extracting the flooded area using equation (15.1) further refinement is necessary, i.e., applying a threshold because some pixels might result in false-positive or negative signals as flooded. Therefore, a threshold of 1.3 is used which is obtained from the field after examining the flooded

TABLE 15.1
List of Datasets

S. No	Data Layer	Resolution (m)	Type of Data	Source
1	Sentinel-1	10	Synthetic Aperture Radar (SAR) data	European Space Agency (2020)
2	Sentinel-2	10	Optical, i.e., multispectral data	European Space Agency (2018)
3	Climate Hazards Group InfraRed Precipitation with Station (CHIRPS)	5,566	Raster data type depicting the rainfall	Funk et al. (2015)
4	HydroSHEDS by World Wildlife Fund (WWF)	92.77	Raster data type depicting the	Lehner, Verdin, and Jarvis (2008)
5	District boundary	—	Vector layer	BNGRC Madagascar (2020)

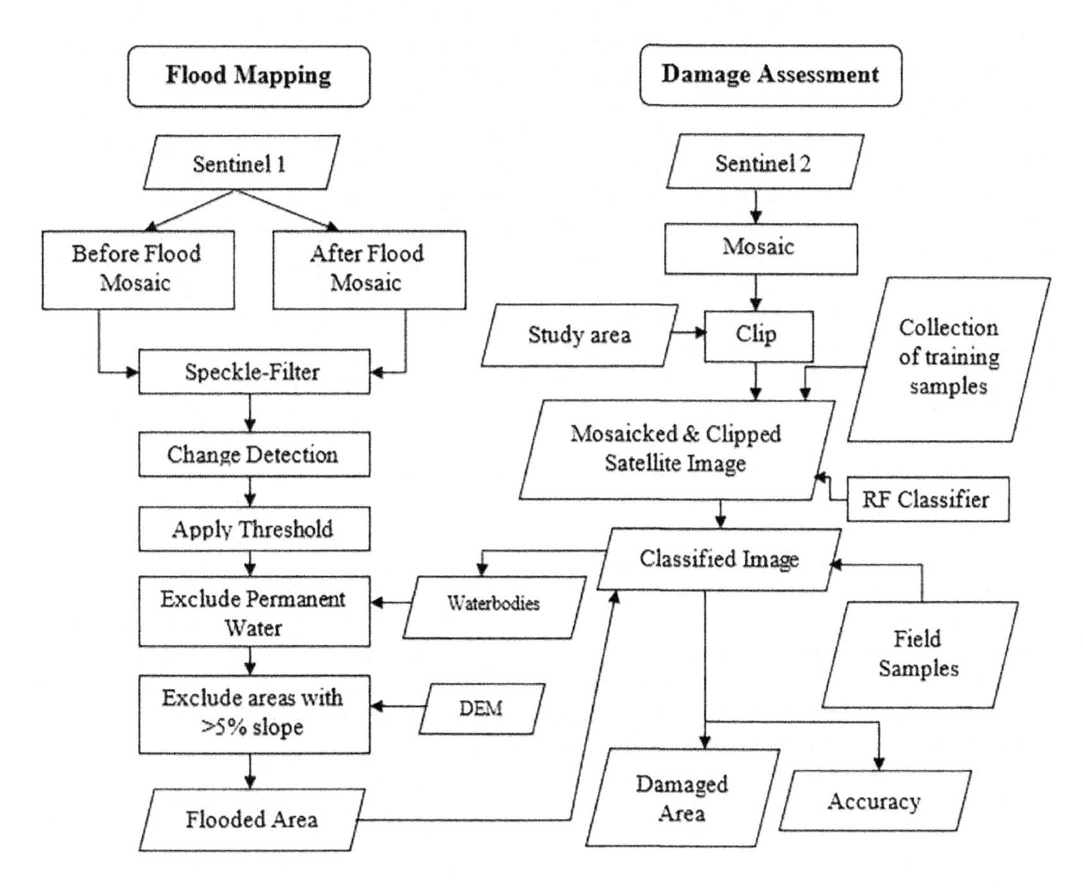

FIGURE 15.2 Flow diagram indicating methodology of the study.

area and pixels with false positives; permanent or seasonal water bodies (e.g., ponds, rivers, lakes, etc.) were required to be masked from extracted flooded areas (CD), say for example if a reservoir is empty before the flood and after the flood is has been filled and on difference image, it may represent as flooded area but in reality, it's not a flooded area, therefore, the classified image which is derived from sentinel-2 satellite imagery is used to differentiate the actual flooded region and final refinement is to exclude areas having a slope >5% (which is obtained from the field after studying the flooded and non-flooded region), since water doesn't remain stagnant. This exclusion is done by using HydroSHEDS data (Lehner, Verdin, and Jarvis 2008) as it provides information on hydrographic, on regional and global-scale which has been created by World Wildlife Fund (WWF) in view of SRTM Digital Elevation Model (DEM). HydroSHEDS data works the same as SRTM but has more derived data on hydrology. It also offers information about layers such as stream aggregations, distances, and information of river topology.

Land Use/Land Cover Map Generation: Supervised classification technique is applied on a False Color Composite (FCC) image for generating LULC maps; 586 simple random training samples were collected for each LULC class. According to level-1 classification (Lillesand et al. 2002), a total of six classes were identified, i.e., barren land, built-up, cropland, sand, forest/vegetation, and waterbody. Finally, the Google Earth Engine Random Forest (RF) algorithm is applied using the training samples. The RF classifier is a collection of the tree-structured classifier which is an advanced form of bagging and creates randomness. Instead of isolating each node using the ideal split of all the variables, RF separates each node utilizing the best of a subset of randomly picked indicators at that node. A new set of training data is constructed with a substitution from the authentic dataset, then, at that point, a tree is grown utilizing random feature selection, and grown trees are not trimmed (Akar and Güngör 2012). The technique makes RF outstanding in precision and can form many trees depending on user requirements, here for this study a total of six trees were used and it is very fast and strong against overfitting (Breiman and Cutler, 2001). Finally, overall, the producer and user accuracies were estimated for the obtained LULC result using another set of 255 samples, i.e., 45 barren land, 39 built up, 43 cropland, 41 sand, 45 forest/vegetation, 42 waterbody samples which were collected from publicly open earth observation and online map tools such as an open street map and google earth. Kappa coefficient is also calculated for testing the interrater reliability.

Flood Damage Assessment: Flood damage is described as the likelihood that a misfortune will occur (Kefi, Mishra, and Kumar 2018). Therefore, flooded areas that were extracted from the sentinel-1 image are overlaid on the LULC map which was generated from sentinel-2, to estimate the damage that occurred due to flood, and finally, the area of each inundated land use/land cover class is calculated.

15.4 RESULTS AND DISCUSSIONS

15.4.1 Flood Inundation Map

The cause of the flood is due to excessive and continuous rainfall in January from the 18th to the 24th of 2020 which was obtained from CHIRPS (Funk et al. 2015) data as shown in Figure 15.3 and mean rainfall map is shown in Figure 15.4. Based on the rainfall data, pre and post-flood datasets were selected from sentinel-1. Therefore, a flood-inundated map was prepared, and a total area of 373.88 km^2, i.e., 3.15% was flooded which is shown in Figure 15.5.

15.4.2 Land Use/Land Cover Map

To estimate damage initially LULC map was prepared using a mosaic of sentinel-2 pre-flood datasets which is shown in Figure 15.6. The total area of both districts is 11,860.37 km^2 where cropland occupies the most and sand occupies the least, i.e., 47.96%, and 0.25% of the total area, respectively, and detailed statistics of each LULC class were shown in Table 15.2.

The overall accuracy estimated using 255 training samples for the 2019 sentinel-2 derived land use/land cover map was found to be 92.89%, and the kappa coefficient is 0.91 which indicates the level of agreement is almost perfect and the complete details of user and producer accuracies were shown in Table 15.3.

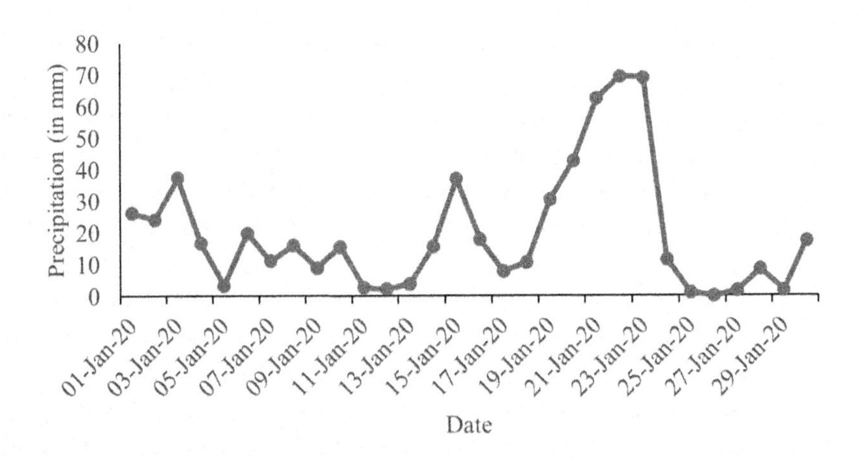

FIGURE 15.3 Rainfall data during January 2020, CHIRPS data.

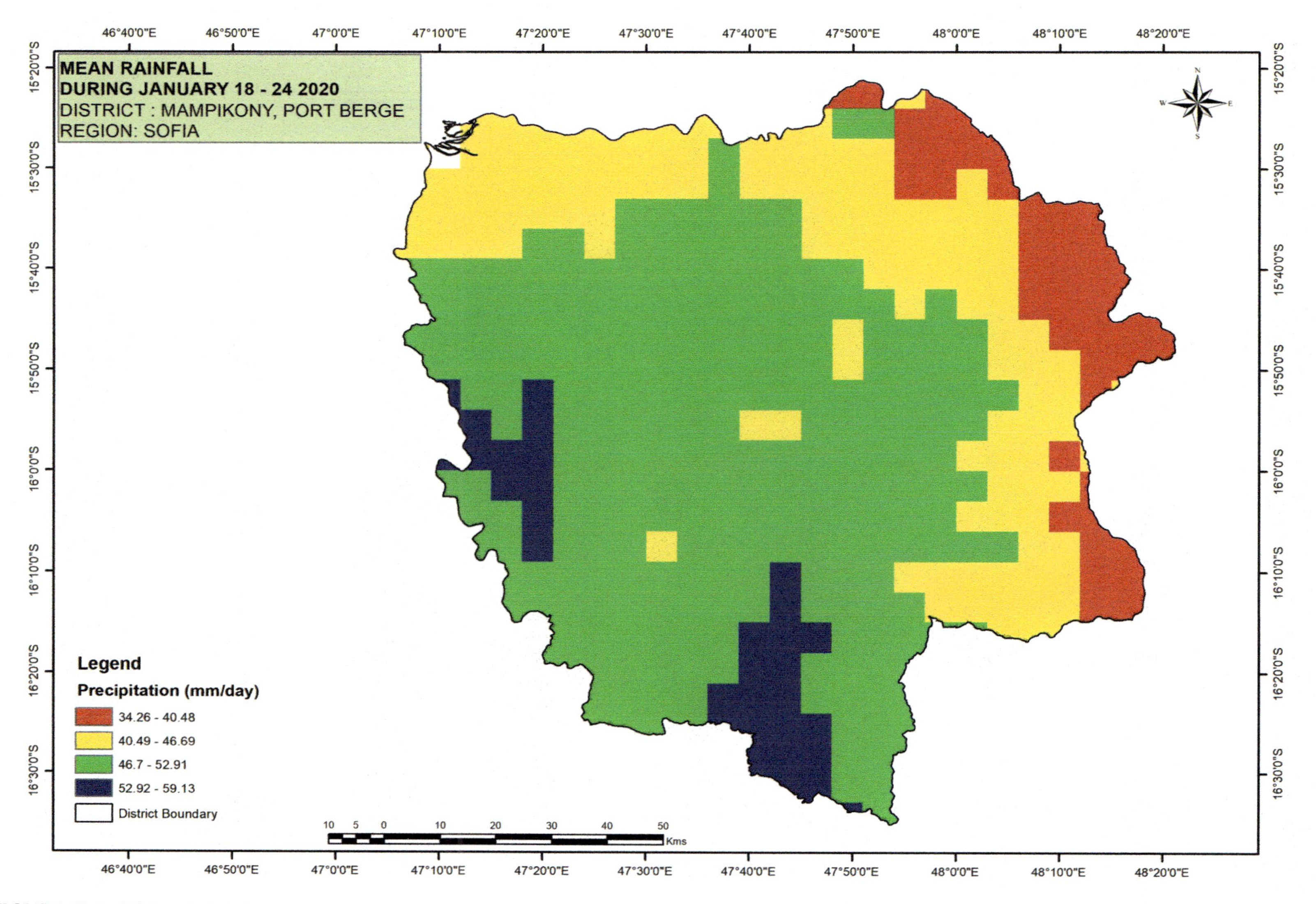

FIGURE 15.4 Mean rainfall map.

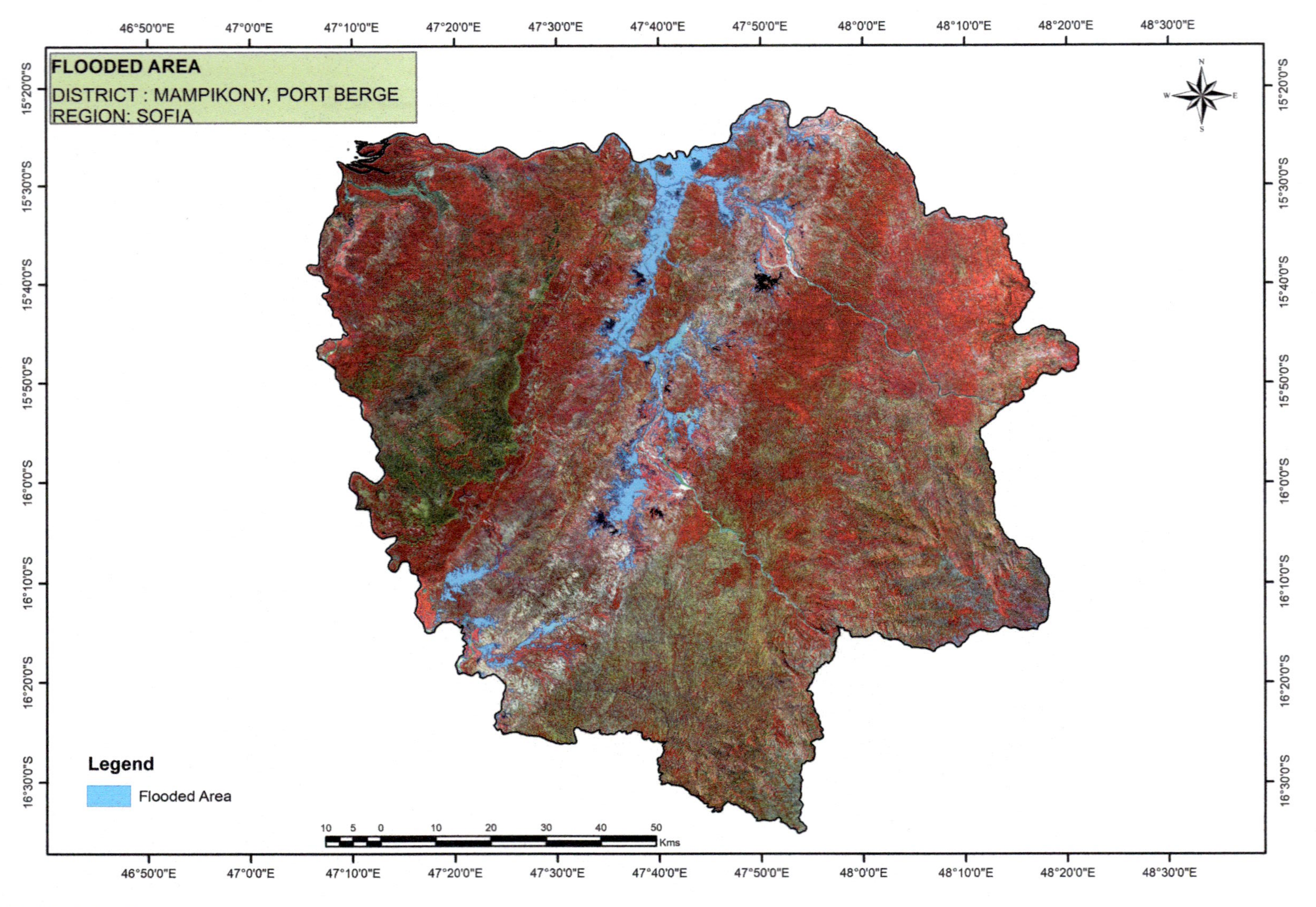

FIGURE 15.5 Flood-inundated map.

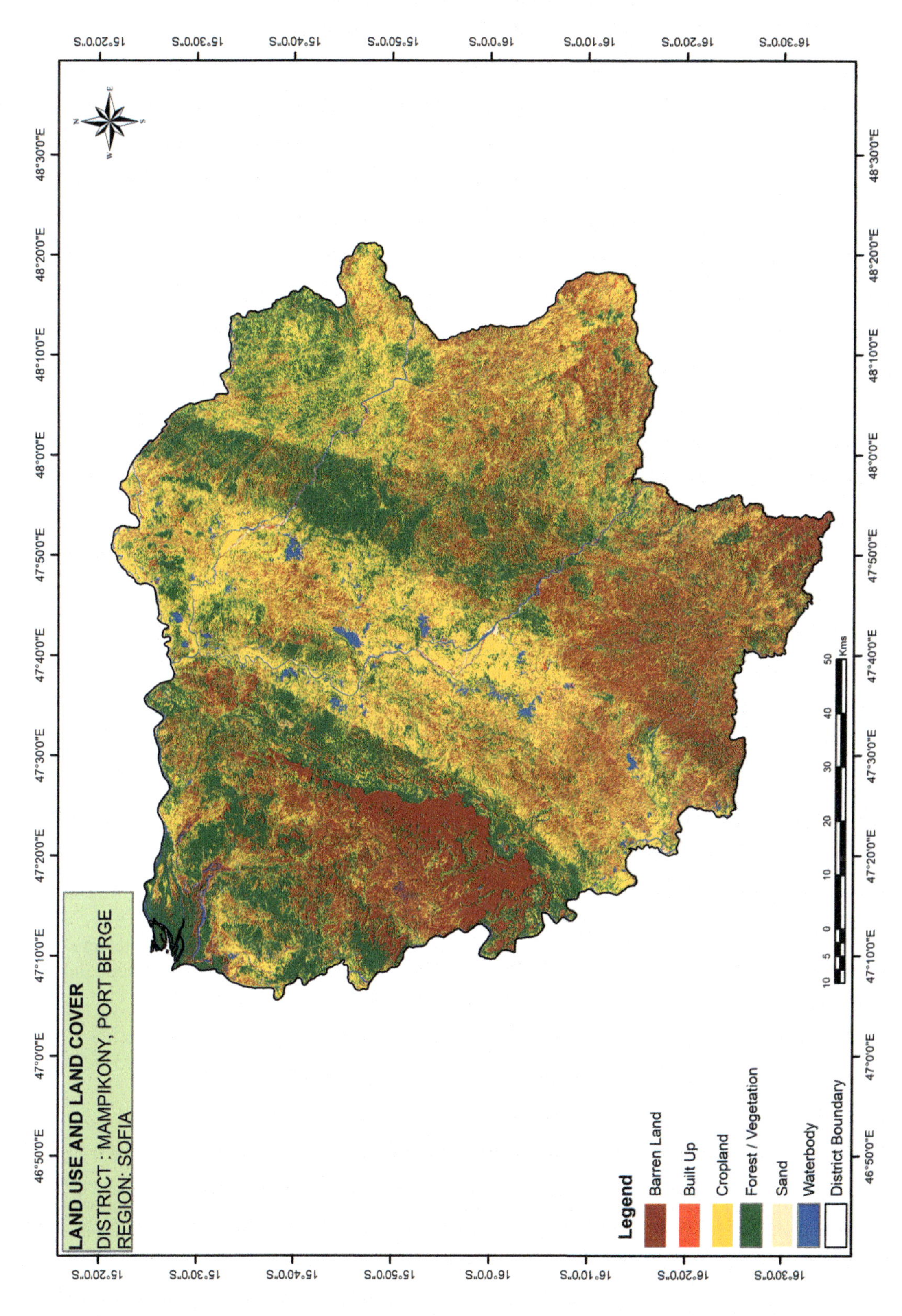

FIGURE 15.6 LULC map of Port Berge and Mampikony.

TABLE 15.2
Area Occupied by Different LULC Classes

LULC Classes	Area (km²)	% of the Total Area
Barren land	3,461.28	29.18
Built up	21.24	0.18
Cropland	5,687.61	47.96
Forest/vegetation	2,509.23	21.16
Sand	30.2	0.25
Waterbody	150.81	1.27
Total	11,860.37	100

TABLE 15.3
Error Matrix of LULC Map for 2019

LULC Classes	Barren Land	Built-up	Cropland	Sand	Forest/ Vegetation	Waterbody	Total	Producer Accuracy (%)
Barren land	53	0	2	0	0	0	55	96.36
Built up	0	89	1	0	0	0	90	98.88
Cropland	0	5	78	2	0	0	85	91.76
Sand	0	0	7	73	3	0	83	87.95
Forest/vegetation	0	0	0	2	42	1	45	93.33
Waterbody	0	0	1	3	0	18	22	81.81
Total	53	94	89	80	45	19	380	N/A
User accuracy (%)	100	94.68	87.64	91.25	93.33	94.74	N/A	N/A
Overall accuracy (%)	92.89							

15.4.3 Flood Damage Assessment

As discussed in the methodology the affected LULC classes were extracted from the flood map which was derived from the sentinel-1 dataset. The affected classes were shown in Figure 15.7, cropland is most affected whereas built-up is least affected and the detailed statistics of each affected LULC classes were shown in Table 15.4.

The entire analysis was performed in Google Earth Engine which is a cloud-based interactive development environment (IDE) platform and the code was written using JavaScript, the code link of the analyzed work, and results.

https://code.earthengine.google.com/bad02cb987a4c838a5e153bed277ff36

15.5 CONCLUSIONS

This approach was developed by utilizing freely available data and analysis is performed using an open-source platform, i.e., GEE, this methodology is particularly useful for underdeveloped countries as commercial platforms cost a lot. Depending on the findings of the study, an area of 373.88 km^2 was affected due to the flood, when compared with the areas of each class of the entire study area of 0.9% of barren land, 4.75% of built-up, 5.47% of cropland, 7.35% of sand, 1.11% of forest/vegetation were affected and overall accuracy of the estimated LULC was 92.89%, therefore we can infer that the geospatial and earth observation technologies furnish timely data for efficient

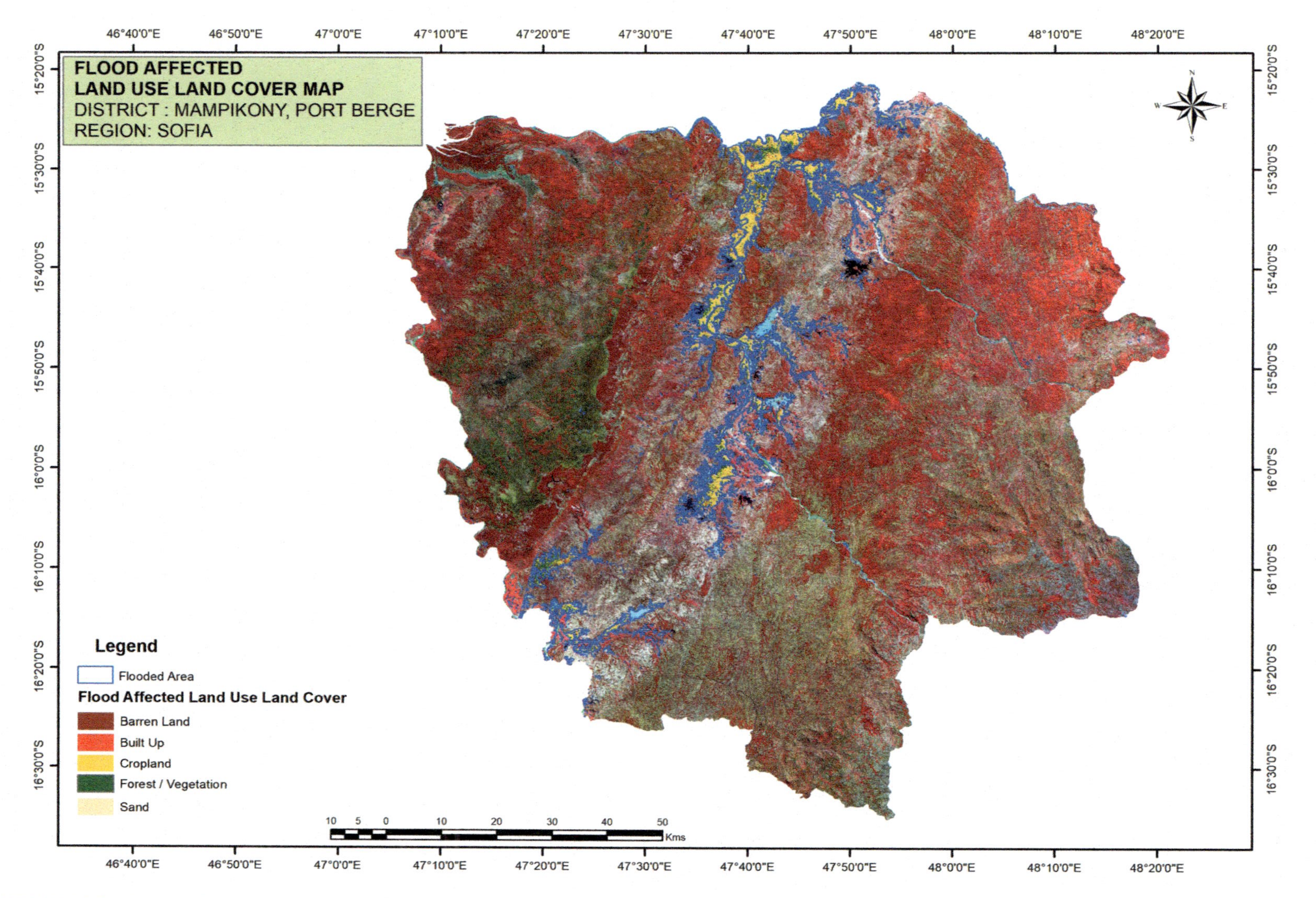

FIGURE 15.7 Affected land use land cover map.

TABLE 15.4
Statistics of Affected Land Use/Land Cover Classes

LULC Classes	Affected Area (km^2)	% of the Total Affected Area
Barren land	31.12	8.32
Built up	1.01	0.27
Cropland	311.52	83.32
Forest/vegetation	28.01	7.49
Sand	2.22	0.6
Total	373.88	100

decisions and detailed management of flood disasters. Due to the weather conditions during flooding, accessing a cloud-less optical data for analysis is difficult in replacement a SAR data can be utilized for estimating the flooded area. Cloud-based platform, i.e., GEE is very useful for users in preparing an emergency response related to flooding and evaluating the damaged area by generating land use land cover map.

ACKNOWLEDGMENTS

We are grateful to the staff of CGARD, Madagascar for their valuable suggestions and assistance in the fieldwork. We'd like to acknowledge National Office for Risk and Disaster Management (BNGRC), Madagascar for providing flood information.

REFERENCES

Akar, Özlem, and Oğuz Güngör. 2012. "Classification of Multispectral Images Using Random Forest Algorithm Classification of Multispectral Images Using Random Forest Algorithm." *Journal of Geodesy and Geoinformation*. doi: 10.9733/jgg.241212.1.

Ali, Haider, Parth Modi, and Vimal Mishra. 2019. "Increased Flood Risk in Indian Sub-Continent under the Warming Climate." *Weather and Climate Extremes* 25. doi: 10.1016/j.wace.2019.100212.

Armenakis, Costas, Erin Xinheng Du, Sowmya Natesan, Ravi Ancil Persad, and Ying Zhang. 2017. "Flood Risk Assessment in Urban Areas Based on Spatial Analytics and Social Factors." *Geosciences (Switzerland)* 7(4). doi: 10.3390/geosciences7040123.

Bureau National de Gestion des Risques et des Catastrophes (BNGRC) à Madagascar. 2020 "District Shapefiles of Madagascar." Accessed January 29, 2020. http://www.bngrc.mg/.

Canty, Morton, and Allan Aasbjerg Nielsen. 2017. "Spatio-Temporal Analysis of Change with Sentinel Imagery on the Google Earth Engine." *Proceeding of the 2017 Conference on Big Data from Space*, Publications Office of the European Union, Luxembourg.

D'Addabbo, Annarita, Alberto Refice, Francesco P. Lovergine, and Guido Pasquariello. 2018. "DAFNE: A Matlab Toolbox for Bayesian Multi-Source Remote Sensing and Ancillary Data Fusion, with Application to Flood Mapping." *Computers & Geosciences* 112: 64–75. doi: 10.1016/j.cageo.2017.12.005.

ESRI. 2016. "ArcGIS 10.5." Redlands, CA: Environmental Systems Research Institute.

European Space Agency. 2018. "Sentinel-2 MSI Level-2A BOA Reflectance." European Space Agency.

European Space Agency. 2020. "Sentinel-1 Synthetic Aperture Radar." European Space Agency.

Funk, Chris, Pete Peterson, Martin Landsfeld, Diego Pedreros, James Verdin, Shraddhanand Shukla, Gregory Husak, et al. 2015. "The Climate Hazards Infrared Precipitation with Stations: A New Environmental Record for Monitoring Extremes." *Scientific Data* 2(1): 150066. doi: 10.1038/sdata.2015.66.

Gorelick, Noel, Matt Hancher, Mike Dixon, Simon Ilyushchenko, David Thau, and Rebecca Moore. 2017. "Google Earth Engine: Planetary-Scale Geospatial Analysis for Everyone." *Remote Sensing of Environment*. doi: 10.1016/j.rse.2017.06.031.

Jeyaseelan, Ayyemperumal. 2004. "Droughts & Floods Assessment and Monitoring Using Remote Sensing and GIS." In: Mannava Sivakumar, Parth Sarathi Roy, K. Harmsen, and Sudip Kumar Saha (Eds.), *Satellite Remote Sensing and GIS Applications in Agricultural Meteorology*. Proceedings of a Training Workshop, Dehra Dun, India.

Kefi, Mohamed, Binaya Kumar Mishra, and Pankaj Kumar. 2018. "Assessment of Tangible Direct Flood Damage Using a Spatial Analysis Approach under the Effects of Climate Change : Case Study in an Urban Watershed in Hanoi, Vietnam." *International Journal of Geo-Information*. doi: 10.3390/ijgi7010029.

Lal, Preet, Aniket Prakash, Amit Kumar, Prashant K. Srivastava, Purabi Saikia, Arvind Chandra Pandey, Parul Srivastava, and Mohammed Latif Khan. 2020. "Evaluating the 2018 Extreme Flood Hazard Events in Kerala, India." *Remote Sensing Letters* 11(5). doi: 10.1080/2150704X.2020.1730468.

Lehner, Bernhard, Kristine Verdin, and Andy Jarvis. 2008. "New Global Hydrography Derived from Spaceborne Elevation Data." *Eos, Transactions, American Geophysical Union* 89(10): 93–94. doi: 10.1029/2008EO100001.

LeoBreiman,andAdeleCutler.2001."RandomForest."https://www.stat.berkeley.edu/~breiman/RandomForests/cc_home.htm.

Lillesand, Thomas, Ralph W. Kiefer, and Jonathan Chipman. 2002. *Remote Sensing and Image Interpretation*, 4th ed. New York: John Wiley and Sons.

Long, Stephanie, Temilola E. Fatoyinbo, and Frederick Policelli. 2014. "Flood Extent Mapping for Namibia Using Change Detection and Thresholding with SAR." *Environmental Research Letters* 9 (2014). doi: 10.1088/1748-9326/9/3/035002.

Maniruzzaman, S.K., S. Balaji, and Sharma S.V. Sivaprasad. 2020. "Flood Inundation Mapping Using Synthetic Aperture Radar (SAR) Data and Its Impact on Land Use /Land Cover (LULC): A Case Study of Kerala Flood 2018, India." *Disaster Advances* 13(3):46–53.

Martinis, Sandro, André Twele, and Stefan Voigt. 2009. "Towards Operational near Real-Time Flood Detection Using a Split-Based Automatic Thresholding Procedure on High Resolution TerraSAR-X Data." *Natural Hazards and Earth System Science* 9(2). doi: 10.5194/nhess-9-303-2009.

Miasnikov, Alexei D., Jayson E. Rome, and Robert M. Haralick. 2004. "A Hierarchical Projection Pursuit Clustering Algorithm." *In Proceedings - International Conference on Pattern Recognition*, Vol. 1. doi: 10.1109/ICPR.2004.1334104.

Moreira, Alberto, Pau Prats-iraola, and Marwan Younis. 2013. "A Tutorial on Synthetic Aperture Radar." *IEEE Geoscience and Remote Sensing Magazine* 1(1):6–43.

Papamarkos, N., and B. Gatos. 1994. "A New Approach for Multilevel Threshold Selection." *CVGIP: Graphical Models and Image Processing* 56(5). doi: 10.1006/cgip.1994.1033.

Peter, Lamovec, Mikoš Matjaž, and Oštir Krištof. 2013. "Detection of Flooded Areas Using Machine Learning Techniques: Case Study of the Ljubljana Moor Floods in 2010." *Disaster Advances* 6(7): 4–11.

Rahman, Rejaur. 2006. "Flood Inundation Mapping and Damage Assessment Using Multi-Temporal RADARSAT and IRS 1C LISS III Image." *Asian Journal of Geoinformatics* 6(2):1–11.

Tiwari, Varun, Vinay Kumar, Mir Abdul Matin, Amrit Thapa, Walter Lee Ellenburg, Nishikant Gupta, and Sunil Thapa. 2020. "Flood Inundation Mapping-Kerala 2018; Harnessing the Power of SAR, Automatic Threshold Detection Method and Google Earth Engine." *PLoS One* 15. doi: 10.1371/journal.pone.0237324.

Uddin, Kabir, Mir A. Matin, and Franz J. Meyer. 2019. "Operational Flood Mapping Using Multi-Temporal Sentinel-1 SAR Images : A Case Study from Bangladesh," 1–19. doi: 10.3390/rs11131581.

Vishnu, Chakrapani Lekha, Kochappi Sathyan Sajinkumar, Thomas Oommen, Richard A. Coffman, K. P. Thrivikramji, V. R. Rani, and S. Keerthy. 2019. "Satellite-Based Assessment of the August 2018 Flood in Parts of Kerala, India." *Geomatics, Natural Hazards and Risk* 10(1). doi: 10.1080/19475705.2018.1543212.

Xue, Jing Hao, and Yu Jin Zhang. 2012. "Ridler and Calvard's, Kittler and Illingworth's and Otsu's Methods for Image Thresholding." *Pattern Recognition Letters* 33(6). doi: 10.1016/j.patrec.2012.01.002.

Zhou, Shilun, and Wanchang Zhang. 2017. "Flood Monitoring and Damage Assessment in Thailand Using Multi-Temporal HJ-1A/1B and MODIS Images." *In IOP Conference Series: Earth and Environmental Science*, Vol. 57. doi: 10.1088/1755-1315/57/1/012016.

Section VI

Case Studies from the Geospatial Domain

16 Fuzzy-Based Meta-Heuristic and Bi-Variate Geo-Statistical Modelling for Spatial Prediction of Landslides

Suvam Das[1,2], *Shubham Chaudhary*[1,2], *Shantanu Sarkar*[1,2], *and Debi Prasanna Kanungo*[1,2]
[1]Academy of Scientific and Innovative Research
[2]CSIR-Central Building Research Institute

CONTENTS

DOI: 10.1201/9781003270928-22

16.1 INTRODUCTION

Landslides are one of the most destructive and recurring hazards in mountainous regions often causes life losses and property damages. Generally, landslides are the typical non-linear process governed by the local geo-environmental conditions and appeared when subjected to any triggering agents (Guzzetti 2002). Increasing population in hilly areas also advanced the landslide process and invites frequent fatal interactions between them. Studies shows in India, landslide caused 2.0 billion USD (0.11% of total GDP) in economic losses annually and during 2007–2015 total 6306, causalities are reported due to this phenomenon (Azimi et al. 2018; Dikshit et al. 2020). Therefore, landslide becomes one of the most expensive and life-endangering hazards in this territory. In terms of landslide counts, the Himalayan region is most vulnerable because of its rugged topography, tectonic instability, complex geology supported by extreme rainfalls with frequent earthquakes (Kanungo et al. 2006; Froude and Petley 2018). In the recent past, many fatal landslides have been witnessed in this region, the most well-known ones being the Malpa landslide (1998); Pithoragarh landslide (2009); Kotropi landslide (2017); Chamoli disaster (2021). Thus, identification of landslide potential areas is one of the primary steps to avoiding landslide induced damages (Martha et al. 2013; Lombardo et al. 2021). To the date, researchers' efforts a long to schedule different approaches in terms of landslide susceptibility zonation (LSZ) mapping for estimating the probability of landslides in a particular area. Such modeling approaches are spanning from qualitative/semi-quantitative methods (Sarkar and Kanungo 2004; Ghosh et al. 2011; Das et al. 2022) to quantitative statistical methods including machine learning approaches (Kanungo et al. 2006; Mandal and Mandal 2018; Chen et al. 2019). With this, the derived LSZ maps can be effectively used for land use planning and are also apt as a good alternative for regional observation where ground-based landside early warning systems (LEWS) are lacking. To forecast landslides in spatial scale, a LSZ map often considers the factors responsible for past landslides and used them as auxiliary information to anticipate the future ones (Sarkar and Kanungo 2004; Kanungo et al. 2006). Although, evaluation of the spatial relationship between past landslides and their responsible factors is very sensitive process and often controls the overall prediction strength (Saha et al. 2005). To recognize and predict landsides in spatial level, application of remote sensing (RS) and geographic information system (GIS) is continuously increasing (Roodposhti et al. 2014; Wu et al. 2020). It is well known that geo-environmental conditions are varies spatially and to represent their spatial heterogeneity, GIS is become the most consistent platform that offers inexpensive solution in a reasonable time budget. In this direction, researchers' extensively relies on RS and GIS for landslide investigation. Keeping in view the severity of landslide process and on-going developmental dynamics in such hilly terrains; information of landslide potential areas definitely helpful for the end users to estimate the vulnerability level and implement preventive measures in the landslide sensitive areas.

However, so far, to forecast landslides different modeling approaches have been proposed. Such methods are either based on subjective judgements and/or statistical approximation assisted by RS and GIS technologies. It is worth mentioning that subjective judgments often suffer from high generalization and similarly, statistical methods constrain researchers' role in the decision-making process. Hence, each modeling approaches have their own limitations and capable to bring out promising results to some extent. Concerning this, the present study includes meta-heuristic fuzzy analytic hierarchy process (FAHP) and bi-variate statistical-based Yule coefficient (YC) models for LSZ mapping. To the best of our knowledge these models are not yet extensively applied in the Himalayan region and thus, provide an opportunity to test them for present study. For implementing these models, an integrated framework has been designed where the FAHP model is used for factor class weights determination and YC is used for defining the subclass weights. Many studies show combining two models can compensate each other by minimizing their limitations and brings more robust predictions (Chen et al. 2019; Wu et al. 2020). The selected models are applied in parts of Kalimpong region of Darjeeling Himalaya. This region was occasionally suffering from the fatal landslides with collateral damages and thus, demands for a scientific and systematic study to find landslide potential areas.

16.2 LSZ MAPPING AND ASSOCIATED MODELING APPROACHES

Considering the process involved in landslide occurrence, it is commonly accepted that in-situ geo-environmental conditions offers the preparatory condition that makes the slope material unstable and often mobilized under the influence of any triggering agents. LSZ mapping aims to find the areas having such unstable situation and depending on state of the condition it ranks or categorized the area in different severity classes. Therefore, to interpret the likelihood of landslides three type of modeling approaches are generally practiced including qualitative or heuristic terrain interpretation, semi-quantitative and quantitative statistical approaches (Sarkar and Kanungo 2004). The qualitative methods are the expert opinion-based techniques often try to establish coarsely scaled subjective judgements to forecast landslides. The semi-quantitative approaches like analytic hierarchy process (AHP) involves the ranking and weighting of factors with sophisticated consistency ratio-based statistical appraisals, but create difficulties for the evaluator during weights assignments to optimize the factors rankings (Ghosh et al. 2011). The quantitative statistical methods consider the statistical relation between landslide and their responsible factors and further can be divided into two types, i.e., bi-variate and multi-variate approaches (Khanna et al. 2021). The bi-variate methods like frequency ratio (FR), information value (InV), weight of evidence (WofE), certainty factor (CF), yule coefficient (YC) evaluates the presence-absence ratio of the past landslides over subclasses of a factor to establish the mutual relationship but they do not examine the individual factor influence in modeling process. Contrastingly, multi-variate methods like logistic regression are more often used to understand the relative importance of each factor considered for LSZ analysis but this method is not suitable for subclass weights determination (Mandal and Mandal 2018). In consideration to the diversity of LSZ modeling approaches and their shortcomings, FHAP and YC models are selected for this study. We thereby try to ensure that FAHP model will allow us to incorporate more meaningful judgments in the decision making and the YC model will maintain the statistical consistency throughout the prediction.

In practice, the traditional AHP method proposed by Saaty and Wind (1980), suffers due to high generalization of weights used in pair-wise comparison and found to be inadequate for representing the complex natural relationship of landslides with their responsible factors (Sur et al. 2020). To address such complex problems, Zadeh (1965) introduced the fuzzy set theory that examines the degree of association for specific attributes in terms of membership function. And in landslide prediction, this fuzzy membership function with AHP allows to achieve complex non-linear decisions in a flexible framework. Over the years FAHP model evolved as a good alternative for multi criteria decision making (MCDA) in a fuzzy environment where all the factors are evaluated in a two-dimensional priority matrix using fuzzy linguistic variables. At the same time, this study also considers the YC model because ease of use and free from any statistical assumptions. A detail description of them is presented below.

16.2.1 Fuzzy Set Theory and FAHP

Fuzzy logic in multi-criteria decision making particularly for LSZ mapping allows researchers to evaluate the non-linearity in factors influence for landslide occurrence by incorporating the fuzzy membership function (MF). In fuzzy set theory, the MF value ranges from 0 to 1, representing the minimum and maximum strength of association between two variables, respectively. Using this auxiliary MF values, AHP method defines the linguistic variables in the pair-wise matrix for performing the comparative evaluation. For the present study, the FAHP method developed by Chang (1996) is used to define the relative importance of the considered landslide causative factors.

16.2.1.1 Extent Analysis on FAHP

In the following, the extent analysis in the FAHP method is given. Let, $M \in F(R)$ called a fuzzy number, if:

1. Exists $x_0 \in R$ such that $\mu_M(x_0) = 1$;
2. For any $\alpha \in [0, 1]$

$A_\alpha = [x, \mu_{A\alpha}(x) \geq a]$ is close interval. Here, $F(R)$ represents all fuzzy sets and R is the set of real numbers.

16.2.1.2 Triangular Fuzzy MF

Chang (1996) suggested a fuzzy number M on R should be a triangular fuzzy number if its membership function $\mu_M(x): R \rightarrow [0, 1]$ is equal to

$$\mu_M(x) = \left\{ \begin{array}{ll} \dfrac{x}{m-l} - \dfrac{l}{m-l}, & |x \in [l, m] \\ \dfrac{x}{m-u} - \dfrac{u}{m-u}, & |x \in [m, u] \\ 0 & \text{otherwise} \end{array} \right\} \tag{16.1}$$

where, l, m, u is the lower, modal and upper limit of triangular fuzzy number on the support of M, respectively. And the support of M is the set of elements $\{x \in R \mid l < x < u\}$. When $l = m = u$, it's a non-fuzzy number by conversion.

16.2.1.3 Fuzzy Operational Laws

According to Chang's method (1996), let consider two triangular fuzzy numbers (TFNs), e.g., M_1 and M_2, where, $M_1 = (l_1, m_1, u_1)$ and $M_2 = (l_2, m_2, u_2)$. Their operational laws are as follows:

$$(l_1, m_1, u_1) \oplus (l_2, m_2, u_2) = (l_1 + l_2, m_1 + m_2, u_1 + u_2) \tag{16.2}$$

$$(l_1, m_1, u_1) \odot (l_2, m_2, u_2) \approx (l_1 l_2, m_1 m_2, u_1 u_2) \tag{16.3}$$

$$\begin{gathered} (\lambda, \lambda, \lambda) \odot (l_1, m_1, u_1) = (\lambda l_1, \lambda m_1, \lambda u_1) \\ \lambda > 0, \lambda \in R \end{gathered} \tag{16.4}$$

$$(l_1, m_1, u_1)^{-1} \approx (1/u_1, 1/m_1, 1/l_1) \tag{16.5}$$

To implement the fuzzy synthetic extent, let $X = \{x_1, x_2, \ldots, x_n\}$ be an object set and $U = \{u_1, u_2, \ldots, u_m\}$ be a goal set. According to the method of extent analysis, we can get m extent values for each object with the following signs:

$$M_{gi}^1, M_{gi}^2, \ldots, M_{gi}^m, \quad i = 1, 2, \ldots, n \tag{16.6}$$

where, all the $M_{gi}^m (i = 1, 2, \ldots, n)$ are the TFNs. The value of fuzzy synthetic extent with respect to the ith object is defined as:

$$S_i = \sum_{j=1}^{m} M_{gi}^j \otimes \left[\sum_{i=1}^{n} \sum_{j=1}^{m} M_{gi}^j \right]^{-1} \tag{16.7}$$

The degree of possibility $M_1 \geq M_2$ is defined as:

$$V(M_1 \geq M_2) = \sup_{x \geq y}\left[\min\left(\mu_{M_1}(x), \mu_{M_2}(y)\right)\right] \tag{16.8}$$

When a pair (x, y) exist such that $x \geq y$ and $\mu M_1(x) = \mu M_2(y)$, then, we have $V(M_1 \geq M_2)$. Since M_1 and M_2 are convex fuzzy numbers, we have that:

$$\begin{aligned} &V(M_1 \geq M_2) = 1 && \xrightarrow{if} m_1 \geq m_2 \\ &V(M_1 \geq M_2) = hgt(M_1 \cap M_2) = \mu M_1(d) && \text{otherwise} \end{aligned} \tag{16.9}$$

where, d is the ordinate of highest intersection point D between μM_1 and μM_2 (Figure 16.1). The ordinate of D is given by:

$$V(M_1 \geq M_2) = hgt(M_1 \cap M_2) = \frac{m_1 - m_3}{(m_2' - m_3') + (m_2 - m_3)} \tag{16.10}$$

To compare M_1 and M_2, we need both the values of $V(M_1 \gg M_2)$ and $V(M_2 \gg M_1)$. The degree of possibility for a convex fuzzy number to be greater than k convex fuzzy number $M_i(i = 1,2,\ldots,k)$ can be defined:

$$V(M \geq M_1, M_2, \ldots, M_k) = \min\left(V(M \geq M_i)\right) \quad i = 1,\ 2, \ldots, k \tag{16.11}$$

Assume that:

$$W'(A_i) = \min\left\{V(S_i \geq S_k)\right\} \quad k = 1,2,\ldots,n;\ k \neq i \tag{16.12}$$

Then, the weight vector is given by:

$$W'(A_i) = \left[W'(A_1), W'(A_2), \ldots, W'(A_n)\right]^T \tag{16.13}$$

where $A_i(i = 1,2,\ldots,n)$ are n elements. Via normalization, the final weights are derived as:

$$W(A_i) = \left[W(A_1), W(A_2), \ldots, W(A_n)\right]^T \tag{16.14}$$

where, W is a non-fuzzy number.

16.2.2 Yule Coefficient

Yule coefficient (YC) also known as Phi-coefficient (Cárdenas & Mera 2016), has been used for determining the strength of association between past landslides with a causative factor subclasses. The main advantage of this model is the trade-offs in multi-class classification of different factors evaluated on the basis of proportional association of one variable to other (Ghosh et al. 2011). It's calculated using the following equation:

$$\text{YC} = \frac{\sqrt{T_{11}/T_{21}} - \sqrt{T_{12}/T_{22}}}{\sqrt{T_{11}/T_{21}} + \sqrt{T_{12}/T_{22}}} \tag{16.15}$$

where, T_{11} = area of positive match, where subclass of a factor and landslide both are present; T_{12} = area of mismatch, where subclass of a factor is present but landslide is absent; T_{21} = area of negative match, where neither attribute is present; T_{22} = area of mismatch, where subclass of a factor

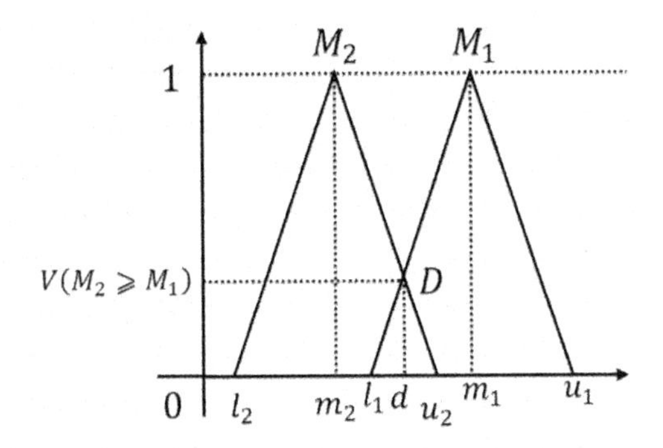

FIGURE 16.1 Triangular fuzzy number.

is absent but landslide is present. The value of YC ranges from −1 to +1. The minimum value of −1 indicates the considered variable has less ability to explain the present landslide scenario and a value close to +1 indicates the most promising variable for explaining landslides in the considered area. Although, a value close to zero indicates null predictiveness and it's difficult to give any certainty about this variable role in landslide occurrences.

16.3 APPLICATION OF RS AND GIS IN LSZ STUDIES

The use of RS and GIS in landslide studies is mainly associated with mapping and modeling aspects. Although, one can use them for monitoring also, as the recent advancement of earth observation (EO) satellites offer high resolution images in short revisit time. Such sort of facilities with GIS helps to prepare a reliable landslide inventory by identifying landslide signatures from satellite images and also allows updating the information with newly available data. Similarly, for susceptibility assessment GIS incorporates the development of thematic layers representing the factors responsible for landslides and use them as input parameter for final modeling. GIS also supports data handling, processing and geo-statistical modeling that enables researchers' to perform LSZ mapping using different approaches and integrate them in a sophisticated framework.

16.4 APPLICATION OF FAHP AND YC MODELS FOR LSZ MAPPING IN PARTS OF KALIMPONG REGION OF DARJEELING HIMALAYA

The methodology adopted for modeling and mapping LSZ in parts of Kalimpong region involves three main steps, as shown in Figure 16.2: (1) Thematic layer development; (2) Configure FHAP model for factor weights determination and apply YC model for subclass weights determination; and (3) Integrate the FAHP and YC models output to generate the final LSZ map and perform the validation measures to check the prediction accuracy.

16.4.1 Description of the Area

Kalimpong region is located in the Eastern part of Darjeeling Himalaya and has long been affected by landslides. For the present study, total 550 km^2 area has been selected in the southern part of Kalimpong district of West Bengal state of India, located between 26°52′30″ N to 27°7′30″ N and 88°25′30″ E to 88°43′ E (Figure 16.3). According to the rainfall data collected form Teesta Low Dam-IV Power Station (TLD-IV PS); the area received an average annual rainfall (last 20 years) of about 3748.86 mm and reached in peak during the months of June-August (Figure 16.4). Apart from torrential rainfall events, the area is also seismically very active and highly susceptible to earthquakes. In 2011, a shallow focus (19.7 km depth) earthquake of 6.9 M occurred in the Sikkim-Darjeeling region that initiated many new slides and also reactivated few old slides (Chakraborty

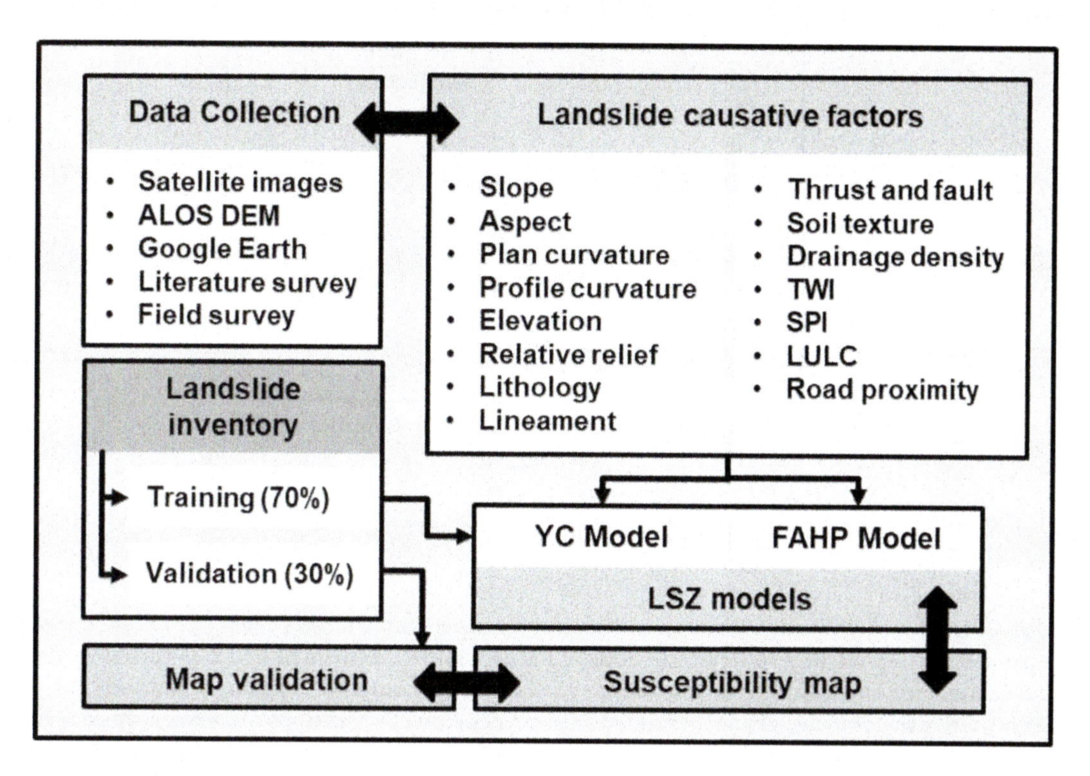

FIGURE 16.2 Methodological flow chart.

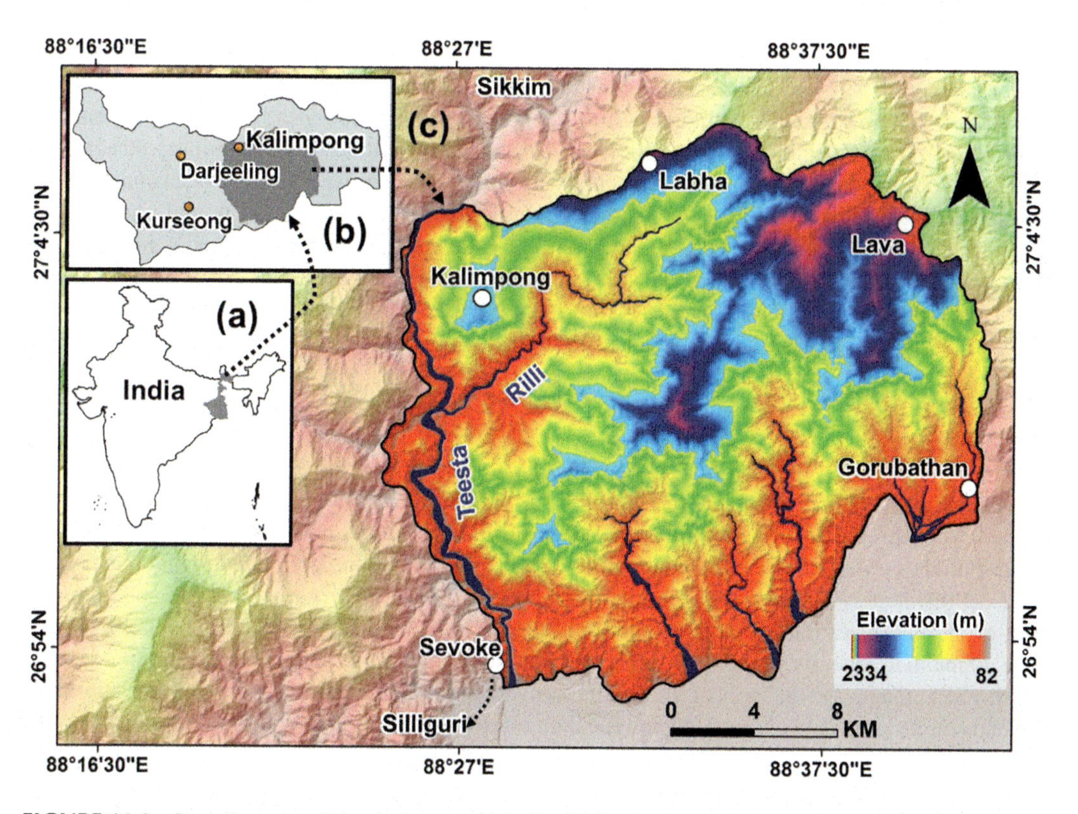

FIGURE 16.3 Location map of the study area. (a) India, (b) Darjeeling Himalaya, and (c) present study area.

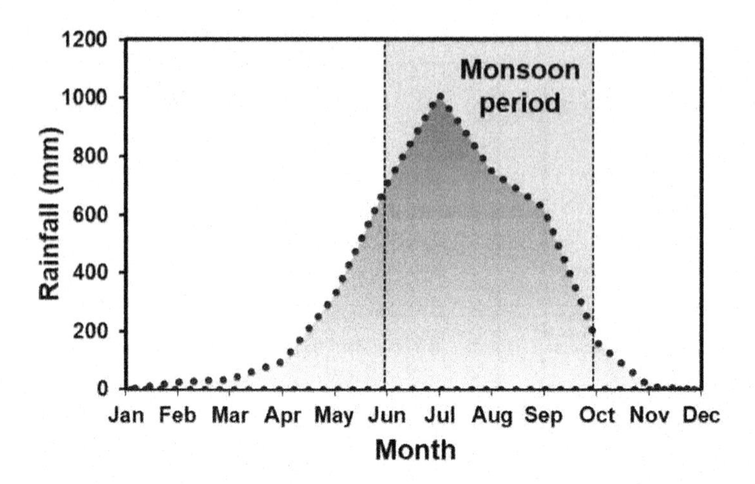

FIGURE 16.4 Ten years (2010–2020) average monthly rainfall in Kalimpong region. (Data collected from Teesta Low Dam – IV Power Station.)

et al. 2011). Topographically, the area is consisted by rugged mountains running in all directions separated by deep drainage valleys. Elevation of the area ranges between 82 and 2,334 m above msl and continuously increased from southern to northerly direction. The slope angle is mostly steep mainly along the drainage channels with an average slope of 24°. The slope orientation shows that most of the past landslides traces are found in the south, southeast and southwest slope directions, wherein the north facing slopes are relatively less suffered by landslides. The geology of the area shows presence of parallel tectonic features like thrust, faults and shear zones where rock structures are mainly ranging from recent Quaternary to Pre-Cambrian age. Alike, thrust and faults presence of closely spaced lineaments along these geological structures increases the propensity for future landslides. The soil texture of the area is mainly governed by fine to coarse loamy and gravelly to coarse loamy groups. Although, other soil textural groups are also found in this area. The river Teesta with its tributaries like Rilli, Lish, Gish, Chel constitutes the preliminary structurally controlled drainage network and is connected with many lower order channels, occasionally rejuvenated during the monsoon period. Such ephemeral streams caused intense fluvial undercutting along the steep slopes and have been found one of the exacerbating agents for the slope instability in this region. The land use pattern of this area is mainly dominated by forest covers with agricultural lands where major settlement units like Kalimpong, Lava, Labha, Sevoke and Gorubathan are connected via NH-10 and SH-12 to the adjacent locations.

16.4.2 ThemSatic Layer Development

16.4.2.1 Landslide Inventory

Landslide inventory maps provide the spatial information of past landslides and considered as one of the key requirements for LSZ modeling and its validation (Aleotti and Chowdhury (1999). Apart from spatial distribution a complete inventory also provides information related to landslide types, their depths, material involved in the movement, date of initiation, total losses, etc. and helps for post-event recoveries (Guzzetti 2002; Ghosh et al. 2011). The advent of RS and GIS helps to prepare a landslide inventory through visual interpretation of satellite images, aerial photos, Google Earth (GE) images or through digitizing past inventories. The success of an inventory initially depends on the accuracy of mapping and the data set used for this purpose. For this study, the landslide inventory has been prepared through on-screen digitization on Google Earth Pro software. The temporal scale ranging from 2018 to 2020 was used for better visual interpretation and the identified landslides have been traced out in polygon-based vector format. Throughout digitization ample care has been taken to delineate the exact boundary of landslides. Total number of 647 landslides are identified (Figure 16.5) from the GE images having minimum and maximum areas of 104 m^2 and 64,617 m^2, respectively, covers a total area of ~1.71 km^2, thus the landslide frequency is about >1 landslide per km^2. The identified landslides are preliminarily categorized as debris slides, debris flows, rock slides with subsidence zones (Cruden and Varnes 1996). A limited field survey has also been carried out along the accessible road tracks to cross-check the identified landslide locations (Figure 16.5). During the field survey, it has been observed landslides in this area are predominantly shallow in depths and mostly are reactivated in the pre-existing deep-seated mass movements may due to the saturation of slope materials during the monsoon period. Some long run-out debris flows are also identified often associated along the drainage channels. Similarly, rock-slides are also identified located mainly in the western higher elevation zones, where rock beds are dipping almost parallel to the slope surface. It is important to note that the present study does not aim to distinguish one type of landslide from another and for this reason, all identified

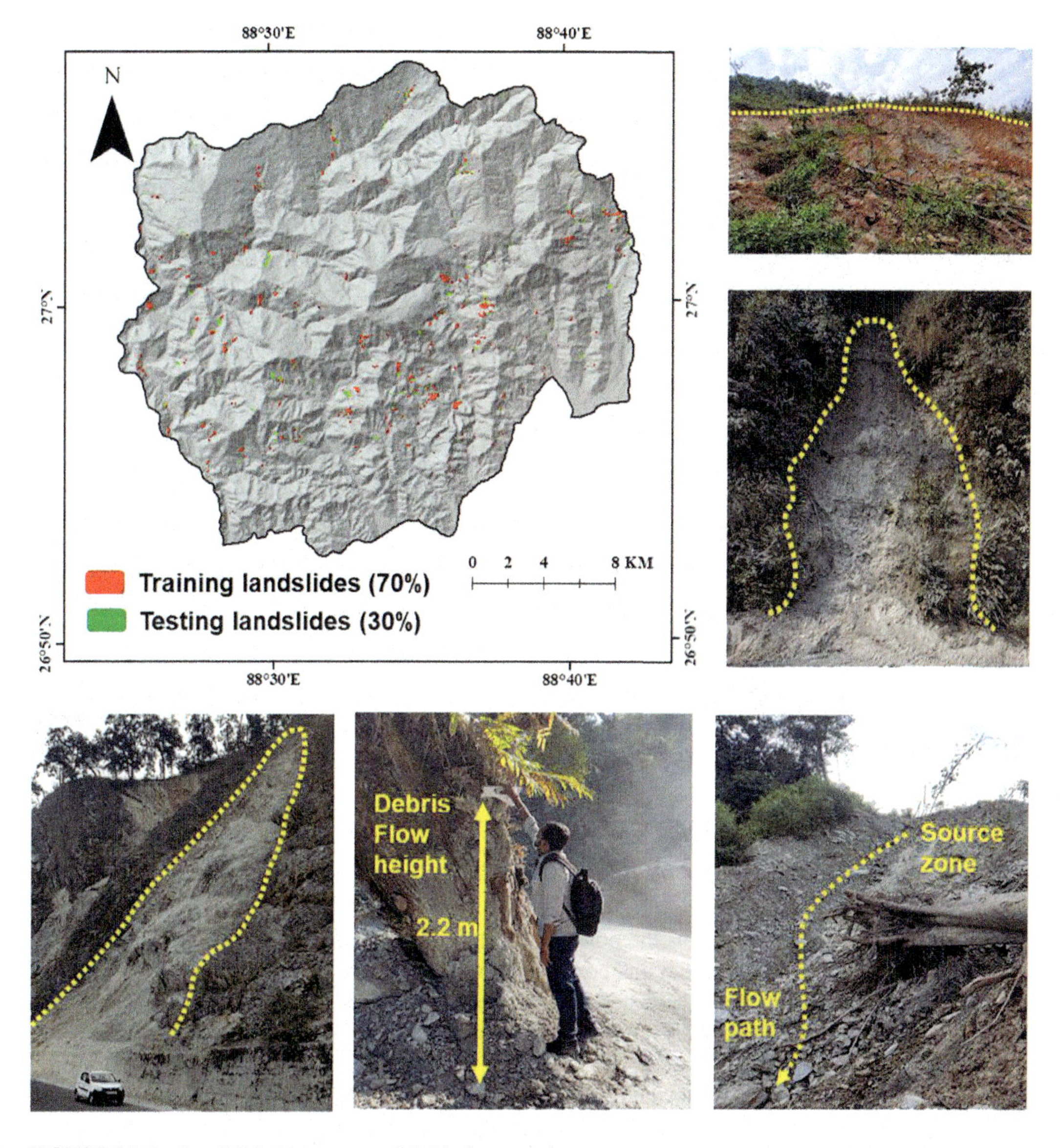

FIGURE 16.5 Landslide inventory and field photographs.

landslides are grouped in a single inventory, contains their location and area only. In India, landslide datasets are very scarce and it's difficult to find out their date of initiation with associated damages. Therefore, RS and GIS derived polygon-based inventory has always become an agile and efficient technique to represent landslides on a spatial scale. After this the prepared inventory is divided into 70:30 ratios for training (453 landslides) and validation (194 landslides) datasets. Here, the training dataset is used for bi-variate YC model configuration and the validation dataset is used for final model validation.

16.4.2.2 Landslide Causative Factors

Occurrences of landslides often attributed by the combined influence of different causative factors and in susceptibility modeling spatial relation of these factors with past landslides are taken as a proxy for estimating the future landslides (Saha et al. 2005; Khanna et al. 2021). Although, no commonly accepted standards available for selecting the landslide causative factors. In this study, depending on the field survey and data availability, total 15 factors are selected belonging to morphological, geological, hydrological and landuse groups. Such kind of selection reduces the biases in the factors' contribution in the landslide process and allows modeling the real-world scenario more effectively. It is worth mentioning that our selected factors do not include the triggering factors like rainfall and earthquake, because such variables are highly dynamic over space and time, and our present landslide inventory does not have any information related to them. Considering these factor groups, different data sources are used including the 12.5 m spatial resolution ALOS PALSAR digital elevation model (DEM) (https://asf.alaska.edu/), 10 m spatial resolution Sentinel-2 satellite images (https://earthexplorer.usgs.gov/), 15 m spatial resolution Landsat-8 OLI satellite images (https://earthexplorer.usgs.gov/), soil map collected from National Bureau of Soil Survey and Land Use Planning (NBSSLUP) and the regional geological information collected from different literature database (after Acharyya 1980; Gangopadhyay 1995; Kellett et al. 2014; Mukul et al. 2017; Patra and Saha 2019). All the collected datasets are processed in a 25 m × 25 m raster resolution using ESRI ArcGIS and for modeling and validation, Microsoft Excel and R Studio platforms are used. A detail description of the selected factors is given below.

16.4.2.2.1 Morphological Factors

Morphological factors include slope, aspect, plan curvature, profile curvature, elevation, relative relief like factors presented in Figure 16.6a–f. Geometry of the slope controls the effective force of gravity and positively correlated with shear stress of slope forming materials and thus, controls the landslide process (Chen et al. 2019). In our study area slope angle ranges from 0° to 79° and based on that the area was categorized into five subclasses at an interval of 15° (Figure 16.6a). Aspect defines the slope orientation measured clockwise in degrees with respect to the north and its orientation controls the insolation amount, rainfall intensity, land use pattern along with the weathering rate and therefore, play an indirect role in landslide process (Sarkar and Kanungo 2004). For this study, the aspect map has been divided into nine classes (Figure 16.6b) namely, north (0°–22.5° and 337.5°–360°), northwest (292.5°–337.5°), northeast (22.5°–67.5°), south (157.5°–202.5°), southeast (112.5°–157.5°), southwest (202.5°–247.5°), west (247.5°–292.5°), east (67.5°–112.5°) and flat (–1) regions. Plan curvature and profile curvature indicate the geometry of slope and used as an indicator for understanding the surface flow characteristics that often determines the ground stability (Wu et al. 2020). Plan curvature controls the flow convergence and divergence, and classified into three classes (Figure 16.6c) as convex, flat and concave. Profile curvature shows the topographic control on the velocity of surface runoff that steer the erosional and depositional processes. Here, we classified this factor into three classes (Figure 16.6d) as convex, flat and concave. Elevation determines the effectiveness of gravitational potential energy and also strengthens the working aptness of secondary relief reliant factors like erosion rate, weathering rate, vegetation cover, rainfall patterns and therefore, considered as an important factor in LSZ studies (Sarkar et al. 2013; Mandal and Mandal 2018). Based on elevation range, the entire area is classified into six zones with 300 m interval

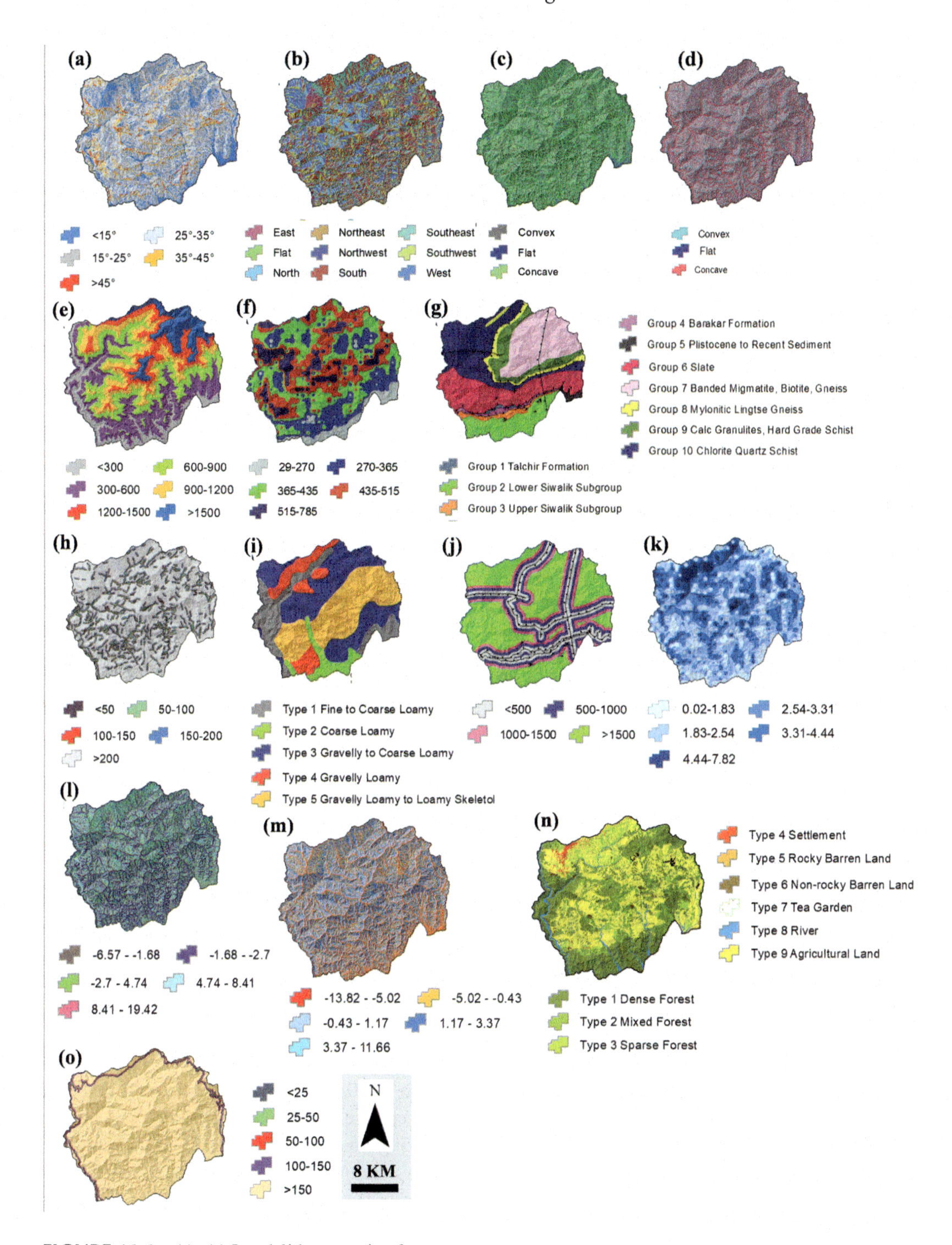

FIGURE 16.6 (a)–(o) Landslide causative factors.

having maximum range of >1,500 m (Figure 16.6e). Relative relief (RR), shows the relief contrast in a particular unit area and high RR values are often associated with steep slope areas indicating the intense erosion by geomorphological agents (Sarkar et al. 2013). To prepare the RR map 1 km × 1 km grid analysis was used to extract the minimum and maximum elevation values from the DEM and based on this relief contrast IDW (Inverse Distance Weighting), interpolation method is used for developing the spatial RR map, which is further reclassified into five zones using jenk's natural break (JNB) method (Figure 16.6f).

16.4.2.2.2 Geological Factors

For this study, different geological factors including lithological setup, lineament proximity, soil texture information, thrust/faults proximity have been used and are presented in Figure 6g-j. In the lithology map spatial extent of rock boundaries are delineated based on the reference map given by Gangopadhyay (1995) and was further updated using other the literature information (Acharyya 1980; Kellett et al. 2014; Mukul et al. 2017; Patra and Saha 2019). Moreover, during filed survey the prepared map was cross-checked and the final map contains six major formations (Figure 16.6g). For lineament identification DEM derived hill shade interpretation was used and to check their influence in landslide occurrence buffer-based proximity analysis was performed in GIS (Figure 16.6h). In general it was commonly anticipated that more landslides will occur nearby the lineaments as they shows the signature of sub-surface deformations (Sarkar and Kanungo 2004). The soil texture information is collected from the district soil map of NBSSLUP and is finalized by incorporating the published literature information (after Roy and Saha 2019). The final map shows five major soil textural groups (Figure 16.6i) including fine to coarse loamy, coarse loamy, gravelly loamy, gravelly loamy to loamy skeletal and gravelly to coarse loamy groups. The distribution of thrust and faults shows that the area is tailored by different thrust systems, like Main Boundary Thrust (MBT 1; MBT 2) and Main Central Thrust (MCT) with some regional faults like Gish Transverse Fault (GTF), Geil Khola Fault (GKF). To show their role in landslide process, buffer-based proximity analysis was performed with an interval of 500 m and extended up to 1,500 m (Figure 16.6j).

16.4.2.2.3 Hydrological Factors

In hydrological group drainage density, topographic wetness index and stream power index factors are used. To prepare the drainage density (DD) map, the identified drainage channels length in per km^2 area has been calculated and using the IDW interpolation the continuous DD raster was developed in GIS having a range of 0–7.82 km/km^2. Distribution of drainage channels shows spatial heterogeneity in their concentration and thus, using the JNB method five classes are prepared (Figure 16.6k) as very low, low, moderate, high and very high DD areas. Topographic wetness index (TWI) shows the areas have potential of high surface runoff without considering the soil transmissivity (Conforti et al. 2014). For generating TWI map, hydrological analyst tool in GIS is used ($TWI = \ln\left(As/\tan\beta\right)$), where As is the upslope contributing area and $\tan\beta$ is slope gradient) and reclassified into five classes using JNB method (Figure 16.6l) as −6.57 to −1.68, −1.68 to −2.70, −2.70 to 4.74, 4.74 to 8.41 and 8.41 to 19.42. Similarly, stream power index (SPI) evaluates the erosive power of surface runoff by assuming the discharge is proportional to the upslope contributing area (Hu et al. 2021). To develop the SPI raster, hydrological analyst tool in GIS is used ($SPI = As \times \tan\beta$, where As is upslope contributing area and $\tan\beta$ is slope gradient) and reclassified into five classes using JNB method (Figure 16.6m) as −13.82 to −5.02, −5.02 to −0.43, −0.43 to 1.17, 1.17 to 3.37 and 3.37 to 11.66.

16.4.2.2.4 Landuse Factors

In this group two factor are included, i.e., land use land cover (LULC) and road proximity. To prepare the LULC map qualitative satellite image analysis of Sentinal-2 and Landsat-8 OLI has been performed in GIS and based on this different LULC categories are identified as forest area, agricultural lands, tea gardens, settlement zones, rocky and non-rocky barren lands. The density of forest varies spatially and depending on that it was further categorized into three classes as dense forest, sparse forest and mixed forest areas (Figure 16.6n). To highlight the influence of roads on landslides, we considered the major roadways including National Highway (NH)-10, and State Highway (SH)-12 and used buffer-based analysis at different intervals like <25 m zone, 25–50 m zone, 50–100 m zone, 100–150 m zone and >150 m zone (Figure 16.6o).

16.4.3 Model Implementation

16.4.3.1 Factors Weights Determination Using FHAP Model

For configuring and constructing the FAHP model, the TFNs (l,m,u) shown in Table 16.1, are assigned in the pair-wise matrix. For defining these numbers subjective as well as field experiences are incorporated and after several trial-and-error iterations, the final ratings are configured, as shown in Table 16.2. Thereafter, by evaluating the pair-wise matrix, eigenvectors are derived to define the relative priority or weights of different causative factors (Table 16.3). Based on this, the ordinate of highest intersection points and the degree of possibility for TFNs are obtained as shown in Table 16.4.

16.4.3.2 Factors Subclasses Weights Determination Using YC Model

To deploy the YC model, four sets of information were prepared including the total pixels in the study area (882,818 pixels), total landslide pixels in the training data (1,994 pixels) and the pixels covered by a particular subclass and the landslide pixels fall in this subclass. For example, let a class have 134,075 pixels, in which landslides covered 420 pixels, thus the T_{11}, T_{12}, T_{21}, and T_{22}, will be 420, 1,574, 133,655 and 747,169, respectively, and according to the YC formula the result will be 0.10 for that subclass. By following this rule YC value of each subclass of all factors are determined, which are shown in Figure 16.7.

16.4.4 Landslide Susceptibility Zonation (LSZ) and Validation

The weights of the factors derived from the FAHP model and their subclass weights derived from the YC model are combined in the GIS platform to develop numerical data layers. Finally, all of

TABLE 16.1
Linguistic Variables for Triangular Fuzzy Membership Function

Code	Linguistic Variables	Fuzzy Triangular Scale		
		l	*m*	*u*
1	Equally important	1	1	1
2	Intermediate value between 1 and 3	1	2	3
3	Slightly important	2	3	4
4	Intermediate value between 3 and 5	3	4	5
5	Important	4	5	6
6	Intermediate value between 5 and 7	5	6	7
7	Strongly important	6	7	8
8	Intermediate value between 7 and 9	7	8	9
9	Extremely important	9	9	9

Source: After Sur et al. (2020).

TABLE 16.2
Pair-Wise Comparison between the Selected Geo-Environmental Factors

	A			B			C			D			E		
	M_1	M_2	M_3	M_1	M_2	M_3	M_1	M_2	M_3	M_1	M_2	M_3	M_1	M_2	M_3
A	1.00	1.00	1.00	0.50	1.00	2.50	0.50	1.00	1.67	0.40	1.00	1.67	0.40	0.50	1.00
B	0.40	1.00	2.00	1.00	1.00	1.00	0.50	1.00	2.50	0.50	0.67	2.50	0.40	0.50	1.67
C	0.60	1.00	2.00	0.40	1.00	2.00	1.00	1.00	1.00	0.67	1.00	2.00	0.67	1.00	2.00
D	0.60	1.00	2.50	0.40	1.50	2.00	0.50	1.00	1.50	1.00	1.00	1.00	1.00	2.00	3.00
E	1.00	2.00	2.50	0.60	2.00	2.50	0.50	1.00	1.50	0.33	0.50	1.00	1.00	1.00	1.00
F	1.00	2.50	3.00	0.60	2.00	2.50	1.00	1.50	2.00	0.33	1.00	1.50	0.33	0.50	1.00
G	1.50	3.00	3.50	1.00	2.50	3.00	1.00	2.00	2.50	0.50	1.00	2.00	0.50	1.00	2.00
H	1.50	3.00	4.00	1.50	2.50	3.00	1.50	2.00	2.50	0.50	1.50	2.00	0.50	1.00	2.00
I	2.00	3.00	4.00	1.50	2.50	3.50	1.50	2.00	3.00	1.00	1.50	2.00	1.00	1.50	2.00
J	2.00	3.50	4.00	2.00	2.50	3.50	2.00	2.50	3.00	1.50	2.00	2.50	1.00	1.50	2.50
K	2.50	4.00	4.50	2.50	3.00	3.50	2.00	3.00	3.50	1.50	2.00	2.50	1.00	2.00	2.50
L	3.00	4.00	4.50	3.00	3.50	4.00	2.50	3.00	3.50	2.00	2.50	3.00	1.50	2.00	2.50
M	3.00	4.50	5.00	3.00	4.00	5.00	3.00	3.50	4.00	2.00	2.50	3.00	2.00	2.50	3.00
N	3.50	5.00	6.00	3.50	4.00	5.00	3.00	4.00	4.50	2.50	3.00	4.00	2.00	2.50	3.50
O	4.00	5.00	6.00	3.50	4.00	5.00	3.00	4.00	4.50	2.50	3.00	4.00	2.00	2.50	3.50

	F			*G*			*H*			*I*			*J*		
	M_1	M_2	M_3	M_1	M_2	M_3	M_1	M_2	M_3	M_1	M_2	M_3	M_1	M_2	M_3
A	0.33	0.40	1.00	0.29	0.33	0.67	0.25	0.33	0.67	0.25	0.33	0.50	0.25	0.29	0.50
B	0.40	0.50	1.67	0.33	0.40	1.00	0.33	0.40	0.67	0.29	0.40	0.67	0.29	0.40	0.50
C	0.50	0.67	1.00	0.40	0.50	1.00	0.40	0.50	0.67	0.33	0.50	0.67	0.33	0.40	0.50
D	0.67	1.00	3.00	0.50	1.00	2.00	0.50	0.67	2.00	0.50	0.67	1.00	0.40	0.50	0.67
E	1.00	2.00	3.00	0.50	1.00	2.00	0.50	1.00	2.00	0.50	0.67	1.00	0.40	0.67	1.00
F	1.00	1.00	1.00	1.00	2.00	3.00	0.50	1.00	2.00	0.50	1.00	2.00	0.50	0.67	1.00
G	0.33	0.50	1.00	1.00	1.00	1.00	1.00	2.00	3.00	1.00	2.00	3.00	1.00	2.00	3.00
H	0.50	1.00	2.00	0.33	0.50	1.00	1.00	1.00	1.00	1.00	2.00	3.00	1.00	2.00	3.00
I	0.50	1.00	2.00	0.33	0.50	1.00	0.33	0.50	1.00	1.00	1.00	1.00	1.00	2.00	3.00
J	1.00	1.50	2.00	0.33	0.50	1.00	0.33	0.50	1.00	0.33	0.50	1.00	1.00	1.00	1.00
K	1.50	2.00	2.50	0.50	1.00	1.50	0.33	0.50	1.00	0.50	1.00	1.50	0.50	1.00	1.50
L	1.50	2.00	2.50	0.50	1.00	1.50	0.33	0.50	1.00	0.50	1.00	1.50	0.50	1.00	1.50
M	2.00	2.50	3.00	1.00	1.50	2.00	0.50	1.00	1.50	0.50	1.00	1.50	0.50	1.00	1.50
N	2.00	2.50	3.00	1.50	2.00	2.50	0.50	1.00	1.50	1.00	1.50	2.00	1.00	1.50	2.00
O	2.00	2.50	3.00	1.50	2.00	2.50	0.50	1.00	1.50	1.00	1.50	2.00	1.00	1.50	2.00

	K			*L*			*M*			*N*			*O*		
	M_1	M_2	M_3	M_1	M_2	M_3	M_1	M_2	M_3	M_1	M_2	M_3	M_1	M_2	M_3
A	0.22	0.25	0.40	0.22	0.25	0.33	0.20	0.22	0.33	0.17	0.20	0.29	0.17	0.20	0.25
B	0.29	0.33	0.40	0.25	0.29	0.33	0.20	0.25	0.33	0.20	0.25	0.29	0.20	0.25	0.29
C	0.29	0.33	0.50	0.29	0.33	0.40	0.25	0.29	0.33	0.22	0.25	0.33	0.22	0.25	0.33
D	0.40	0.50	0.67	0.33	0.40	0.50	0.33	0.40	0.50	0.25	0.33	0.40	0.25	0.33	0.40
E	0.40	0.50	1.00	0.40	0.50	0.67	0.33	0.40	0.50	0.29	0.40	0.50	0.29	0.40	0.50
F	0.40	0.50	0.67	0.40	0.50	0.67	0.33	0.40	0.50	0.33	0.40	0.50	0.33	0.40	0.50
G	0.67	1.00	2.00	0.67	1.00	2.00	0.50	0.67	1.00	0.40	0.50	0.67	0.40	0.50	0.67
H	1.00	2.00	3.00	1.00	2.00	3.00	0.67	1.00	2.00	0.67	1.00	2.00	0.67	1.00	2.00
I	0.67	1.00	2.00	0.67	1.00	2.00	0.67	1.00	2.00	0.50	0.67	1.00	0.50	0.67	1.00
J	0.67	1.00	2.00	0.67	1.00	2.00	0.67	1.00	2.00	0.50	0.67	1.00	0.50	0.67	1.00
K	1.00	1.00	1.00	0.67	1.00	2.00	0.67	1.00	2.00	0.50	0.67	1.00	0.50	0.67	1.00

(Continued)

TABLE 16.2 (*Continued*)
Pair-Wise Comparison between the Selected Geo-Environmental Factors

	K			*L*			*M*			*N*			*O*		
	M_1	M_2	M_3	M_1	M_2	M_3	M_1	M_2	M_3	M_1	M_2	M_3	M_1	M_2	M_3
L	0.50	1.00	1.50	1.00	1.00	1.00	0.67	1.00	2.00	0.50	0.67	1.00	0.50	0.67	1.00
M	0.50	1.00	1.50	0.50	1.00	1.50	1.00	1.00	1.00	0.50	0.67	1.00	0.50	0.67	1.00
N	1.00	1.50	2.00	1.00	1.50	2.00	1.00	1.50	2.00	1.00	1.00	1.00	1.00	1.00	1.00
O	1.00	1.50	2.00	1.00	1.50	2.00	1.00	1.50	2.00	1.00	1.00	1.00	1.00	1.00	1.00

Note: A, SPI; B, TWI; C, Road proximity; D, Profile curvature; E, Plan curvature; F, Thrust and fault proximity; G, Aspect; H, Elevation; I, Lithology; J, Soil texture; K, Relative relief; L, Land use land cover; M, Drainage density; N, Lineament proximity; O, Slope.

TABLE 16.3
FAHP Derived Weights

Factor	Weights
SPI	0.004
TWI	0.016
Road proximity	0.013
Profile curvature	0.043
Plan curvature	0.044
Thrust and fault proximity	0.050
Aspect	0.078
Elevation	0.090
Lithology	0.076
Soil texture	0.076
Relative relief	0.086
Land use land cover	0.089
Drainage density	0.101
Lineament proximity	0.116
Slope	0.117

them are integrated in GIS using arithmetic overlay approach to obtain the landslide susceptibility index (LSI) values, as shown in equation (16.16).

$$\begin{aligned}\text{LSI} = \{&(\text{slope} \times 0.117) + (\text{lineament proximity} \times 0.116) + (\text{drainage density} \times 0.101)\\ &+ (\text{landuse} \times 0.089) + (\text{relative relief} \times 0.086) + (\text{soil texture} \times 0.076) +\\ &(\text{lithology} \times 0.075)\\ &+ (\text{elevation} \times 0.090) + (\text{aspect} \times 0.078) + (\text{thrust and fault proximity} \times 0.050)\\ &+ (\text{plancurvature} \times 0.044) + (\text{profile curvature} \times 0.043) + (\text{road proximity} \times 0.013)\\ &+ (\text{topographic wetness index} \times 0.016) + (\text{stream power index} \times 0.004)\}\end{aligned} \tag{16.16}$$

The final LSI values ranges between −0.434 and 0.178 with a mean (μ) and standard deviation (σ) of −0.054 and 0.076, respectively. For better spatial representation, we select five susceptibility

TABLE 16.4
The Ordinate of Highest Intersection and Degree of Possibility of TFNs

S_1 (SPI)		**S_2 (TWI)**		**S_3 (Road Proxi)**		**S_4 (Profile Cur)**		**S_5 (Plan Cur)**	
$S_1 \geq S_2$	0.96	$S_2 \geq S_1$	1.00	$S_3 \geq S_1$	1.00	$S_4 \geq S_1$	1.00	$S_5 \geq S_1$	1.00
$S_1 \geq S_3$	0.88	$S_2 \geq S_3$	0.92	$S_3 \geq S_2$	1.00	$S_4 \geq S_2$	1.00	$S_5 \geq S_2$	1.00
$S_1 \geq S_4$	0.72	$S_2 \geq S_4$	0.78	$S_3 \geq S_4$	0.83	$S_4 \geq S_3$	1.00	$S_5 \geq S_3$	1.00
$S_1 \geq S_5$	0.66	$S_2 \geq S_5$	0.72	$S_3 \geq S_5$	0.76	$S_4 \geq S_5$	0.93	$S_5 \geq S_4$	1.00
$S_1 \geq S_6$	0.61	$S_2 \geq S_6$	0.68	$S_3 \geq S_6$	0.71	$S_4 \geq S_6$	0.89	$S_5 \geq S_6$	0.95
$S_1 \geq S_7$	0.45	$S_2 \geq S_7$	0.52	$S_3 \geq S_7$	0.54	$S_4 \geq S_7$	0.73	$S_5 \geq S_7$	0.77
$S_1 \geq S_8$	0.37	$S_2 \geq S_8$	0.45	$S_3 \geq S_8$	0.46	$S_4 \geq S_8$	0.66	$S_5 \geq S_8$	0.69
$S_1 \geq S_9$	0.43	$S_2 \geq S_9$	0.51	$S_3 \geq S_9$	0.53	$S_4 \geq S_9$	0.74	$S_5 \geq S_9$	0.79
$S_1 \geq S_9$	0.40	$S_2 \geq S_{10}$	0.49	$S_3 \geq S_{10}$	0.50	$S_4 \geq S_{10}$	0.72	$S_5 \geq S_{10}$	0.76
$S_1 \geq S_{11}$	0.31	$S_2 \geq S_{11}$	0.40	$S_3 \geq S_{11}$	0.40	$S_4 \geq S_{11}$	0.63	$S_5 \geq S_{11}$	0.66
$S1 \geq S_{12}$	0.26	$S_2 \geq S_{12}$	0.36	$S_3 \geq S_{12}$	0.35	$S_4 \geq S_{12}$	0.59	$S_5 \geq S_{12}$	0.62
$S1 \geq S_{13}$	0.18	$S_2 \geq S_{13}$	0.28	$S_3 \geq S_{13}$	0.27	$S_4 \geq S_{13}$	0.51	$S_5 \geq S_{13}$	0.53
$S1 \geq S_{14}$	0.04	$S_2 \geq S_{14}$	0.15	$S_3 \geq S_{14}$	0.13	$S_4 \geq S_{14}$	0.38	$S_5 \geq S_{14}$	0.39
$S1 \geq S_{15}$	0.03	$S_2 \geq S_{15}$	0.14	$S_3 \geq S_{15}$	0.12	$S_4 \geq S_{15}$	0.37	$S_5 \geq S_{15}$	0.38
min{V(Si>Sj)}	0.03	min{V(Si>Sj)}	0.14	min{V(Si>Sj)}	0.12	min{V(Si>Sj)}	0.37	min{V(Si>Sj)}	0.38
S_6 (Thrust and Fault)		**S_7 (Aspect)**		**S_8 (Elevation)**		**S_9 (Lithology)**		**S_{10} (Soil)**	
$S_6 \geq S1$	1.00	$S_7 \geq S1$	1.00	$S_8 \geq S1$	1.00	$S_9 \geq S1$	1.00	$S_{10} \geq S1$	1.00
$S_6 \geq S_2$	1.00	$S_7 \geq S_2$	1.00	$S_8 \geq S_2$	1.00	$S_9 \geq S_2$	1.00	$S_{10} \geq S_2$	1.00
$S_6 \geq S_3$	1.00	$S_7 \geq S_3$	1.00	$S_8 \geq S_3$	1.00	$S_9 \geq S_3$	1.00	$S_{10} \geq S_3$	1.00
$S_6 \geq S_4$	1.00	$S_7 \geq S_4$	1.00	$S_8 \geq S_4$	1.00	$S_9 \geq S_4$	1.00	$S_{10} \geq S_4$	1.00
$S_6 \geq S_5$	1.00	$S_7 \geq S_5$	1.00	$S_8 \geq S_5$	1.00	$S_9 \geq S_5$	1.00	$S_{10} \geq S_5$	1.00
$S_6 \geq S_7$	0.82	$S_7 \geq S_6$	1.00	$S_8 \geq S_6$	1.00	$S_9 \geq S_6$	1.00	$S_{10} \geq S_6$	1.00
$S_6 \geq S_8$	0.73	$S_7 \geq S_8$	0.92	$S_8 \geq S_7$	1.00	$S_9 \geq S_7$	0.98	$S_{10} \geq S_7$	0.99
$S_6 \geq S_9$	0.84	$S_7 \geq S_9$	1.00	$S_8 \geq S_9$	1.00	$S_9 \geq S_8$	0.90	$S_{10} \geq S_8$	0.91
$S_6 \geq S_{10}$	0.81	$S_7 \geq S_{10}$	1.00	$S_8 \geq S_{10}$	1.00	$S_9 \geq S_{10}$	0.98	$S_{10} \geq S_9$	1.00
$S_6 \geq S_{11}$	0.71	$S_7 \geq S_{11}$	0.91	$S_8 \geq S_{11}$	0.99	$S_9 \geq S_{11}$	0.89	$S_{10} \geq S_{11}$	0.90
$S_6 \geq S_{12}$	0.67	$S_7 \geq S_{12}$	0.88	$S_8 \geq S_{12}$	0.97	$S_9 \geq S_{12}$	0.86	$S_{10} \geq S_{12}$	0.87
$S_6 \geq S_{13}$	0.58	$S_7 \geq S_{13}$	0.80	$S_8 \geq S_{13}$	0.89	$S_9 \geq S_{13}$	0.78	$S_{10} \geq S_{13}$	0.78

(Continued)

TABLE 16.4 (*Continued*)
The Ordinate of Highest Intersection and Degree of Possibility of TFNs

S_6 (Thrust and Fault)		**S_7 (Aspect)**		**S_8 (Elevation)**		**S_9 (Lithology)**		**S_{10} (Soil)**	
$S_6 \geq S_{14}$	0.44	$S_7 \geq S_{14}$	0.67	$S_8 \geq S_{14}$	0.77	$S_9 \geq S_{14}$	0.66	$S_{10} \geq S_{14}$	0.66
$S_6 \geq S_{15}$	0.43	$S_7 \geq S_{15}$	0.66	$S_8 \geq S_{15}$	0.77	$S_9 \geq S_{15}$	0.65	$S_{10} \geq S_{15}$	0.65
min{V(Si>Sj)}	0.43	min{V(Si>Sj)}	0.66	min{V(Si>Sj)}	0.77	min{V(Si>Sj)}	0.65	min{V(Si>Sj)}	0.65
S_{11} (Relative Relief)		**S_{12} (LULC)**		**S_{13} (Drainage)**		**S_{14} (Lineament)**		**S_{15} (Slope)**	
$S_{11} \geq$ S1	1.00	$S_{12} \geq$ S1	1.00	$S_{13} \geq$ S1	1.00	$S_{14} \geq$ S1	1.00	$S_{15} \geq$ S1	1
$S_{11} \geq S_2$	1.00	$S_{12} \geq S_2$	1.00	$S_{13} \geq S_2$	1.00	$S_{14} \geq S_2$	1.00	$S_{15} \geq S_2$	1
$S_{11} \geq S_3$	1.00	$S_{12} \geq S_3$	1.00	$S_{13} \geq S_3$	1.00	$S_{14} \geq S_3$	1.00	$S_{15} \geq S_3$	1
$S_{11} \geq S_4$	1.00	$S_{12} \geq S_4$	1.00	$S_{13} \geq S_4$	1.00	$S_{14} \geq S_4$	1.00	$S_{15} \geq S_4$	1
$S_{11} \geq S_5$	1.00	$S_{12} \geq S_5$	1.00	$S_{13} \geq S_5$	1.00	$S_{14} \geq S_5$	1.00	$S_{15} \geq S_5$	1
$S_{11} \geq S_6$	1.00	$S_{12} \geq S_6$	1.00	$S_{13} \geq S_6$	1.00	$S_{14} \geq S_6$	1.00	$S_{15} \geq S_6$	1
$S_{11} \geq S_7$	1.00	$S_{12} \geq S_7$	1.00	$S_{13} \geq S_7$	1.00	$S_{14} \geq S_7$	1.00	$S_{15} \geq S_7$	1
$S_{11} \geq S_8$	1.00	$S_{12} \geq S_8$	1.00	$S_{13} \geq S_8$	1.00	$S_{14} \geq S_8$	1.00	$S_{15} \geq S_8$	1
$S_{11} \geq S_9$	1.00	$S_{12} \geq S_9$	1.00	$S_{13} \geq S_9$	1.00	$S_{14} \geq S_9$	1.00	$S_{15} \geq S_9$	1
$S_{11} \geq S_{10}$	1.00	$S_{12} \geq S_{10}$	1.00	$S_{13} \geq S_{10}$	1.00	$S_{14} \geq S_{10}$	1.00	$S_{15} \geq S_{10}$	1
$S_{11} \geq S_{12}$	0.97	$S_{12} \geq S_{11}$	1.00	$S_{13} \geq S_{11}$	1.00	$S_{14} \geq S_{11}$	1.00	$S_{15} \geq S_{11}$	1
$S_{11} \geq S_{13}$	0.88	$S_{12} \geq S_{13}$	0.90	$S_{13} \geq S_{12}$	1.00	$S_{14} \geq S_{12}$	1.00	$S_{15} \geq S_{12}$	1
$S_{11} \geq S_{14}$	0.74	$S_{12} \geq S_{14}$	0.77	$S_{13} \geq S_{14}$	0.87	$S_{14} \geq S_{13}$	1.00	$S_{15} \geq S_{13}$	1
$S_{11} \geq S_{15}$	0.74	$S_{12} \geq S_{15}$	0.76	$S_{13} \geq S_{15}$	0.86	$S_{14} \geq S_{15}$	0.99	$S_{15} \geq S_{14}$	1
min{$V(S_i > S_j)$}	0.74	min{$V(S_i > S_j)$}	0.76	min{$V(S_i > S_j)$}	0.86	min{$V(S_i > S_j)$}	0.99	min{$V(S_i > S_j)$}	1

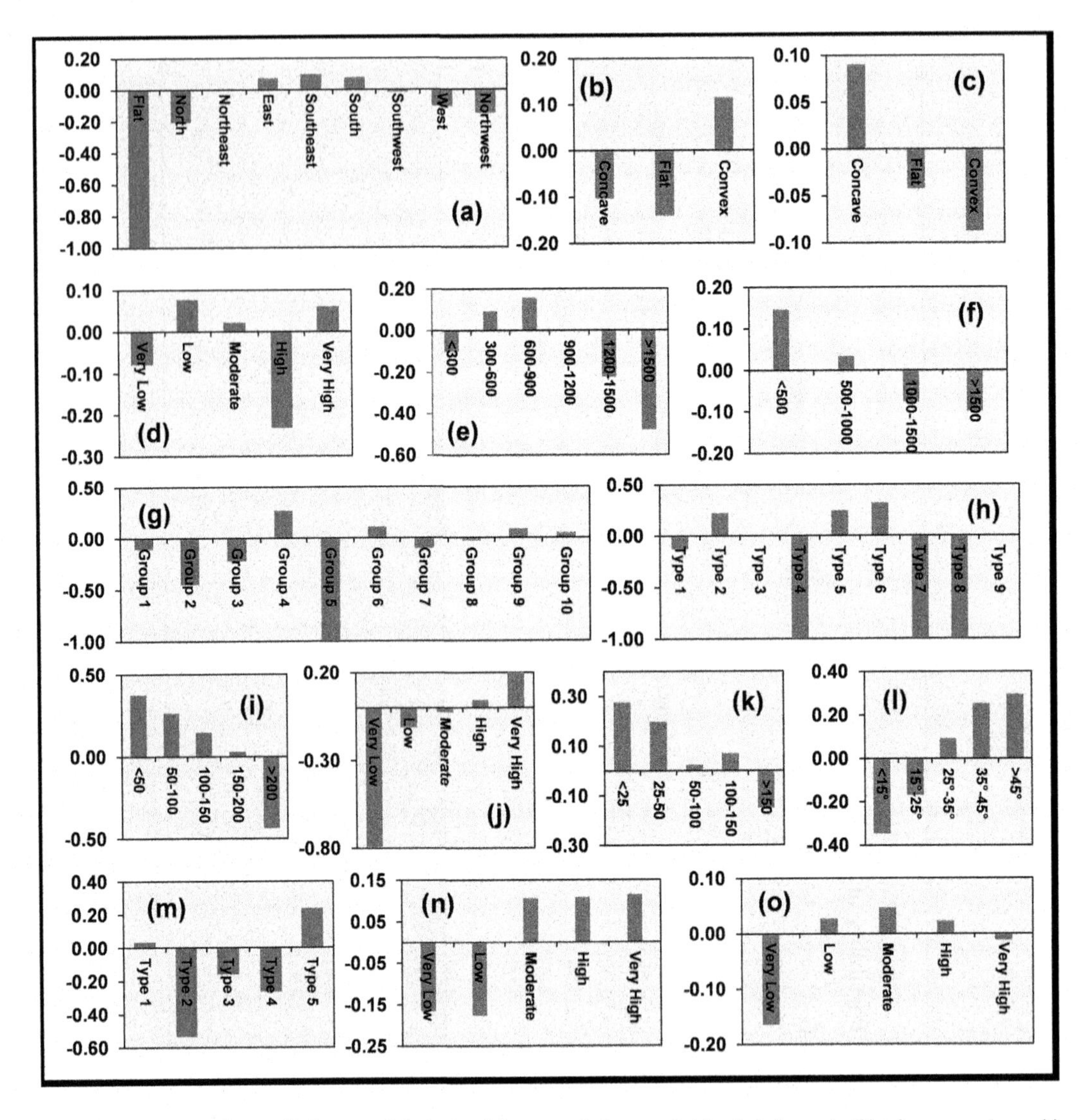

FIGURE 16.7 Yule coefficient model derived factor subclass weights. (a.) Aspect, (b) plan curvature, (c) profile curvature, (d) drainage density, (e) elevation, (f) thrust and fault proximity, (g) lithology, (h) landuse landcover, (i) lineament proximity, (j) relative relief, (k) road proximity, (l) slope, (m) soil texture, (n) stream power index, and (o) topographic wetness index.

classes, i.e., very low, low, moderate, high and very high to present the LSZ map. Therefore, considering the obtained LSI information, the four class boundaries need to be defined to separate five consecutive susceptibility classes. For this purpose, the success rate curve (SRC) method proposed by Saha et al. (2005) has been utilized to specify the class boundaries at $(\mu - 1.5m\sigma)$, $(\mu - 0.5m\sigma)$, $(\mu + 0.5m\sigma)$ and $(\mu + 1.5m\sigma)$. In this SRC method m is a positive non-zero value and initially depends on the statistical distribution of the data. For our case to define the optimum *m-value*, a series of trials has been made using different combinations like 0.7, 0.8, 0.9, 1.0, 1.1 and 1.2 (Figure 16.8). Finally, based on their comparative assessment in view of maximum number of landslides covered by the areas classified as very high susceptible zone, $m = 0.8$ is found as optimum for this study (Figure 16.8). Using this *m-value*, the class boundaries are assigned over the LSI range at −0.15 $(\mu - 1.5m\sigma)$, −0.08 $(\mu - 0.5m\sigma)$, −0.02 $(\mu + 0.5m\sigma)$ and 0.04 $(\mu + 1.5m\sigma)$ values, and five susceptibility

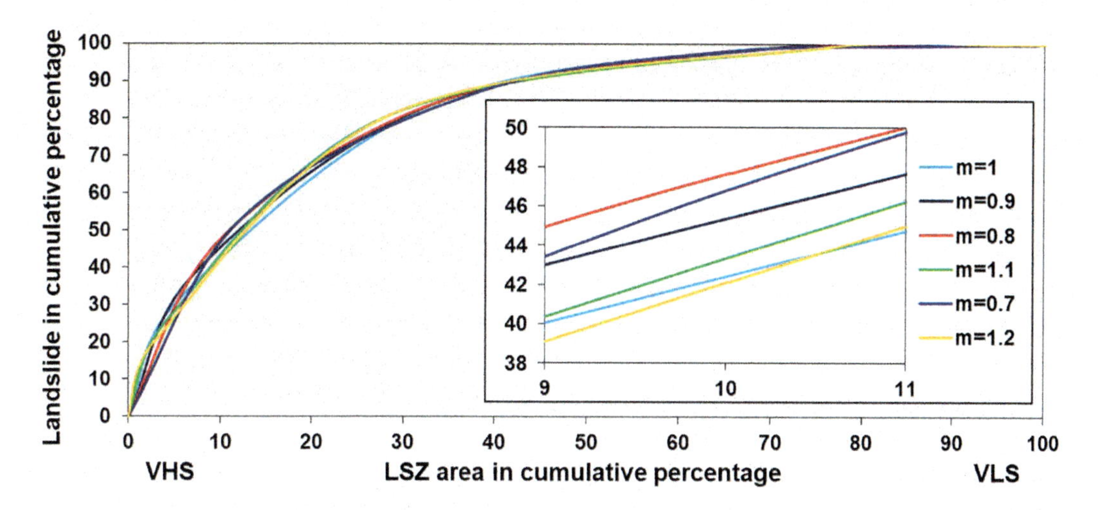

FIGURE 16.8 Success rate curve plot.

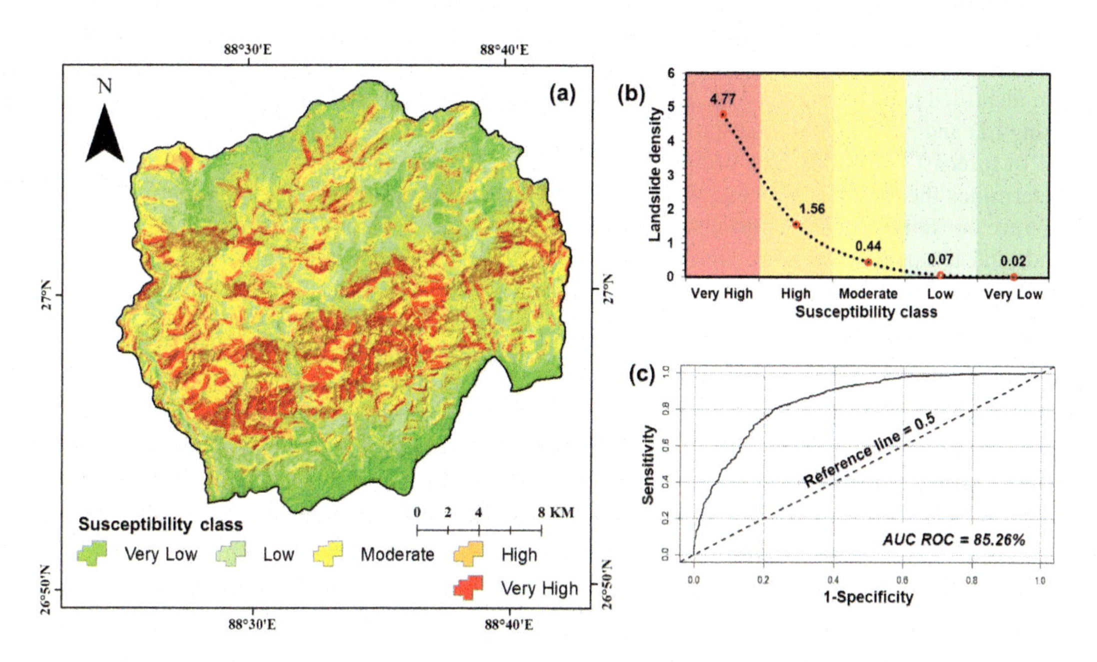

FIGURE 16.9 (a) Landslide susceptibility map, (b) landslide density in different susceptibility classes, and (c) AUC ROC.

classes are finalized. The spatial distribution of LSZ classes show very low, low, moderate, high and very high susceptibility classes covers 9.13%, 25.89%, 31.59%, 23.44% and 9.96% area, respectively, (Figure 16.9a).

To validate the results, landslide density (LD) in different susceptibility classes has been considered. In a general assumption, it is expected that maximum LD should be found in the areas predicted as high susceptible and the values become lower toward the low susceptible areas (Sur et al. 2020). Apart from this, we also include the area under curve of receiver operating characteristic (AUC ROC) curve to examine the overall prediction strength in terms of *sensitivity* and *specificity* results. The statistical summary of AUC value ranges between 0 to 1 and maximum value indicates perfect goodness-of-fit (Sarkar et al. 2013; Chen et al. 2019).

Therefore, we calculate the LD for each susceptibility class considering the testing landslides (30%). The obtained LD values for very high, high, moderate, low and very low susceptibility classes are 4.77, 1.56, 0.44, 0.07 and 0.02, respectively, (Figure 16.9b). The decreasing trend of LD values from very high susceptible areas to very low susceptible areas indicates a good agreement regarding the acceptability of the developed LSZ map. To construct the AUC ROC, we select the testing landslides contains 886 pixels and similar number of non-landslide pixels are randomly generated from the entire study area. Based on this the *sensitivity* and *1-specificity* values are calculated using '*pROC*' package (Robin et al. 2011) in RStudio (http://www.rstudio.com/) platform. The obtained AUC ROC value is 85.26% for this study which indicates the good prediction results (Figure 16.9c).

16.5 DISCUSSION AND CONCLUSION

The preset study highlights the development of LSZ map using geo-statistical models supported by RS and GIS technologies. Here, to establish the susceptibility level, 15 landslide causative factors are considered and based on their spatial association with past landslides; the likelihood of future landslides is predicted. The implementation strategy shows RS and GIS offered an immense endorsement in preparation of landslide inventory, development of thematic layers, LSI integration and LSZ categorization. Thus, RS and GIS should be considered as an integral part of landslide studies. Depending on the selected modeling approaches, different landslide susceptibility zones are identified considering the SRC method derives class boundaries. Finally, the map is validated using LD and AUC ROC-based statistical measures.

To forecast landslides, it is often difficult to select the optimum parameters those intricately determines the landslide process. Therefore, to reduce the biases regarding the parameter selection we considered different groups of factors including morphological, hydrological, geological and land use group. Thereby, we tried to ensure that our selected factors will provide a more deliberation of the real-world scenario. Although, it's very true that all the selected factors are not equally contributing to the landslide process but are accountable to some extent. Similarly, if we consider a single factor, its subclasses are not altogether endorsed the landslide process. Keeping these conditions, we used parallel combination theory in which the FAHP is used to determine the relative importance of the selected causative factors and the subclasses weights are assimilate using YC model. During the priority vector or TFN assignment in the pair-wise matrix, we considered the field experiences along with some trial-error steps to critically determine the final weights. Our derived weight vectors show slope, lineament, drainage density, LULC are most prominent factors controls the landslide process in the Kalimpong region with respect to other factors. The results of the YC models shows the subclass have value ≥ 0 are support the landslide process and the values ≤ 0 are not significantly contribute in the sliding process. Thus, by evaluating both these weighted scenarios one can easily estimate a factor importance and also which section of the factor is more influential for landslide occurrence. To simulate this complete situation for the entire study area, we used arithmetic overlay approach and thereby, obtained the probability of landslide, i.e., the LSI values of each pixel. To generalize this result in a more readable format, we further categorized the LSI values in five classes using SRC method that shows the certainty of landslide ranging from very high to very low classes. The final LSZ map show 33.39% area is covered by very high and high susceptible zones mainly located along the areas have steep slopes and shows abundance of lineament concentration with thrust and faults. Although, moderately steep slope areas have sparse vegetation with high relative relief are predicted as moderate susceptible zones covered by 31.59% area. The low elevation zones in south part and the northern higher elevation zones have dense vegetation are predicted as low and very low susceptible zones covered by 35.02% area. It's observed that factors like TWI, SPI, road proximity, aspect, plan and profile curvature are not significantly contributing in the final LSZ map may because of their low weights. It is important to note here that there are some limitations inherently associated in our modeling process. Like the inventory we used only represent the cartographic information and all the slope instability signatures are directly used as landslide without any classification. Similarly, most of the factors we used here are derived on small scale that limits the detail ground information. Despite these limitations, the LD of 4.77 in the very high susceptibility zone with AUC ROC value of 85.26% significantly justifies our develop model is able to forecast satisfactory results and provides the information of landslide hot spots in parts of Kalimpong region. Similar to this study, earlier Das et al. (2022) has been investigated the same area considering the same group of geo-environmental factors by utilizing the analytic hierarchy process (AHP) method and achieved an prediction accuracy of 79.5%. Thus, by considering the earlier study, the present investigation yields more robust predictions and thereby, ensures that the present findings are more promising for end users. Finally, we conclude that such studies will be helpful for the Govt. agencies to schedule their development projects and land use management in such high landslide-sensitive areas.

ACKNOWLEDGMENTS

The authors are thankful to the Director, CSIR-Central Building Research Institute, Roorkee, India for granting permission to publish this work. The first author acknowledges University Grants Commission (New Delhi, India) for providing the fellowship under Junior Research Fellowship (JRF) Scheme [**UGC-Ref. No. 3511/(NET-JULY 2018)**] and AcSIR (Ghaziabad, India) for providing an opportunity to carry out this doctoral research.

FUNDING

Not applicable.

Availability of Data and Material

The datasets generated during and/or analyzed in this study are available from the corresponding author on reasonable request.

Code Availability

Not applicable

DECLARATIONS

Conflicts of Interest

The authors declare no conflicts of interest.

REFERENCES

Acharyya, S. K. (1980). Structural framework and tectonic evolution of the eastern Himalaya. *Himalayan Geology*, *10*, 412–439.

Aleotti, P., & Chowdhury, R. (1999). Landslide hazard assessment: Summary review and new perspectives. *Bulletin of Engineering Geology and the Environment*, *58*(1), 21–44. doi: 10.1007/s100640050066.

Azimi, S. R., Nikraz, H., & Yazdani-Chamzini, A. (2018). Landslide risk assessment by using a new combination model based on a fuzzy inference system method. *KSCE Journal of Civil Engineering*, *22*(11), 4263–4271.

Cárdenas, N. Y., & Mera, E. E. (2016). Landslide susceptibility analysis using remote sensing and GIS in the western Ecuadorian Andes. *Natural Hazards*, *81*(3), 1829–1859. doi: 10.1007/s11069-016-2157-8.

Chakraborty, I., Ghosh, S., Bhattacharya, D., & Bora, A. (2011). Earthquake induced landslides in the Sikkim Darjeeling Himalayas–An aftermath of the 18th September 2011 Sikkim earthquake. Kolkata: Geological Survey of India.

Chang, D. Y. (1996). Applications of the extent analysis method on fuzzy AHP. *European Journal of Operational Research, 95*(3), 649–655.

Chen, W., Shahabi, H., Shirzadi, A., Hong, H., Akgun, A., Tian, Y., ... & Li, S. (2019). Novel hybrid artificial intelligence approach of bivariate statistical-methods-based kernel logistic regression classifier for landslide susceptibility modeling. *Bulletin of Engineering Geology and the Environment, 78*(6), 4397–4419. doi: 10.1007/s10064-018-1401-8.

Conforti, M., Pascale, S., Robustelli, G., & Sdao, F. (2014). Evaluation of prediction capability of the artificial neural networks for mapping landslide susceptibility in the Turbolo River catchment (northern Calabria, Italy). *Catena, 113*, 236–250. doi: 10.1016/j.catena.2013.08.006.

Cruden, D. M., & Varnes, D. J. (1996). Landslide types and processes. In: Turner, A. K., & Schuster, R. L. (Eds.), *Landslides: Investigation and Mitigation (Special Report 247).* Washington, DC: National Research Council, Transportation and Research Board, pp. 36–75.

Das, S., Sarkar, S., & Kanungo, D. P. (2022). GIS-based landslide susceptibility zonation mapping using the analytic hierarchy process (AHP) method in parts of Kalimpong Region of Darjeeling Himalaya. *Environ Monit Assess, 194*, 234. doi: 10.1007/s10661-022-09851-7.

Dikshit, A., Sarkar, R., Pradhan, B., Segoni, S., & Alamri, A. M. (2020). Rainfall induced landslide studies in Indian Himalayan region: A critical review. *Applied Sciences, 10*(7), 2466. doi: 10.3390/app10072466.

Froude, M. J., & Petley, D. N. (2018). Global fatal landslide occurrence from 2004 to 2016. *Natural Hazards and Earth System Sciences, 18*(8), 2161–2181. doi: 10.5194/nhess-18-2161-2018.

Gangopadhyay, P. K. (1995). Intrafolial folds and associated structures in a progressive strain environment of Darjeeling-Sikkim Himalaya. *Proceedings of the Indian Academy of Sciences-Earth and Planetary Sciences, 104*(3), 523.

Ghosh, S., Carranza E. J. M., van Westen, C. J., Jetten, V. G., & Bhattacharya, D. N. (2011). Selecting and weighting spatial predictors for empirical modeling of landslide susceptibility in the Darjeeling Himalayas (India). *Geomorphology, 131*(1–2), 35–56. doi: 10.1016/j.geomorph.2011.04.019.

Guzzetti, F. (2002). Landslide hazard assessment and risk evaluation: Limits and prospectives. *In Proceedings of the 4th EGS Plinius Conference*, Mallorca, Spain, pp. 2–4.

Hu, X., Huang, C., Mei, H., & Zhang, H. (2021). Landslide susceptibility mapping using an ensemble model of Bagging scheme and random subspace–based naïve Bayes tree in Zigui County of the Three Gorges Reservoir Area, China. *Bulletin of Engineering Geology and the Environment*, 1–15. doi: 10.1007/s10064-021-02275-6.

Kanungo, D. P., Arora, M. K., Sarkar, S., & Gupta, R. P. (2006). A comparative study of conventional, ANN black box, fuzzy and combined neural and fuzzy weighting procedures for landslide susceptibility zonation in Darjeeling Himalayas. *Engineering Geology, 85*(3–4), 347–366. doi: 10.1016/j.enggeo.2006.03.004.

Kellett, D., Grujic, D., Mot-tram, C., Mukul, M., & Larson, K. P. (2014). Virtual field guide for the Darjeeling-Sik-kim Himalaya, India. In: Montomoli, C., Carosi, R., Law, R., Singh, S., & Rai, S. M. (Eds.), *Geological Field Trips in the Himalaya, Kar-akoram and Tibet, Journal of the Virtual Explorer*, Electronic Edition, vol. 47, ISSN, 1441-8142.

Khanna, K., Martha, T. R., Roy, P., & Kumar, K. V. (2021). Effect of time and space partitioning strategies of samples on regional landslide susceptibility modelling. *Landslides, 18*(6), 2281–2294. doi: 10.1007/s10346-021-01627-3.

Lombardo, L., Tanyas, H., Huser, R., Guzzetti, F., & Castro-Camilo, D. (2021). Landslide size matters: A new data-driven, spatial prototype. *Engineering Geology, 293*, 106288. doi: 10.1016/j.enggeo.2021.106288.

Mandal, S., & Mandal, K. (2018). Modeling and mapping landslide susceptibility zones using GIS based multivariate binary logistic regression (LR) model in the Rorachu river basin of eastern Sikkim Himalaya, India. *Modeling Earth Systems and Environment, 4*(1), 69–88. doi: 10.1007/s40808-018-0426-0.

Martha, T. R., van Westen, C. J., Kerle, N., Jetten, V., & Kumar, K. V. (2013). Landslide hazard and risk assessment using semi-automatically created landslide inventories. *Geomorphology, 184*, 139–150. doi: 10.1016/j.geomorph.2012.12.001.

Mukul, M., Srivastava, V., & Mukul, M. (2017). Out-of-sequence reactivation of the Munsiari thrust in the Relli River basin, Darjiling Himalaya, India: Insights from Shuttle Radar Topography Mission digital elevation model-based geomorphic indices. *Geomorphology, 284*, 229–237. doi: 10.1016/j.geomorph.2016.10.029.

Patra, A., & Saha, D. (2019). Stress regime changes in the Main Boundary Thrust zone, Eastern Himalaya, decoded from fault-slip analysis. *Journal of Structural Geology, 120*, 29–47. doi: 10.1016/j.jsg.2018.12.010.

Robin, X., Turck, N., Hainard, A., Tiberti, N., Lisacek, F., Sanchez, J. C., & Müller, M. (2011). PROC: An open-source package for R and S+ to analyze and compare ROC curves. *BMC Bioinformatics*, *12*, 77.

Roodposhti, M. S., Rahimi, S., & Beglou, M. J. (2014). PROMETHEE II and fuzzy AHP: An enhanced GIS-based landslide susceptibility mapping. *Natural Hazards*, *73*(1), 77–95. doi: 10.1007/s11069-012-0523-8.

Roy, J., & Saha, S. (2019). Landslide susceptibility mapping using knowledge driven statistical models in Darjeeling District, West Bengal, India. *Geoenvironmental Disasters*, *6*(1), 1–18. doi: 10.1186/s40677-019-0126-8.

Saaty, T. L., & Wind, Y. (1980). Marketing applications of the analytic hierarchy process. *Management Science*, *26*(7), 641–658. doi: 10.1287/mnsc.26.7.641.

Saha, A. K., Gupta, R. P., Sarkar, I., Arora, M. K., & Csaplovics, E. (2005). An approach for GIS-based statistical landslide susceptibility zonation—with a case study in the Himalayas. *Landslides*, *2*(1), 61–69. doi: 10.1007/s10346-004-0039-8.

Sarkar, S., & Kanungo, D. P. (2004). An integrated approach for landslide susceptibility mapping using remote sensing and GIS. *Photogrammetric Engineering & Remote Sensing*, *70*(5), 617–625. doi: 10.14358/PERS.70.5.617.

Sarkar, S., Roy, A. K., & Martha, T. R. (2013). Landslide susceptibility assessment using information value method in parts of the Darjeeling Himalayas. *Journal of the Geological Society of India*, *82*(4), 351–362. doi: 10.1007/s12594-013-0162-z.

Sur, U., Singh, P., & Meena, S. R. (2020). Landslide susceptibility assessment in a lesser Himalayan road corridor (India) applying fuzzy AHP technique and earth-observation data. *Geomatics, Natural Hazards and Risk*, *11*(1), 2176–2209. doi: 10.1080/19475705.2020.1836038.

Wu, Y., Ke, Y., Chen, Z., Liang, S., Zhao, H., & Hong, H. (2020). Application of alternating decision tree with AdaBoost and bagging ensembles for landslide susceptibility mapping. *Catena*, *187*, 104396. doi: 10.1016/j.catena.2019.104396.

Zadeh, L. A. (1965). Fuzzy sets. *Information and Control*, *8*, 338–353.

17 Understanding the Dynamics of the City through Crowdsourced Datasets

A Case Study of Indore City

Vipul Parmar
Research Scholar, Centre for Urban Science & Engineering, Indian Institute of Technology, Bombay, India

Anugrah Anilkumar Nagaich
Assistant Professor, Department of Architecture and Planning, Maulana Azad National Institute of Technology, Bhopal, India

CONTENTS

17.1 INTRODUCTION, BACKGROUND AND NEED OF STUDY

Urban planning is the field of study which focuses on the effective management of land and designing the urban environment for its citizens. Effective management of land requires to assess the ongoing activities and identify the best possible use of that land parcel. In urban planning, the land use plan is one of the important legal documents which gives details regarding the existing use of land parcels as well as regulates activity on the future developable land. Generally, this plan is

DOI: 10.1201/9781003270928-23

prepared for more than 20+ years, but in the fast-growing cities, the landuse pattern is changing frequently. Various kinds of activities are mixing altogether [1]. It is required to capture the dynamics for the management and development of the land for the betterment of the city.

Cities are collection of the places to live, work and play [2]. These places have spatial attribute attached along with them. By superimposing various spatial attribute like the demographic profile, socio-economic profile, infrastructure layers all together can be used to identify, which area has access to all the infrastructure facilities and whether distribution is inclusive or not. Geographical information system (GIS) software enables to analysis, manage, and visualize and converts this spatial dataset into location information.

With the advancement in the information and communication technology (ICT) sector, the mode of interaction between people has been changed completely. Increase in the number of smartphone users and cheaper internet facilities boosted the digital communication channel. Social Media is the platform to interact with people, share thoughts, and emotions in textual format, images or videos along with the geographical location [3]. People, now generating gigabytes of dataset related to places where they work, live, and enjoy, imparting their digital footprints all over. This leads to the rise of information urbanism where information is continuously flowing in the environment [4,5]. Collecting these digital foot-prints and analyzing it with the help of GIS software could help planners understand more truthful insights about the city, spatially.

Different researchers have used crowdsourced data content, to identify any cultural event, emergency event, people behavior, etc. The location dataset can be utilized to analyze the movement of people and help to identifies the interaction point in cities [6]. The emergence of new data processing technologies can act as powerful tools for gathering urban knowledge. Big data, data extraction or data analysis along with GIS can provide vital information about the dynamics of the city [7–10].

17.2 LITERATURE REVIEW

17.2.1 Location Intelligence

Location intelligence (LI) can be defined as gathering and analyzing geospatial data to overcome a variety of business challenges [11]. This provides insights to the consumers, their socio-economic profile, and behavior which can be used to target a specific set of audiences for a product.

Initially, LI was mainly used for the expansion of business but the introduction of the smart city concept globally emphasized using it to empower inclusive and holistic planning, decision making and prediction [12]. This helps urban planners create a better understanding regarding interaction point of people, or the level of activity in the public space. It can be complementary to the existing planning processes, providing a fine picture of a city at a local level.

GIS technology helps to collect, analyze, and visualize the location data into more meaningful information. LI, powered by GIS, can effectively improve the quality of life along with boosting the local economy.

17.2.2 Rise of Social Media and Urban Datasets

Due to the growth of ICT, the number of people using the internet is exponentially increasing. In India, there are 462 million active internet users (Figure 17.1), out of 3.63 billion people all around the world [13]. Also, there are 2.43 billion people who use social media platforms, out of which 326.1 million people are from India. From Figure 17.1, it can be seen that after 2016 the number of internet users had a sharp increase from 295 million to 437 million in 2017. This is because, Reliance Jio 4G was introduced in 2016, which provides high speed internet facilities at a cheaper rate. Currently, the average Indian consumes 9.8 GB per month of data and it will be double up to 18 GB per month by 2024 [14].

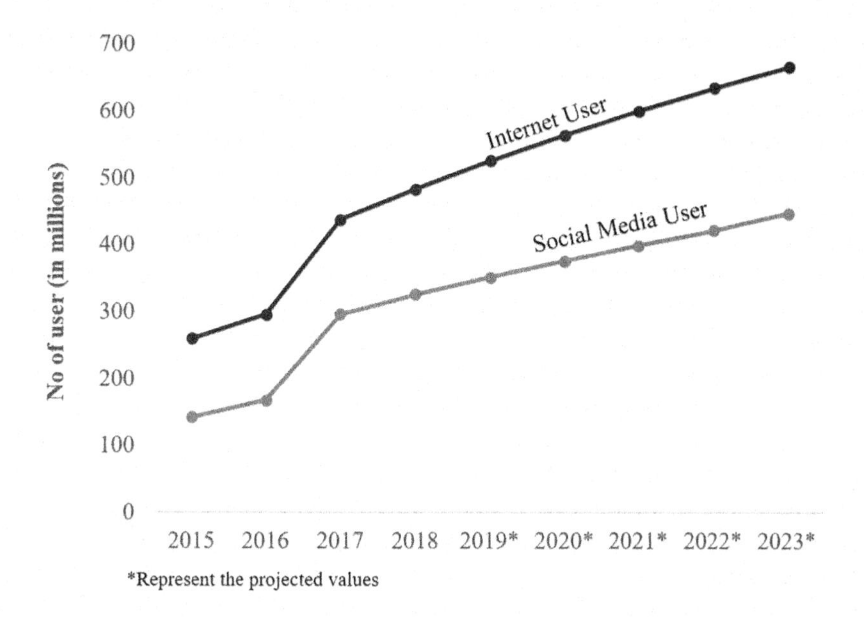

FIGURE 17.1 Growth trend of Internet and Social Media User in India. (Data from Statista (2020).)

The availability of good internet facilities and smartphones increased the use of social media. Also, around 90% of people in urban and rural areas use a smartphone for accessing internet facilities. As per the Ericsson report 2018 [14], the number of smartphone users will reach 1.1 billion by 2024. Furthermore, out 2 in 3 Indians who have smartphones are on some kind of social media platform. Most of the time, people spend accessing social media websites over the internet. In India, average person spends 2.4 hours a day on social media which is almost near the global average of 2.5 hours a day [15].

17.2.3 Using the Social Media/Urban Datasets for Various Urban Studies

Social media is not only just a place for interacting with new people but also acts as a source of information in the context of urban planning. For years, these platforms have been used for participation and are considered as a communication tool for study [16]. Campagna [17] suggests using these platforms to "listen" to the local community's opinions and feelings and knowing the feasibility of the planning proposals. Another common goal for using social media platform is to increase participation.

Therefore, considering crowdsourced data as a reliable source of data and analyzing it using various algorithms can provide an efficient way of understanding the views of the public [16]. In a study, city planners are identified as a possible class for using social media geo-tagged data in planning practices of using social media [18]. They conclude that "geotagged social media dataset can be used to enhance urban planning and quality of life" incite by providing granular level data about socioeconomic profile, mobility pattern and quality of life, and extracting growth patterns and comparisons between cities structures. The use of the crowdsourced dataset, big data analytics are prominent issues in urban e-planning research [19].

17.2.4 Different Approaches and Clustering-Based Algorithms

In fast-growing cities, land-use patterns have changed rapidly, especially in urban centers in developing nations and mixing different types of activity. It is difficult to evaluate timely and accurately the pattern of land use for these fast-growing cities. To tackle this problem, multi-source data mining techniques can be used to identify the complex trends of urban land use.

To estimate the types and patterns of urban land use in Beijing [1], a regular grid of 400 m was established to divide the urban core. Through analyzing the temporal frequency patterns of social media messages in each cell using the K-means clustering algorithm, seven types of land use clusters are established in Beijing: residential areas, college dormitories, commercial areas, workplace, transportation hubs, and two types of mixed land use areas. For successful verification of the estimated land use types, text mining, word clouds, and point of interest (POI) distribution analysis were used. This helps in resource mobilization and enhances the decision-making process of the local administrative bodies.

17.2.5 Incorporation of Text-Based Classification

Social media posts generally consist of images, videos or text. The text associated with the post can be a single word or line. This text helps to identify the feelings associated with the post [20]. There are hashtags which are one word that describes things in non-hierarchical form. These tags can be classified through a text-based classification method [21].

In Ljubljana, Slovenia, an experiment was performed to reveal perceptions of this kind. It is based on photography analytic with crowdsourcing and descriptions attached. Initially, developed as part of the EU Human Cities project to provide Ljubljana citizens with a tool to express their concerns about living environments. It focuses on revealing local communities shared values, which are seen as a starting point for the development of participatory strategies for urban regeneration. There are images, submitted to the category "The nicest place of my neighbourhood". There is small corner in the neighborhood where whole community gathers from mothers with their babies to youth, observing the transition in lifestyle. There are some attached shared values: conviviality, imagination, leisure. Shared values associated with images align the perception of other people with its owners analyzing these keywords and spatially mapping the activity of the city helps to understand the dynamics of the city [22].

17.3 FRAMEWORK AND METHODOLOGY

The study is carried to understand how social media dataset can be used to explore the dynamics of various places inside the city. Research is carried out in three steps. Firstly, to identifies which social media platform can be used for extraction of dataset on the basis of prominently used platform. Secondly, to select case study area on the basis of its population and other prominent characteristic like smart city initiatives and achievements. Furthermore, data is extracted for the selected city and spatially mapped using GIS software. Lastly, analysis is performed using various algorithms and spatially mapped to extract micro level details.

17.3.1 Identification of Platform for Extraction of Data

Facebook and Instagram are the two most popular social media platforms in India. They have lots of activities and posts related to the city. Generally, a post consists of images, video, text or a combination of them. These two platforms were selected for collecting and extraction of the data for city along with location data associated with the post. This research also requires the land-use plan of the city which has been collected from the respective departments.

17.3.2 Case Study – Indore City

Crowdsources dataset having location information can be used for expansion of business, distribution of resources or understanding the behavior of people. This research aims to explore the potential of these datasets to understand the city dynamics. In this digitally transformed era, the city has digital footprints over the internet which can be analyzed to know how people feel about their city.

For the study, economic capital of Madhya Pradesh, Indore city, has been chosen. This is having vibrant characteristics in terms of economy, social life, etc. Also, it has been ranked as 1st in list of India's cleanest city from last 5 years as per the Swachh Survekshan. During smart city mission, corporation has used different social media platform for collecting responses from the people for preparation of smart city proposal.

Indore Smart city is also well recognized at national level for its command-and-control center for traffic monitoring, weather, etc. In the study, various levels of city are considered, i.e., data is extracted for complete city and further assessment are done at ward level.

17.3.3 Data Collection/Extraction

For the collection of the dataset, the application programming interface of Facebook and Instagram has been used. Due to privacy concerns, dataset has been extracted from the open profiles only. Approximately 40,000 posts were extracted from these platforms. Once data is extracted from this platform, the cleaning of data is done using the python language. Since based on the extracted dataset, it was difficult to conclude the result, so later a Ward was selected for further study. Additional datasets were collected through digital tool.

17.3.4 Data Analysis

In this research, three kinds of analysis were performed using different algorithms. Each analysis specifically focuses on the spatial aspect of the particular location. Using clustering algorithms, different activities are clubbed together in the largest domain, then the mapping of location points is done through Google Maps and GIS software. Later, the activity was superimposed on the land-use plan of the selected ward and changes were identified.

For the selected area, Point of Interest (POI) of the people was identified by calculating the frequency of posts for a particular location. Using text-based classification algorithms, text associated with the posts in the form of hashtags was analyzed to identify the issues in the particular locality.

Sentiment mapping was done using valence aware dictionary for sentiment reasoning (VADER) algorithms which help to identify the polarity of the statement. The textual contain associated with the post is categorized as positive, neutral and negative and spatially mapped out. Further, the sentiment was linked with the activity associated with the post to identify which category has a positive impact on the environment.

17.3.5 Limitations of the Study

This research mainly focuses on how the crowdsourced dataset in the form of images and hashtags can be used for understanding the city at the micro-level. This research mainly used a text-based classification method for analysis. Due to the availability of a limited dataset, intermediate points for sentiment mapping are interpolated with the help of GIS. Also, hashtags are considered as the base for classification which may exclude a certain age group because senior citizens don't use such kinds of tags with their posts. In this study, intermediate gaps of data are fulfilled through qualitative survey.

17.4 OBSERVATION AND INFERENCES

17.4.1 Activity Mapping

Activities are identified through the hashtags which are associated with the post. There are different sets of hashtags, so using K-mean algorithms, activity mapping was done by making a cluster of the different activities, i.e., Commercial activities like shopping, salons, gyms, and similar kinds of words are searched among the complete data set and exported into a single category. During extraction, it was identified that the city has the maximum number of food posts, so all posts containing food as a keyword were extracted and a separate classification was made.

From Figure 17.2, it is clear that most of the commercial activity is concentrated at the center of Indore city. It can also be seen that the cluster of food shops is also much higher in some particular areas. Vijay Nagar is also one of the clusters of commercial activity as well as food activity apart from the core of the city. From the qualitative interaction with the residents of Indore city, it is known that this area has developed a lot in the last 4–5 years. Similar to this, ward-74 has also shown variation. Therefore, for further study at the micro-level ward-74 has been selected.

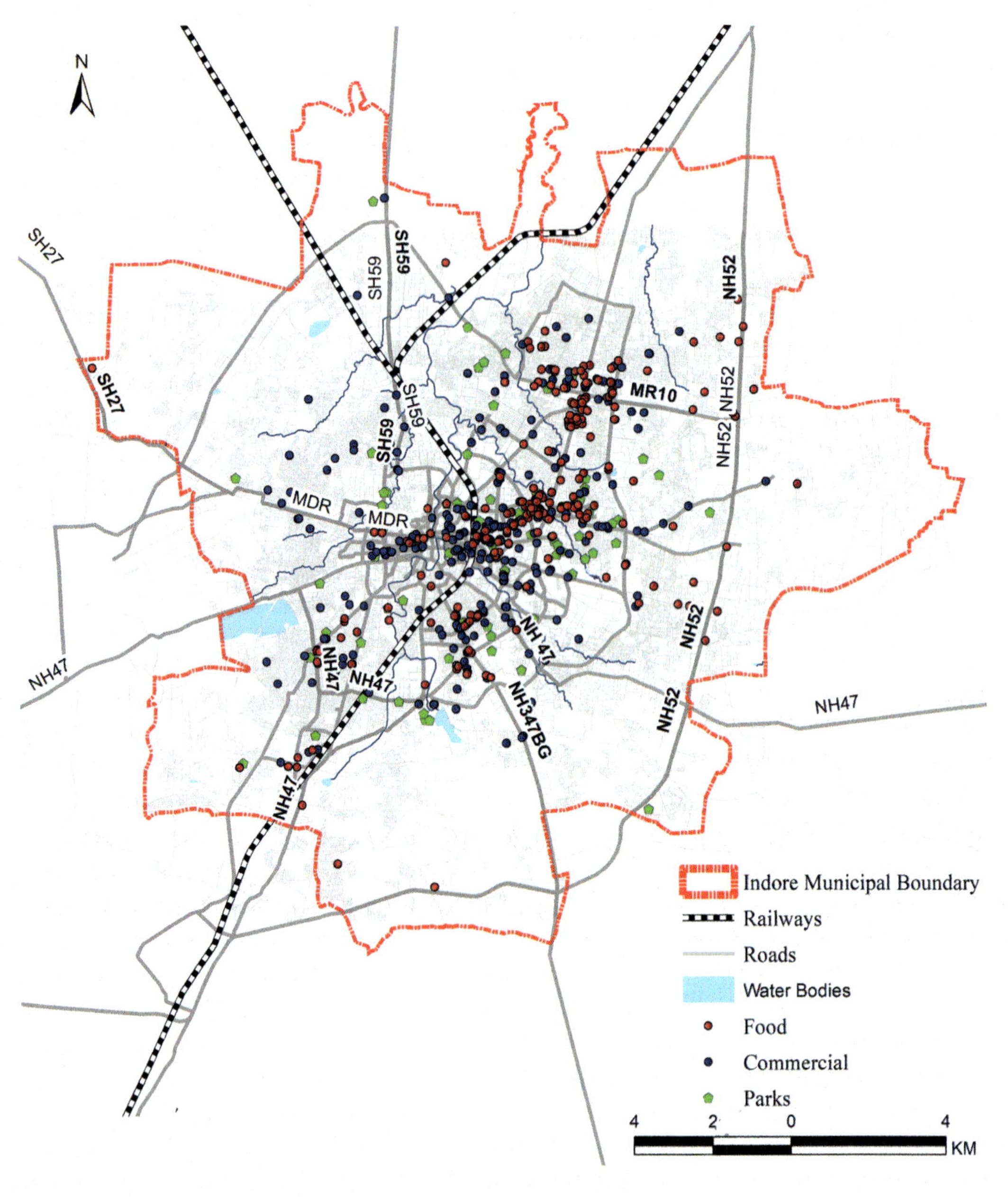

FIGURE 17.2 Activity mapping for Indore city (Source: Author).

17.4.2 Landuse Change Detection

Land use of ward-74 has a mainly residential character as well as public and semi-public uses as shown in Figure 17.3. But when we superimpose the activity mapping over the land use plan, then it can be seen that there are commercial activities taking place and residential land use has changed to commercial.

It can be seen from Figure 17.3, Bhawarkua Chouraha has become one of the commercial points including activities like coaching classes, food/café houses and other stationery related shops. It acts as the center of attraction for other commercial activities to change the character of the ward. Along the road, most of the residential land use has been changed to commercial. These changes have been seen in the last 2–3 years. Due to an increase in the coaching institutes in this area, it has become a hub for students. The facilities which are required by students are now available in the nearby vicinity. Some of the residences have become hostels and PGs for students. Regional Park is also located in this ward. This area is allocated for public and semi-public uses, but the municipal corporation of Indore has changed it to a recreational area.

It can easily be understood that due to certain specific type of activity at a particular location, characteristics of that place changes. This area has now become the student hub of Indore. Due to a coaching institution and youth crowds day by day, the number of food shops are also increasing in this area. It can be concluded from Figure 17.3, land use has changed from the prescribed plan and lots of commercial activities are ongoing in this particular area as well as it is acting as the center of attraction for other kind of activities which are in the multiplier of the needs of that particular area.

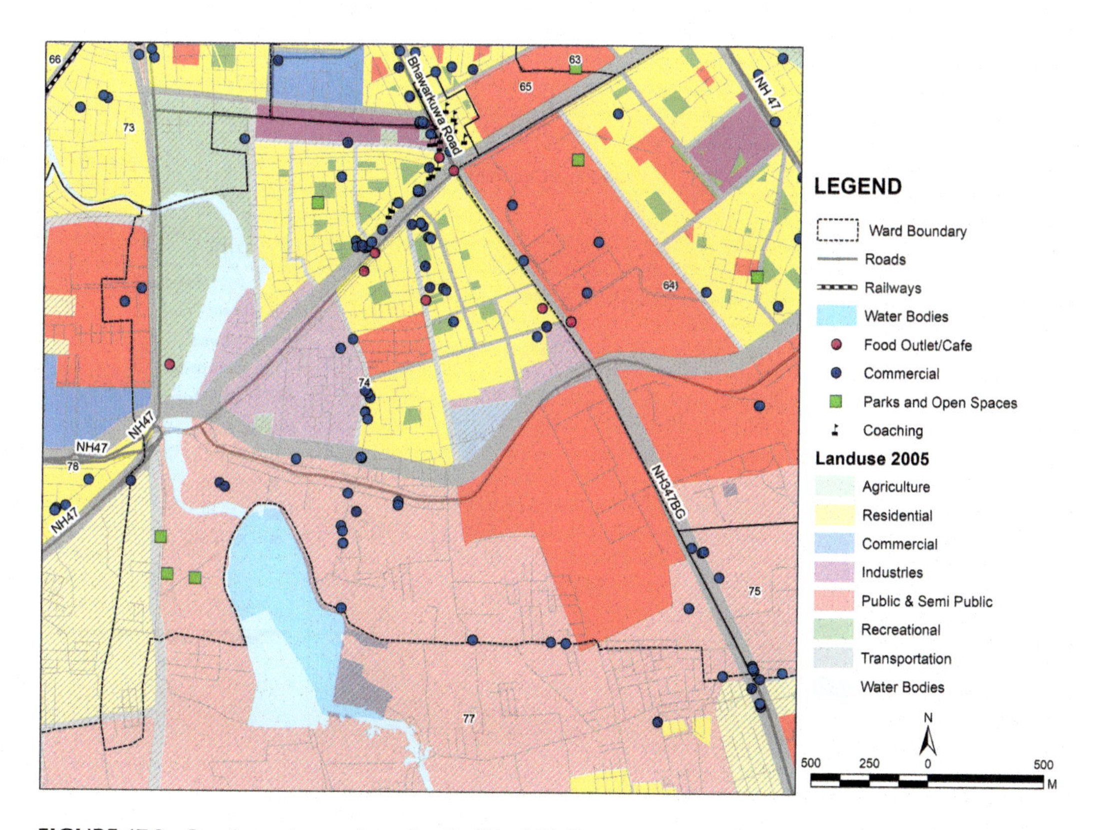

FIGURE 17.3 Land use change detection for Ward-74 (Source: Author).

17.4.3 Point of Interest (POI)

POI of the people is considered as the most socially active area of any neighbourhood. For POI identification at a granular level, data extracted from these platforms were not enough; therefore, a primary survey was conducted for more accurate and effective results. Based on the frequency of occurrence of a particular location in the cumulative dataset, POI is identified.

From Figure 17.4, it can be seen that Bhawarkua Chouraha is one of the important POI. This is because most of the coaching/training institutes for preparation of competitive examinations are situated along with this. Furthermore, other additional facilities for students are also available at this point. However, some hashtags talk about traffic congestion and accidents at this location. Tinku's cafe is another POI. This place is famous for its delicious food. For most of the students, this café is the first choice. Also, cafe has a multiplier effect which leads setting up of other cafes. During the field visit, it was understood that around 10–15 new food outlets have come up within a few years after this cafe. Regional Park is a recreational area developed by Municipal Corporation. This park has various fun activities, like boating where people spend their leisure time. This place consists of tags related to happiness, hangout, greenery, fun, boating, etc.

17.4.4 Sentiment Mapping

It is visible from Figure 17.5 that an activity like food outlets has more positive reaction or sentiments concerning another area. Also, recreational area like the regional park has more positive tags as compared to the nearby area. Due to the unavailability of intermediate data points, it is difficult to get the complete scenario, but this can be very helpful when planning that particular area. It also helps in planning other areas with similar characteristics.

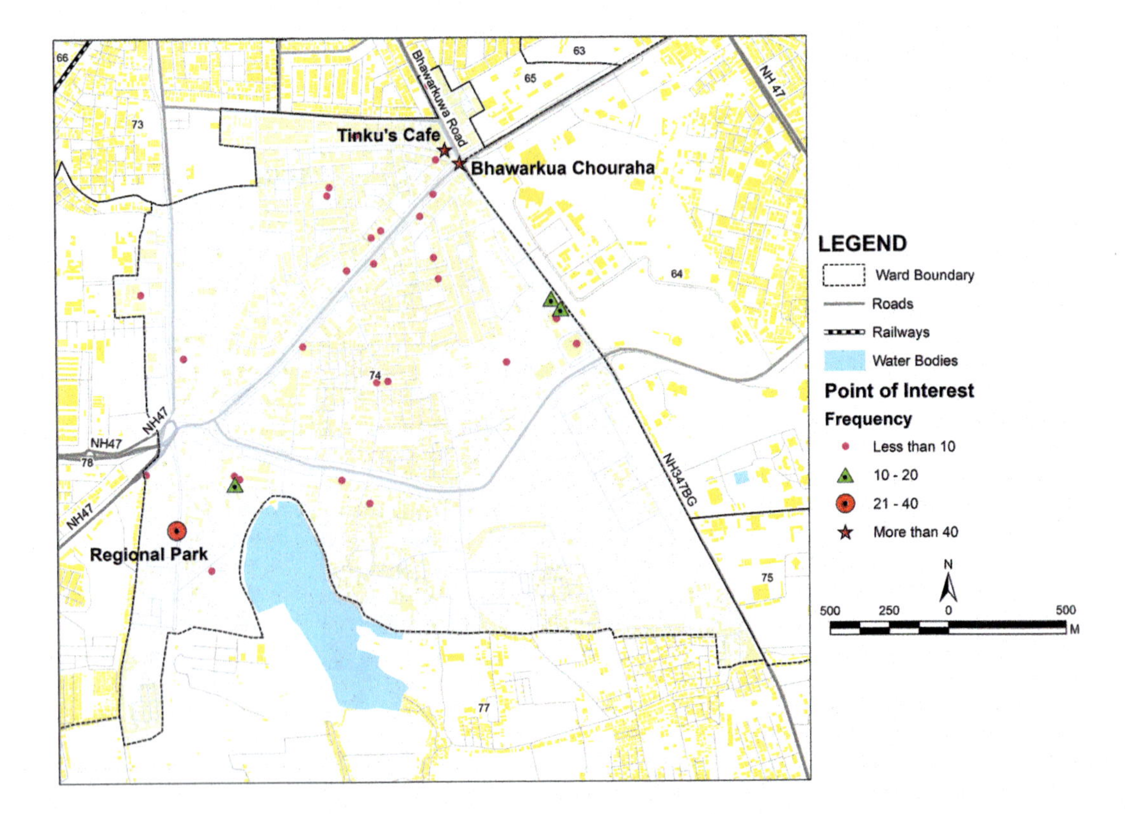

FIGURE 17.4 Point of interest for Ward-74 (Source: Author).

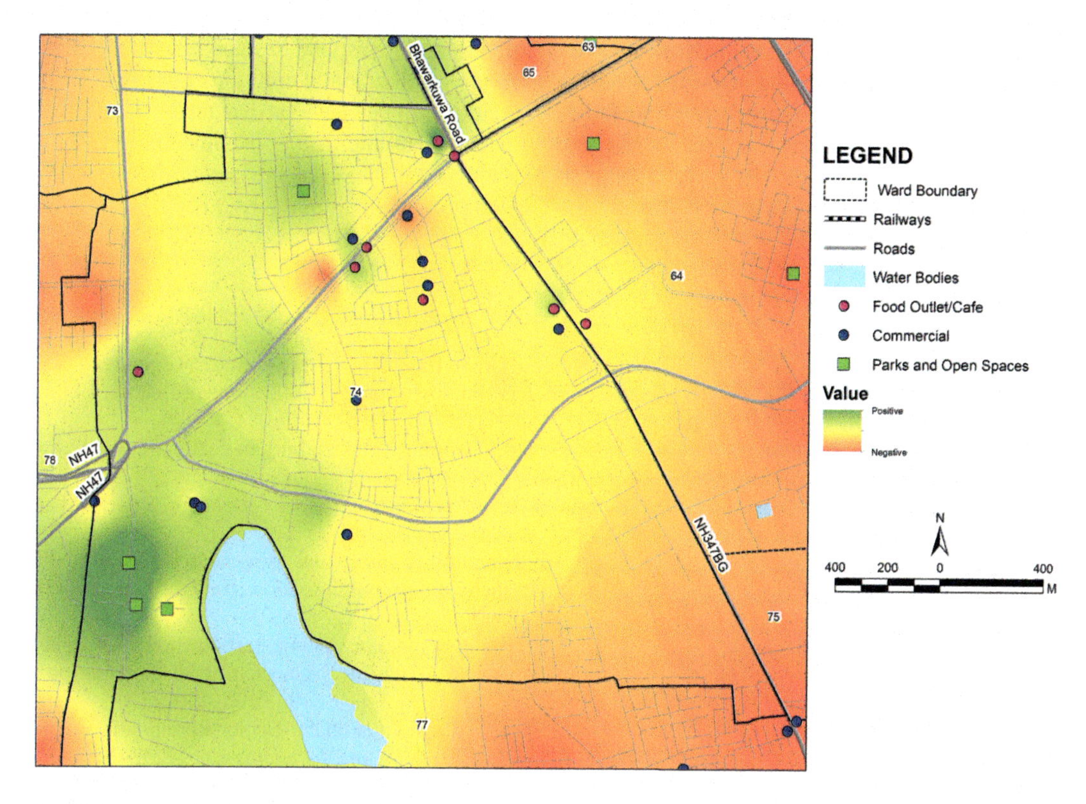

FIGURE 17.5 Sentiment mapping (Source: Author).

17.5 CONCLUSION AND WAY FORWARD

LI can be effectively used for understanding the urban dynamics. Crowdsourced dataset with location information can be utilized as raw data for extracting spatial information using advance geospatial technology. This information can effectively be used by urban local bodies to plan as per the needs of the people. Currently, the non-traditional technique focuses on active collection of datasets like primary surveys, but crowdsourced dataset enriches the passive dataset. It is an indirect way of participation. Mapping of this dataset regularly can answer the various question like which area is developing very rapidly, and what the characteristics of that area are. This can also act as complementary layer for preparing the master plan for the city. The location-based crowdsourced dataset has lots of potential that can help the authorities to identify the needs of the people, monitor the city activity and plan for the local area.

People's daily lives are increasingly reliant on social networking sites. These mediums are used to express their ideas, invitations, and feelings in the form of photographs, videos, and text. Different stakeholders use these datasets to help them expand their businesses. A dataset like this may be useful in the domain of urban planning. As can be observed, crowdsourced geotagged photos with textual captions or hashtags can be used to explore granular details of neighborhood, implying that there is an opportunity or issue that needs to be solved. A micro-level strategy like this makes spatial planning easier. A more people-centered methodology raises the likelihood of societal acceptance of subsequent planning interventions. It further bridges the divide between the metropolitan city government and the general public

A crowdsourced dataset can also be used to explain how people respond to some kind of incident, product, or location. It should be remembered that spatial visualization of emotions associated with various places assists developers in recognizing the relationship between various land uses and the feelings associated with them. It will assist in the development and design of areas that promote positivity in the community, thus increasing people's efficiency. The frequent gathering of crowdsourced data from various social media platforms will add dynamism to the traditionally stagnant nature of planning.

REFERENCES

1. Yandong Wang, T. W.-H. (2016). Mapping Dynamic Urban Land Use Pattern with Crowd Sourced Geo Tagged Social Media (Sina - Weibo) and Commercial Points of Interest Collections in Beijing, China. *Sustainability*, 8, 1202.
2. Huston, R. (2020). People, places, and cultures, 8 6.1 defining cities and urban centers, https://open.library.okstate.edu/culturalgeography\/chapter/6-1/.
3. Higinio Mora, R. P.-D.-P.-S. (2018). Analysis of Social Networking Service Data for Smart Urban Planning. *Sustainability,* 10, 4732.
4. Marsal-Llacuna, M.-L., & Gesa, R. F. (2016). Modeling citizen's time- use using adaptive hypermedia surveys to obtain an urban planning, citizen-centric, methodological reinvention. *Time & Society*, 25(2), 272–294.
5. Wilson, A., Tewdwr-Jones, M., & Comber, R. (2019). Urban planning, public participation, and digital technology: App development as a method of generating citizen involvement in local planning processes. *Environment and Planning B: Urban Analytics and City Science*, 46(2), 286–302.
6. Liu, B., Yuan, Q., Cong, G., & Xu, D. (2014). Where your photo is taken: Geolocation prediction for social images. *Journal of the Association for Information Science and Technology*, 65, 1232–1243.
7. Bollier, D. (2016). The city as platform-How digital networks are changing urban life and governance. The ASPEN Institute.
8. Rathore, M. M., Paul, A., Ahmad, A., & Rho, S. (2015). Urban planning and building smart cities based on using inter of things using big data analytics. *Computer Networks*, 101, 63–80.
9. Batty, M. (2013). Big data, smart cities and city planning. *Dialogues in Human Geography*, 3(3), 274–279.
10. Frias-Martinez, V., & Frias-Martinez, E. (2014). Spectral Clustering for sensing urban land use using twitter activity. *Engineering Application of Artificial Intelligence* 35, 237–245.
11. Geoblink (2022). What is location Intelligence, https://www.geoblink.com/what-is-location-intelligence/.
12. esri (2022). Location Intelligence Driving digital transformation, https://www.esri.com/en-us/location-intelligence.
13. Statista. (2019). Retrieved November 2019, from https://www.statista.com/statistics/278407/number-of-social-network-users-in-india.
14. Ericsson Mobility Report (2017). https://www.ericsson.com/en.
15. Noshir Kaka, A. M. (2019). Digital India: Technology to transform a connected nation.
16. Evans-Cowley, Jennifer & Griffin, Greg. (2011). Micro-Participation: The Role of Microblogging in Planning. SSRN Electronic Journal. 10.2139/ssrn.1760522.
17. Campagna, M. (2014). The geographic turn in social media: Opportunities for spatial planning and geodesign. In: Murgante, B., Misra, S., Rocha, A.M.A.C., Torre, C.M., Rocha, J.G., Falcão, M.I., Taniar, D., Apduhan, B.O., & Gervasi, O. (Eds.), *Computational Science and Its Applications – ICCSA 2014* (pp. 598–610). Berlin: Springer.
18. Tasse, D. & Hong, J. I. (2014). Using social media data to understand cities. In: Thakuriah, P., Tilahun, N., & Zellner, M., (Eds.) *Proceedings of NSF Workshop on Big Data and Urban Informatics* (pp. 64–79). Chicago, IL: University of Illinois Press.
19. Lopez-Ornelas, E., Abascal-Menaa R., & Zepeda-Hernández, S. (2017). Social media participation in urban planning: A new way to interact and make decisions. In: *The International Archives of the Photogrammetry, Remote Sensing and Spatial Information Science, 2nd International Conference on Smart Data and Smart Cities*, 4–6 October 2017, Puebla, Mexico, vol. XLII/W3, pp. 59–64.

20. Bo Pang, L. L. (2008). Opinion mining and sentiment analysis. *Foundations and Trends in Information Retrieval*, 2, 1–135.
21. Fathullah, A., & Willis, K. (2018). Engaging the senses: The potential of emotional data for participation in urban planning. *Urban Science*, 2(4), 98. https://doi.org/10.3390/urban-sci2040098.
22. Niksic, M., Tominc, B., & Gorsic, N. (2018). Revealing resident's shared value through crowdsourced photography: Experimental approach in participatory urban regeneration. *Urbani izziv*, 29, 1–14.

18 A Hybrid Model for the Prediction of Land Use/Land Cover Pattern in Kurunegala City, Sri Lanka

Mohamed Haniffa Fathima Hasna
University of Peradeniya, Sri Lanka

Mathanraj Seevarethnam and Vasanthakumary Selvanayagam
Eastern University, Sri Lanka

CONTENTS

18.1 INTRODUCTION

Urban land use is a complex process that is influenced by various drivers, including physical and human factors (Huang et al., 2008). Land use change is the process of how human activities alter the natural environment, which refers to how the land has been utilised, emphasising its functional role in economic activity (Paul & Rashid, 2016). Land use/land cover (LULC) changes are more dynamic in space and time (Rathnayake, Jones, & Soto-Berelov, 2020), which arise as an effect of interactions between socioeconomic and natural elements on various magnitudes (Han, Yang, & Song, 2015). Urbanisation and urban growth are major issues in land use change that are usually accompanied by poor planning (Deep & Saklani, 2014; Samat, Hasni, & Elhadary, 2011; Shahraki et al., 2011). The rapid increase in the urban population becomes one of the main confrontations in developing cities that accelerates the pace of change in LULC. Thus, increasing socioeconomic demands continue to increase pressure on LULC (Uduporuwa, 2020).

DOI: 10.1201/9781003270928-24

Economic development has led to massive changes in the earth's surface, which have been visible in recent decades (Vinayak, Lee, & Gedem, 2021). Haphazard modification of LULC in cities of developing countries has many adverse implications for both urban dwellers and the environment (Uduporuwa, 2020). The recent trend implies that the change rate will be significant in the future (Vinayak et al., 2021). As a result of increased pressure, unplanned and uncontrolled changes occur in LULC (Uduporuwa, 2020). These spontaneous changes in an urban area are the source of many socioeconomic and environmental challenges (Ranagalage et al., 2020). The lack of planning leads to several negative effects on environmental balance, such as deforestation, reduction of infiltration and destruction of fertile farmland, and conversion of bare and forest land into a man-made environment (Quan et al., 2015; Yu, Zang, Wu, Liu, & Na, 2011).

Comparatively, the high urban population growth in Asian cities has become a major concern in the world (Ranagalage et al., 2020). LULC and its consequences are evident in almost all sectors associated with population growth. Thus, as a developing nation, Sri Lanka has been experiencing a similar pattern of emerging global urbanisation, characterised by rapid growth in cities in developing countries (UN-Habitat, 2018). The country encounters many challenges due to LULC changes and urban population growth. However, statistics on the trend of Sri Lanka's urban population indicate different dimensions. Rapid urban sprawl is a significant feature of urban expansion in many Sri Lankan cities (UN-Habitat, 2018). This causes cities to increase in both physical size and number faster than earlier, exacerbating future sprawling issues (Amarawickrama, Singhapathirana, & Rajapaksha, 2015). Accordingly, Kurunegala city, the capital of the North Western Province, has also undergone substantial changes in LULC in recent decades due to urbanisation in conjunction with the rapid urbanisation process in Sri Lanka.

Kurunegala is densely developed with commercial and administrative capitals in Sri Lanka, and it has direct connections to other major capital cities and towns (Cho, 2020). Its contribution to the national economy is immense. From a development perspective, since Kurunegala is a historic city, its significance becomes a major concern. Compared to the national and district levels, Kurunegala has a high population growth rate, and an agglomeration of work opportunities and services in the core city area attracts more people into the area. Thus, population growth, daily mobility (Kurunegala has a high level of daily mobility compared to other major parallel capital cities), commercial tendency, and national development initiatives are the key characteristics that have caused dynamics in LULC change in Kurunegala (UDA, 2019). The spatial pattern indicates that out of the total built-up area, the proportion of land use for municipal council (MC) activities is higher in Kurunegala city (UN-Habitat, 2018). With the population increase, Kurunegala also shows a growth in the land size of the urban areas. It is noted that urban area, which was 0.51% in 1995, has increased to 3.54% in 2017. A similar pattern is also evident in semi-urban areas (0.77% in 1995 – 3.57% in 2017) (UN-Habitat, 2018). It is clear from the LULC of Kurunegala city limit from 2001 to 2017, that significant growth has been identified in commercial usage rather than residential development (UDA, 2019). Since Kurunegala city has been identified as an economic hub along the Colombo-Trincomalee economic corridor, economic development strategies such as expanding the urban and industrial town areas are important priorities in this plan (UDA, 2019), the consideration of LULC changes and their consequences would be more significant. Therefore, concern on LULC proportional to the growing population becomes a development priority; thus, this area should be concentrated to study the LULC patterns, changes and predictions for the sustainable growth of the city.

While research on LULC is critical to revealing the changes and their consequences, the studies have sometimes been hampered in Sri Lanka by the lack of available data (Rathnayake et al., 2020). However, the urban population scenario for Sri Lanka is underestimated (UN-Habitat, 2018), and the urban expansion outside the municipal and urban councils is not well quantified so far. This is one of the major causes in assessing LULC and its changes in the urban areas of Sri Lanka. Furthermore, ambiguity in defining urban areas, their sizes, and the changes that have occurred over time are major limitations in the Sri Lankan urban planning process (UN-Habitat, 2018). Therefore, to understand and assess the various social, economic, and environmental consequences of these changes, LULC information from past to present and possible future changes is crucial (Parsa & Salehi, 2016). It is vital in several sectors to achieve sustainable development (Vinayak et al., 2021). Thus, environmental change analysis and natural resource management require accurate and up-to-date LULC change information for proper planning (Rathnayake et al., 2020). Identifying and mapping spatio-temporal LULC changes in cities is crucial for various applications in urban studies, such as urban sprawl monitoring, urban management analysis, environmental and land use planning, and mitigating urban disasters (Ranagalage et al., 2020; Uduporuwa, 2020). The Landsat Archive for Sri Lanka has been available since 1988 (Rathnayake et al., 2020), and advances in spatial analysis and remote sensing techniques have provided scholars with excellent tools for map-making and investigating LULC changes (Suthakar & Bui, 2008).

Land use modelling has evolved from a spatial mathematical specification of linear relations to spatially explicit dynamic simulations (Meiyappan, Dalton, O'Neill, & Jain, 2014). Furthermore, they allow feedback between the subsystems of the model and account for a diverse set of institutional and ecological influences, making it a useful tool for understanding the various drivers of urbanisation (Aronoff, 2004). Understanding and predicting a pattern are the two main goals of modelling (Gleicher, 2016). Predictive models absorb a collection of facts to simulate possible outcomes, whereas descriptive models reflect what is already available (Verburg, Kok, Pontius, & Veldkamp, 2006). Temporal models build a bridge between the two by defining transition processes that have been seen over time and provide a basis for future predictions (Aronoff, 2004). Further, land use models are currently used to analyse the causes and implications of land use dynamics. There are different modelling traditions, such as optimisation models, spatial interaction models, integrated models, econometric and statistical models, and other types of models, like natural science-oriented modelling and GIS modelling (Briassoulis, 2020).

The prediction models for LULC can be categorised into three groups: GIS-based models, machine learning models, and hybrid models (Márquez, Guevara, & Rey, 2019). The LULC patterns are generally predicted with GIS-based models. For instance, the CLUE (Verburg et al., 2002), LTM (Pijanowski et al., 2014), STSM (Daniel, Frid, Sleeter, & Fortin, 2016) and SLEUTH (Saxena & Jat, 2019) models are some of the GIS-based models in land use and urban studies. GIS-based models have encoded the datasets to produce spatial layers for predictor variables created from a series of LULC maps. A time series image is generated based on the quantity of changing scale with future land use (Márquez et al., 2019). Machine learning models use computer algorithms for simulating and predicting LULC. They are an important component of artificial intelligence. Logistic regression (Park, Jeon, & Choi, 2012), Markov chain (Huang et al., 2008), multi-agent model (Rui & Ban, 2010), cellular automata (Deep & Saklani, 2014), SVM (Thakur, Kumar, & Gosavi, 2020) and ANN (Silva, Xavier, Silva, & Santos, 2020) are often used models in urban studies. These models can be run without human assistance, and they have the feature of using drivers of LULC in the prediction (Aburas, Ahamad, & Omar, 2019). Hybrid models are systems-based theories with multi-scale features. LR-MC-CA (Arsanjani, Helbich, Kainz, & Boloorani, 2013), CLUE-S-MC (Han et al., 2015), MLP-MC (Mirici, Berberoglu, Akin, & Satir, 2018), ANN-CA (Yang, Chen, & Zheng, 2016) and CA-MC (Mathanraj, Rusli, & Ling, 2021) are some hybrid models identified from past studies. Different LULC scenarios were investigated at different spatial scales with these models. These models are well-fit with other models and are used to find the association between LULC change and driving factors (Márquez et al., 2019). Understanding the relationship between LULC change and its driving forces is critical in land use planning and management, as is being able to reasonably estimate future land demand (Han et al., 2015).

Various models have been used in urban studies of Sri Lankan cities. A GIS-based logistic regression model was applied to depict the urban fringe growth of Colombo (Weerakoon, 2017), and a cellular automaton model was used to assess urbanisation trends in the Colombo Metropolitan Region (Divigalpitiya, Ohgai, Tani, Watanabe, & Gohnai, 2007). These studies mainly focused on a large city in the western part of Sri Lanka, which is a rapidly developing region, and much more

LULC is the evidence. Simultaneously, the CA-Markov model and the Markovian transition estimate were used to predict LULC patterns in Batticaloa Municipal Council, Sri Lanka (Mathanraj et al., 2021; Zahir et al., 2021). However, the LULC in the Kurunegala city area was studied to check urban heat islands, mainly using spatio-temporal analysis (Ranagalage et al., 2020). However, this study was not considered a prediction of LULC for future planning. Thus, the present study focused on a hybrid model, such as MLP neural networks and CA-MC, which was not previously considered in the studies of Sri Lankan cities. Therefore, this study aims to apply a hybrid model to predict the LULC pattern in Kurunegala city, Sri Lanka.

18.2 METHOD AND MATERIALS

18.2.1 The Study Area

Kurunegala is the capital of the north-western province of Sri Lanka (see Figure 18.1). It has been bordered by Anuradhapura in the north, Matale in the east, Gampaha and Kegalle in the south, and Puttalam in the west. While the city serves as the administrative capital of the north-western province, educational and health services are provided within a 50 Kilometre radius, including other urban services (UDA, 2019). Kurunegala belongs to the intermediate climatic zone in the lowland region with a tropically hot and humid climate with an elevation of nearly 300 meters. The climate of the city remains hot throughout the year, and air temperature shows a general increase, parallel to the national trend. This situation is aggravated by the rock outcrops in and around the city during the day (Cho, 2020). The average rainfall in this area varies between 1,750 and 2,500 mm, and the

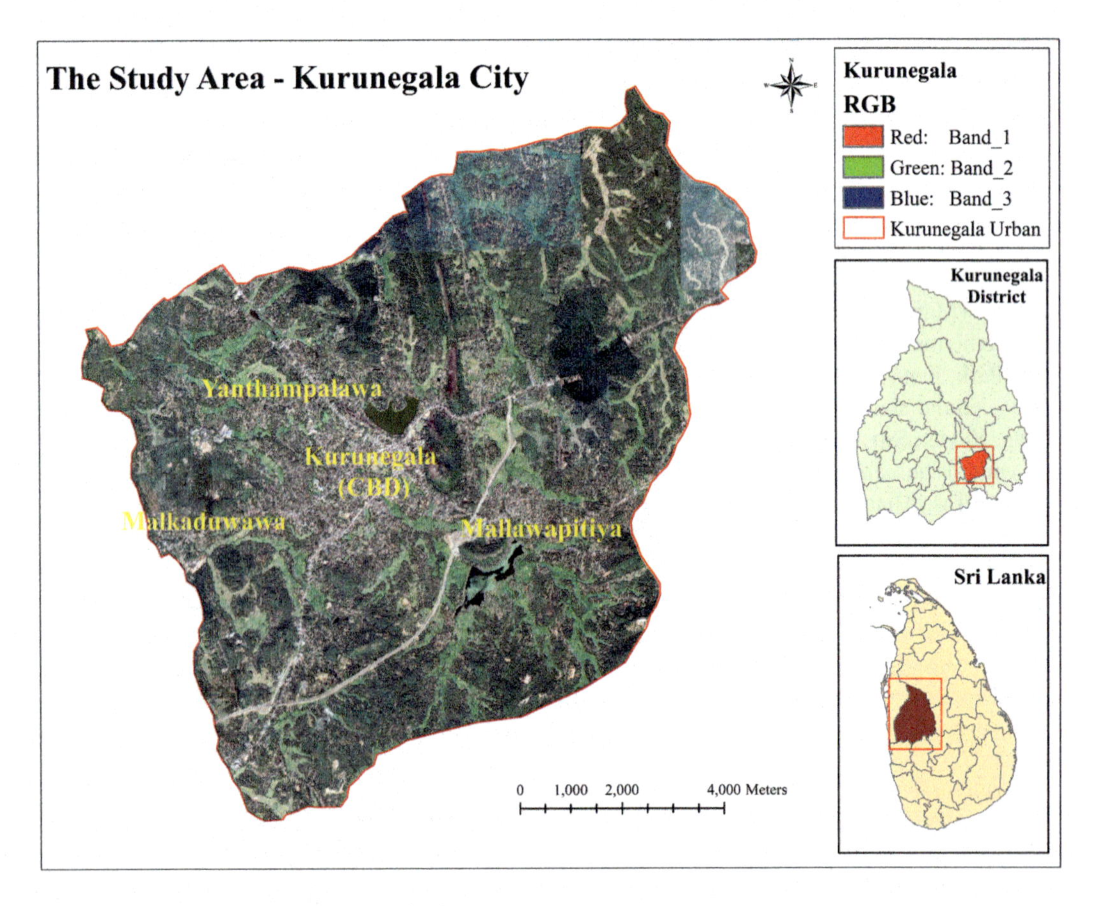

FIGURE 18.1 Study area—Kurunegala city.

average temperature is 27°C (UDA, 2019). The total population of the study area was ~119,120 in 2019 (Divisional-Secretariat, 2019).

The LULC of Kurunegala mainly comprises of built-up land (commercial, residential, religious, and recreational places), agricultural lands with major cultivation (rice and coconut), forest land, water bodies and rock outcroppings. Corresponding to the LULC changes of Kurunegala, a rapid change can be seen over the last two decades (UDA, 2019). As an important urban centre in Sri Lanka, it would be a valiant effort to study the LULC patterns and changes in the city and surrounding areas. Thus, the major urban areas in Mallawapitya, Maspotha and Kurunegala divisional secretariat divisions were considered in this study.

18.2.2 Data Source

Landsat satellite images were used for the classification of LULC patterns for 2001, 2011 and 2021 in Kurunegala city, which was downloaded from the Earth Explorer, United States Geological Survey (see Table 18.1).

18.2.3 Data Pre-Processing

The downloaded bands of satellite images were overlaid using a composite band tool in ArcGIS 10.6.1 to get an RGB (red, green, blue) natural colour image. The composited images which were in WGS 84 world geodetic coordinate system that was converted to Kandawala Sri Lanka local coordinate system using project coordinate tool. These images employed contrast stretching, noise reduction, and histogram adjustment to enhance the image quality. Further, Kurunegala urban area was digitised with an existing city map and delineated the study area boundary. All three (3) satellite images were clipped based on this boundary for the LULC classification. In addition, the elevation for the city was generated from Google Earth Pro and converted to a GPX file using the GPS visualiser. Then, the GPX file was imported to DEM in ArcGIS to map the elevation of the city. This elevation map was used to find the slope of the areas, which were used for the simulation of LULC. Also, the major roads (highway and secondary roads) of the city were digitised using Google Earth Pro and the existing road network map of the city. These road maps were employed in the Euclidean distance tool to find the distance from roads in order to understand the impact of roads on LULC changes.

18.2.4 Data Analysis

18.2.4.1 Supervised Classification

LULC patterns for 2001, 2011 and 2021 were classified with ArcGIS's Maximum Likelihood classification tool. The clipped Landsat images were converted to false-colour images (at least one invisible wavelength is used in the image) since the band is signified in RGB. Then, these images were utilised for collecting the training samples to classify the LULC. These training samples were evaluated with the histogram, statistics, and scatter plot tools available in the Training

TABLE 18.1
Details of Satellite Images

Name	Date Acquired	Path/Row	Resolution (m)	Bands Used
Landsat 7 ETM+	2001/12/11	141/055	30	Band 1–5, 7
Landsat 7 ETM+	2011/05/13	141/055	30	Band 1–5, 7
Landsat 8 OLI/TIRS	2021/10/23	141/055	30	Band 1–7

Sample Manager in ArcGIS. These tools were used to find if the samples overlapped with others or not. Meanwhile, it reveals the normality of the collected samples, whether they are distributed as normal or non-normal, which is more useful for accurate classification results. A signature file was created for each period using Create Signatures tool, which was used in the maximum likelihood classification analysis. The signature file is the multivariate statistics for the LULC classes that include the mean, number of cells, and variance-covariance matrix. Accordingly, five (5) major LULC classes were obtained in Kurunegala city, such as built-up, forest, agriculture, water bodies and rocks.

Accuracy assessment is the quality of the results of the classified images taken from the remote sensing data. The accuracy of the classified images was evaluated using ground truth samples acquired from Google Earth Pro and Land use images from the Department of Survey, Sri Lanka. Producer accuracy, User accuracy, Kappa accuracy and Overall accuracy were computed with 263 ground samples distributed as follows: 84 for built-up, 69 for agriculture, 57 for the forest, 31 for water bodies and 22 for rocks. The overall accuracy of about 85% is good enough for further analysis (Bhatta, 2010). Also, the Kappa accuracy varies as excellent (0.80 and 1.00), good (0.60 and 0.80), moderate (0.40 and 0.60), fair (0.20 and 0.40) and poor (<0.20). The index above 0.80 is considered for validating classified LULC maps (Mathanraj, Ratnayake, & Rajendram, 2019). These validated images were used for the simulation of LULC patterns integrating with several explanatory variables.

18.2.4.2 Selection of Drivers Variables

Driving forces have been contributing to a major change in the LULC patterns in the city. LULC classes are the dependent variables, and driving forces are the independent variables. Multiple spatial forces (independents variables), such as elevation, slope, distance to highway and distance to secondary road (see Figure 18.2), were used in the analysis for the prediction of LULC pattern in 2031. Elevation and slope are essential biophysical variables to predict the LULC pattern. From the Cramer's V test, the selected drivers were identified as significant forces of LULC changes in the city. The overall and individual value of Cramer's V test is presented in Table 18.2. The value of Cramer's V is >0.15 is useful, while a value higher than 0.4 is good (Hamdy, Zhao, Osman, Salheen, & Eid, 2016; Hamdy, Zhao, Salheen, & Eid, 2017). Thus, the variables >0.15 were used in the MLP model for the simulation.

18.2.4.3 Multi-Layer Perceptron Neural Network and CA-Markov Model

The transition potential for LULC was obtained using the multi-layer perceptron neural network (MLP), which is a type of artificial neural network (ANN) method. The MLP is a non-parametric technique with more advantages for modelling a complex functional association of LULC classes. This model is more appropriate for LULC prediction than other models. Many types of research indicate that the ability of the MLP model is higher than other empirical models in the Land Change Modeller (Wang et al., 2016). The MLP model is integrated with the CA-Markov model in the module of Land Change Modeller (LCM) of IDRISI 17.0. This module allows incorporating the driving forces of LULC in the prediction. Elevation, slope, distance to highway and distance to secondary road were some explanatory variables included in the MLP model to develop the transition potential with the earlier (2011) and later (2021) LULC maps. To determine the influences of the variables, the major transitions were processed in the transition potential with the MLP model. This transition map was used in the CA-Markov model to simulate the LULC patterns.

CA-Markov is a widely used model for LULC simulation and prediction (Al-Hameedi et al., 2021). Markov model defines the change of LULC patterns from one period to another to simulate future patterns. A transition probability matrix was computed for two LULC maps from different periods in this area. The transition matrixes for the periods of 2001–2011 and 2011–2021 were obtained with the LULC images for 2001, 2011 and 2021. Of these, a transition matrix for 2001–2011 was utilised to simulate the LULC pattern in 2021 for model validation. Then, the transition

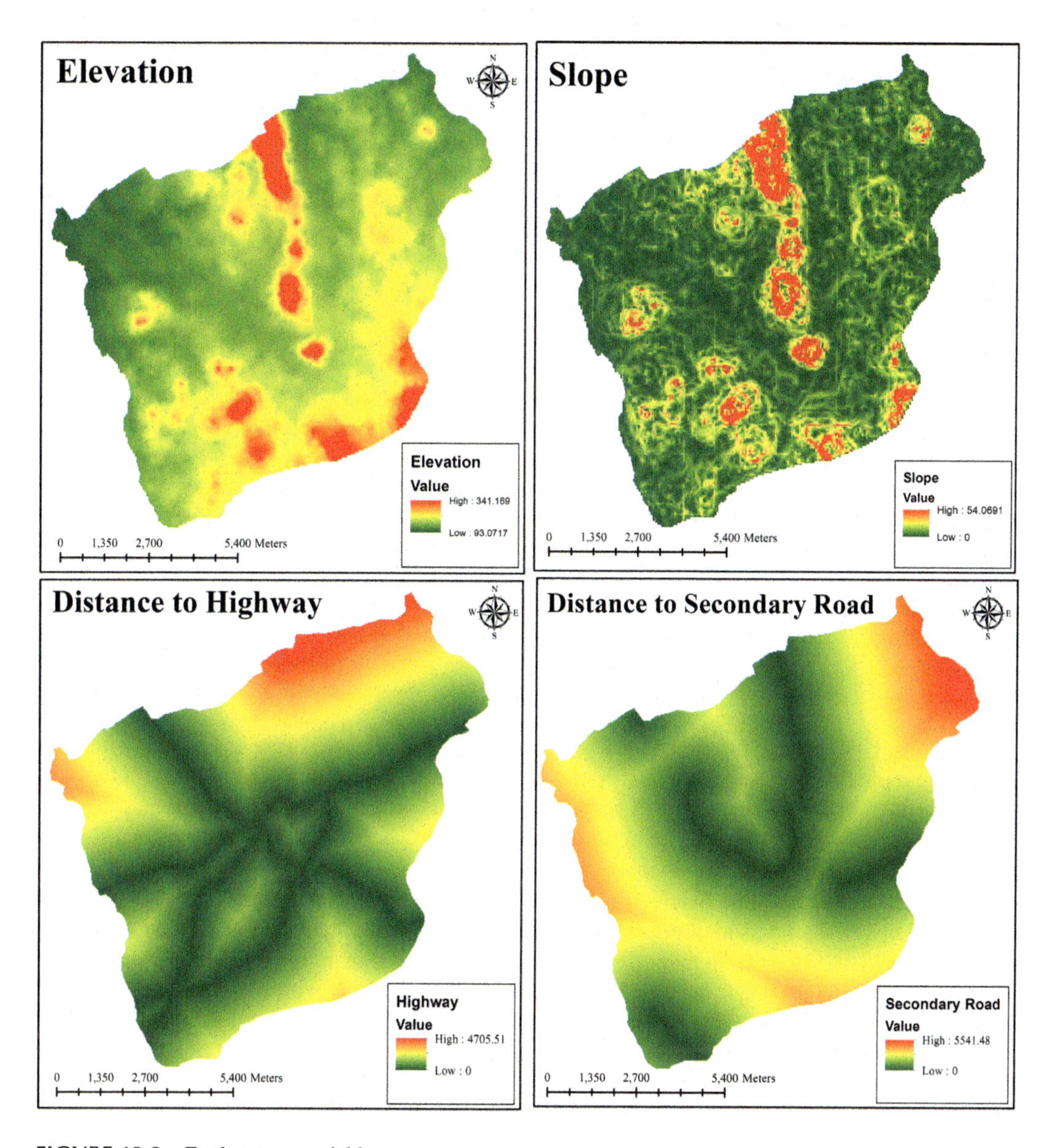

FIGURE 18.2 Explanatory variables.

TABLE 18.2
Cramer's Value for Potential Driving Forces

Driving Forces	Overall V	Built-up	Agriculture	Forest	Water Bodies	Rocks
Elevation	0.2547	0.2001	0.2842	0.2474	0.3138	0.2154
Slope	0.1136	0.1003	0.1193	0.1085	0.1156	0.1042
Distance to highway	0.1669	0.2937	0.1033	0.2699	0.0715	0.0707
Distance to secondary road	0.1589	0.2474	0.0699	0.2667	0.1004	0.0998

matrix of 2011–2021 was deployed to predict the LULC pattern in 2031. A Markov model is calculated as follows (Hamad, Balzter, & Kolo, 2018; Mathanraj et al., 2021):

$$S(t,t+1) = P_{ij} \times S(t) \tag{18.1}$$

Where $S(t)$ is the system status at a period of t, $S(t+1)$ is the system status at a period of $t+1$; P_{ij} is the probability matrix for transition in a state. It is computed as follows the equation (18.2):

$$\|P_{ij}\| = \begin{Vmatrix} P_{1,1} & P_{1,2} & \dots & P_{1,N} \\ P_{2,1} & P_{2,2} & \dots & P_{2,N} \\ \vdots & \vdots & \vdots & \vdots \\ P_{N,1} & P_{N,2} & \dots & P_{N,N} \end{Vmatrix} \tag{18.2}$$

$$\left(0 \le P_{ij} \le 1\right)$$

P is the probability for transition; P_{ij} is converting probability from existing state i to other state j in subsequent time; P_N is the state probability for any period. A less transition is possible near (0), while a great transition has possibilities near (1). Three indicators, such as K_{no}, $K_{location}$ and $K_{standard}$, were utilised to validate the model prediction of LULC in 2021. K_{no} (Kappa for no ability) affords the overall accuracy for the simulation, which is a strongly recommended evaluation method. Further, $K_{location}$ and $K_{standard}$ provide the quantity and location of existing and simulated LULC images (Mathanraj et al., 2021).

The overall method is presented in Figure 18.3. Landsat images were employed with pre-processing tools that were used to classify the LULC patterns for 2001, 2011 and 2021 in the city. These classified images were evaluated with an accuracy assessment to obtain precise maps. These maps were used to find the spatial changing patterns of LULC between 2001–2011 and 2011–2021. Based on these changes, a transition matrix was developed using a Markovian transition estimator. In addition, the driver variables were selected using the Cramer's V test available from IDRISI. The potentiality of variables was identified based on the LULC patterns, which were added to the MLP model to produce the transition map. According to the transition map, the CA-Markov model was performed to create the LULC map for 2021. This map was evaluated with the existing LULC map of the city of 2021 for validation purpose. Since the map for 2021 confirmed an excellent accuracy of simulation, the prediction for the LULC patterns for 2031 was achieved to the city.

18.3 RESULTS AND DISCUSSION

18.3.1 Land Use/Land Cover Pattern

Spatial LULC pattern has primarily been classified as built-up, agriculture, forest, water bodies and rocks for 2001, 2011 and 2021. Accuracy levels were estimated to determine the accuracy of these classified images. Producer accuracy (PA) and user accuracy (UA) were greater than 80% for most LULC classes. However, producer accuracy for agriculture showed 78% in 2001. Simultaneously, producer and user accuracy for water bodies were indicated 78% in 2011. But, the overall accuracy for 2001, 2011 and 2021 was 89.19%, 90.16% and 93.44%, respectively, (see Table 18.3), which indicates a very good precision for all periods. Similarly, the accuracy of the Kappa coefficient showed an excellent agreement with 0.86, 0.87, and 0.91 for 2001, 2011 and 2021, respectively. Therefore, the classified LULC images are significant for further analysis in this study.

Table 18.4 presents the extent of LULC patterns from 2001 to 2021. In 2001, forest area was the dominant pattern with an area of 6,396.31 ha, and the built-up land substituted it in 2021. Concurrently, the built-up area was 19% in 2001, which increased to 44% in 2021. Agricultural

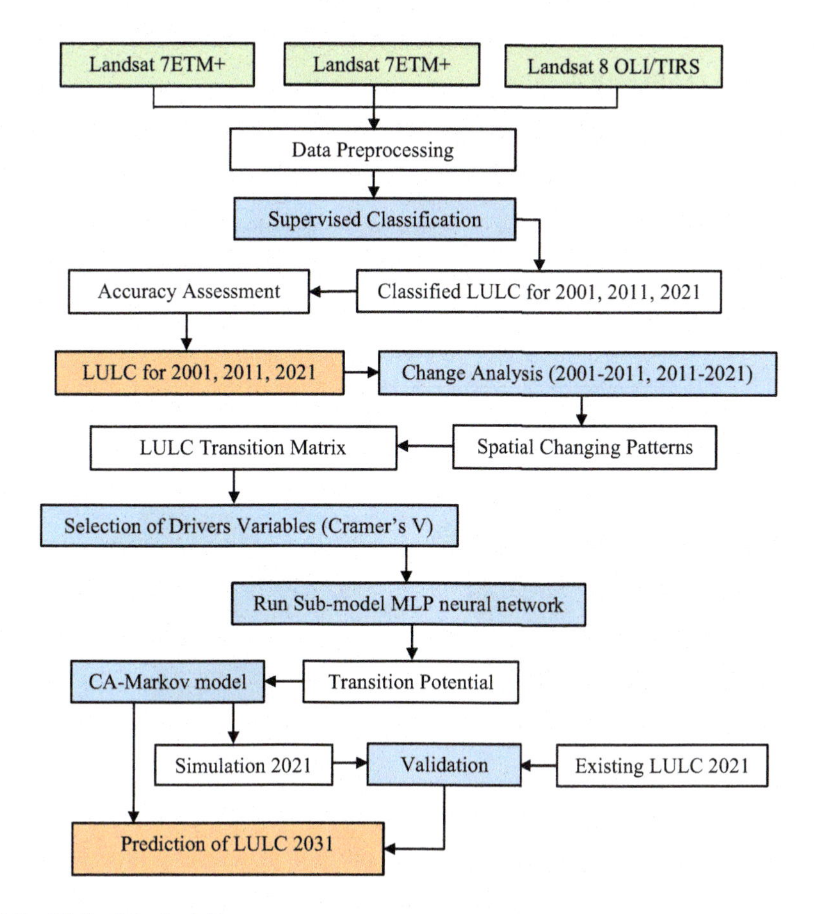

FIGURE 18.3 Methodological diagram.

TABLE 18.3
The Accuracy for LULC from 2001 to 2021

Class Name	2001 PA	2001 UA	2011 PA	2011 UA	2021 PA	2021 UA
Built-up	93	93	92	100	92	100
Agriculture	78	82	87	87	93	88
Forest	92	92	100	86	100	100
Water bodies	89	89	78	78	89	89
Rocks	100	86	100	83	100	83
Overall accuracy	**89.19%**		**90.16%**		**93.44%**	
Kappa accuracy	**0.86**		**0.87**		**0.91**	

land showed a drastic decline in these periods. It had ~2,048.66 ha in 2001, which was reduced to 1,406.20 ha in 2021 (see Table 18.4). Correspondingly, the forest area was 60% in 2001, which has decreased to 40% in 2021. The rocks remain the same, accounts for 1% of total land use over all three periods. Meantime, the water bodies showed a slight change in extent depending on seasonal climatic factors, as Kurunegala receives comparatively more rainfall from the southwest and the second inter-monsoons. During the northeast monsoon, the temperature exceeds 30°C, sometimes rising to 36°C in the central area of the city; therefore, the high-temperature level is a severe concern

TABLE 18.4
Extent of LULC Patterns (in Hectares)

Value	Class Name	2001	2011	2021
1	Built-up	2045.77	3378.01	4699.76
2	Agriculture	2048.66	2269.05	1406.20
3	Forest	6396.31	4874.53	4321.42
4	Water bodies	158.49	130.02	223.87
5	Rocks	75.34	72.95	73.32
Total Land		**10724.57**	**10724.57**	**10724.57**

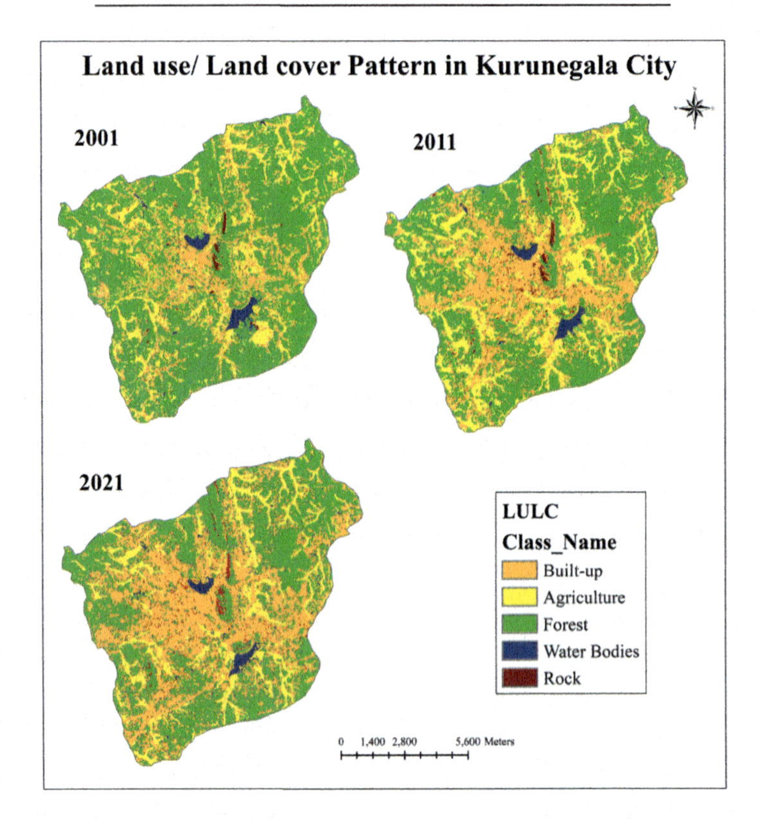

FIGURE 18.4 LULC pattern in Kurunegala City.

in the city that influences the water bodies (UDA, 2019). The satellite image used for the classification of LULC changes was taken from the months of December 2001 (northeast monsoon), May 2011 (southwest monsoon) and October 2021 (second inter-monsoon). Therefore, extent of water bodies in the area have been shown to be higher during 2021 than in previous years, and several small tanks in the region have also been renovated in recent years, considering their economic importance.

Figure 18.4 illustrates the LULC pattern for 2001, 2011 and 2021. The results indicate that the core city had undergone a major transformation from forest and agricultural area to built-up land. According to the spatial trends of the LULC pattern, it is apparent that the built-up area was increasing dramatically during the study periods, and the trend showed an enhancing pattern in the city. A considerable increase in commercial buildings and residential land is the most important cause of the built-up growth in this area. It is widely spread from the city centre towards the southern and western parts of the main city, especially on the Kandy, Colombo, Puttalam and Negombo highways. Further, physical features constrain the built-up area's extension in the north and northeast of the main city. The continuous increase in urban built-up areas is a common feature of Kurunegala, which is similarly identified in several cities in Sri Lanka, such as Kaduwela (Ratnayake & Ranaweera, 2017), Colombo Municipality (Pushpakumara & Ranga, 2020), Kandy (Uduporuwa, 2020) and Batticaloa (Mathanraj et al., 2021; Zahir et al., 2021).

Meanwhile, the decline or conversion of agricultural land into other types of land use, especially built-up, is evident in the city. In particular, more than half of the agricultural area was reduced or converted to another type of land use. Field observations confirmed that speculative land conversion by residents and suburban development are the primary modes for LULC changes. The high demand for land for commercial purposes has led traditional farmers to seek other livelihood alternatives for survival. According to UDA (2019), agriculture occupied a significantly larger share of the land; however, its contribution to the city's economy is relatively low. Ranagalage et al. (2020) stated that growing urbanisation had transformed naturally vegetated areas into impervious surfaces in Kurunegala municipality. The decline in green space has caused the loss of ecosystem services and an increase in the urban heat index in the city (Asmone, Conejos, Chew, & Piyadasa, 2016; Kafy et al., 2021).

18.3.2 Land Use/Land Cover Changes

The gains and losses of LULC are shown between 2001 and 2011(see Figure 18.5) and 2011 and 2021(see Figure 18.6). The highest gain of 2,362 ha was recorded in the built-up land, and the highest loss of 2,199 ha was noticed in the forest land between 2001 and 2011. Built-up land had a gain of 1,848 ha in 2021, which is slightly lower than in 2011. Agricultural land showed a loss of 1,105 ha, compared to 2001, it is somewhat lower. A loss of 1,182 ha found in forest land. Thus, the area indicated significant changes between these periods.

LULC changes with the transition probability matrix are presented in Tables 18.5 and 18.6, obtained between 2001–2011 and 2011–2021 for all classes. Gains and losses explained the probability of changing LULC classes for both periods. The built-up, agricultural and forest area probably increased by 114.13%, 88.58% and 51.56%, respectively, between 2001 and 2011. On the other hand, water bodies and rocks probably reduced by 40.91% and 55.47%, respectively. Some human

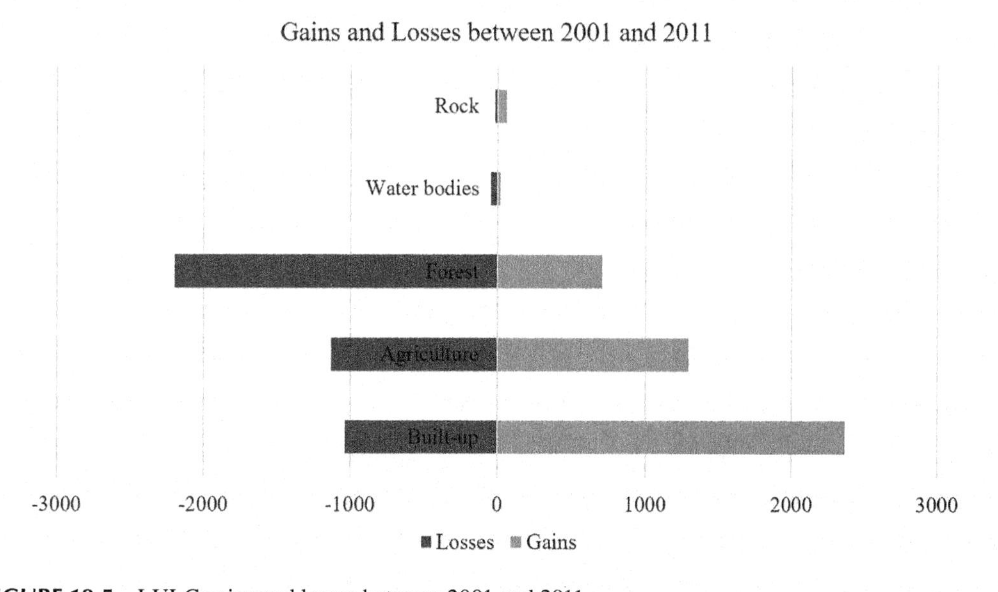

FIGURE 18.5 LULC gains and losses between 2001 and 2011.

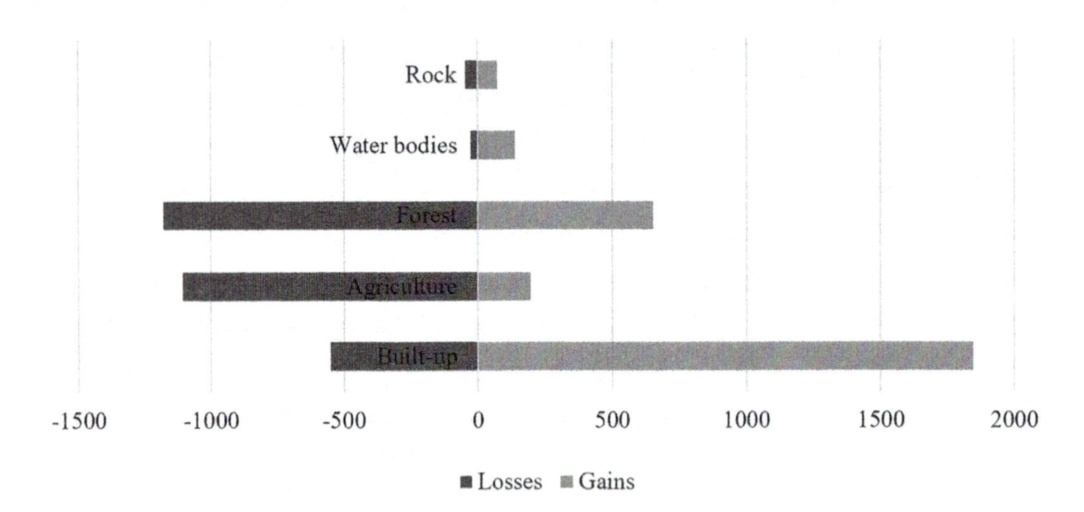

FIGURE 18.6 LULC gains and losses between 2011 and 2021.

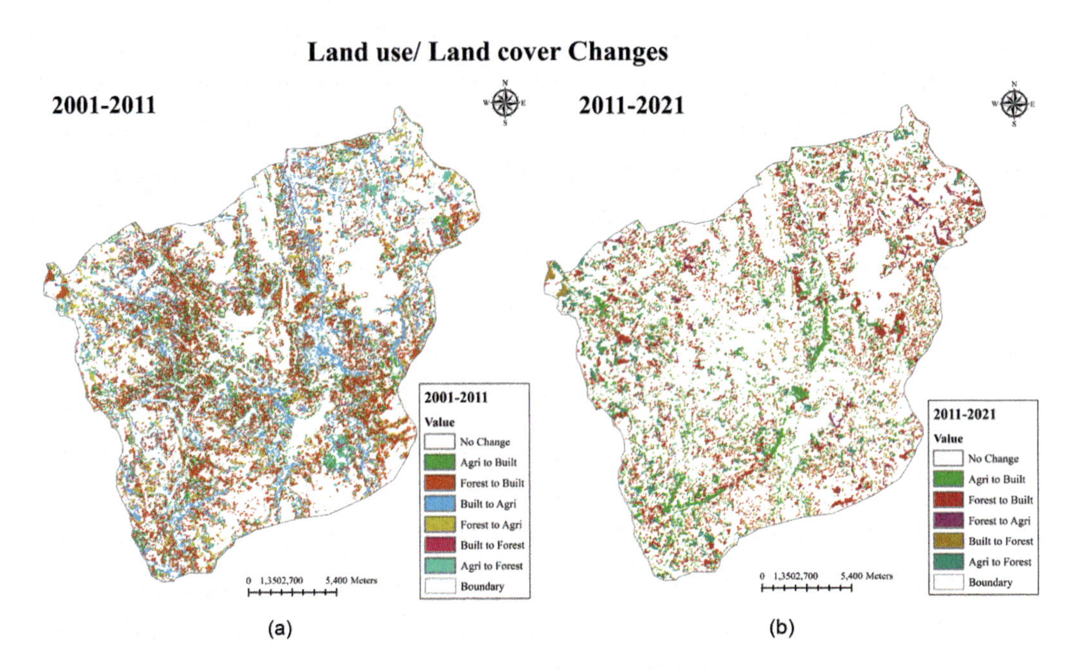

FIGURE 18.7 Land use/land cover changes between 2001–2011 (a) and 2011–2021 (b).

influences over the rock outcrops are the main cause for the decline in rock exposure in recent years. Further, the probability of built-up land can be gained by nearly 40.56% from an agricultural area, 15.30% from forest land and a smaller proportion from the remaining land uses. Correspondingly, the conversion of agricultural area to built-up land can be 37.1%, while the possibility of change as forest land is 22.05%. Approximately 19.03% of water bodies can be transformed into built-up land and 8.47% into agricultural land (see Figure 18.7).

In addition, the built-up area could be increased over this period by 135.36%, whereas the agricultural area could be reduced by 56.05% between 2011 and 2021 (see Table 18.6). Almost 39.07% of the agricultural area can be converted into built-up areas and 14.79% forest land. Likewise, there

TABLE 18.5
Probability Matrix of Transition between 2001 and 2011

	Probability of Transformation: 2011						
Given: 2001	**Built-up**	**Agriculture**	**Forest**	**Water Bodies**	**Rocks**	**Total**	**Loss**
Built-up	**0.4239**	0.4056	0.1530	0.0025	0.0151	1	0.5761
Agriculture	0.3710	**0.3972**	0.2205	0.0038	0.0075	1	0.6028
Forest	0.3305	0.1104	**0.5546**	0.0013	0.0032	1	0.4454
Water bodies	0.1903	0.0847	0.1221	**0.5909**	0.012	1	0.4091
Rocks	0.2495	0.2851	0.0200	0.0000	**0.4453**	1	0.5547
Total	1.5652	1.283	1.0702	0.5985	0.4831	5	
Gain	1.1413	0.8858	0.5156	0.0076	0.0378		

TABLE 18.6
Probability Matrix of Transition between 2011 and 2021

	Probability of Transformation: 2021						
Given: 2011	**Built-up**	**Agriculture**	**Forest**	**Water Bodies**	**Rocks**	**Total**	**Loss**
Built-up	**0.7122**	0.0441	0.1859	0.0257	0.0321	1	0.2878
Agriculture	0.3907	**0.4395**	0.1479	0.0188	0.0031	1	0.5605
Forest	0.3112	0.0292	**0.6421**	0.0160	0.0015	1	0.3579
Water bodies	0.1613	0.1312	0.0387	**0.6687**	0.0000	1	0.3313
Rocks	0.4904	0.0751	0.0436	0.0133	**0.3777**	1	0.6223
Total	2.0658	0.7191	1.0582	0.7425	**0.4144**	5	
Gain	1.3536	0.2796	0.4161	0.0738	**0.0367**		

is a possibility for transformation of forest land at 29.2% and built-up land around 31.12% into agricultural areas. In addition, 3.87% of the water bodies can be converted into forests, while 16.13% into built-up land. However, the probability of loss in the built-up is the lowest conversion in this period.

18.3.3 Land Use/Land Cover Simulation and Prediction

LULC for 2021 was simulated based on the classified maps of 2001 and 2011. The simulated map was confirmed with the existing LULC map of 2021. The values of K_{no}, $K_{location}$ and $K_{standard}$ were calculated to the simulated map with a validation tool in IDRISI 17.0. K_{no} at 0.9550, $K_{standard}$ at 0.9459 and $K_{location}$ at 0.9572 were obtained for this map, all of which indicated an excellent agreement with the simulated map of 2021 (see Figure 18.8). Therefore, the prediction for 2031 can be adequate based on the simulation results.

Table 18.7 shows the transition probability matrix between 2021 and 2031. The built-up area has the potential to achieve the highest gain of 98.53%, while the agricultural area could be subsidised by 55.98%. The probability of the forest cover showed a reduction of 27.73%, and the water bodies could be increased to 23.8%. Approximately 41.22% of the agricultural area can be converted to built-up land and 13.49% to forest land. In addition, 26.92% of forest land would be transformed into built-up areas.

Figure 18.9 presents the predicted LULC pattern for Kurunegala city in 2031. The results indicate that the largest land could be occupied with a built-up area of 6,418.26 ha. However, the agricultural area will be covered by 842.71 ha, which is a great loss compared to the previous periods.

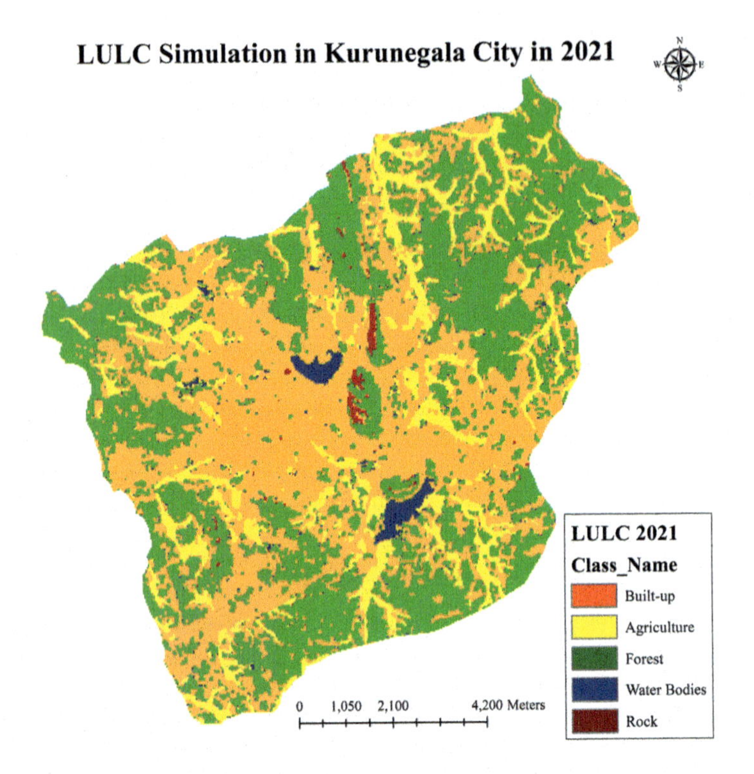

FIGURE 18.8 Land use/land cover simulation in Kurunegala city in 2021.

TABLE 18.7
Transitional Probability Matrix between 2021and 2031

	Probability of Changing to: 2031						
Given: 2021	**Built-up**	**Agriculture**	**Forest**	**Water Bodies**	**Rocks**	**Total**	**Loss**
Built-up	**0.6870**	0.2242	0.0681	0.0098	0.0109	1	0.313
Agriculture	0.4122	**0.4402**	0.1349	0.0084	0.0042	1	0.5598
Forest	0.2692	0.0000	**0.7227**	0.0065	0.0016	1	0.2773
Water bodies	0.1471	0.0452	0.0445	**0.7620**	0.0011	1	0.238
Rocks	0.1568	0.1387	0.0091	0.0016	**0.6938**	1	0.3062
Total	1.6723	0.8483	0.9793	0.7943	0.7116		
Gain	0.9853	0.4081	0.2566	0.0323	0.0178		

The forest cover would extend to 3,145.77 ha, which still showed a declining pattern in the city. In addition, the water bodies will account for 242.3 ha, while rocks nearly 75.53 ha in the area, which showed an almost constant pattern compared to 2021. The non-built-up area has sharply declined over time. The city's non-built-up area of 9.14% in 1995 was diminished to 3.31% in 2017 (UN-Habitat, 2018) due to the rapid construction development. This trend indicates that the challenges arising from LULC changes will be severe in the future. In particular, since the city consists of various environmentally essential sites, including wetlands, the systematic evaluation for the impacts of LULC changes is essential (UDA, 2019). Further, the city has been identified as an important economic hub that expand the urban and industrial areas in the city, thus, it has important priorities in the development (UDA, 2019). Hence, considering LULC changes and their consequences would

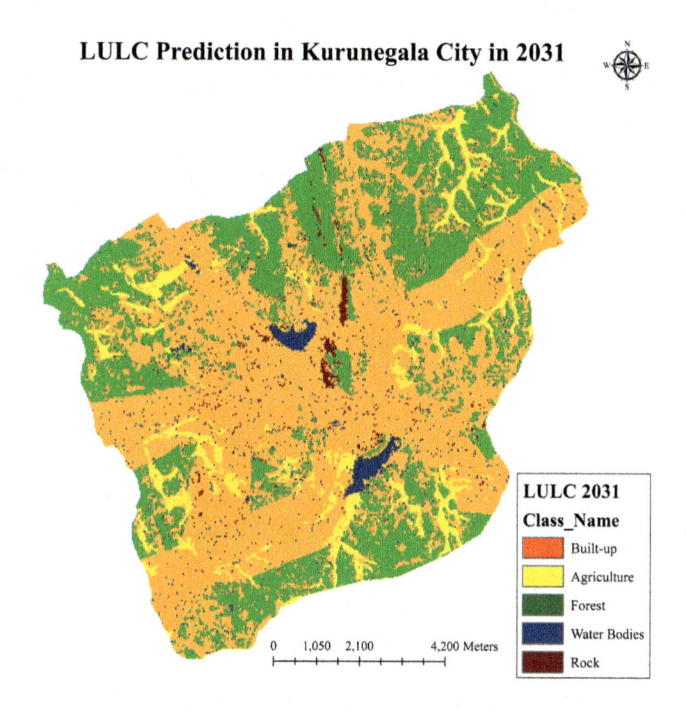

FIGURE 18.9 Land use/land cover prediction Kurunegala city in 2031.

be more significant in space and time. Rapid loss of agricultural and forest land could be accelerated environmental challenges, such as loss of biodiversity, urban heat islands, and pollution. Thus, these findings should be considered in the spatial planning process of the city.

Overall, the study applied a hybrid model, which is Multi-Layer Perceptron neural networks and Cellular Automata-Markov is more appropriate to the prediction of LULC in Kurunegala city. This model facilitates useful tools for the suitability maps, which is a crucial step in future modelling. Meantime, it offers tools for incorporating the explanatory variables, which cause the LULC change in the city. The potential variables were tested using Cramer's V for a suitability map. Especially, highway and major road development significantly caused the LULC change, which is empirically identified based on the 30 years of changes. Several models applied in the developed and developing cities are not capable to integrate the explanatory variables. However, the hybrid models such as logistic regression -Markov chain, CLUE- Markov chain are some well-accepted models for the LULC prediction. Logistic regression efficiently handles the binary dependent variables, which are useful for the urban sprawl prediction. The CLUE model is most widely applied for LULC studies in the world. However, the MLP model was rarely applied around the cities worldwide that encourages to apply in LULC prediction for Kurunegala city. Meantime, the model provided a very good accuracy for the prediction of 2031, which is more useful for future development.

18.4 CONCLUSION

The study aimed to apply a hybrid model to predict the LULC pattern in Kurunegala city, Sri Lanka. LULC spatial pattern was classified for the years 2001, 2011 and 2021. The built-up pattern showed a gradual increase since 2001, while the forest cover exposed a decreasing pattern. Built-up land stretched much of the distance from the city centre to the southern and western part of the city. Besides, the growth in other directions is limited to physical features, such as forest and rock outcroppings. Further, rocks indicated almost the same pattern during the study period. However, the water bodies identified a fluctuating pattern due to monsoonal rainfall changes because the Landsat images used in the study were obtained in various season.

In addition, the probability of the LULC pattern explains that the agricultural area had a substantial loss, and the built-up had a sharp increase between 2001 and 2021. Rapid urban development in recent decades has resulted a high built-up growth. This pattern is reflected on agricultural and forest land. Therefore, a prediction model was used to simulate the LULC pattern, which is essential for the sustainable development of the city.

The model used in the study was more appropriate to present the prediction of LULC in 2031. First, the model was validated with the existing and simulated LULC in 2021, indicating an excellent prediction accuracy. Hence, the prediction for 2031 can be relied upon for future planning. The prediction in 2031 revealed that the built-up area would have a high proportion compared with other land use classes. Meanwhile, the agricultural area is expected to lose large extent of land between the classified LULC. Further, built-up land will be remained as a prominent class in 2031. Therefore, the LULC prediction in 2031 is useful for spatial planning and decision-making process. In addition, these findings can be incorporated into the future planning and development of the city.

REFERENCES

Aburas, M. M., Ahamad, M. S. S., & Omar, N. Q. (2019). Spatio-temporal simulation and prediction of land-use change using conventional and machine learning models: A review. *Environmental Monitoring and Assessment, 191*(4), 205.

Al-Hameedi, W. M. M., Chen, J., Faichia, C., Al-Shaibah, B., Nath, B., Kafy, A.-A., ... Al-Aizari, A. (2021). Remote sensing-based urban sprawl modeling using multilayer perceptron neural network Markov chain in Baghdad, Iraq. *Remote Sensing, 13*(20), 4034.

Amarawickrama, S., Singhapathirana, P., & Rajapaksha, N. (2015). Defining urban sprawl in the Sri Lankan context: With special reference to the Colombo Metropolitan Region. *Journal of Asian and African studies, 50*(5), 590–614.

Aronoff, S. (2004). *Remote Sensing for GIS Managers*. Redlands, CA: ESRI Press, Environmental Systems Research.

Arsanjani, J. J., Helbich, M., Kainz, W., & Boloorani, A. D. (2013). Integration of logistic regression, Markov chain and cellular automata models to simulate urban expansion. *International Journal of Applied Earth Observation and Geoinformation, 21*, 265–275.

Asmone, A., Conejos, S., Chew, M., & Piyadasa, R. (2016). Urban green cover protocol to reduce urban heat island in Sri Lanka. *Paper Presented at the 4th International Conference on Countermeasures to Urban Heat Island, National University of Singapore*, Singapore.

Bhatta, B. (2010). *Analysis of Urban Growth and Sprawl from Remote Sensing Data*. Berlin/Heidelberg: Springer Science & Business Media.

Briassoulis, H. (2020). *Analysis of Land Use Change: Theoretical and Modeling Approaches* (Loveridge, S. & Jackson, S. Eds., 2nd ed.). Morgantown, WV: WVU Research Repository.

Cho, H. (2020). Climate change risk assessment for Kurunegala, Sri Lanka: Water and heat waves. *Climate, 8*(12), 140.

Daniel, C. J., Frid, L., Sleeter, B. M., & Fortin, M. J. (2016). State-and-transition simulation models: A framework for forecasting landscape change. *Methods in Ecology and Evolution, 7*(11), 1413–1423.

Deep, S., & Saklani, A. (2014). Urban sprawl modeling using cellular automata. *The Egyptian Journal of Remote Sensing and Space Science, 17*(2), 179–187.

Divigalpitiya, P., Ohgai, A., Tani, T., Watanabe, K., & Gohnai, Y. (2007). Modeling land conversion in the Colombo Metropolitan Area using cellular automata. *Journal of Asian Architecture and Building Engineering, 6*(2), 291–298.

Divisional-Secretariat. (2019). Population by Grama Niladhari divisions.

Gleicher, M. (2016). A framework for considering comprehensibility in modeling. *Big Data, 4*(2), 75–88.

Hamad, R., Balzter, H., & Kolo, K. (2018). Predicting land use/land cover changes using a CA-Markov model under two different scenarios. *Sustainability, 10*(10), 3421.

Hamdy, O., Zhao, S., Osman, T., Salheen, M. A., & Eid, Y. Y. (2016). Applying a hybrid model of Markov chain and logistic regression to identify future urban sprawl in Abouelreesh, Aswan: A case study. *Geosciences, 6*(4), 43.

Hamdy, O., Zhao, S., Salheen, M. A., & Eid, Y. (2017). Analyses the driving forces for urban growth by using IDRISI® Selva models Abouelreesh-Aswan as a case study. *International Journal of Engineering and Technology, 9*(3), 226.

Han, H., Yang, C., & Song, J. (2015). Scenario simulation and the prediction of land use and land cover change in Beijing, China. *Sustainability, 7*(4), 4260–4279.

Huang, W., Liu, H., Luan, Q., Jiang, Q., Liu, J., & Liu, H. (2008). Detection and prediction of land use change in Beijing based on remote sensing and GIS. *The International Archives of the Photogrammetry, Remote Sensing and Spatial Information Sciences (ISPRS Archives), 37*, 75–82.

Kafy, A.-A., Al Rakib, A., Akter, K. S., Rahaman, Z. A., Mallik, S., Nasher, N. R., ... Ali, M. Y. (2021). Monitoring the effects of vegetation cover losses on land surface temperature dynamics using geospatial approach in Rajshahi city, Bangladesh. *Environmental Challenges, 4*, 100187.

Márquez, A. M., Guevara, E., & Rey, D. (2019). Hybrid model for forecasting of changes in land use and land cover using satellite techniques. *IEEE Journal of Selected Topics in Applied Earth Observations and Remote Sensing, 12*(1), 252–273.

Mathanraj, S., Ratnayake, R., & Rajendram, K. (2019). A GIS-based analysis of temporal changes of land use pattern in Batticaloa MC, Sri Lanka from 1980 to 2018. *World Scientific News, 137*, 210–228.

Mathanraj, S., Rusli, N., & Ling, G. (2021). Applicability of the CA-Markov model in land-use/land cover change prediction for urban sprawling in Batticaloa Municipal Council, Sri Lanka. *IOP Conference Series: Earth and Environmental Science, 620*(1), 012015.

Meiyappan, P., Dalton, M., O'Neill, B. C., & Jain, A. K. (2014). Spatial modeling of agricultural land use change at global scale. *Ecological Modelling, 291*, 152–174.

Mirici, M., Berberoglu, S., Akin, A., & Satir, O. (2018). Land use/cover change modelling in a mediterranean rural landscape using multi-layer perceptron and Markov chain (mlp-mc). *Applied Ecology and Environmental Research, 16*, 467–486.

Park, S., Jeon, S., & Choi, C. (2012). Mapping urban growth probability in South Korea: Comparison of frequency ratio, analytic hierarchy process, and logistic regression models and use of the environmental conservation value assessment. *Landscape and Ecological Engineering, 8*(1), 17–31.

Parsa, V. A., & Salehi, E. (2016). Spatio-temporal analysis and simulation pattern of land use/cover changes, case study: Naghadeh, Iran. *Journal of Urban Management, 5*(2), 43–51.

Paul, B., & Rashid, H. (2016). Climatic hazards in coastal Bangladesh: Non-structural and structural solutions. In *Land Use Change and Coastal Management*. Oxford: Butterworth-Heinemann. pp. 183–186.

Pijanowski, B. C., Tayyebi, A., Doucette, J., Pekin, B. K., Braun, D., & Plourde, J. (2014). A big data urban growth simulation at a national scale: Configuring the GIS and neural network based land transformation model to run in a high performance computing (HPC) environment. *Environmental Modelling & Software, 51*, 250–268.

Pushpakumara, T., & Ranga, K. (2020). Land use management in Colombo Municipal Council Area. *International Journal of Advanced Remote Sensing and GIS, 9*(1), 3438–3445.

Quan, B., Bai, Y., Römkens, M., Chang, K.-t., Song, H., Guo, T., & Lei, S. (2015). Urban land expansion in Quanzhou city, China, 1995–2010. *Habitat International, 48*, 131–139.

Ranagalage, M., Ratnayake, S. S., Dissanayake, D., Kumar, L., Wickremasinghe, H., Vidanagama, J.,... Simwanda, M. (2020). Spatiotemporal variation of urban heat islands for implementing nature-based solutions: A case study of Kurunegala, Sri Lanka. *ISPRS International Journal of Geo-Information, 9*(7), 461.

Ratnayake, R., & Ranaweera, D. (2017). Urban Landuse Changes in Sri Lanka with Special Reference to Kaduwela Town from 1975 to 2016. *International Journal of Innovative Research and Development, 6*(6), 1–13.

Rathnayake, C. W., Jones, S., & Soto-Berelov, M. (2020). Mapping land cover change over a 25-year period (1993–2018) in Sri Lanka using Landsat time-series. *Land, 9*(1), 27.

Rui, Y., & Ban, Y. (2010). Multi-agent simulation for modeling urban sprawl in the greater Toronto area. *Paper Presented at the 13th AGILE International Conference on Geographic Information Science 2010*, Guimarães, Portugal, 10–14 May 2010.

Samat, N., Hasni, R., & Elhadary, Y. A. E. (2011). Modelling land use changes at the Peri-Urban areas using geographic information systems and cellular automata model. *Journal of Sustainable Development 4*(6), 1–13.

Saxena, A., & Jat, M. K. (2019). Capturing heterogeneous urban growth using SLEUTH model. *Remote Sensing Applications: Society and Environment, 13*, 426–434.

Shahraki, S. Z., Sauri, D., Serra, P., Modugno, S., Seifolddini, F., & Pourahmad, A. (2011). Urban sprawl pattern and land-use change detection in Yazd, Iran. *Habitat International, 35*(4), 521–528.

Silva, L. P. E., Xavier, A. P. C., Silva, R. M. D., & Santos, C. A. G. (2020). Modeling land cover change based on an artificial neural network for a semiarid river basin in northeastern Brazil. *Global Ecology and Conservation, 21*, e00811.

Suthakar, K., & Bui, E. N. (2008). Land use/cover changes in the war-ravaged Jaffna Peninsula, Sri Lanka, 1984–early 2004. *Singapore Journal of Tropical Geography, 29*(2), 205–220.

Thakur, P. K., Kumar, M., & Gosavi, V. E. (2020). Monitoring and modelling of urban sprawl using geospatial techniques: A case study of Shimla City, India. In: Sahdev, S., Singh, R. B., & Kumar, M. (Eds.), *Geoecology of Landscape Dynamics* (pp. 263–294). Berlin/Heidelberg: Springer.

UDA. (2019). Kurunegala Town Development Plan. Retrieved from UDA, Battaramulla.

Uduporuwa, R. (2020). Spatial and temporal dynamics of land use/land cover in Kandy City, Sri Lanka: An analytical investigation with geospatial techniques. *American Academic Scientific Research Journal for Engineering, Technology, and Sciences, 69*(1), 149–166.

UN-Habitat. (2018). The State of Sri Lankan Cities 2018. Retrieved from Colombo.

Verburg, P. H., Soepboer, W., Veldkamp, A., Limpiada, R., Espaldon, V., & Mastura, S. S. (2002). Modeling the spatial dynamics of regional land use: The CLUE-S model. *Environmental Management, 30*(3), 391–405.

Verburg, P. H., Kok, K., Pontius, R. G., & Veldkamp, A. (2006). Modeling land-use and land-cover change. In: Lambin, E. F., & Geist, H. J. (Eds.), *Land-Use and Land-Cover Change* (pp. 117–135). Berlin/Heidelberg: Springer.

Vinayak, B., Lee, H. S., & Gedem, S. (2021). Prediction of land use and land cover changes in Mumbai City, India, using remote sensing data and a multilayer perceptron neural network-based Markov chain model. *Sustainability, 13*(2), 471.

Wang, W., Zhang, C., Allen, J. M., Li, W., Boyer, M. A., Segerson, K., & Silander, J. A. (2016). Analysis and prediction of land use changes related to invasive species and major driving forces in the state of Connecticut. *Land, 5*(3), 25.

Weerakoon, P. (2017). GIS integrated spatio-temporal urban growth modelling: Colombo Urban Fringe, Sri Lanka. *Journal of Geographic Information System, 9*, 372–389.

Yang, X., Chen, R., & Zheng, X. (2016). Simulating land use change by integrating ANN-CA model and landscape pattern indices. *Geomatics, Natural Hazards and Risk, 7*(3), 918–932.

Yu, W., Zang, S., Wu, C., Liu, W., & Na, X. (2011). Analyzing and modeling land use land cover change (LUCC) in the Daqing City, China. *Applied Geography, 31*(2), 600–608.

Zahir, I. L. M., Thennakoon, S., Sangasumana, R., Herath, J., Madurapperuma, B., & Iyoob, A. L. (2021). Spatiotemporal land-use changes of batticaloa municipal council in Sri Lanka from 1990 to 2030 using land change modeler. *Geographies, 1*(3), 166–177.

19 Spatio-Temporal Dynamics of Tropical Deciduous Forests under Climate Change Scenarios in India

Rajit Gupta and Laxmi Kant Sharma
Central University of Rajasthan

CONTENTS

19.1 INTRODUCTION

Climate change will impact every part of the world (IPCC 2021). Climate change directly or indirectly affects forest growth and productivity, disturbing the plant's physiological processes, composition and distribution patterns (Gupta and Sharma 2021; del Río et al. 2021). Due to climate change, significant forest changes have been observed in the last half-century (Duo et al. 2016). Climatic conditions such as temperature and precipitation are closely related to forest growth and distribution (Kong et al. 2017). Also, climatic conditions are commonly interpreted to observe the response of forests to the changing climate (He et al. 2015; Sun et al. 2021). Therefore, there is some mutual relation between climate change and forest dynamics, which need to be critically investigated for sustainable forest management and climate change mitigation (Hansen and Loveland 2012; Nguyen et al. 2018).

Climatic conditions and topographical structures are dominant environmental factors that control vegetation distribution patterns locally and globally (Boonpragob and Santisirisomboon 1996). Bao et al. (2021) stated that it is crucial to investigate the forests sensitivity to climate change to identify how

DOI: 10.1201/9781003270928-25

forest cover would distribute under future climate change scenarios. According to the Forest Survey of India report, the entire area of moist and dry tropical deciduous forest in India will fall under climate hotspots in mid-century. Previous studies conducted to determine the impact of climate change in different forest ecosystems had used dynamic vegetation models. In India, very limited research demonstrated the impacts of climate change on tropical deciduous forest cover. For instance, Ravindranath et al. (2011) explored the role of India's forests for potential mitigation of climate changes and compared net primary productivity derived from dynamic vegetation models such as BIOME4 and Integrated Biosphere Simulator (IBIS). Gopalakrishnan et al. (2011) studied the impact of the changing climate on tropical deciduous *Tectona grandis* trees using IBIS and climatic data of the Hadley Regional Model (HadRM3) model. The issue with dynamic vegetation models is that they require several inputs, complexity in model structures, and their parameters are not easy to obtain (Bao et al. 2021).

Over the previous two decades, geospatial technology and machine learning have shown considerable progress in many Earth observation applications (Lary et al. 2016), including forests spatial mapping and prediction. Supervised machine learning algorithms-based models are the most widely used machine learning models (Osisanwo et al. 2017). The machine learning algorithms are "universal approximators," meaning they learn the underlying behavior and identify system patterns from training datasets (Lary et al. 2016). Machine learning is valuable where conventional approaches such as simple regression methods have limited applicability (Patil et al. 2017). Further, machine learning-based models do not need prior information about the state of the relationships between the datasets (Lary et al. 2016). Machine learning algorithms like Random Forest (RF) can learn complex patterns and non-linear relations between a target variable and predictors data (Evans et al. 2011). In addition, integrating machine learning and remotely sensed datasets unlocks new opportunities to study the forest cover dynamics and distribution mapping at a large scale, with faster speed and accuracy.

According to Kirilenko and Sedjo (2007), the changing temperature and precipitation patterns under climate change scenarios will directly impact natural forests and plantations. Therefore, the objective of the current study is to predict the spatio-temporal dynamics of India's tropical deciduous forest cover under climate change scenarios. A set of bioclimatic variables related to temperature and precipitation in the near current and future climate change scenarios, along with topographic predictors were used as input in the RF model. Pearson correlation test was performed to assess the collinearity among predictors. Further, we identified which predictors are the most to least important for prediction. The model evaluation was performed to determine the RF model classification accuracy. This study is important as very limited studies determine the impacts of climate change on India's tropical deciduous forest cover using a machine learning approach.

19.2 MATERIALS AND METHODS

19.2.1 Study Area

The current study region is Indian tropical moist and dry deciduous forests (Figure 19.1). India is a mega-biodiversity country with rich floral species diversity (Reddy et al. 2015). According to (ISFR 2021), the total forest cover is 7,13,789 km^2, constituting about 21.71% of India's geographical area. According to Champion and Seth (1968) forest classification types, India comprises 16 major forest types and 221 sub-types. The dominant forest types of India are tropical dry deciduous and tropical moist deciduous, covering an area of 3,13,617 km^2 and 1,35,492 km^2, respectively (ISFR 2021).

19.2.2 Data Used

19.2.2.1 Forest Cover Data

India's tropical dry and moist deciduous forest cover spatial data in raster format was freely downloaded from the Bhuvan portal (https://bhuvan.nrsc.gov.in). This data is based on a nationwide forest

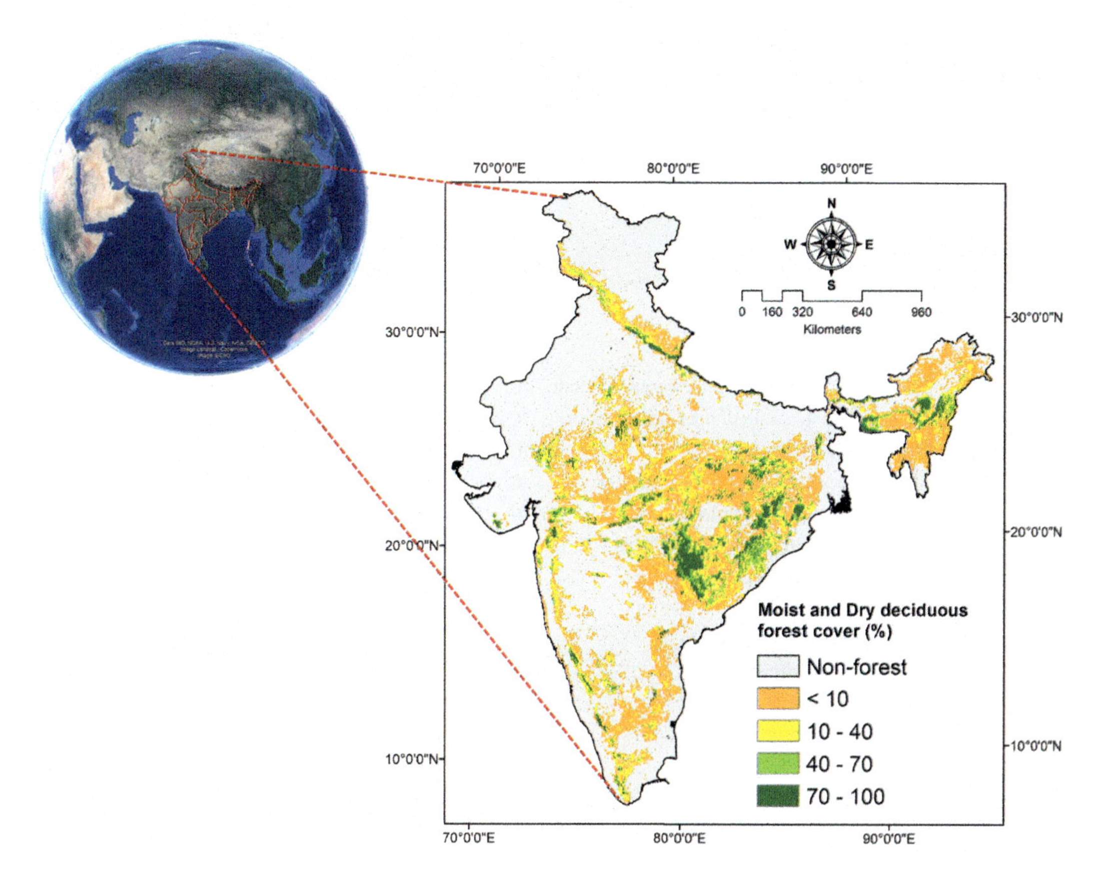

FIGURE 19.1 Spatial distribution of the tropical moist and dry deciduous forest cover in India. (Data source: https://bhuvan.nrsc.gov.in.)

cover assessment using a hybrid classification on a multi-seasonal satellite dataset Resourcesat-2 Advanced Wide Field Sensor (AWiFS) in the year 2013, having a spatial resolution of 56 m (Reddy et al. 2015). They categorized data based on ecological principles and followed Champion and Seth (1968) classification system of forest types in India (Reddy et al. 2015). The tropical deciduous dry and moist forest cover data is available at a 5×5 km grid. Further, it is arranged in the forest cover density class from <10% to 100% density in an interval of 10%.

19.2.2.2 Predictors Used

We used different climatic and topographical predictors as shown in Table 19.1. The climatic predictors datasets for the near-current and future scenarios were acquired from WorldClim version 2.1 (https://www.worldclim.org/), having a spatial resolution of 2.5 minutes (Fick and Hijmans 2017). We downloaded the general circulation model (GCM), the sixth version of the Model for Interdisciplinary Research on Climate (MIROC6) (Tatebe et al. 2019) data for three Shared Socio-economic Pathways (SSPs): SSP1–2.6, SSP2–4.5, and SSP5–8.5. Future climate scenarios data were based on Coupled Model Intercomparison Projects (CMIP6), downscaled and bias-corrected with WorldClim version 2.1 as baseline climate. The monthly data were averaged over 20 years for the 2030s (2021–2040), 2050s (2041–2060), and 2070s (2061–2080) (Fick and Hijmans 2017). Further, topographical predictors such as DEM, slope, and aspect were the used in the current study. Shuttle Radar Topography Mission (SRTM) elevation data (Farr et al. 2007), was used to obtain DEM, while slope and aspect map was generated from DEM in QGIS version 3.18.

TABLE 19.1
List of Predictors, Including Climatic and Topographic Variables (Grey Shaded Predictors Were Removed)

Category	Predictors	Units	Abbreviation	Data Source
Climatic	Annual mean temperature	°C	Bio1	WorldClim version 2.1 (https://www.worldclim.org/)
	Mean diurnal range	°C	Bio2	
	Isothermality	°C	Bio3	
	Temperature seasonality	°C	Bio4	
	Maximum temperature of the warmest month	°C	Bio5	
	Minimum temperature of the coldest month	°C	Bio6	
	Temperature annual range	°C	Bio7	
	Mean temperature of the wettest quarter	°C	Bio8	
	Mean temperature of the driest quarter	°C	Bio9	
	Mean temperature of warmest quarter	°C	Bio10	
	Mean temperature of coldest quarter	°C	Bio11	
	Annual precipitation	mm	Bio12	
	Precipitation of wettest month	mm	Bio13	
	Precipitation of driest month	mm	Bio14	
	Precipitation seasonality	mm	Bio15	
	Precipitation of wettest quarter	mm	Bio16	
	Precipitation of driest quarter	mm	Bio17	
	Precipitation of warmest quarter	mm	Bio18	
	Precipitation of coldest quarter	mm	Bio19	
Topographic	Elevation	m	DEM	SRTM (https://www.worldclim.org/)
	Slope	%	Slope	
	Aspect	°	Aspect	

19.2.3 Data Processing

Figure 19.2 shows the outline of the methodology used in the current study. After downloading and extracting dry and moist deciduous forest cover data, it was merged into a single category of deciduous forest. Then, the deciduous forest cover data was converted into point features and reclassified into four forest cover density classes, including <10%, 10%–40%, 40%–70% and 70%–100%. Both bioclimatic and topographical predictors were clipped using the shapefile of India.

The extent of all predictors was matched to use in the machine learning model. Then, these point features forest cover data were used to extract the pixel value from each bioclimatic and topographic predictor.

19.2.4 Random Forest Model Building

RF is an ensemble decision tree classification and regression machine learning algorithm (Breiman 2001). RF is a combination of classifiers in which a single vote is assigned from each classifier to the most recurrent class to the input vector (x) in equation (19.1),

$$\hat{C}_{rf}^{B} = \text{Majority vote}\left\{\hat{C}_b(x)\right\}_1^B \quad (1)$$

where $\hat{C}_b(x)$ denotes class prediction of the bth RF tree.

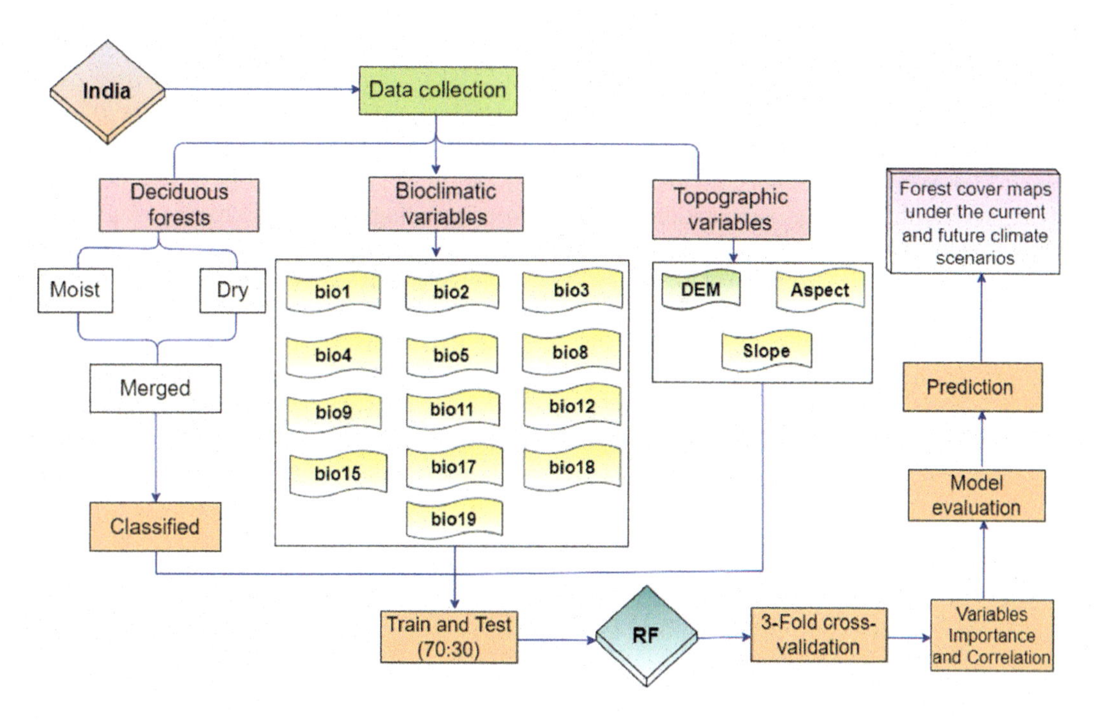

FIGURE 19.2 Flowchart shows the methodological framework used in the current study.

RF use decisions trees as a base classifiers in the form $\{h(x, U_k), k=1,\ldots,\}$ where the x is an input pattern and $\{U_k\}$ are independent identically distributed random vectors and (Breiman 2001; Rodriguez-Galiano et al. 2012). RF can capture the complex relationship between a target and a set of predictors. Moreover, the RF algorithm has high accuracy, avoids over-fitting, inbuilt variable importance function, has fewer hyperparameters, has lower sensitivity to the parameters tuning, has fast training, and is robust to noise (Shen et al. 2018; Li et al. 2019). RF is widely used in forest variables prediction studies, such as above-ground biomass estimation, forest cover and canopy height mapping. In this study, we were trained RF algorithm to predict tropical deciduous forest cover in the near current and future climatic conditions. The extracted data of target variable and predictors was divided into the train (70%) and test (30%) datasets. We used the caret package (Kuhn 2008) available in RStudio version 1.4.1717 (RStudio Team 2021) to build the RF model. We used the random forest 'rf' algorithm from the caret package, applied 3-fold cross-validation, and data pre-processing steps such as center and scale were used to normalize the data. A tune length value of ten was applied for tuning hyperparameter "number of variables randomly sampled (mtry)." The number of random trees (ntree) ware taken as 500. Variables importance was derived to assess the most to least important predictor variables. Pearson correlation test was performed in RStudio version 1.4.1717 using the "cor" function. The Pearson correlation test determines the multicollinearity relation among the predictor variables.

19.2.5 Model Evaluation and Spatial Prediction

RF model evaluation was performed using an accuracy metric based on a confusion matrix. The optimal model was selected based on the largest accuracy value. After achieving desired model accuracy, the model was tested on test data. The trained RF model was used to predict forest cover of the study area using the "predict" function in the RStudio version 1.4.1717. Spatial prediction of forest cover density was made in the near current and future climatic scenarios (SSP1–2.6, SSP2–4.5 and SSP5–8.5) for the 2030s, 2050s and 2070s periods. Further, we calculated the area changes of forest cover density classes in future climatic conditions with respect to the near current.

19.3 RESULTS

19.3.1 Pearson Correlation

Figure 19.3 demonstrates the collinearity among the predictors used in the current study. The use of highly correlated variables does not always increase the accuracy of the prediction model; however, it can increase the machine computation time. Therefore, excluding the variables having a high positive correlation (>0.85) helps to reduce the number of variables used in our model.

It is observed that precipitation of the wettest month (Bio13) and the wettest quarter (Bio16) correlate at 0.98, while Bio13 has a correlation of 0.88 with annual precipitation (Bio12). Therefore, we excluded Bio13 and Bio16 and retained only Bio12 in our model. Also, the precipitation of the driest month (Bio14) and the driest quarter (Bio17) correlate at 0.88, and we removed Bio14 and retained Bio17. Similarly, the annual temperature range (Bio7) was removed as it correlates 0.87 with temperature seasonality (Bio4). The topographical predictor DEM and slope were retained, whereas the aspect was removed from our analysis as it does not indicate any significant positive or negative correlation with other predictors.

19.3.2 Accuracy and Predictors Importance

Figure 19.4 shows the variations in the training accuracy of RF with the hyperparameter 'mtry.' The accuracy metrics selected the optimal model using the largest value. Resampling results across

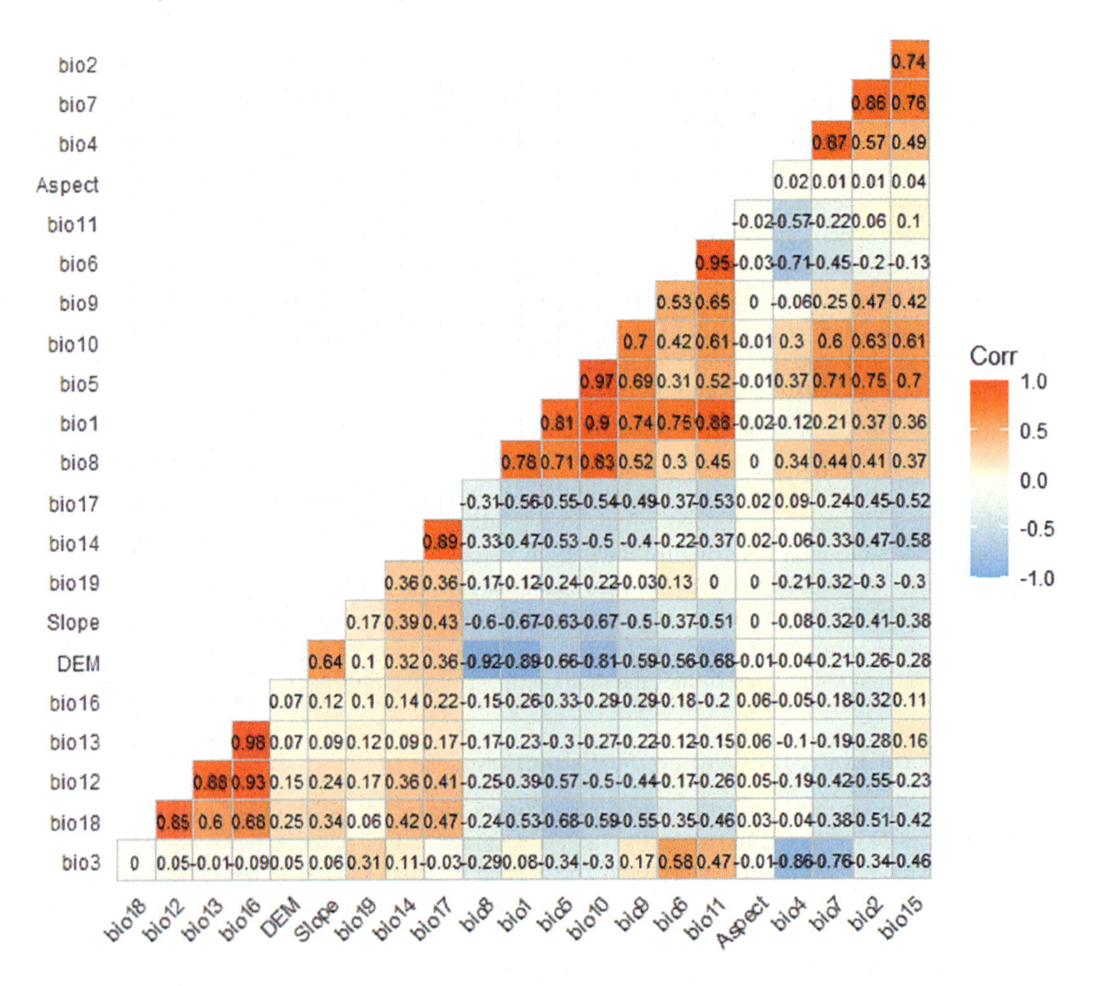

FIGURE 19.3 Pearson correlation matrix among the predictor variables used in the current study.

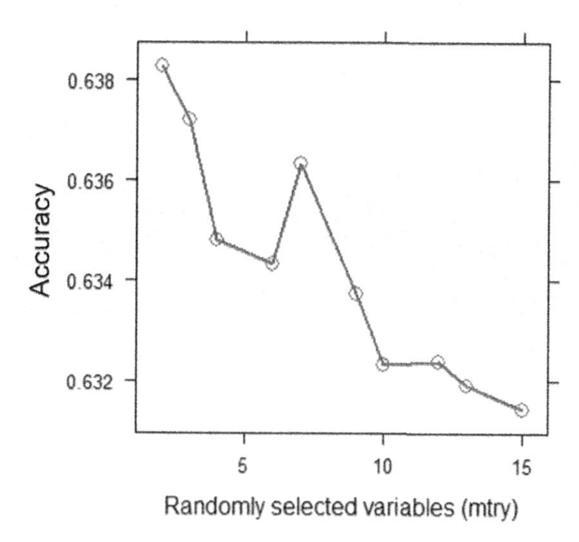

FIGURE 19.4 Variations in the training accuracy of RF with the hyperparameter 'mtry.'

tuning parameters demonstrated that the final value used for the model was mtry=2, at which the maximum classification training accuracy of RF achieved was 0.638. From Figure 19.4, it is observed that the accuracy generally goes lower as the 'mtry' values increase. The RF model classification testing accuracy achieved was 0.615, representing moderate RF performance.

Figure 19.5 shows the relative importance of different predictors. It is observed that annual precipitation (Bio12) is the dominating, while slope is the second main predictor of tropical deciduous forest cover classes. The mean diurnal range (Bio2) has an importance score of 68.20 for classes 10% to 40% and 70% to 100%. DEM, Bio11, and Bio17 are the least important predictor variables, as shown in Figure 19.5.

19.3.3 Spatial Prediction of Tropical Deciduous Forest Cover

Figure 19.6 shows the predicted map of tropical deciduous forest cover over India in the near current period. In the near current, the area covered under the forest cover class <10%, 10%–40%, 40%–70% and 70%–100% would be 0.67, 0.26, 0.107, and 0.098 M km^2, respectively.

Figure 19.7a–i shows the predicted maps of tropical deciduous forest cover over India in future climate change scenarios. Under the future climate scenarios of SSP5–8.5 in the 2070s, the forest cover class (<10%) has been increased to 0.783 M km^2, while the forest cover class (10%–40%) would be reduced to 0.193 M km^2. Forest cover class (70%–100%) shows improvement under the SSP1–2.6 scenarios in the 2030s and 2050s than in the near current period.

Figure 19.8 demonstrates that under the warmer conditions in the period 2070s, the forest cover class (<10%) shows an increasing trend, while the forest cover class (10%–40% and 70%–100%) shows a decreasing trend for SSP1–2.6, SSP2–4.5 and SSP5–8.5 climate change scenarios. Therefore, the moderate and dense tropical deciduous forest cover would be adversely affected under the warmer conditions of future climate change scenarios.

Table 19.2 shows an area percent difference (future and near current) covered under different forest cover classes. It is observed that under the SSP2–4.5 in the period 2030s than the near current, the forest cover class (70%–100%) would be increased by 3.59%. Meanwhile, the forest cover class (<10%) would be decreased by 4.99% under the SSP2–4.5 in the period 2030s. In the scenarios of SSP2–4.5 and SSP5–8.5 in the 2070s, the forest cover class (<10%) would be increased, indicating the deteriorating conditions of deciduous forest cover in the future.

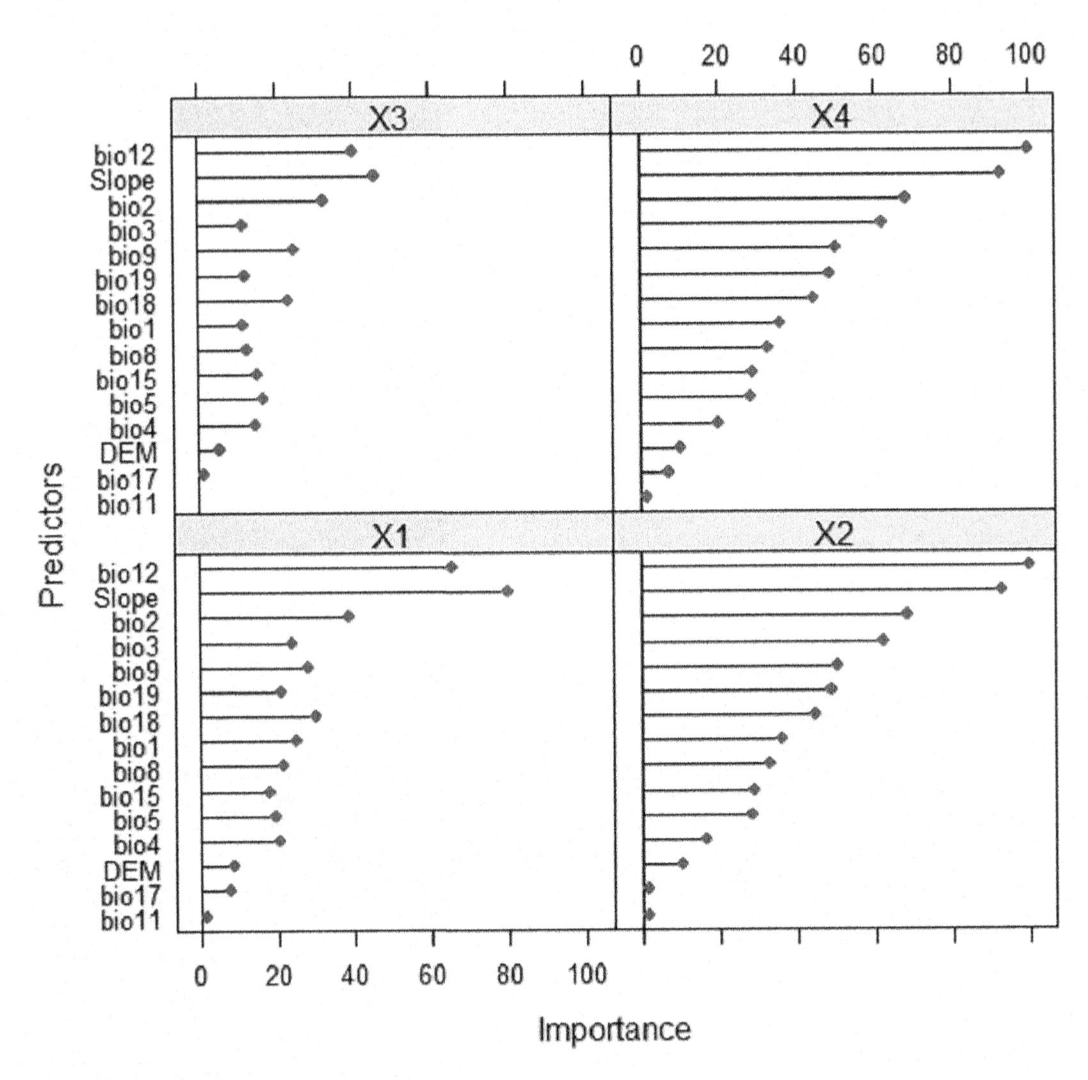

FIGURE 19.5 Relative importance of predictors for the prediction of tropical deciduous forest cover density classes (X_1=<10%, X_2=10%–40%, X_3=40%–70%, X_4=70%–100%).

19.4 DISCUSSION

Tropical deciduous forest cover in India is predominant and susceptible to climate change. Moreover, climate change impacts tropical deciduous forest distribution patterns and density. This study attempted to assess the impacts of climate change on the distribution and area changes in the tropical deciduous forest of India using a machine learning RF model. Generally, dynamic vegetation models are used for the detailed investigation and to assess the growth parameters, including net primary productivity and biomass dynamics under the impacts of climate change on forests. However, this study only deals with the spatial mapping and area changes based on trained data under the current and future climate change scenarios; therefore, a non-linear supervised machine learning and robust RF model is suitable for this objective.

Our trained and tested model showed an accuracy of 0.63 and 0.61, respectively, which is moderate. Moderate accuracy showed by RF is possibly due to the use of low-resolution datasets and the non-inclusion of anthropogenic and physical predictors. Also, the use of coarser resolution data can be one reason for the moderate accuracy of RF. Further, forest cover has seasonal variations, and it is difficult to create a threshold for forest cover density based on the climatic and topographical

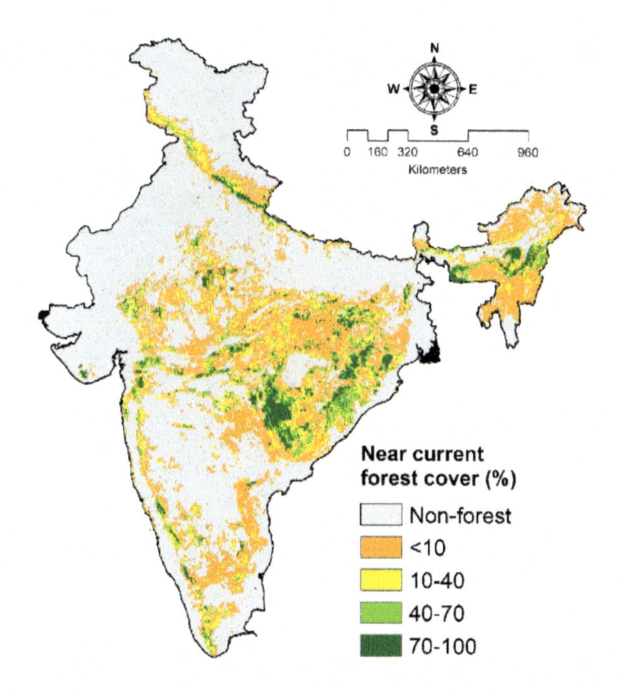

FIGURE 19.6 Predicted map of tropical deciduous forest cover over India in the near current period.

predictors. The relative importance of different predictors shows that annual precipitation (Bio12) is the dominating predictor. The slope is the second most important predictor variable. The mean diurnal range (Bio2) has an importance score of 68.20 for classes 10%–40% and 70%–100%. Therefore, precipitation and temperature would show a dominant role in the future deciduous forest cover distribution and forest cover density changes. DEM, mean temperature of coldest quarter (Bio11), and precipitation of driest quarter (Bio17) are the least important predictors variables. Many studies revealed that temperature and precipitation change is the prime driving factor of forest dynamics and their ecosystems.

Generally, dynamic vegetation models are commonly used to evaluate the impact of climate change on forests. Previously, only limited work has assessed the impacts of climate change on the Indian tropical deciduous forests. A study by Gopalakrishnan et al. (2011) found that about 30% of deciduous *Tectona grandis* species grids in India are susceptible to climate scenarios under A2 and B2 SRES scenarios using IBIS and projected HadRM3 model climatic data. Ravindranath et al. (2006) revealed that around 77% and 68% of the forest grids in India are estimated to face an alteration due to climate change under A2 and B2 scenarios, respectively. According to Zhang et al. (2014), 60% of woody tree species in southwest China will likely lose >30% of their current habitat range under the warmer climatic conditions by the year 2080. Our results also indicate an increase in forest density class (<10%) under the warmer condition of SSP2–4.5 and SSP5–8.5 in the 2070s. Under the SSP2–4.5 scenarios in the 2030s and 2050s, there would be an improvement in the forest density class (70%–100%). In previous studies, it is seen that in addition to climate change parameters (temperature, precipitation and CO_2), anthropogenic activities are also the primary driving force in changing the forest's structure, abundance and composition (Kirilenko and Sedjo 2007; Wu et al. 2013; Cai et al. 2014, Zheng et al. 2019; Gupta and Sharma 2020).

In general, the results showed that India's tropical deciduous forest cover shows non-linear variations with climate change in the future period. The distribution of these forests would also be affected, and climate change would also affect the forest cover density. Further, work using ensemble GCMs models data and adding additional predictors of anthropogenic factors would improve

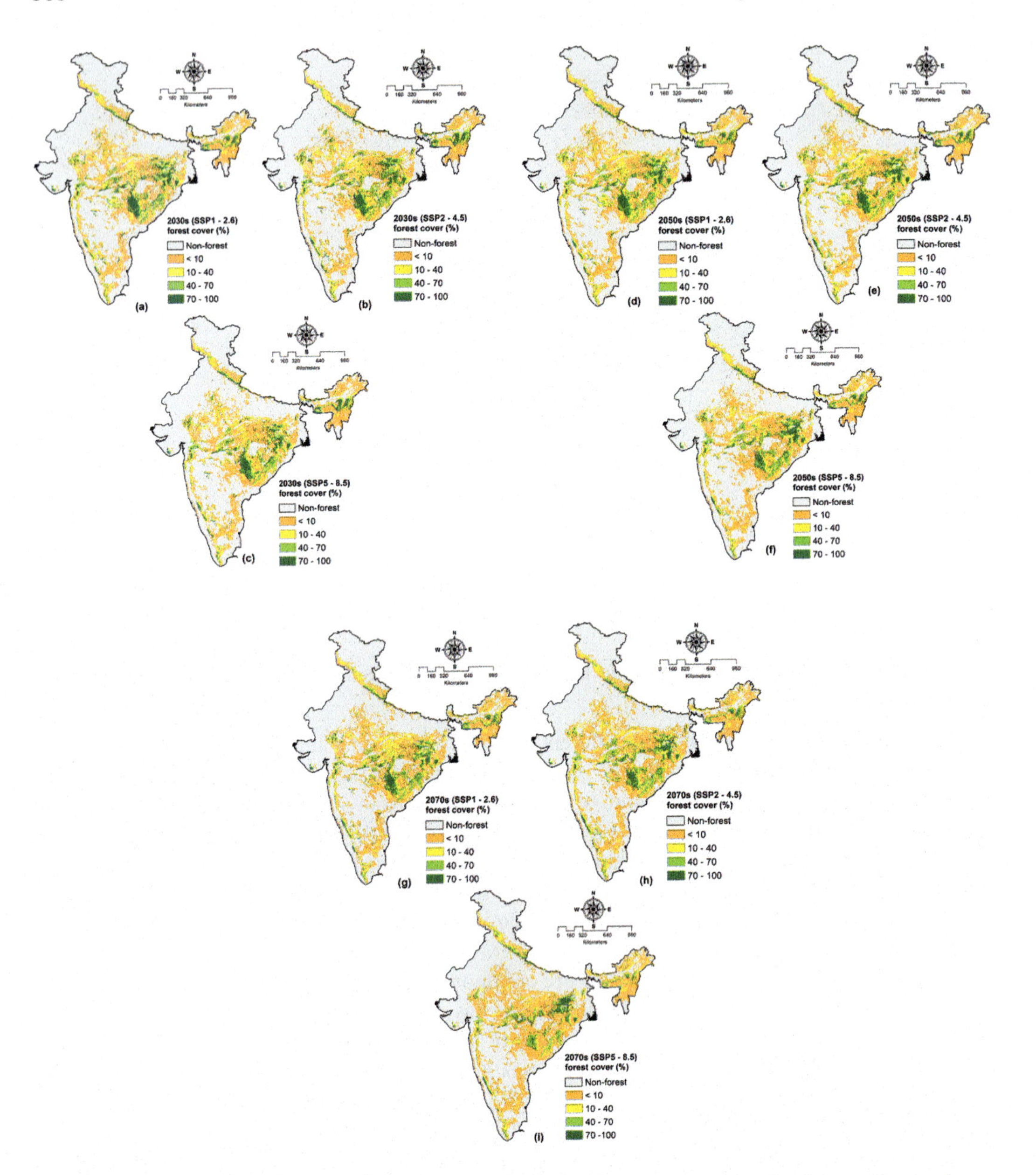

FIGURE 19.7 (a)–(i) Predicted maps of tropical deciduous forest cover under the future climate change scenarios (SSP1–2.6, SSP2–4.5, SSP5–8.5) in the period (2030s, 2050s and 2070s).

prediction accuracy. Further, using high-resolution remote sensing datasets would clear the potential impacts of climate change on India's tropical deciduous forests.

19.5 CONCLUSION

The current study attempted to predict the spatial dynamics of Indian tropical deciduous forest cover under the climate change scenarios using a machine learning RF model. The RF model shows a moderate training (0.63) and testing (0.61) classification accuracy in the spatial prediction of deciduous forest cover classes in the near-current climatic conditions. Under the SSP2–4.5 scenario, the forest cover class (70%–100%) shows an improvement of 3.59% than the near current time. However, under

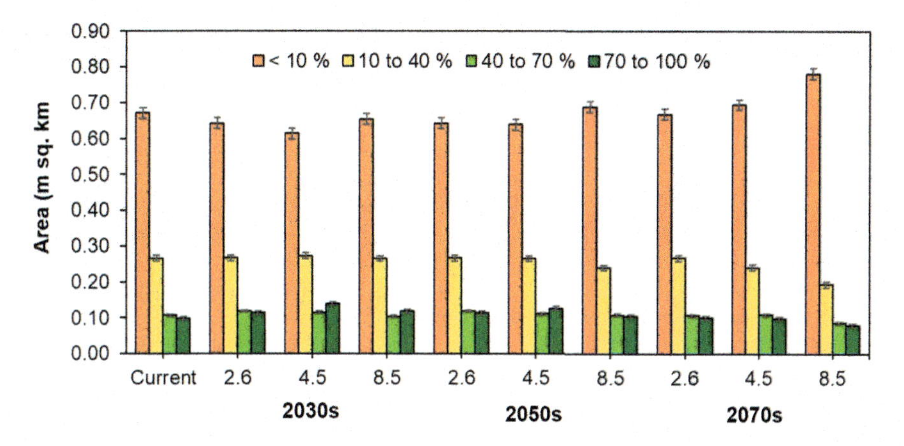

FIGURE 19.8 Area occupied by tropical deciduous forest under the near current and future climate change scenarios.

TABLE 19.2
Percent Area Difference of Deciduous Forest under the Future Climate Scenarios than the Near Current

	2030s – Near Current (%)			2050s – Near Current (%)			2070s – Near Current (%)		
Class	**2.6**	**4.5**	**8.5**	**2.6**	**4.5**	**8.5**	**2.6**	**4.5**	**8.5**
<10%	−2.48	−4.99	−1.46	−2.48	−2.81	1.53	−0.25	2.05	9.76
10%–40%	0.13	0.65	−0.01	0.13	−0.06	−2.23	0.13	−2.19	−6.32
40%–70%	1.04	0.75	−0.29	1.04	0.37	0.11	0.01	0.17	−1.84
70%–100%	1.30	3.59	1.76	1.30	2.50	0.60	0.11	−0.03	−1.59

high emission scenarios of SSP5–8.5 in the 2070s, there would be an increase in the forest density class (<10%) than the near current time, which indicates that climate change could adversely affect the tropical deciduous forests under extreme climatic conditions. This study only considered bioclimatic and topographic variables; however, forest cover is also affected by anthropogenic activities and hazards such as forest fires and droughts. Therefore, including predictor variables from these categories would help improve the model's prediction accuracy. This study helps employ the conservation practices, management and restoration of the tropical deciduous forest cover under the changing climate.

ACKNOWLEDGMENT

We are grateful to the Central University of Rajasthan for research facilities. The first author is grateful the University Grants Commission (UGC) for the UGC NET-JRF fellowship (Ref no. 3551/(NET-JAN2017) for financial support. We are thankful to the anonymous reviewers and editors for their constructive comments.

REFERENCES

Bao, Z., Zhang, J., Wang, G., Guan, T., Jin, J., Liu, Y., Li, M., and T. Ma. 2021. The sensitivity of vegetation cover to climate change in multiple climatic zones using machine learning algorithms. *Ecological Indicators* 124:107443. doi: 10.1016/j.ecolind.2021.107443.

Boonpragob, K., and J. Santisirisomboon. 1996. Modeling potential changes of forest area in Thailand under climate change. *Water, Air, and Soil Pollution* 92(1):107–117. doi: 10.1007/BF00175557.

Breiman, L. 2001. Random forests. *Machine Learning* 45:5–32. doi: 10.1023/A:1010933404324.

Cai, H., Yang, X., Wang, K., and L. Xiao. 2014. Is forest restoration in the southwest China Karst promoted mainly by climate change or human-induced factors?. *Remote Sensing* 6(10): 9895–9910. doi: 10.3390/rs6109895.

Champion, H.G., and S.K. Seth. 1968. *A Revised Survey of the Forest Types of India*. Delhi: Manager of Publications.

del Río, S., Canas, R., Cano, E., Cano-Ortiz, A., Musarella, C., Pinto-Gomes, C., and A. Penas. 2021. Modelling the impacts of climate change on habitat suitability and vulnerability in deciduous forests in Spain. *Ecological Indicators* 131:108202. doi: 10.1016/j.ecolind.2021.108202.

Duo, A., Zhao, W., Qu, X., Jing, R., and K. Xiong. 2016. Spatio-temporal variation of vegetation coverage and its response to climate change in North China plain in the last 33 years. *International Journal of Applied Earth Observation and Geoinformation* 53:103–117. doi: 10.1016/j.jag.2016.08.008.

Evans, J.S., Murphy, M.A., Holden, Z.A., and S.A. Cushman. 2011. Modeling species distribution and change using Random forest. In Drew, C.A., Wiersma, Y.F., and F. Huettmann (Eds.), *Predictive Species and Habitat Modeling in Landscape Ecology*, pp. 139–159. New York: Springer.

Farr, T.G., Rosen, P.A., Caro, E., Crippen, R., Duren, R., Hensley, S., Kobrick, M., Paller, M., Rodriguez, E., Roth, L., and D. Seal. 2007. The shuttle radar topography mission. *Reviews of Geophysics* 45(2). doi: 10.1029/2005RG000183.

Fick, S.E., and R.J. Hijmans. 2017. WorldClim 2: New 1km spatial resolution climate surfaces for global land areas. *International Journal of Climatology* 37(12):4302–4315. doi: 10.1002/joc.5086.

Gopalakrishnan, R., Jayaraman, M., Swarnim, S., Chaturvedi, R.K., Bala, G., and N.H. Ravindranath. 2011. Impact of climate change at species level: A case study of teak in India. *Mitigation and Adaptation Strategies for Global Change* 16(2):199–209. doi: 10.1007/s11027-010-9258-6.

Gupta, R., and L.K. Sharma. 2020. Efficacy of spatial land change modeler as a forecasting indicator for anthropogenic change dynamics over five decades: A case study of Shoolpaneshwar Wildlife Sanctuary, Gujarat, India. *Ecological Indicators* 112:106171. doi: 10.1016/j.ecolind.2020.106171.

Gupta, R., and L. Sharma. 2021. Modelling the growth response to climate change and management of *Tectona grandis* L. f. using the 3-PGmix model. *Annals of Forest Science* 78(4):1–17. doi: 10.1007/s13595-021-01102-y.

Hansen, M.C., and T.R. Loveland. 2012. A review of large area monitoring of land cover change using Landsat data. *Remote Sensing of Environment* 122:66–74. doi: 10.1016/j.rse.2011.08.024.

He, B., Chen, A., Wang, H., and Q. Wang. 2015. Dynamic response of satellite-derived vegetation growth to climate change in the Three North Shelter Forest Region in China. *Remote Sensing* 7(8):9998–10016. doi: 10.3390/rs70809998.

IPCC, 2021. *Climate Change 2021: The Physical Science Basis. Contribution of Working Group I to the Sixth Assessment Report of the Intergovernmental Panel on Climate Change* Masson-Delmotte, V., et al. (Eds.) Cambridge: Cambridge University Press.

ISFR, 2021. Dehradun: Ministry of Environment and Forests, Forest Survey of India (FSI).

Kirilenko, A.P., and R.A. Sedjo. 2007. Climate change impacts on forestry. *Proceedings of the National Academy of Sciences* 104(50):19697–19702. doi: 10.1073/pnas.070142410.

Kong, D., Zhang, Q., Singh, V.P., and P. Shi. 2017. Seasonal vegetation response to climate change in the Northern Hemisphere (1982–2013). *Global and Planetary Change* 148:1–8. doi: 10.1016/j.gloplacha.2016.10.020.

Kuhn, M. 2008. Building predictive models in R using the caret package. *Journal of Statistical Software* 28:1–26. doi: 10.18637/jss.v028.i05.

Lary, D.J., Alavi, A.H., Gandomi, A.H., and A.L. Walker. 2016. Machine learning in geosciences and remote sensing. *Geoscience Frontiers* 7(1):3–10. doi: 10.1016/j.gsf.2015.07.003.

Li, Y., Li, C., Li, M., and Z. Liu. 2019. Influence of variable selection and forest type on forest aboveground biomass estimation using machine learning algorithms. *Forests* 10(12):1073. doi: 10.3390/f10121073.

Nguyen, T.H., Jones, S.D., Soto-Berelov, M., Haywood, A., and S. Hislop. 2018. A spatial and temporal analysis of forest dynamics using Landsat time-series. *Remote Sensing of Environment* 217:461–475. doi: 10.1016/j.rse.2018.08.028.

Osisanwo, F.Y., Akinsola, J.E.T., Awodele, O., Hinmikaiye, J.O., Olakanmi, O., and J. Akinjobi. 2017. Supervised machine learning algorithms: Classification and comparison. *International Journal of Computer Trends and Technology* 48(3):128–138. doi: 10.14445/22312803/IJCTT-V48P126.

Patil, S.D., Gu, Y., Dias, F.S.A., Stieglitz, M., and G. Turk. 2017. Predicting the spectral information of future land cover using machine learning. *International Journal of Remote Sensing* 38(20):5592–5607. doi: 10.1080/01431161.2017.1343512.

Ravindranath, N.H., Aaheim, A., and J. Sathaye. 2011. Climate change and forests in India: Note from the guest editors. *Mitigation and Adaptation Strategies for Global Change* 16(2):117–118. doi: 10.1007/s11027-010-9280-8.

Ravindranath, N.H., Joshi, N.V, Sukumar, R., and A. Saxena. 2006. Impact of climate change on forests in India. *Current Science* 90(3):354–361.

Reddy, C.S., Jha, C.S., Diwakar, P.G., and V.K. Dadhwal. 2015. Nationwide classification of forest types of India using remote sensing and GIS. *Environmental Monitoring and Assessment* 187(12):1–30. doi: 10.1007/s10661-015-4990-8.

Rodriguez-Galiano, V.F., Ghimire, B., Rogan, J., Chica-Olmo, M., and, J.P. Rigol-Sanchez. 2012. An assessment of the effectiveness of a random forest classifier for land-cover classification. *ISPRS Journal of Photogrammetry and Remote Sensing* 67:93–104 doi: 10.1016/j.isprsjprs.2011.11.002.

RStudio Team, 2021. RStudio: Integrated Development for R. RStudio, PBC, Boston, MA. http://www.rstudio.com/.

Shen, W., Li, M., Huang, C., Tao, X., and A. Wei. 2018. Annual forest aboveground biomass changes mapped using ICESat/GLAS measurements, historical inventory data, and time-series optical and Radar imagery for Guangdong province, China. *Agricultural and Forest Meteorology* 259:23–38. doi: 10.1016/j.agrformet.2018.04.005.

Sun, R., Chen, S., and H. Su. 2021. Climate dynamics of the spatiotemporal changes of vegetation NDVI in Northern China from 1982 to 2015. *Remote Sensing* 13:187. doi: 10.3390/rs13020187.

Tatebe, H., Ogura, T., Nitta, T., Komuro, Y., Ogochi, K., Takemura, T., Sudo, K., Sekiguchi, M., Abe, M., Saito, F. and M. Chikira. 2019. Description and basic evaluation of simulated mean state, internal variability, and climate sensitivity in MIROC6. *Geoscientific Model Development* 12(7):2727–2765. doi: 10.5194/gmd-12-2727-2019.

Wu, Z., Wu, J., Liu, J., He, B., Lei, T., and Q. Wang. 2013. Increasing terrestrial vegetation activity of ecological restoration program in the Beijing–Tianjin Sand Source Region of China. *Ecological Engineering* 52:37–50. doi: 10.1016/j.ecoleng.2012.12.040.

Zhang, M.G., Zhou, Z.K., Chen, W.Y., Cannon, C.H., Raes, N., and J. Slik. 2014. Major declines of woody plant species ranges under climate change in Yunnan, China. *Diversity and Distributions* 20:405–415 doi: 10.1111/ddi.12165.

Zheng, K., Wei, J.Z., Pei, J.Y., Cheng, H., Zhang, X.L., Huang, F.Q., Li, F.M. and J.S. Ye. 2019. Impacts of climate change and human activities on grassland vegetation variation in the Chinese Loess Plateau. *Science of the Total Environment* 660:236–244. doi: 10.1016/j.scitotenv.2019.01.022.

20 A Survey of Machine Learning Techniques in Forestry Applications Using SAR Data

Naveen Ramachandran
Indian Institute of Technology Kanpur

K.K. Sarma and Dibyajyoti Chutia
North Eastern Space Applications Centre

Onkar Dikshit
Indian Institute of Technology Kanpur

CONTENTS

20.1 INTRODUCTION

Forest is a crucial part of the terrestrial ecosystem. The role of forest ecosystems in understanding the global carbon cycle (GCC) has drawn the immense attention of the scientific community over the last few decades. Forest stores 70%–90% of the terrestrial biomass (Houghton, Hall, and Goetz 2009). Forest absorbs the carbon during photosynthesis and acts as a major carbon sink. However, the carbon is released back into the atmosphere during forest loss. Hence, monitoring the state of the forest ecosystem is essential for understanding GCC and its role in climate change. Apart from its role in GCC, the forest ecosystem plays a major role in different aspects. Forest covers around 4.06 billion hectares, i.e., ~30.8% of the earth's land area (FAO 2020), and serves as the home for more than 80% of animals, insects, and plants (UNDP 2022). They assist the earth's surface in controlling runoff, prevent flood, soil and wind erosion, and provide sloped surface stability. They also help in reducing air and noise pollution. Further, they provide food, fiber, timber, and a source of bioenergy. However, natural and manmade activities such as deforestation, degradation, forest fire, improper forest management, over-exploitation of natural resources, and forest diseases have led to the current state of the forest, making it one of the most dynamic environments. Over the last three-decade, the global forest has seen a decline of 178 million hectares area (FAO 2020). Particularly in boreal and temperate forests, the primary disturbance was due to wildfire and forestry, as agriculture expansion and deforestation dominated the tropical forests (Curtis et al. 2018). Also, the impact of climate change on forest disturbance dynamics and its feedback remains

DOI: 10.1201/9781003270928-26

unclear. Hence, understanding the forest dynamics is an essential component for improving global models to accurately predict future scenarios (Purves and Pacala 2008), planning mitigation strategies for climate change (Canadell and Raupach 2008; Bolte et al. 2009), and forest management (Gauthier, Leduc, and Bergeron 1996; Björse and Bradshaw 1998).

Monitoring forest ecosystems over a regional and global scale over time by field-based measurements is a tedious and unfeasible process. RS provides suitable tools for monitoring global forests and supporting forest inventory updation but with lesser accuracy than field measurements. Among the existing RS techniques, SAR and Light Detection and Ranging (LiDAR) are the only techniques that can measure the three-dimensional (3D) structure of forests at regional and global scales. At the local scale, airborne LiDAR sensors have been widely used to map forest structure and predict forest parameters such as canopy volume, canopy height model, plant leaf area index, and so on (Peterson et al. 2007; Wulder et al. 2008; Kelly and Tommaso 2015), but constrained by the cost of data acquisition. Spaceborne missions like Global Ecosystem Dynamics Investigation (GEDI) and Ice, Cloud and land Elevation SATellite (ICESat) provide global coverage but are limited by sampling and cloud coverage (Schutz et al. 2005; Dubayah et al. 2020). SAR techniques have emerged as a pivotal tool for monitoring and characterizing forest environments due to its ability to penetrate under-foliage along with sensitivity to the size, orientation, and moisture content of targets (Tiner, Lang, and Klemas 2015). Since the implementation of the first fully polarimetric SAR system, AIRSAR, in 1985, numerous techniques and applications have been developed, especially for forest type classification (Rignot et al. 1994; Lapini et al. 2020; Bjerreskov, Nord-Larsen, and Fensholt 2021; Udali, Lingua, and Persson 2021), forest species classification (Wollersheim, Collins, and Leckie 2011; Bjerreskov, Nord-Larsen, and Fensholt 2021; Udali, Lingua, and Persson 2021), forest/-non-forest mapping (Shimada et al. 2014; Martone et al. 2018) and forest parameter estimation, such as tree height and biomass (Fransson, Walter, and Ulander 2000; Ranson and Sun 1994).

In addition, the rapid evolution of interferometric techniques allowed measurement of the phase difference between two SAR images which found a wide range of applications in mapping land topography (Nico et al. 2005; Geymen 2014), stem volume (Askne et al. 1997; Fransson et al. 2001), forest height (Balzter, Rowland, and Sacih 2007; Lei, Treuhaft, and Gonçalves 2021), and many others. Later, in the late 1990s, the inclusion of polarimetry to interferometric techniques constituted a watershed moment in SAR forestry applications. It is now possible to map the 3D structure of a forest and retrieve forest parameters such as height and biomass with acceptable precision using interferometric techniques such as Polarimetric SAR interferometry (PolInSAR) (Mette et al. 2002; Liao et al. 2018) and tomography (TomoSAR) (Ramachandran et al. 2021). However, the SAR polarimetric and interferometric data are limited due to saturation and decorrelation problems, respectively, which affect the accuracy of the derived products (Ghasemi, Tolpekin, and Stein 2018b; Lavalle and Hensley 2015). As a result of such hindrance, the physical or semi-empirical models may not exhibit the true relationships between the parameter and the observable. In such scenarios, the ML approach can be considered an excellent alternative to perform different applications from SAR data. Also, with the number of current and future SAR missions, handling such a large volume of data demands computationally efficient processing strategies. While ML has shown its capability to handle big data problems (Zhou et al. 2017; Sun and Scanlon 2019), the use of ML with SAR data is still in the early days.

This article aims to survey the forestry applications using SAR and ML-based algorithms, highlight its advantages and limitations along with current knowledge gaps, and propose future trends. This review focuses on four major forest applications using SAR data, i.e., forest degradation mapping, forest type/species classification, forest height, and biomass estimation. Even though numerous literatures can be found for these applications using SAR data and ML with optical datasets, the use of SAR data with ML is limited and less found. This chapter is organized as follows: The fundamental principles and techniques in SAR and their role in forestry applications are discussed in Section 20.2. This chapter doesn't intend to provide an exhaustive cover of SAR techniques or ML algorithms but rather to present the reader with an overall idea of the topics, emphasizing their application in forestry. Sections 20.3–20.6 review key forestry applications based on SAR and ML algorithms. Finally, the chapter concludes with a future perspective and conclusion.

20.2 SAR AND MACHINE LEARNING

The radar sensors operated in the microwave wavelengths, ranging from a few millimeters to meters, of the electromagnetic (EM) spectrum. All-illumination, all-weather, and cloud penetrating capabilities (Oliver and Quegan 2004; Cumming and Wong 2005) make it a suitable choice for monitoring the earth's surface. The sensitivity of microwave signals to moisture, dielectric properties, target roughness, and soil and vegetation penetration provide complementary information compared to other wavelengths used for RS. The penetration in the volumetric media depends upon the wavelength used. The longer the wavelength, the deeper the penetration. Hence, longer wavelengths, such as L- or P-band, are more suitable for forest studies. The measurement, processing, and analysis of SAR data at different polarization allow to develop a better understanding of the polarimetric properties of the target under investigation. This technique is called SAR polarimetry (PolSAR). The SAR antenna is designed to transmit and receive EM waves in a defined polarization. Now, varying the input and output polarization state of transmitting and receiving signals, images can be acquired in different polarization. This allows us to study the size, shape, orientation and dielectric properties of the target under investigation making PolSAR a powerful tool for classification, target identification and detection, and parameter estimation. Further, the SAR interferometry (InSAR) technique demonstrated by Graham (1974) exploits the absolute phase difference between two or more SAR images acquired at slightly shifted orbits and/or time. This allows us to estimate any difference between the range values of observations, even to a centimeter to sub-centimeter levels. These techniques have demonstrated their capabilities as powerful techniques for a wide range of applications such as ground deformation studies and Digital Elevation Model (DEM) generation, forest height estimation, velocity mapping, and target detections. Another recent advancement deals with reconstructing the 3D reflectivity of volumetric media under observation, known as TomoSAR (Reigber and Moreira 2000). It utilizes multiple SAR images acquired at different orbital paths shifted in vertical and/or horizontal directions to synthesize an aperture in the vertical direction. This allows the estimation of reflectivity profiles in the vertical direction. This means that to achieve higher resolution in a vertical direction, a large number of SAR data is required.

ML is a disciple of artificial intelligence that is designed to mimic human intelligence, using statistical techniques and a rule set by learning from data. The ML algorithms are primarily classified into four categories (1) supervised, (2) unsupervised, (3) semi-supervised, and (4) reinforcement learning (Mohammed, Khan, and Bashier 2016). The supervised learning algorithm tries to understand the relationship between the input and output variables and develop a function or mapping to obtain the desired output or class. The classification and regression problems are examples of supervised learning. Unsupervised learning is a data-driven approach where the unlabeled data is analyzed to find hidden patterns. Some common examples of unsupervised learning are clustering, feature learning, and dimensionality reduction. A semi-supervised approach is a hybrid approach combining supervised and unsupervised learning. Here the data are grouped into labeled and unlabeled data, and the label portion is used to predict the unlabeled data allowing an increased accuracy than the supervised learning. Text classification and machine translation are some examples of semi-supervised learning. Finally, reinforcement learning is an environment-driven approach that allows the model to automatically assess the best possible solution in a particular environment to improve model accuracy. Self-driving cars, gaming, and robotics are some examples of reinforcement learning. Before understanding ML algorithms, it is essential to understand the ML process, which involves four key steps (1) defining a problem, (2) data pre-processing, (3) model building and evaluation, and (4) prediction. The first step involves the identification and framing of the objectives of the project undertaken and the collection of data. It requires discussions with experts and stakeholders in that domain, selection and availability of appropriate data to address these objectives, and understanding data using statistics and data visualization. If the data is poorly recorded, erroneous, repetitive, uncorrelated, and not properly structured, this may degrade model accuracy. Thus, the *data pre-processing* step involves adding, removing, or transforming data into a well-structured dataset. The *model building and evaluation* stage begins by splitting the data into training and testing sets. Then the training data is used to build the model and predict the accuracy. The -cross-validation and hyperparameter tuning are used to obtain a robust model and results. The testing data can be used to indicate the model's generalizability and then improve the model further, if necessary. The selection of ML algorithms depends on the problem and data complexity. The final step involves the prediction of labels or output depending on whether it is a classification or regression problem.

In recent years, the acceleration and success of ML algorithms have attracted the SAR community to evaluate ML in comparison to conventional approaches. However, the way SAR data looks in comparison to optical data makes it difficult to decode the information. Further, the presence of SAR geometric distortions such as slant-range scale distortion, layover, foreshortening, and shadow, especially over terrain with high relief, makes it more complicated. Hence, extreme care must be taken when exploring the relationship between SAR datasets and ML. Although ML algorithms have been used for different SAR applications (Zhu et al. 2020a), here we focus on the application of SAR and ML in four major applications related to forestry.

20.3 FOREST CLASSIFICATION

The forest type and species are important ecological traits of the forest community, which define the functioning and services the forest provides. Forest species accumulate different amounts of carbon, water, and other nutrients differently. Higher diversity results in a more efficient intake of resources (Barry et al. 2019) across space, time, chemical form, or across all these, which usually yields higher above-ground biomass (AGB) productivity. Thus, it contributes differently to the carbon and water cycle and climate change mitigations. Further, different socio-economic forest products are closely related to the spatial distribution of different species and yield different economic profiles. However, the effect of climate change on different tree species distribution is of key importance for understanding the future supply of these goods since climate change may result in the redistribution of tree species (McKenney et al. 2007). Also, understanding species distribution is essential for forest management activities (Falk and Mellert 2011; Pecchi et al. 2019). Thus, monitoring the spatial distribution of forest types and species is essential.

The interaction of forest components with SAR signals mainly depends upon frequency or wavelength. Low-frequency signals, such as the L- or P-band, penetrate deep into the trees and interact largely with structural components like branches or trunks, whereas high-frequency signals, such as the X- or C-bands, engage substantially with crown features like leaves, needles, and twigs (Le Toan et al. 1992). Based on previous literature, forest classification using high-frequency SAR data has shown higher potential than low-frequency data (Rignot et al. 1994; Ranson, Saatchi, and Sun 1995). In contradiction, Saatchi and McDonald (1997) reported the ability of L- & P-band to differentiate the coniferous from broadleaved species. Further, the sensor parameters, such as look angle, polarization, and resolution, along with the environmental parameters, such as terrain slope and weather conditions, also play a critical role. The high-frequency SAR data have low sensitivity to terrain variation than low-frequency SAR (Wollersheim, Collins, and Leckie 2011). Rignot et al. (1994) found HV polarization data acquired at different frequencies more suitable for classification. The seasonal variation of forest structure and environmental conditions must be accounted for in forest classification. Combining data obtained during leaf-off and leaf-on was observed to provide better accuracy than standalone (Ranson and Sun 1994, 2000; Maghsoudi, Collins, and Leckie 2012; Ortiz et al. 2012). The role of phenology using high-frequency bands has been extensively addressed (Maghsoudi, Collins, and Leckie 2012; Ortiz et al. 2012; Rutsches, Schaepman, and Small 2018; Tran et al. 2021). Also, using data acquired during the arid and snowy conditions decreases the texture information, hampering the classification accuracy (Kurvonen and Hallikainen 1999).

PolSAR-based classifier algorithms use the richness in polarimetric information to boost the differentiability between classes. Ferro-Famil, Pottier, and Lee (2001) compared two unsupervised classification approaches based on the iterative maximum likelihood and polarimetric cross-correlation information. It was observed that, single-frequency data alone could not provide high descriptive values. Maghsoudi, Collins, and Leckie (2012) compared two approaches based on support vector machines (SVM) and Wishart's classifier. They observed an ensemble of SVM classifiers based on class-based feature selection provided better accuracy than a conventional SVM and traditional Wishart's classifier. Mazza et al. (2019) integrated different observables such as backscatter, coherence, volume decorrelation, and local incidence angle from TanDEM-X InSAR data to extract the forest maps. They assessed the capability of different Convolutional neural network (CNN) architectures and found that the U-Net performed better than all architectures. The feasibility of using multi-temporal C-band SAR data for forest type classification at a continental scale was explored, and it was noted that an accuracy almost within the range of Copernicus High-Resolution Layers was achieved (Dostálová et al. 2021). Similarly, Haarpaintner and Hindberg (2019) compared multi-temporal observations from the C-and L-band data and their synergies for mapping land cover over the tropical area of the Democratic Republic of Congo. They found that multi-temporal observations achieved better accuracies than global forest maps produced by Landsat and ALOS. Udali, Lingua, and Persson (2021) used the random forest (RF) approach and the multi-temporal SAR data acquired at C-band to classify forest type and species information. They observed that even though good accuracy was achieved for forest type classification, the tree species classification resulted in degraded accuracy. The use of multiple winter information improved the estimates compared to single-season images. Another possibility to enhance the estimates is based on combining information from multiple frequencies (Ranson, Saatchi, and Sun 1995). Lapini et al. (2020) compared the potentials of different ML algorithms to classify forest/non-forest and forest type over Mediterranean forests using multi-frequency-multi-polarization data acquired at L-, C-, and X-bands. They observed combining different frequencies improved the accuracy of the RF algorithm. Numerous studies combining multiple sensors have recently shown the possibility of improving classification accuracies by compensating for each other's weaknesses. Wolter and Townsend (2011) used the partial least squares regression (PLSR) approach to combine multi-frequency SAR data with Landsat and forest structure information derived from SPOT-5 data to achieve forest species-level classification. They found PLSR to be a valuable tool for integrating multi-sensor data for forest-type discrimination. The spectral neighborhood information from SPOT-5 data assisted SAR data in discriminating many species, while the structural information helped identify coniferous species. Attarchi and Gloaguen (2014) compared three different ML classifications using combined L-band SAR with Landsat 7 data under rough terrain conditions and compared it with the traditional classifier. They observed that in the case of Landsat, the terrain-corrected data improved the accuracy, while for the SAR sensor, texture-based information enhanced the mapping accuracy. The combined usage resulted in slightly improved performance, with ML proving to provide more robust mapping compared to the conventional classifier. Yu, Li, and Fu (2018) performed forest-type classification based on RF, combining SPOT-5 and multi-temporal RADARSAT-2 data, and showed improved accuracy compared to single-source classification. Bjerreskov, Larsen, and Fensholt (2021) propounded a multi-temporal, multi-sensor-based approach for classifying forest/-non-forest, type, and species. The cloud-based approach combined Sentinel-1 and 2A sensors data using RF. Although the forest/non-forest and type classification resulted in good accuracy, species classification resulted in comparatively lower accuracy, with broadleaf groups with higher error rates. (Dobrinić, Gašparović, and Medak 2021) combined Sentinel-1 and Sentinel-2 time-series data, texture features, spectral indices, and DEM, which were used for vegetation classification using RF. Combining Sentinel -1 (S1) and Sentinel -2 (S2) provided the highest overall accuracy of all classification scenarios. Similarly, Ruiz et al. (2021) compared the performance of RF and *k*-nearest neighbor (*k*-NN) for vegetation species classification in the tropical wetland by combining S1 and S2A data and observed RF performed better. A recent study by Minh, Ngo, and Lê (2021) demonstrated an approach based on TomoSAR backscattered power for forest density classification. The results were compared with the conventional SAR backscatter approach, demonstrating that incorporating forest vertical structure information from TomoSAR resulted in an improved estimate. Both approaches were trained and validated using a traditional RF algorithm.

The integrated use of multi-temporal, multi-sensor data provides encouraging results. However, from a SAR point of view, it is observed that using multi-temporal data improved classification accuracies compared to a single date or season data. Further, in the absence of multi-temporal data, multi-frequency data at low frequency allows better mapping, especially over tropical forests. HV polarization assists in better classification compared to co-polar channels. From the ML perspective, RF and SVM are two widely used approaches. However, the use of CNN and Recurrent neural network (RNN) or a combined model must be critically explored as it can recognize spatial, structural, and temporal patterns, respectively.

20.4 FOREST DEGRADATION/DEFORESTATION MAPPING

Forest degradation refers to the reduction of forest attribute quality of the site or stands, resulting in a reduction in the capacity of forest supply or products because of man-made and environmental change (Thompson et al. 2013). In contrast, deforestation refers to the loss of forest area to any other land-use class (FAO 2005). While deforestation is a rapid process, degradation is usually gradual and eventually leads to deforestation. The main causes for these changes are improper logging management, forest fires, shifting cultivation, fuelwood collection, invasive species, and cattle grazing. In efforts to mitigate climate change, the Reduction of Emission due to Deforestation and Degradation and Sustainable Forest Management (REDD+) framework by the United Nations Framework Convention on Climate Change (UNFCCC) requires accurate reporting of both forest degradation and deforestation for policy implementation and potential financial structure and incentive programs. However, there is still a large uncertainty about carbon emissions due to deforestation and degradation due to the inability to precisely monitor and map larger areas over a temporal scale. A few review papers address deforestation and forest degradation mapping (Goetz et al. 2015; Hirschmugl et al. 2017; Mitchell, Rosenqvist, and Mora 2017; Dupuis et al. 2020; Gao et al. 2020). However, forest degradation/deforestation is primarily discussed based on optical data. Here, we extensively review approaches to map forest degradation and deforestation using SAR data. Deforestation usually results in a change from volume to ground scattering in cross-polarized backscatter data, while degradation diminishes volume scattering depending on the level of degradation (Flores-Anderson et al. 2019). However, the soil roughness and moisture conditions can introduce ambiguities, especially at higher frequencies. The use of time-series data allows us to overcome such situations.

The use of SAR backscatter data to estimate deforestation/degradation is primarily based on classification (Hoekman, Vissers, and Wielaard 2010; Dong et al. 2015) or time-series analysis (Antropov et al. 2015; Reiche et al. 2015b). Zhang et al. (2012a) analyzed multi-temporal polarimetric RADARSAT-2 images for forest classification and deforestation mapping from three different perspectives. The temporal behavior analysis found winter to be the most optimum for forest type and deforestation classification. Polarimetric decomposition comparison showed Freeman-Durden decomposition was useful in identifying forest type and deforestation. Finally, fusing multi-temporal information enhanced forest information, and the SVM-based classification allowed the identification of forests with an accuracy of 87.63%. Over the main mining district in central Guyana, an approach using the feature-based fusion of multi-temporal medium resolution ALOS PALSAR and Landsat subpixel information was performed using a decision tree classifier. It was found that the forest classification and change map achieved an overall accuracy of 88% and 89.3%, respectively. This fusion approach shows improved performance compared to individual sensor performance (Reiche et al. 2013). Pantze, Santoro, and Fransson (2014) proposed a two-stage change detection approach, combining a polynomial-based histogram matching and Markov random field, using bi-temporal ALOS PALSAR L-band data. The approach was validated over the boreal forest, and a pixel-wise accuracy of 90% was achieved while estimating clear-cut. Similarly, L-band data acquired using ALOS PALSAR during the dry season was used by Joshi et al. (2015) to propose an approach to estimate fast- and slow-recovering changes. The time-series data was calibrated using local moving-window filtering, and multiple thresholding was manually selected from the training data to estimate the disturbed area and observed that the SAR-based approach could provide much improved results compared to optical-based products. Barreto et al. (2016) proposed a change detection approach based on object detection using multi-temporal X-band data acquired using BRADAR airborne OrbiSAR-2 sensor over the heterogeneous Brazilian Atlantic forest. SAR images were first segmented into superpixels, followed by feature extraction using gray-level cooccurrence matrix (GLCM) and object correlation images (OCI) methods before being classified into deforested, unchanged, and other changed regions using the multi-layer perceptron (MLP). The results show a 10% improvement in accuracy compared to state-of-the-art object-based approaches. Another approach based on expectation maximization was proposed by Mermoz and Le Toan (2016) and validated using L-band ALOS PALSAR data over Vietnam, Cambodia, and Lao PDR. They found the forest disturbance and growth rates to be $-1.23\pm0.03\%$ and 0.61%, with more than 90% of the user's accuracy. Ma et al. (2019) presented a SAR change detection approach based on multi-grained cascade forest (gcForest) and multi-scale image fusion. The algorithm uses gradient information and a probability map from gcForest to detect the pixels that change gray values abruptly and reduce the effect of speckle compared to other change detection approaches. Further, reduced parameter space makes it easy to train compared to conventional deep learning approaches. Tarazona et al. (2021) mapped tropical forests using multi-sensor time-series data and ML algorithms to identify deforestation spots with acceptable accuracy. The impact of seasonality was reduced using Landsat-derived Photosynthetic Vegetation (PV) index, and then SAR images were fused with Landsat using Principal Component Analysis (PCA). Comparing the ML tools, RF-NN and PVts-b showed similar results, with PVts-b having the highest commission and lowest omission errors. Hethcoat et al. (2021) validated the ability of two approaches, using RF and the Breaks For Additive Season and Trend (BFAST) algorithm for analyzing time-series data, to map and monitor selecting logging. The data was acquired at C- (Sentinel -1 and RADARSAT-2) and L-band (ALOS-2 PALSAR-2) over Brazilian Amazon tropical forests. The RF classification using the SAR datasets performed poorly. However, the BFAST approach using S1 time-series data was able to detect areas with selective logging $>20\,m^3/ha$. Kuck et al. (2021) compared the performance of three ML algorithms to estimate selective logging using multi-temporal COSMO-SkyMed images over the Brazilian Amazon. They observed that MLP slightly improved the performance compared to RF and AdaBoost. Ortega et al. (2021) compared the capability of the different sensors to map deforestation using three different Fully Convolutional network architectures over the Amazon rainforest. Although S1-based results were slightly inferior to Landsat and S2, it was sufficient to map deforestation under cloudy conditions. Zhao et al. (2022) proposed a deep learning-based approach using U-net architecture for mapping forest harvesting over two deforestation hotspots: California, USA and Rondônia, Brazil, which outperformed the conventional object-based method and could reveal significant changes in monthly harvesting. Some studies exploited the geometric artifact, shadow, occurring at the borders of the deforested patch using S1 time series (Bouvet et al. 2018a; Hirschmugl et al. 2020; Ballère et al. 2021). The approach outperformed the UMD-GLAD Forest Alert dataset in spatial and temporal estimation with more than a 95% detection rate for a sample size larger than 4 ha (Bouvet et al. 2018a). Further, combining results based on this approach with optical-based results improved the accuracy (Hirschmugl et al. 2020). Another key approach based on Bayesian to combine multi-sensor information to extract near real-time deforestation was propounded by Reiche et al. (2015a, 2018) and showed detection at high spatial and temporal accuracies. A similar approach was adopted using Sentinel-1 data alone, which allowed mapping disturbance and providing alerts, of which more than 80% of events were <0.5 ha (Reiche et al. 2021). Further, different approaches based on texture (Rauste et al. 2013), physical modeling (Hoekman et al. 2020), and the recurrence quantification analysis (RQA) Trend (Cremer et al. 2020) were also explored for forest change mapping.

Numerous studies on forest change based on interferometric data have also shown good capabilities but are not explored the full potential. Wang, Ouchi, and Jin (2011) utilized two single-pass SAR data acquired at X-band to access the damage caused by the typhoon over Tomakomai, Hokkaido, Japan. Surface height change detection was performed and found to be in agreement with the ground data. Solberg, Astrup, and Weydahl (2013) estimated forest clearcut between 2000 and 2011 using digital surface models (DSM) estimated from Shuttle Radar Topography Mission (SRTM) and Tandem-X. Based on the thresholding approach, the clear-cut areas were demarcated from the DSMs. Using SRTM X-band data as the reference provided slightly better results compared to C-band. Similarly, an approach to generate forest disturbance based on the DSM derived from COSMO-SkyMed X-band data using an integrated approach that combines interferometric processing with radargrammetry was adopted (Deutscher et al. 2013). The study explored two different approaches, (1) the height variance, where the difference between DSM and mean flattened DSM is used (2) the SRTM Difference approach, where SRTM DSM replaces SRTM Difference, to detect forest degradation. Although both approaches performed relatively well, the prior was not useful in hilly regions. The combination provided the best accuracy of all. Lei et al. (2018) developed a disturbance index, a forest disturbance indicator (DI), from interferometric data to detect selective logging. Time series of InSAR from TerraSAR-X was used to generate DI over the Tapajos National Forest in Brazil and validate it against ground data. Compared with ALOS L-band data, it was noted that X-band results were much more stable and accurate for mapping disturbance. Carstairs et al. (2022) estimated biomass change using InSAR height over selectively logged plots in the hilly terrain of central Gabon. It was observed that the selection of apt pass-direction at the pixel level could improve degradation results over hilly terrain.

The backscatter-based approaches are generally preferred to the interferometric-based methods to map forest change. Most studies have used L-band data to map deforestation. However, the recent availability of time-series backscatter data at the C-band has shown promising results even in tropical forests. Time-series-based approaches offer an excellent way to monitor forest change detection compared to classification and bi-temporal-based approaches. With different missions such as NISAR and sentinel-1 series, freely available data at different frequencies will open new possibilities to combine multi-frequency time-series data to better monitor and map forest changes and deforestation. The ML-based algorithms have not been exploited to their full potential to monitor forest change and possess good potential. However, the ML architecture must be designed to internally adopt multi-source, multi-scale features and reduce computational complexity.

20.5 FOREST TREE HEIGHT ESTIMATION

Forest height is a key geophysical parameter in forest inventories. Having a quantitative assessment of forest height allows us to assist prediction of forest growth (O'Neill and Nigh 2011), stand volume (Rodil, Aranda, and Burkhart 2017), and monitor activities such as degradation, unauthorized logging, windfall, and many others, which alters the state of the forest. Apart from these, it can be directly linked via an allometric equation to the forest AGB and carbon stock estimation (Sullivan et al. 2018; Hunter et al. 2013; Chave et al. 2014). Thus, a better estimation of forest tree height allows for a better characterization of forest environments. Although the SAR data provides an excellent tool for estimating tree height, literature using ML for tree height is majorly based on the combined use of optical and LiDAR data (Chen, Hay, and Zhou 2010; Chen, Hay, and St-Onge 2012; Stojanova et al. 2010; Gu et al. 2018; Zhu et al. 2020b; Potapov et al. 2021).

From the perspective of SAR, backscattered power and interferometric approaches allow the estimation of mean tree height, either empirically or using physical models. Backscattered power provides an indirect measurement of tree height, as the backscattered power depends on the number of scatterers within a resolution cell, which varies with tree height and density (Flores-Anderson et al. 2019).

Limited studies have explored the use of backscattered power for tree height estimation (Fransson, Walter, and Ulander, 2000; Kovacs et al. 2008). Garestier et al., (2009) explored the potentials of polarimetric data at the L- and P-band and observed a linear correlation between anisotropy and mean tree height at P-band. Even though polarimetric data provides valuable information about the scattering process, its lack of ability to establish a direct relationship to the vertical structure has prompted its use along with other data sources, such as LiDAR, which provides information on a vertical structure. García et al. (2018) demonstrated a multi-sensor approach, integrating LiDAR, SAR, and multispectral data using Least-Squares SVM to estimate tree height. Li et al. (2020) compared the ability of a deep neural network and RF to estimate tree height over mountainous regions and obtained satisfactory results using S1, S2, and ICESat-2 data. Another RF study used backscattered power, derived texture information from S1 data, and integrated it with ICESat-2 data to achieve 4.5% RMSE for tree height (Nandy, Srinet, and Padalia 2021). Pourshamsi et al. (2021) compared SVM, RF, rotation forest (RoF), and canonical correlation forest (CCF) for tree height estimation from airborne polarimetric variables in combination with LiDAR measurements and found that a reasonable accuracy can be achieved even in dense tropical forests.

The concept of acquiring interferometric data at different polarization remarked a decisive step in the retrieval of forest height and the Digital Terrain Model (DTM). The sensitivity of polarimetric data to differentiate different scattering processes and the interferometric ability to measure the distribution of these elements along the vertical direction allowed PolInSAR to characterize the vertical structure and estimate different phase centers. However, the direct measurement of vegetation height is impossible as the phase centers are not usually located at the top of the volume, i.e., located somewhere between ground and volume, depending on the wavelength. The backscattering forward models relate forest parameters to interferometric observables and allow inversion to estimate precise parameters. In view of this, numerous models, such as Random Volume over Ground (RVoG) (Treuhaft et al. 1996; Treuhaft and Siqueira 2000), the interferometric water cloud model (IWCM) (Askne et al. 1997; Askne, Soja, and Ulander 2017), the two-level method (TLM) (Soja, Persson, and Ulander 2015a, b) and many others. The RVoG model relates the interferometric coherence to four forest parameters, including tree height. This model is one of the most successful and extensively used model and is validated at different frequencies (Neumann et al. 2012; Kugler et al. 2015; Khati, Singh, and Kumar 2018) and regimes (Liao et al. 2018; Schlund et al. 2019; Praks et al. 2007; Hensley et al. 2012; Hajnsek et al. 2009). A modified form of the RVoG model was presented by Lu et al. (2013) to account for the effect of sloped terrain. Further, to achieve optimal inversion over a range of tree height, multiple PolInSAR observation is required (Kugler et al. 2015). However, in a repeat-pass system, the effect of temporal decorrelation is prominent, resulting in height estimations being overestimated (Lee et al., 2013). Different models such as the RVoG model with vertical temporal decorrelation (Papathanassiou and Cloude 2003), Random Motion over Ground (RMoG) (Lavalle and Hensley 2012, 2015), modified RMoG (Ghasemi, Tolpekin, and Stein 2018a, c) were developed to minimize the effect of temporal decorrelation on tree height estimation from a multi-baseline observation. Ghasemi, Tolpekin, and Stein (2018a) compared RVoG, RMoG, and modified RMoG models over boreal forests at the L- and P-band and found that modified RMoG outperformed other models at P-band and provided similar results to RMoG at L-band. Even with multi-baseline data, inversion of models is mostly an ill-poised problem, as adding each baseline introduces temporal decorrelation terms for ground and volume, and the ground-to-volume ratio is polarization-dependent. Achieving a balanced inversion usually relies on the assumption or estimation of parameters from external sources. For example, the conventional RVoG assumes the unit ground coherence magnitude, which is not valid in the presence of temporal decorrelation (Simard and Denbina 2018). Similarly, the ground-to-volume ratio is assumed to be zero for volumetric coherence but fails when the signal reaches the surface and has ground contributions. The approaches based on fixed extinction (Hajnsek et al. 2009) and temporal decorrelation (Simard and Denbina 2018) values to estimate the remaining parameters have been discussed. However, the forest ecosystem's heterogeneous nature and temporal behavior don't justify the usage of constant parameter values. Another possibility is to use the LiDAR data to estimate the temporal decorrelation (Simard and Denbina 2018) and/or other model parameters such as extinction coefficient and/or ground-to-volume ratio (Qi et al. 2019) and constrain the inversion. Also, it was observed that using ground phase information estimated from external DTMs, such as LiDAR DTM, significantly improved the results of inversion rather than estimating it from the model (Praks et al. 2009; Qi et al. 2019). Different merging approaches have been adopted to retrieve the best height estimates from a multi-baseline dataset. Lee et al. (2011) discuss incoherent merging approaches based on coherence region eccentricity and height variance, while (Denbina, Simard, and Hawkins 2018) proposed a method that selects baselines with high coherence magnitude and larger phase diversity.

Now, combining ML and an interferometric approach, Denbina, Simard, and Hawkins (2018) used SVM to select the interferometric baseline for tree height estimation along with LiDAR data. The selection of baseline was treated similarly to the classification problem. Pourshamsi et al. (2018) introduced an approach to merge multi-baseline interferometric data using the SVM model by minimizing the absolute difference between LiDAR-derived heights and height estimated from the PolInSAR approach. The studies showed that the estimated SVM merged height provides better results than single-baseline and conventional PolInSAR merging approaches. Similarly, the capability of RF and MLP to combine LiDAR and interferometric data to estimate forest structure parameters, including forest height, was investigated by Brigot et al. (2019). A study focused on selecting a dual-baseline from a multi-baseline dataset was proposed using SVM. An unsupervised deep learning-based approach, known as a tailored Generative Adversarial Network (GAN), was adopted by Zhang et al. (2022) to estimate forest height using high-resolution multi-pass PolInSAR and low-resolution LiDAR datasets.

TomoSAR allows retrieval of forest height from the tomographic cube based on the backscatter power distribution function analysis along the vertical direction. The assumption here is that the shape of the distribution can be divided into three zones: backscattering from the canopy layer, especially the phase center, power loss, and noise zone (Tebaldini and Rocca 2012). Once we estimate the phase center, the power loss parameter (K) must be estimated above the phase center region, allowing us to retrieve the forest height. Usually, to estimate the optimum K parameter, we use CHM derived from LiDAR data. For a detailed understanding, the reader may refer to Minh et al. (2016) and Ramachandran et al. (2021). Apart from this, TomoSAR approaches have tried to approximate the vertical structure in terms of orthogonal function basis (Cloude 2006; Zhang, Ma, and Wang 2012b; Pardini and Papathanassiou 2018; Roman Guliaev, Matteo Pardini, and Papathanassiou 2021) and estimate the tree height values (Zhang et al. 2018; Cloude 2006). Praks et al. (2008) provided a ground phase estimated using LiDAR DTM as an input to Polarization Coherence Tomography (PCT) rather than calculating it from the RVoG model and achieved improved estimation.

Interferometric-based approaches have been widely used to estimate tree height from repeat pass SAR data. The RVoG Model has been used extensively and shown acceptable accuracy. However, the accuracy is diminished in the presence of temporal decorrelation. The modified RMoG model has demonstrated improved accuracy over RVoG and RMoG models (Ghasemi, Tolpekin, and Stein 2018a, c) but needs to be validated extensively. Similarly, the TomoSAR approach integrated with LiDAR data provides an alternative but suffers from high computational costs. An ML-based approach combining SAR backscatter and LiDAR data has also shown good potential and is a way to look forward due to the availability of the datasets in the same timeframe and as they reduce the complexity of physically-based models.

20.6 FOREST BIOMASS ESTIMATION

From a forestry perspective, biomass refers to the total amount of living organic matter, including both living and dead organic matter from vegetation. The biomass density is usually expressed in tones or Megagram per unit area. AGB is a portion of biomass above the soil and is defined as "*All living biomass above the soil including stem, stump, branches, bark, seeds, and foliage*" by IPCC (Penman et al. 2003). Once the AGB is quantified, it can be used as a proxy for carbon values based on the corresponding forest regimes' carbon fraction of dry matter (Eggleston et al. 2006). However, continuous flux change due to forest fire, logging, and land cover change makes it a dynamic component that requires periodic monitoring. This may allow us to understand the exchange of carbon between the terrestrial biosphere and the atmosphere and hence was identified as an Essential Climate variable by UNFCCC (Mason et al. 2003). Apart from its importance in the GCC (Houghton 2005; Houghton, Hall, and Goetz 2009), AGB also plays a crucial role in radiation balance and microclimate (Parker et al. 2004), hydrological cycle (Tague, Moritz, and Hanan 2019), socioeconomic development and forest characterization and management. Despite its importance, there is a large amount of uncertainty in forest AGB estimation, especially in the tropics (Pan et al. 2011, 2013).

SAR techniques have shown great potential for estimating AGB over different forest regimes. Over the last three decades, several approaches have been adopted to estimate AGB from SAR backscattered measurements based on empirical (Dobson et al. 1992; Le Toan et al. 1992; Luckman 1997; Sandberg et al. 2011; Ningthoujam et al. 2016; Yu and Saatchi 2016), semi-empirical (Cartus, Santoro, and Kellndorfer 2012; Bouvet et al. 2018b), physically-based model (Loi, Saatchi, and Jaruwatanadilok 2015; Soja et al. 2021) and ML (Antropov et al. 2017; Vafaei et al. 2018; Santi et al. 2020). Although the empirical and semi-empirical approaches have been frequently used, they are limited by saturation, site, environmental, and forest type dependency problems. For every frequency-polarization combination above a certain AGB value, the sensitivity of the backscatter decreases, a phenomenon known as saturation resulting from attenuation of signals (Woodhouse 2017) and interaction with forest structure properties (Joshi et al. 2017). This imposes a constraint on the maximum limit of AGB values that can be estimated within a required accuracy. Based on the previous studies, saturation was observed usually at 30–50 Mg/ha for X- band (Englhart, Keuck, and Siegert 2011), 30–160 Mg/ha for C-band (Ranson and Sun 1994; Imhoff 1995), 70–250 Mg/ha for L-band (Yu and Saatchi 2016; Sandberg et al. 2011) and ~120–400 Mg/ha for P-band (Sandberg et al. 2011), which depends upon the density and type of forest under observation. S-band saturation was found to be <100 Mg/ha based on limited studies (Ningthoujam et al. 2016, 2017). From the polarization point, HV polarization was found to be highly correlated to AGB (Le Toan et al. 1992; Dobson et al. 1992), while VV showed the weakest correlation. HH polarization shows a good correlation to AGB, but under conditions such as sparse forest and/or lower frequency, where signals interaction with the ground is significant, the correlation is weakened by dihedral scattering. Also, it was observed that combining information at different polarization could improve the accuracy of AGB estimates (Saatchi et al. 2007; Soja, Sandberg, and Ulander 2013). Further, for robust AGB model development, the effect of terrain slope (Soja, Sandberg, and Ulander 2013; Villard and Le Toan 2015), incidence angle (Mladenova et al. 2013), and environmental condition (Lucas et al. 2010) must be removed/compensated during model building. To overcome these limitations and retrieve precise AGB estimates, model-based and ML-based approaches were introduced. The Model-based approaches exploit the physics behind the scattering process and can be grouped into distorted born model approximation (DBA) and radiative transfer (RT) models. The RT is an incoherent modeling approach, the most common of which is the Michigan microwave canopy scattering model (MIMICS) (Ulaby et al. 1990). It has consistently predicted scattering mechanisms at different frequencies (Ningthoujam et al. 2016, 2017). Another popular model based on RT is the water cloud model (Attema and Ulaby 1978), which models the canopy as a water cloud with identical water droplets distributed randomly within the canopy. In contrast to RT, the DBA is a coherent modeling approach (Chauhan, Lang, and Ranson 1991; Lang et al. 1994; Saatchi and McDonald 1997). However, model-based approaches are constrained by many model parameters and may sometimes result in ill-posed inversion problems from a single SAR observation. Some studies recently explored the use of ML to retrieve AGB from SAR data. Vafaei et al. (2018) compared individual and combined use of ALOS-2 PALSAR-2 and S2 data for AGB estimation using four different ML. It was observed that the best accuracy was observed using a combination of SAR-Optical data, while PALSAR-2 alone showed poor performance. Of different ML algorithms, SVR outperformed Gaussian Processes (GP), RF, and MPL with r^2 values of 0.73. Similarly, Ghosh and Behera (2018) combined S1 & S2 data with its derived variables using RF and Stochastic gradient boosting algorithms. The RF-based ($r^2 = 0.71$) approach outperformed the SGB ($r^2 = 0.6$) model. Another study compared SVR and artificial neural network (ANN) potentials to estimate AGB over boreal and tropical forests using P-band data acquired using airborne sensors during BioSAR-2007 & -2008, TropiSAR-2009, and AfriSAR-2015–2016 campaigns. It was observed both algorithm performed good, with ANN slightly edging SVR with r^2 values of 0.88 compared to 0.84 for SVR. Apart from backscatter-based estimation, some studies explored the possibilities of using SAR texture-based information to estimate AGB at different frequencies (Kuplich, Curran, and Atkinson 2005; Sarker et al. 2012, 2013). It was observed that adding texture information along with backscatter improved the accuracy of AGB estimates. Further, combing the SAR texture features with optical data also showed positive improvements (Cutler et al. 2012). The relationship between texture and AGB is significantly influenced by stand structure, ground topography, and soil (Champion et al. 2013).

Forest AGB is usually estimated indirectly from the interferometric data (Mette et al. 2002) by retrieving tree height first and then relating it to AGB via allometric equations (Mette et al. 2002, 2004a, b). However, few approaches directly estimate AGB from interferometric coherence. Treuhaft et al. (2015) estimated AGB at X-band data over tropical forests using linear regression and achieved a root-mean-square error (RMSE) of 29%–35% of average AGB. Similarly, Schlund et al. (2015) used Tandem-X coherence to estimate AGB over tropical peat forests and achieved moderate to high correlation. Further, combining C- & L-band coherence provided the best AGB estimates (Tsui et al. 2012). Zinab, Maghsoudi, and Sayedain (2020) applied three-component Freeman–Durden decomposition on PolSAR and PolInSAR data and used derived features to estimate AGB based on linear regression and SVR at P- and L-band data. It was noted that double bounce contribution derived from PolInSAR showed the highest correlation to AGB at both bands, with the best RMSE of 12.73% achieved using SVR at the P-band. Neumann et al. (2012) retrieved vertical structure decomposing PolInSAR into surface and volume contributions and used these variables along with other PolInSAR variables to estimate AGB using LR, SVR, and RF. The RMSE of 17–25% and 5–27% at L- & P-band, respectively, is achieved depending upon the combination of variables used. Ghosh and Behera (2021) compared the IWCM with the deep learning (DL) technique over Indian tropical mangrove forests. They found DL based approach could represent AGB much better in comparison to IWCM. The interferometric water cloud model was used to estimate AGB using X-band SAR with and without using local field data and was found to provide AGB estimates with good accuracies (Askne et al. 2013; Askne, Soja, and Ulander 2017; Askne, Persson, and Ulander 2019). Liao et al. (2019) combined multiple information estimated from single-baseline PolInSAR data to estimate AGB. A volume backscatter was derived and combined with the tree height at different baselines to estimate an RMSE of <10%. Solberg et al. (2010) used X- band data to estimate AGB over Norway Spruce and Scots pine by estimating tree height from interferometric data and achieved an RMSE of 19%. Ghasemi, Tolpekin, and Stein (2018b) assessed the effect of temporal decorrelation on biomass estimation at the P-band and observed that accounting for temporal decorrelation when retrieving the tree height, which is then used for AGB estimation, resulted in significant improvement of AGB estimates with relative error reduction from 47% to 30%. Treuhaft et al. (2017) exploited the phase-height relationship to AGB change to understand dynamics over tropical forests using Tandem-X data. Similarly, forest biomass change was estimated based on height change using InSAR based on Tandem-X data overall boreal forest (Solberg et al. 2014). D'Alessandro et al. (2020) demonstrated that removing ground contribution from interferometric observations can significantly improve correlation with AGB. A model-based retrieval of AGB from ground-canceled interferometric data was shown by Banda et al. (2020). Numerous studies have combined interferometric data with optical and/or LiDAR data to estimate AGB at different frequencies and regimes (Hyde et al. 2007; Nelson et al. 2007; Sun et al. 2011; Mitchard et al. 2012; Kaasalainen et al. 2015).

Recently, the use of TomoSAR has shown great potential for estimating AGB (Minh et al. 2014, 2016; Tebaldini et al. 2019). It has been observed that the HV polarization-based 30 m layer of the tomographic cube provided the best correlation with AGB estimates. Also, the need for slope compensation to improve AGB estimation from tomographic focusing was demonstrated. However, for Capon-based estimation, slope compensation resulted in the degradation of AGB estimates (Ramachandran et al. 2019). Further AGB estimates were improved by combining different polarization and height from a tomographic estimator (Ramachandran et al. 2019). The texture-based information from different tomographic layers also showed a signification correlation to AGB estimates, especially from top layers, and possesses significant potential along with backscatter power to improve AGB estimates further (Liao, He, and Quan 2020). A volumetric backscatter quantity integrating backscatter values between 10 and 30 m height was used to estimate AGB values over the boreal forest and showed significant results at L- and P-bands (Blomberg et al. 2018, 2021). Another set of studies used PCT to estimate parameters from reflectivity profiles and then related it to AGB (Luo et al. 2011; Zhang et al. 2018). Later, segmented polarimetric information was added along with the parameters to improve the accuracy (Li et al. 2012, 2015). Similarly, a structure ratio derived from Legendre coefficients were used to estimate AGB to account for contribution from the different compartment of the forest (Caicoya et al. 2014, 2015).

TomoSAR has the potential to be an excellent tool for mapping AGB. With BIOMASS (P-band) sensor with tomographic capability available in the future, it provides an active area of research. However, some crucial issues, such as the impact of temporal decorrelation and approaches for dealing with it, must be investigated. The availability of PolInSAR data allows the estimation of tree heights, as discussed in Section 20.4.3, which is then used to estimate AGB. However, it is observed that the height alone is insufficient to estimate AGB, especially over dense tropical forests (Ramachandran et al. 2019). Hence, new approaches integrating interferometric observables and derived variables must be employed. The parametric-based (Yu and Saatchi 2016) and ML-based (Santi et al. 2020) approaches have also shown promise in estimating AGB from backscatter data. However, using time series backscatter data for AGB estimation has to be explored with these models.

20.7 FUTURE PERSPECTIVE AND CONCLUSION

In this chapter, we review the current state of the art using SAR techniques and ML algorithms for forestry applications. The fundamental ideas of SAR and ML are covered. Due to its relevance in climatic and socio-economic balance, several important research focused on forest classification, forest change detection, and retrieval of forest tree height and AGB were examined in depth. From a data point of view, using a multi-sensor data approach is the key to attaining a robust solution. With several missions in an overlapping timeframe, the integration of complementing datasets should be encouraged to achieve improved and robust results. For example, the use of SAR sensors in synergy with optical data for biomass estimation (Zhao et al. 2016; Urbazaev et al. 2018) or species classification (Nagendra et al. 2013; Rajah, Odindi, and Mutanga 2018) or combining SAR with LiDAR variables to estimate tree heights (Ramachandran et al. 2021; Pourshamsi et al. 2021) have shown great potential. Even if optical and LiDAR have limitations, integrating multiple SAR observations and derived variables will lead to a potential improvement in estimates. Further, with the current understanding of how SAR signals interact with the different forest components at a particular frequency and polarization, a multi-frequency integration should allow a better illustration of the forest scenario and parameter retrieval. With various space-borne missions operating at different frequencies (X-, to -P band) over similar time in the future, identifying the ways to integrate these datasets and counterbalancing their limitations is the way to look forward toward achieving a complete characterization of forests. In addition, the use of multi-temporal datasets to account for environmental effects is highly encouraged. Although SAR techniques have been extensively used over the last three decades in forestry applications, the use of ML algorithms with SAR data for forestry applications is still relatively young, and rapid advancement in the future is expected. Hence, a few challenges/questions have yet to be addressed to assess the potential to monitor forests globally.

a. The physical-based models are complicated and require the inversion of several parameters, sometimes resulting in ill-poised problems. Further, any deviation from assumptions made by the physical models results in a deterioration of accuracy. The ML-based approach provides a simple solution to estimate the parameter of interest and establish a much better relationship between input and output. However, with limited training samples across the global regime, the transferability of ML-based models must be addressed in detail to be proven as an efficient tool. The ML algorithms should be designed so that the architecture can deal with multi-source, multi-scale data with reduced computational complexity.
b. Given many existing algorithms and the pace of development, benchmarking algorithms for a particular application must be focused. Extensive analysis and comparison of strengths and weakness of algorithms, hyperparameter tuning, and optimization approaches for a particular application must be visited.
c. The availability of multi-temporal and multi-sensor data provides an opportunity to integrate and improve the estimation and mapping procedures. However, it poses challenges due to unstructured, inconsistent, heterogeneous non-stationary datasets and may result in complex and ambiguous outcomes.

The scientific community should make efforts to collaborate with experts from forestry, RS, and data science to provide a better understanding and standardization of approaches for forestry applications and urge data sharing to develop robust and precise models to work across global regimes.

REFERENCES

D'Alessandro, Mariotti, Mauro, Stefano Tebaldini, Shaun Quegan, Maciej J. Soja, Lars M. H. Ulander, and Klaus Scipal. 2020. "Interferometric Ground Cancellation for above Ground Biomass Estimation." *IEEE Transactions on Geoscience and Remote Sensing* 58 (9): 6410–19. doi: 10.1109/TGRS.2020.2976854.

Antropov, Oleg, Yrjo Rauste, Frank M. Seifert, and Tuomas Hame. 2015. "Selective Logging of Tropical Forests Observed Using L- and C-Band SAR Satellite Data." 3870–73.

Antropov, Oleg, Yrjö Rauste, Tuomas Häme, and Jaan Praks. 2017. "Polarimetric ALOS PALSAR Time Series in Mapping Biomass of Boreal Forests." *Remote Sensing* 9 (10): 999. doi: 10.3390/rs9100999.

Arias-Rodil, Manuel, Ulises Diéguez-Aranda, and Harold E. Burkhart. 2017. "Effects of Measurement Error in Total Tree Height and Upper-Stem Diameter on Stem Volume Prediction." *Forest Science* 63 (3): 250–60. doi: 10.5849/FS-2016-087.

Askne, I. H. Jan, Patrik. B. G. Dammert, Lars. M. H. Ulander, and Gray Smith. 1997. "C-Band Repeat-Pass Interferometric SAR Observations of the Forest." *IEEE Transactions on Geoscience and Remote Sensing* 35 (1): 25–35. doi: 10.1109/36.551931.

Askne, I. H. Jan, Johan Fransson, Maurizio Santoro, Maciej J Soja, and Lars M.H Ulander. 2013. "Model-Based Biomass Estimation of a Hemi-Boreal Forest from Multitemporal TanDEM-X Acquisitions." *Remote Sensing* 5 (11): 5574–97. doi: 10.3390/rs5115574.

Askne, I. H. , Jan ., Maciej J. Soja, and Lars M.H. Ulander. 2017. "Biomass Estimation in a Boreal Forest from TanDEM-X Data, LiDAR DTM, and the Interferometric Water Cloud Model." *Remote Sensing of Environment* 196: 265–78 doi: 10.1016/j.rse.2017.05.010.

Askne,I.H. Jan , Henrik J. Persson, and Lars M. H. Ulander. 2019. "On the Sensitivity of TanDEM-X-Observations to Boreal Forest Structure." *Remote Sensing* 11 (14): 1644. doi: 10.3390/rs11141644.

Attarchi, Sara, and Richard Gloaguen. 2014. "Classifying Complex Mountainous Forests with L-Band SAR and Landsat Data Integration: A Comparison Among Different Machine Learning Methods in the Hyrcanian Forest." *Remote Sensing* 6 (5): 3624–47. doi: 10.3390/rs6053624.

Attema, P. W. Evert, and Fawwaz T. ULABY. 1978. "Vegetation Modeled as a Water Cloud." *Radio Science* 13 (2): 357–64. doi: 10.1029/RS013i002p00357.

Ballère, Marie, Alexandre Bouvet, Stéphane Mermoz, Thuy Le Toan, Thierry Koleck, Caroline Bedeau, Mathilde André, Elodie Forestier, Pierre-Louis Frison, and Cédric Lardeux. 2021. "SAR Data for Tropical Forest Disturbance Alerts in French Guiana: Benefit over Optical Imagery." *Remote Sensing of Environment* 252: 112159. doi: 10.1016/j.rse.2020.112159.

Balzter, Heiko., Clare. Rowland, and P. Sacih. 2007. "Forest Canopy Height and Carbon Estimation at Monks Wood National Nature Reserve, UK, Using Dual-Wavelength SAR Interferometry." *Remote Sensing of Environment* 108 (3): 224–39. doi: 10.1016/j.rse.2006.11.014.

Banda, Francesco, Davide Giudici, Thuy Le Toan, Mauro M. Mariotti d'Alessandro, Kostas Papathanassiou, Shaun Quegan, Guido Riembauer et al. 2020. "The BIOMASS Level 2 Prototype Processor: Design and Experimental Results of Above-Ground Biomass Estimation." *Remote Sensing* 12 (6): 985. doi: 10.3390/rs12060985.

Barreto, Thiago L. M., Rafael A. S. Rosa, Christian Wimmer, Joao R. Moreira, Leonardo S. Bins, Fabio A. M. Cappabianco, and Jurandy Almeida. 2016. "Classification of Detected Changes from Multitemporal High-Res Xband SAR Images: Intensity and Texture Descriptors from SuperPixels." *IEEE Journal of Selected Topics in Applied Earth Observations and Remote Sensing* 9 (12): 5436–48. doi: 10.1109/JSTARS.2016.2621818.

Barry, Kathryn E., Liesje Mommer, Jasper van Ruijven, Christian Wirth, Alexandra J. Wright, Yongfei Bai, John Connolly et al. 2019. "The Future of Complementarity: Disentangling Causes from Consequences." *Trends in Ecology & Evolution* 34 (2): 167–80. doi: 10.1016/j.tree.2018.10.013.

Bjerreskov, Kristian S., Thomas Nord-Larsen, and Rasmus Fensholt. 2021. "Classification of Nemoral Forests with Fusion of Multi-Temporal Sentinel-1 and 2 Data." *Remote Sensing* 13 (5): 950. doi: 10.3390/rs13050950.

Björse, Gisela, and Richard Bradshaw. 1998. "2000 Years of Forest Dynamics in Southern Sweden: Suggestions for Forest Management." *Forest Ecology and Management* 104 (1–3): 15–26. doi: 10.1016/S0378-1127(97)00162-X.

Blomberg, Erik, Laurent Ferro-Famil, Maciej J. Soja, Lars M. H. Ulander, and Stefano Tebaldini. 2018. "Forest Biomass Retrieval from L-Band SAR Using Tomographic Ground Backscatter Removal." *IEEE Geoscience and Remote Sensing Letters* 15 (7): 1030–34. doi: 10.1109/LGRS.2018.2819884.

Blomberg, Erik, Lars M. H. Ulander, Stefano Tebaldini, and Laurent Ferro-Famil. 2021. "Evaluating P-Band TomoSAR for Biomass Retrieval in Boreal Forest." *IEEE Transactions on Geoscience and Remote Sensing* 59 (5): 3793–3804. doi: 10.1109/TGRS.2020.3020775.

Bolte, Andreas, Christian Ammer, Magnus Löf, Palle Madsen, Gert-Jan Nabuurs, Peter Schall, Peter Spathelf, and Joachim Rock. 2009. "Adaptive Forest Management in Central Europe: Climate Change Impacts, Strategies and Integrative Concept." *Scandinavian Journal of Forest Research* 24 (6): 473–82. doi: 10.1080/02827580903418224.

Bouvet, Alexandre, Stéphane Mermoz, Marie Ballère, Thierry Koleck, and Thuy Le Toan. 2018a. "Use of the SAR Shadowing Effect for Deforestation Detection with Sentinel-1 Time Series." *Remote Sensing* 10 (8): 1250. doi: 10.3390/rs10081250.

Bouvet, Alexandre, Stéphane Mermoz, Thuy Le Toan, Ludovic Villard, Renaud Mathieu, Laven Naidoo, and Gregory P. Asner. 2018b. "An Above-Ground Biomass Map of African Savannahs and Woodlands at 25 M Resolution Derived from ALOS PALSAR." *Remote Sensing of Environment* 206: 156–73 doi: 10.1016/j.rse.2017.12.030.

Brigot, Guillaume, Marc Simard, Elise Colin-Koeniguer, and Alexandre Boulch. 2019. "Retrieval of Forest Vertical Structure from PolInSAR Data by Machine Learning Using LIDAR-Derived Features." *Remote Sensing* 11 (4): 381. doi: 10.3390/rs11040381.

Caicoya, Astor T., Florian Kugler, Matteo Pardini, Irena Hajnsek, and Konstantinos Papathanassiou. 2014. "Vertical Forest Structure Characterization for the Estimation of Above Ground Biomass: First Experimental Results Using SAR Vertical Reflectivity Profiles." *In 2014 IEEE Geoscience and Remote Sensing Symposium,* Quebec, Canada, 1045–48: IEEE.

Caicoya, Astor T., Matteo Pardini, Irena Hajnsek, and Konstantinos Papathanassiou. 2015. "Forest Above-Ground Biomass Estimation from Vertical Reflectivity Profiles at L-Band." *IEEE Geoscience and Remote Sensing Letters* 12 (12): 2379–83. doi: 10.1109/LGRS.2015.2477858.

Canadell, Josep G., and Michael R. Raupach. 2008. "Managing Forests for Climate Change Mitigation." *Science (New York, N.Y.)* 320 (5882): 1456–57. doi: 10.1126/science.1155458.

Carstairs, Harry, Edward T. A. Mitchard, Iain McNicol, Chiara Aquino, Andrew Burt, Médard O. Ebanega, Anaick M. Dikongo, José-Luis Bueso-Bello, and Mathias Disney. 2022. "An Effective Method for InSAR Mapping of Tropical Forest Degradation in Hilly Areas." *Remote Sensing* 14 (3): 452. doi: 10.3390/rs14030452.

Cartus, Oliver, Maurizio Santoro, and Josef Kellndorfer. 2012. "Mapping Forest Aboveground Biomass in the Northeastern United States with ALOS PALSAR Dual-Polarization L-Band." *Remote Sensing of Environment* 124: 466–78 doi: 10.1016/j.rse.2012.05.029.

Champion, Isabelle, Jean-Pierre Da Costa, Adrien Godineau, Ludovic Villard, Patricia Dubois-Fernandez, and Thuy Le Toan, 2013. "Canopy Structure Effect on SAR Image Texture Versus Forest Biomass Relationships." *EARSeL eProceedings* 12 (1), 25–32. https://hal.inrae.fr/hal-02642852/document.

Chauhan, S. Narinder , Roger. H. Lang, and Kenneth. J. Ranson. 1991. "Radar Modeling of a Boreal Forest." *IEEE Transactions on Geoscience and Remote Sensing* 29 (4): 627–38. doi: 10.1109/36.135825.

Chave, Jérôme, Maxime Réjou-Méchain, Alberto Búrquez, Emmanuel Chidumayo, Matthew S. Colgan, Welington B. C. Delitti, Alvaro Duque et al. 2014. "Improved Allometric Models to Estimate the Aboveground Biomass of Tropical Trees." *Global Change Biology* 20 (10): 3177–90. doi: 10.1111/gcb.12629.

Chen, Gang, Geoffrey J. Hay, and Yanlian Zhou. 2010. "Estimation of Forest Height, Biomass and Volume Using Support Vector Regression and Segmentation from Lidar Transects and Quickbird Imagery." *In 2010 18th International Conference on Geoinformatics*, Beijing, China, 1–4: IEEE.

Chen, Gang, Geoffrey J. Hay, and Benoît St-Onge. 2012. "A GEOBIA Framework to Estimate Forest Parameters from LiDAR Transects, Quickbird Imagery and Machine Learning: A Case Study in Quebec, Canada." *International Journal of Applied Earth Observation and Geoinformation* 15: 28–37 doi: 10.1016/j.jag.2011.05.010.

Cloude, Shane R. 2006. "Polarization Coherence Tomography." *Radio Science* 41 (4). doi: 10.1029/2005RS003436.

Cremer, Felix, Mikhail Urbazaev, Jose Cortes, John Truckenbrodt, Christiane Schmullius, and Christian Thiel. 2020. "Potential of Recurrence Metrics from Sentinel-1 Time Series for Deforestation Mapping." *IEEE Journal of Selected Topics in Applied Earth Observations and Remote Sensing* 13: 5233–40 doi: 10.1109/JSTARS.2020.3019333.

Cumming, Ian G., and Frank H. Wong. 2005. *Digital Processing of Synthetic Aperture Radar Data: Algorithms and Implementation*. Boston, MA, London: Artech House.

Curtis, Philip G., Christy M. Slay, Nancy L. Harris, Alexandra Tyukavina, and Matthew C. Hansen. 2018. "Classifying Drivers of Global Forest Loss." *Science (New York, N.Y.)* 361 (6407): 1108–11. doi: 10.1126/science.aau3445.

Cutler, E. J. Mark, Doreen. S. Boyd, Giles. M. Foody, and Anand. Vetrivel. 2012. "Estimating Tropical Forest Biomass with a Combination of SAR Image Texture and Landsat TM Data: An Assessment of Predictions Between Regions." *ISPRS Journal of Photogrammetry and Remote Sensing* 70: 66–77 doi: 10.1016/j.isprsjprs.2012.03.011.

Denbina, Michael, Marc Simard, and Brian Hawkins. 2018. "Forest Height Estimation Using Multibaseline PolInSAR and Sparse Lidar Data Fusion." *IEEE Journal of Selected Topics in Applied Earth Observations and Remote Sensing* 11 (10): 3415–33. doi: 10.1109/JSTARS.2018.2841388.

Deutscher, Janik, Roland Perko, Karlheinz Gutjahr, Manuela Hirschmugl, and Mathias Schardt. 2013. "Mapping Tropical Rainforest Canopy Disturbances in 3D by COSMO-SkyMed Spotlight InSAR-Stereo Data to Detect Areas of Forest Degradation." *Remote Sensing* 5 (2): 648–63. doi: 10.3390/rs5020648.

Dobrinić, Dino, Mateo Gašparović, and Damir Medak. 2021. "Sentinel-1 and 2 Time-Series for Vegetation Mapping Using Random Forest Classification: A Case Study of Northern Croatia." *Remote Sensing* 13 (12): 2321. doi: 10.3390/rs13122321.

Dobson, M. C., F. T. Ulaby, T. LeToan, A. Beaudoin, E. S. Kasischke, and N. Christensen. 1992. "Dependence of Radar Backscatter on Coniferous Forest Biomass." *IEEE Transactions on Geoscience and Remote Sensing* 30 (2): 412–15. doi: 10.1109/36.134090.

Dong, Xichao, Shaun Quegan, Uryu Yumiko, Cheng Hu, and Tao Zeng. 2015. "Feasibility Study of C- and L-Band SAR Time Series Data in Tracking Indonesian Plantation and Natural Forest Cover Changes." *IEEE Journal of Selected Topics in Applied Earth Observations and Remote Sensing* 8 (7): 3692–99. doi: 10.1109/JSTARS.2015.2400439.

Dostálová, Alena, Mait Lang, Janis Ivanovs, Lars T. Waser, and Wolfgang Wagner. 2021. "European Wide Forest Classification Based on Sentinel-1 Data." *Remote Sensing* 13 (3): 337. doi: 10.3390/rs13030337.

Dubayah, Ralph, James B. Blair, Scott Goetz, Lola Fatoyinbo, Matthew Hansen, Sean Healey, Michelle Hofton et al. 2020. "The Global Ecosystem Dynamics Investigation: High-Resolution Laser Ranging of the Earth's Forests and Topography." *Science of Remote Sensing* 1: 100002. doi: 10.1016/j.srs.2020.100002.

Dupuis, Chloé, Philippe Lejeune, Adrien Michez, and Adeline Fayolle. 2020. "How Can Remote Sensing Help Monitor Tropical Moist Forest Degradation?: A Systematic Review." *Remote Sensing* 12 (7): 1087. doi: 10.3390/rs12071087.

Eggleston, H. Simon, Leandro. Buendia, Kyoko. Miwa, Ngara, Todd., and Kiyoto. Tanabe. 2006. *2006 IPCC Guidelines for National Greenhouse Gas Inventories*. https://www.osti.gov/etdeweb/biblio/20880391.

Eini-Zinab, Sajjad, Yasser Maghsoudi, and Seyed A. Sayedain. 2020. "Assessing the Performance of Indicators Resulting from Three-Component Freeman–Durden Polarimetric SAR Interferometry Decomposition at P-and L-Band in Estimating Tropical Forest Aboveground Biomass." *International Journal of Remote Sensing* 41 (2): 433–54. doi: 10.1080/01431161.2019.1641761.

Englhart, Sandra., Vanessa. Keuck, and Florian . Siegert. 2011. "Aboveground Biomass Retrieval in Tropical Forests: The Potential of Combined X- and L-Band SAR Data Use." *Remote Sensing of Environment* 115 (5): 1260–71. doi: 10.1016/j.rse.2011.01.008.

Falk, Wolfgang, and Karl H. Mellert. 2011. "Species Distribution Models as a Tool for Forest Management Planning Under Climate Change: Risk Evaluation of Abies Alba in Bavaria." *Journal of Vegetation Science* 22 (4): 621–34. doi: 10.1111/j.1654-1103.2011.01294.x.

FAO. 2005. Global Forest Resources Assessment 2005. https://www.fao.org/3/ah887e/ah887e.pdf.

FAO. 2020. Global Forest Resources Assessment 2020.

Ferro-Famil, Laurent., Eric. Pottier, and Jong-Sen Lee. 2001. "Unsupervised Classification of Multifrequency and Fully Polarimetric SAR Images Based on the H/A/Alpha-Wishart Classifier." *IEEE Transactions on Geoscience and Remote Sensing* 39 (11): 2332–42. doi: 10.1109/36.964969.

Flores-Anderson, Africa Ixmuca, Kelsey E. Herndon, Rajesh Bahadur Thapa, and Emil Cherrington. 2019. *The SAR Handbook: Comprehensive Methodologies for Forest Monitoring and Biomass Estimation*. No. MSFC-E-DAA-TN67454. https://scholar.google.com/citations?user=w7_ecnmaaaaj&hl=en&oi=sra.

Fransson, E. S. Johan, Fredrik. Walter, and Lars, M. H. Ulander. 2000. "Estimation of Forest Parameters Using CARABAS-II VHF SAR Data." *IEEE Transactions on Geoscience and Remote Sensing* 38 (2): 720–27. doi: 10.1109/36.842001.

Fransson, E. S. Johan, Gray Smith, Jan Askne, and H. Olsson. 2001. "Stem Volume Estimation in Boreal Forests Using ERS-1/2 Coherence and SPOT XS Optical Data." *International Journal of Remote Sensing* 22 (14): 2777–91. doi: 10.1080/01431160010006872.

Gao, Yan, Margaret Skutsch, Jaime Paneque-Gálvez, and Adrian Ghilardi. 2020. "Remote Sensing of Forest Degradation: A Review." *Environmental Research Letters* 15 (10): 103001. doi: 10.1088/1748-9326/abaad7.

García, Mariano, Sassan Saatchi, Susan Ustin, and Heiko Balzter. 2018. "Modelling Forest Canopy Height by Integrating Airborne LiDAR Samples with Satellite Radar and Multispectral Imagery." *International Journal of Applied Earth Observation and Geoinformation* 66: 159–73 doi: 10.1016/j.jag.2017.11.017.

Garestier, , P. C. Franck, Pascale ,Dubois-Fernandez, Dominique. Guyon, and Thuy. Le Toan. 2009. "Forest Biophysical Parameter Estimation Using L- and P-Band Polarimetric SAR Data." *IEEE Transactions on Geoscience and Remote Sensing* 47 (10): 3379–88. doi: 10.1109/TGRS.2009.2022947.

Gauthier, Sylvie, Alain Leduc, and Yves Bergeron. 1996. "Forest Dynamics Modelling Under Natural Fire Cycles: A Tool to Define Natural Mosaic Diversity for Forest Management." In *Global to Local: Ecological Land Classification*, edited by Richard A. Sims, Ian G. W. Corns, and Karel Klinka, 417–34. Dordrecht: Springer Netherlands.

Geymen, Abdurrahman. 2014. "Digital Elevation Model (DEM) Generation Using the SAR Interferometry Technique." *Arabian Journal of Geosciences* 7 (2): 827–37. doi: 10.1007/s12517-012-0811-3.

Ghasemi, Nafiseh, Valentyn Tolpekin, and Alfred Stein. 2018a. "A Modified Model for Estimating Tree Height from PolInSAR with Compensation for Temporal Decorrelation." *International Journal of Applied Earth Observation and Geoinformation* 73: 313–22 doi: 10.1016/j.jag.2018.06.022.

Ghasemi, Nafiseh, Valentyn Tolpekin, and Alfred Stein. 2018b. "Assessment of Forest Above-Ground Biomass Estimation from PolInSAR in the Presence of Temporal Decorrelation." *Remote Sensing* 10 (6): 815. doi: 10.3390/rs10060815.

Ghasemi, Nafiseh, Valentyn Tolpekin, and Alfred Stein. 2018c. "Estimating Tree Heights Using Multibaseline PolInSAR Data with Compensation for Temporal Decorrelation, Case Study: AfriSAR Campaign Data." *IEEE Journal of Selected Topics in Applied Earth Observations and Remote Sensing* 11 (10): 3464–77. doi: 10.1109/JSTARS.2018.2869620.

Ghosh, Sujit Madhab., and Mukunda Dev. Behera. 2018. "Aboveground Biomass Estimation Using Multi-Sensor Data Synergy and Machine Learning Algorithms in a Dense Tropical Forest." *Applied Geography* 96: 29–40 doi: 10.1016/j.apgeog.2018.05.011.

Ghosh, Sujit Madhab., and Mukunda Dev. Behera. 2021. "Aboveground Biomass Estimates of Tropical Mangrove Forest Using Sentinel-1 SAR Coherence Data - the Superiority of Deep Learning over a Semi-Empirical Model." *Computers & Geosciences* 150: 104737. doi: 10.1016/j.cageo.2021.104737.

Goetz, Scott J., Matthew Hansen, Richard A. Houghton, Wayne Walker, Nadine Laporte, and Jonah Busch. 2015. "Measurement and Monitoring Needs, Capabilities and Potential for Addressing Reduced Emissions from Deforestation and Forest Degradation Under REDD+." *Environmental Research Letters* 10 (12): 123001. doi: 10.1088/1748-9326/10/12/123001.

Graham, L. C. 1974. "Synthetic Interferometer Radar for Topographic Mapping." *Proceedings of IEEE* 62 (6): 763–68. doi: 10.1109/PROC.1974.9516.

Gu, Chengyan, Jan G. Clevers, Xiao Liu, Xin Tian, Zhouyuan Li, and Zengyuan Li. 2018. "Predicting Forest Height Using the GOST, Landsat 7 ETM+, and Airborne LiDAR for Sloping Terrains in the Greater Khingan Mountains of China." *ISPRS Journal of Photogrammetry and Remote Sensing* 137: 97–111 doi: 10.1016/j.isprsjprs.2018.01.005.

Haarpaintner, Jörg, and Heidi Hindberg. 2019. "Multi-Temporal and Multi-Frequency SAR Analysis for Forest Land Cover Mapping of the Mai-Ndombe District (Democratic Republic of Congo)." *Remote Sensing* 11 (24): 2999. doi: 10.3390/rs11242999.

Hajnsek, Irena, Florian Kugler, Seung-Kuk Lee, and Konstantinos P. Papathanassiou. 2009. "Tropical-Forest-Parameter Estimation by Means of Pol-InSAR: The INDREX-II Campaign." *IEEE Transactions on Geoscience and Remote Sensing* 47 (2): 481–93. doi: 10.1109/TGRS.2008.2009437.

Hensley, Scott, Thierry Michel, Maxim Nuemann, Marco Lavalle, Ron Muellerschoen, Bruce Chapman, Cathleen Jones, Razi Ahmed, Fabrizio Lombardini, and Paul Siqueira. 2012. "Some First Polarimetric-Interferometric Multi-Baseline and Tomographic Results at Harvard Forest Using UAVSAR." 5202–5.

Hethcoat, Matthew G., João M. Carreiras, David P. Edwards, Robert G. Bryant, and Shaun Quegan. 2021. "Detecting Tropical Selective Logging with C-Band SAR Data May Require a Time Series Approach." *Remote Sensing of Environment* 259: 112411. doi: 10.1016/j.rse.2021.112411.

Hirschmugl, Manuela, Heinz Gallaun, Matthias Dees, Pawan Datta, Janik Deutscher, Nikos Koutsias, and Mathias Schardt. 2017. "Methods for Mapping Forest Disturbance and Degradation from Optical Earth Observation Data: A Review." *Current Forestry Reports* 3 (1): 32–45. doi: 10.1007/s40725-017-0047-2.**

Hirschmugl, Manuela, Janik Deutscher, Carina Sobe, Alexandre Bouvet, Stéphane Mermoz, and Mathias Schardt. 2020. "Use of SAR and Optical Time Series for Tropical Forest Disturbance Mapping." *Remote Sensing* 12 (4): 727. doi: 10.3390/rs12040727.

Ho Tong Minh, Dinh, Thuy Le Toan, Fabio Rocca, Stefano Tebaldini, Mauro M. D'Alessandro, and Ludovic Villard. 2014. "Relating P-Band Synthetic Aperture Radar Tomography to Tropical Forest Biomass." *IEEE Transactions on Geoscience and Remote Sensing* 52 (2): 967–79. doi: 10.1109/TGRS.2013.2246170.

Ho Tong Minh, Dinh, Thuy Le Toan, Fabio Rocca, Stefano Tebaldini, Ludovic Villard, Maxime Réjou-Méchain, Oliver L. Phillips et al. 2016. "SAR Tomography for the Retrieval of Forest Biomass and Height: Cross-Validation at Two Tropical Forest Sites in French Guiana." *Remote Sensing of Environment* 175: 138–47 doi: 10.1016/j.rse.2015.12.037.

Ho Tong Minh, Dinh, Yen-Nhi Ngo, and Thu T. Lê. 2021. "Potential of P-Band SAR Tomography in Forest Type Classification." *Remote Sensing* 13 (4): 696. doi: 10.3390/rs13040696.

Hoekman, Dirk H., Martin A. M. Vissers, and Niels Wielaard. 2010. "PALSAR Wide-Area Mapping of Borneo: Methodology and Map Validation." *IEEE Journal of Selected Topics in Applied Earth Observations and Remote Sensing* 3 (4): 605–17. doi: 10.1109/JSTARS.2010.2070059.

Hoekman, Dirk, Boris Kooij, Marcela Quiñones, Sam Vellekoop, Ita Carolita, Syarif Budhiman, Rahmat Arief, and Orbita Roswintiarti. 2020. "Wide-Area Near-Real-Time Monitoring of Tropical Forest Degradation and Deforestation Using Sentinel-1." *Remote Sensing* 12 (19): 3263. doi: 10.3390/rs12193263.

Houghton, A. Richard 2005. "Aboveground Forest Biomass and the Global Carbon Balance." *Global Change Biology* 11 (6): 945–58. doi: 10.1111/j.1365-2486.2005.00955.x.

Houghton, A. Richard , Forrest Hall, and Scott J. Goetz. 2009. "Importance of Biomass in the Global Carbon Cycle." *Journal of Geophysical Research* 114 (G2). doi: 10.1029/2009JG000935.

Hunter, O. Maria, Michael. Keller, Daniel. Victoria, and Douglas. C. Morton. 2013. "Tree Height and Tropical Forest Biomass Estimation." *Biogeosciences* 10 (12): 8385–99. doi: 10.5194/bg-10-8385-2013.

Hyde, Peter, Ross Nelson, Dan Kimes, and Elissa Levine. 2007. "Exploring LiDAR–RaDAR Synergy—Predicting Aboveground Biomass in a Southwestern Ponderosa Pine Forest Using LiDAR, SAR and InSAR." *Remote Sensing of Environment* 106 (1): 28–38. doi: 10.1016/j.rse.2006.07.017.

Imhoff, Marc L. 1995. "Radar Backscatter and Biomass Saturation: Ramifications for Global Biomass Inventory." *IEEE Transactions on Geoscience and Remote Sensing* 33 (2): 511–18. doi: 10.1109/TGRS.1995.8746034.

Joshi, Neha, Edward T. A. Mitchard, Natalia Woo, Jorge Torres, Julian Moll-Rocek, Andrea Ehammer, Murray Collins, Martin R. Jepsen, and Rasmus Fensholt. 2015. "Mapping Dynamics of Deforestation and Forest Degradation in Tropical Forests Using Radar Satellite Data." *Environmental Research Letters* 10 (3): 34014. doi: 10.1088/1748-9326/10/3/034014.

Joshi, Neha, Edward T. A. Mitchard, Matthew Brolly, Johannes Schumacher, Alfredo Fernández-Landa, Vivian K. Johannsen, Miguel Marchamalo, and Rasmus Fensholt. 2017. "Understanding 'Saturation' of Radar Signals over Forests." *Scientific Reports* 7 (1): 3505. doi: 10.1038/s41598-017-03469-3.

Kaasalainen, Sanna, Markus Holopainen, Mika Karjalainen, Mikko Vastaranta, Ville Kankare, Kirsi Karila, and Batuhan Osmanoglu. 2015. "Combining LiDAR and Synthetic Aperture Radar Data to Estimate Forest Biomass: Status and Prospects." *Forests* 6 (12): 252–70. doi: 10.3390/f6010252.

Kelly, Maggi, and Stefania Di Tommaso. 2015. "Mapping Forests with LiDAR Provides Flexible, Accurate Data with Many Uses." *California Agriculture* 69 (1): 14–20. doi: 10.3733/ca.v069n01p14.

Khati, Unmesh, Gulab Singh, and Shashi Kumar. 2018. "Potential of Space-Borne PolInSAR for Forest Canopy Height Estimation over India: A Case Study Using Fully Polarimetric L -, C -, and X -Band SAR Data." *IEEE Journal of Selected Topics in Applied Earth Observations and Remote Sensing* 11 (7): 2406–16. doi: 10.1109/JSTARS.2018.2835388.

Kovacs, John M., Casey V. Vandenberg, Jinfei Wang, and Francisco Flores-Verdugo. 2008. "The Use of Multipolarized Spaceborne SAR Backscatter for Monitoring the Health of a Degraded Mangrove Forest." *Journal of Coastal Research* 241: 248–54 doi: 10.2112/06-0660.1.

Kuck, Tahisa N., Edson E. Sano, Polyanna C. Da Bispo, Elcio H. Shiguemori, Paulo F. F. Silva Filho, and Eraldo A. T. Matricardi. 2021. "A Comparative Assessment of Machine-Learning Techniques for Forest Degradation Caused by Selective Logging in an Amazon Region Using Multitemporal X-Band SAR Images." *Remote Sensing* 13 (17): 3341. doi: 10.3390/rs13173341.

Kugler, Florian, Seung-Kuk Lee, Irena Hajnsek, and Konstantinos P. Papathanassiou. 2015. "Forest Height Estimation by Means of Pol-InSAR Data Inversion: The Role of the Vertical Wavenumber." *IEEE Transactions on Geoscience and Remote Sensing* 53 (10): 5294–5311. doi: 10.1109/TGRS.2015.2420996.

Kuplich, Tatiana. Mora., PaulJ. Curran, and Peter. M. Atkinson. 2005. "Relating SAR Image Texture to the Biomass of Regenerating Tropical Forests." *International Journal of Remote Sensing* 26 (21): 4829–54. doi: 10.1080/01431160500239107.

Lang, H. Roger , Narinder. S. Chauhan, Kenneth. J. Ranson, and Ozlem. Kilic. 1994. "Modeling P-Band SAR Returns from a Red Pine Stand." *Remote sensing of environment*, 47(2), pp.132-141. doi: 10.1016/0034-4257(94)90150-3.

Kurvonen, Lauri, and Martti. T. Hallikainen. 1999. "Textural Information of Multitemporal ERS-1 and JERS-1 SAR Images with Applications to Land and Forest Type Classification in Boreal Zone." *IEEE Transactions on Geoscience and Remote Sensing* 37 (2): 680–89. doi: 10.1109/36.752185.

doi: 10.1016/0034-4257(94)90150-3.

Lapini, Alessandro, Simone Pettinato, Emanuele Santi, Simonetta Paloscia, Giacomo Fontanelli, and Andrea Garzelli. 2020. "Comparison of Machine Learning Methods Applied to SAR Images for Forest Classification in Mediterranean Areas." *Remote Sensing* 12 (3): 369. doi: 10.3390/rs12030369.

Lavalle, Marco, and Scott Hensley. 2012. "Demonstration of Repeat-Pass POLINSAR Using UAVSAR: The RMOG Model." 5876–79.

Lavalle, Marco, and Scott Hensley. 2015. "Extraction of Structural and Dynamic Properties of Forests from Polarimetric-Interferometric SAR Data Affected by Temporal Decorrelation." *IEEE Transactions on Geoscience and Remote Sensing* 53 (9): 4752–67. doi: 10.1109/TGRS.2015.2409066.

Le Toan, Thuy., André. Beaudoin, J. Riom, and Dominique. Guyon. 1992. "Relating Forest Biomass to SAR Data." *IEEE Transactions on Geoscience and Remote Sensing* 30 (2): 403–11. doi: 10.1109/36.134089.

Lee, Seung-Kuk, Florian Kugler, Konstantinos Papathanassiou, and Irena Hajnsek. 2011. "Multibaseline Polarimetric Sar Interferometry Forest Height Inversion Approaches." *Proceedings of ESA Polinsar Workshop*. https://elib.dlr.de/74010/.

Lee, Seung-Kuk, Florian Kugler, Konstantinos P. Papathanassiou, and Irena Hajnsek. 2013. "Quantification of Temporal Decorrelation Effects at L-Band for Polarimetric SAR Interferometry Applications." *IEEE Journal of Selected Topics in Applied Earth Observations and Remote Sensing* 6 (3): 1351–67. doi: 10.1109/JSTARS.2013.2253448.

Lei, Yang, Robert Treuhaft, and Fabio Gonçalves. 2021. "Automated Estimation of Forest Height and Underlying Topography over a Brazilian Tropical Forest with Single-Baseline Single-Polarization TanDEM-X SAR Interferometry." *Remote Sensing of Environment* 252: 112132. doi: 10.1016/j.rse.2020.112132.

Lei, Yang, Robert Treuhaft, Michael Keller, Maiza dos-Santos, Fabio Gonçalves, and Maxim Neumann. 2018. "Quantification of Selective Logging in Tropical Forest with Spaceborne SAR Interferometry." *Remote Sensing of Environment* 211: 167–83. doi: 10.1016/j.rse.2018.04.009.

Li, Wenmei, Erxue Chen, Zengyuan Li, Huanmin Luo, Wei Zhou, Qi Feng, and Xinshuang Wang. 2012. "Combing Polarization Coherence Tomography and PoLInSAR Segmentation for Forest above Ground Biomass Estimation." 3351–54.

Li, Wenmei, Erxue Chen, Zengyuan Li, Yinghai Ke, and Wenfeng Zhan. 2015. "Forest Aboveground Biomass Estimation Using Polarization Coherence Tomography and PolSAR Segmentation." *International Journal of Remote Sensing* 36 (2): 530–50. doi: 10.1080/01431161.2014.999383.

Li, Wang, Zheng Niu, Rong Shang, Yuchu Qin, Li Wang, and Hanyue Chen. 2020. "High-Resolution Mapping of Forest Canopy Height Using Machine Learning by Coupling ICESat-2 LiDAR with Sentinel-1, Sentinel-2 and Landsat-8 Data." *International Journal of Applied Earth Observation and Geoinformation* 92: 102163. doi: 10.1016/j.jag.2020.102163.

Liao, Zhanmang, Binbin He, Albert I. van Dijk, Xiaojing Bai, and Xingwen Quan. 2018. "The Impacts of Spatial Baseline on Forest Canopy Height Model and Digital Terrain Model Retrieval Using P-Band PolInSAR Data." *Remote Sensing of Environment* 210: 403–21 doi: 10.1016/j.rse.2018.03.033.

Liao, Zhanmang, Binbin He, Xingwen Quan, Albert I. van Dijk, Shi Qiu, and Changming Yin. 2019. "Biomass Estimation in Dense Tropical Forest Using Multiple Information from Single-Baseline P-Band PolInSAR Data." *Remote Sensing of Environment* 221: 489–507 doi: 10.1016/j.rse.2018.11.027.

Liao, Zhanmang, Binbin He, and Xingwen Quan. 2020. "Potential of Texture from SAR Tomographic Images for Forest Aboveground Biomass Estimation." *International Journal of Applied Earth Observation and Geoinformation* 88: 102049. doi: 10.1016/j.jag.2020.102049.

Lu, Hongxi, Zhiyong Suo, Rui Guo, and Zheng Bao. 2013. "S-RVoG Model for Forest Parameters Inversion over Underlying Topography." *Electronics Letters* 49 (9): 618–20. doi: 10.1049/el.2012.4467.

Lucas, Richard, John Armston, Russell Fairfax, Rod Fensham, Arnon Accad, Joao Carreiras, Jack Kelley et al. 2010. "An Evaluation of the ALOS PALSAR L-Band Backscatter: Above Ground Biomass Relationship Queensland, Australia: Impacts of Surface Moisture Condition and Vegetation Structure." *IEEE Journal of Selected Topics in Applied Earth Observations and Remote Sensing* 3 (4): LuoLuo576–593. doi: 10.1109/JSTARS.2010.2086436.

Luckman, A. 1997. "A Study of the Relationship between Radar Backscatter and Regenerating Tropical Forest Biomass for Spaceborne SAR Instruments." *Remote Sensing of Environment* 60 (1): 1–13. doi: 10.1016/S0034-4257(96)00121-6.

Luo, Huanmin, Erxue Chen, Zengyuan Li, and Chunxiang Cao. 2011. "Forest above Ground Biomass Estimation Methodology Based on Polarization Coherence Tomography." *Journal of Remote Sensing* 15 (6): 1138–1155. http://gissky.net/paper/uploadfiles_4495/201207/2012072415070781.pdf.

Ma, Wenping, Hui Yang, Yue Wu, Yunta Xiong, Tao Hu, Licheng Jiao, and Biao Hou. 2019. "Change Detection Based on Multi-Grained Cascade Forest and Multi-Scale Fusion for SAR Images." *Remote Sensing* 11 (2): 142. doi: 10.3390/rs11020142.

Maghsoudi, Yasser, Michael Collins, and Donald G. Leckie. 2012. "Polarimetric Classification of Boreal Forest Using Nonparametric Feature Selection and Multiple Classifiers." *International Journal of Applied Earth Observation and Geoinformation* 19: 139–50 doi: 10.1016/j.jag.2012.04.015.

Martone, Michele, Paola Rizzoli, Christopher Wecklich, Carolina González, José-Luis Bueso-Bello, Paolo Valdo, Daniel Schulze, Manfred Zink, Gerhard Krieger, and Alberto Moreira. 2018. "The Global Forest/Non-Forest Map from TanDEM-X Interferometric SAR Data." *Remote Sensing of Environment* 205: 352–73 doi: 10.1016/j.rse.2017.12.002.

Mason, J. Paul , Michael. Manton, D. E. Harrison, Alan Belward, and Alan. R Thomas, and D. K. Dawson. 2003. The Second Report on the Adequacy of the Global Observing Systems for Climate in Support of the UNFCCC.

Mazza, Antonio, Francescopaolo Sica, Paola Rizzoli, and Giuseppe Scarpa. 2019. "TanDEM-X Forest Mapping Using Convolutional Neural Networks." *Remote Sensing* 11 (24): 2980. doi: 10.3390/rs11242980.

McKenney, Daniel W., John H. Pedlar, Kevin Lawrence, Kathy Campbell, and Michael F. Hutchinson. 2007. "Potential Impacts of Climate Change on the Distribution of North American Trees." *BioScience* 57 (11): 939–48. doi: 10.1641/B571106.

Mermoz, Stéphane, and Thuy Le Toan. 2016. "Forest Disturbances and Regrowth Assessment Using ALOS PALSAR Data from 2007 to 2010 in Vietnam, Cambodia and Lao PDR." *Remote Sensing* 8 (3): 217. doi: 10.3390/rs8030217.

Mette, Tobias, K. P. Papathanassiou, Irena Hajnsek, and Reiner Zimmermann, 2002. Forest Biomass Estimation Using Polarimetric SAR Interferometry. *Geoscience and Remote Sensing Symposium, 2002 (IGARSS'02)*, Toronto.

Mette, Tobias, Konstantinos P. Papathanassiou, Irena Hajnsek, Hans Pretzsch, and Peter Biber. 2004a. Applying a Common Allometric Equation to Convert Forest Height from Pol-InSAR Data to Forest Biomass. *Proceedings of IGARSS'04*, Anchorage, Alaska.

Mette, Tobias, Konstantinos P. Papathanassiou, and Irena Hajnsek. 2004b. "Biomass Estimation from Polarimetric SAR Interferometry over Heterogeneous Forest Terrain." *In IEEE International IEEE International IEEE International Geoscience and Remote Sensing Symposium, 2004 (IGARSS '04)*, Anchorage, Alaska, 511–14: IEEE.

Mitchard, E. T. A., S. S. Saatchi, L. J. T. White, K. A. Abernethy, K. J. Jeffery, S. L. Lewis, M. Collins et al. 2012. "Mapping Tropical Forest Biomass with Radar and Spaceborne LiDAR in Lopé National Park, Gabon: Overcoming Problems of High Biomass and Persistent Cloud." *Biogeosciences* 9 (1): 179–91. doi: 10.5194/bg-9-179-2012.

Mitchell, Anthea L., Ake Rosenqvist, and Brice Mora. 2017. "Current Remote Sensing Approaches to Monitoring Forest Degradation in Support of Countries Measurement, Reporting and Verification (MRV) Systems for REDD." *Carbon Balance and Management* 12 (1): 9. doi: 10.1186/s13021-017-0078-9.

Mladenova, Iliana E., Thomas J. Jackson, Rajat Bindlish, and Scott Hensley. 2013. "Incidence Angle Normalization of Radar Backscatter Data." *IEEE Transactions on Geoscience and Remote Sensing* 51 (3): 1791–1804. doi: 10.1109/TGRS.2012.2205264.

Mohammed, Mohssen, Mohammad B. Khan, and Eihab B. M. Bashier. 2016. *Machine Learning: Algorithms and Applications*. Boca Raton, FL, London, New York: CRC Press, Taylor & Francis Group.

Nagendra, Harini, Richard Lucas, João P. Honrado, Rob H. Jongman, Cristina Tarantino, Maria Adamo, and Paola Mairota. 2013. "Remote Sensing for Conservation Monitoring: Assessing Protected Areas, Habitat Extent, Habitat Condition, Species Diversity, and Threats." *Ecological Indicators* 33: 45–59 doi: 10.1016/j.ecolind.2012.09.014.

Nandy, Subrata, Ritika Srinet, and Hitendra Padalia. 2021. "Mapping Forest Height and Aboveground Biomass by Integrating ICESat-2, Sentinel-1 and Sentinel-2 Data Using Random Forest Algorithm in Northwest Himalayan Foothills of India." *Geophysical Research Letters* 48 (14). doi: 10.1029/2021GL093799.

Nelson, Ross F., Peter Hyde, Patrick Johnson, Bomono Emessiene, Marc L. Imhoff, Robert Campbell, and Wilson Edwards. 2007. "Investigating RaDAR–LiDAR Synergy in a North Carolina Pine Forest." *Remote Sensing of Environment* 110 (1): 98–108. doi: 10.1016/j.rse.2007.02.006.

Neumann, Maxim, Sassan S. Saatchi, Lars M. H. Ulander, and Johan E. S. Fransson. 2012. "Assessing Performance of L- and P-Band Polarimetric Interferometric SAR Data in Estimating Boreal Forest Above-Ground Biomass." *IEEE Transactions on Geoscience and Remote Sensing* 50 (3): 714–26. doi: 10.1109/TGRS.2011.2176133.

Nico, Giovanni., Davide. Leva, Joaquim. Fortuny-Guasch, Giuseppe. Antonello, and Dario. Tarchi. 2005. "Generation of Digital Terrain Models with a Ground-Based SAR System." *IEEE Transactions on Geoscience and Remote Sensing* 43 (1): 45–49. doi: 10.1109/TGRS.2004.838354.

Ningthoujam, Ramesh, Heiko Balzter, Kevin Tansey, Keith Morrison, Sarah Johnson, France Gerard, Charles George et al. 2016. "Airborne S-Band SAR for Forest Biophysical Retrieval in Temperate Mixed Forests of the UK." *Remote Sensing* 8 (7): 609. doi: 10.3390/rs8070609.

Ningthoujam, Ramesh, Heiko Balzter, Kevin Tansey, Ted Feldpausch, Edward Mitchard, Akhlaq Wani, and Pawan Joshi. 2017. "Relationships of S-Band Radar Backscatter and Forest above ground Biomass in Different Forest Types." *Remote Sensing* 9 (11): 1116. doi: 10.3390/rs9111116.

O'Neill, Gregory A., and Gordon Nigh. 2011. "Linking Population Genetics and Tree Height Growth Models to Predict Impacts of Climate Change on Forest Production." *Global Change Biology* 17 (10): 3208–17. doi: 10.1111/j.1365-2486.2011.02467.x.

Oliver, Chris, and Shaun Quegan. 2004. *Understanding Synthetic Aperture Radar Images*. The SciTech Radar und Defense Series. Raleigh, NC: SciTech Publishing Inc.

Ortega, Mabel X., Raul Q. Feitosa, Jose D. Bermudez, Patrick N. Happ, and Claudio A. de Almeida. 2021-2021. "Comparison of Optical and SAR Data for Deforestation Mapping in the Amazon Rainforest with Fully Convolutional Networks." *In 2021 IEEE International Geoscience and Remote Sensing Symposium IGARSS*, Brussels, Belgium, 3769–72: IEEE.

Ortiz, Sonia M., Johannes Breidenbach, Ralf Knuth, and Gerald Kändler. 2012. "The Influence of DEM Quality on Mapping Accuracy of Coniferous- and Deciduous-Dominated Forest Using TerraSAR-X Images." *Remote Sensing* 4 (3): 661–81. doi: 10.3390/rs4030661.

Pan, Yude, Richard A. Birdsey, Jingyun Fang, Richard Houghton, Pekka E. Kauppi, Werner A. Kurz, Oliver L. Phillips et al. 2011. "A Large and Persistent Carbon Sink in the World's Forests." *Science (New York, N.Y.)* 333 (6045): 988–93. doi: 10.1126/science.1201609.

Pan, Yude, Richard A. Birdsey, Oliver L. Phillips, and Robert B. Jackson. 2013. "The Structure, Distribution, and Biomass of the World's Forests." *Annual Review of Ecology, Evolution, and Systematics* 44 (1): 593–622. doi: 10.1146/annurev-ecolsys-110512-135914.

Pantze, Andreas, Maurizio Santoro, and Johan E. Fransson. 2014. "Change Detection of Boreal Forest Using Bi-Temporal ALOS PALSAR Backscatter Data." *Remote Sensing of Environment* 155: 120–28 doi: 10.1016/j.rse.2013.08.050.

Papathanassiou, P. Konstantinos, and Shane. R. Cloude. 2003. The Effect of Temporal Decorrelation on the Inversion of Forest Parameters from Polinsar Data. *In, IEEE International Geoscience and Remote Sensing Symposium,* Toulouse, France *2003* 1429-1431.

Pardini, Matteo, and Konstantinos Papathanassiou. 2018. "Linking Sar Tomography and Polarization Coherence Tomography in Forest Scenarios." *In IEEE International Geoscience and Remote Sensing Symposium,* Valencia, Spain,2018, 8671–74: 2018.

Parker, Geoffrey G., Mark E. Harmon, Michael A. Lefsky, Jiquan Chen, Robert van Pelt, Stuart B. Weis, Sean C. Thomas, William E. Winner, David C. Shaw, and Jerry F. Frankling. 2004. "Three-Dimensional Structure of an Old-Growth Pseudotsuga-Tsuga Canopy and Its Implications for Radiation Balance, Microclimate, and Gas Exchange." *Ecosystems* 7 (5). doi: 10.1007/s10021-004-0136-5.

Pecchi, Matteo, Maurizio Marchi, Vanessa Burton, Francesca Giannetti, Marco Moriondo, Iacopo Bernetti, Marco Bindi, and Gherardo Chirici. 2019. "Species Distribution Modelling to Support Forest Management: A Literature Review." *Ecological Modelling* 411: 108817. doi: 10.1016/j.ecolmodel.2019.108817.

Penman, J., M. Gytarsky, T. Hiraishi, T. Krug, D. Kruger, R. Pipatti, L. Buendia et al. 2003. Good Practice Guidance for Land Use, Land-Use Change and Forestry. https://inis.iaea.org/search/search.aspx?orig_q=rn:36103537.

Peterson, Birgit, Ralph Dubayah, Peter Hyde, Michelle Hofton, J. Bryan Blair, and JoAnn Fites-Kaufman, eds. 2007. *Use of LIDAR for Forest Inventory and Forest Management Application*, 77 vols. Washington, DC: U.S. Department of Agriculture, Forest Service.

Potapov, Peter, Xinyuan Li, Andres Hernandez-Serna, Alexandra Tyukavina, Matthew C. Hansen, Anil Kommareddy, Amy Pickens et al. 2021. "Mapping Global Forest Canopy Height Through Integration of GEDI and Landsat Data." *Remote Sensing of Environment* 253: 112165. doi: 10.1016/j.rse.2020.112165.

Pourshamsi, Maryam, Mariano Garcia, Marco Lavalle, and Heiko Balzter. 2018. "A Machine-Learning Approach to PolInSAR and LiDAR Data Fusion for Improved Tropical Forest Canopy Height Estimation Using NASA AfriSAR Campaign Data." *IEEE Journal of Selected Topics in Applied Earth Observations and Remote Sensing* 11 (10): 3453–63. doi: 10.1109/JSTARS.2018.2868119.

Pourshamsi, Maryam, Junshi Xia, Naoto Yokoya, Mariano Garcia, Marco Lavalle, Eric Pottier, and Heiko Balzter. 2021. "Tropical Forest Canopy Height Estimation from Combined Polarimetric SAR and LiDAR Using Machine-Learning." *ISPRS Journal of Photogrammetry and Remote Sensing* 172: 79–94 doi: 10.1016/j.isprsjprs.2020.11.008.

Praks, Jaan, Florian Kugler, Konstantinos P. Papathanassiou, Irena Hajnsek, and Martti Hallikainen. 2007. "Height Estimation of Boreal Forest: Interferometric Model-Based Inversion at L- and X-Band Versus HUTSCAT Profiling Scatterometer." *IEEE Geoscience and Remote Sensing Letters* 4 (3): 466–70. doi: 10.1109/LGRS.2007.898083.

Praks, Jaan, Florian Kugler, Juha Hyyppa, Konstantinos Papathanassiou, and Martti Hallikainen. 2008. "SAR Coherence Tomography for Boreal Forest with Aid of Laser Measurements." *In IGARSS 2008 IEEE International Geoscience and Remote Sensing Symposium,* Boston, MA, II-469-II-472: IEEE.

Praks, Jaan, Martti Hallikainen, Jaakko Seppanen, and Juha Hyyppa. 2009. "Boreal Forest Height Estimation with SAR Interferometry and Laser Measurements." *In 2009 IEEE International Geoscience and Remote Sensing Symposium*, Cape Town, South Africa, V-308-V-311: IEEE.

Purves, Drew, and Stephen Pacala. 2008. "Predictive Models of Forest Dynamics." *Science (New York, N.Y.)* 320 (5882): 1452–53. doi: 10.1126/science.1155359.

Qi, Wenlu, Seung-Kuk Lee, Steven Hancock, Scott Luthcke, Hao Tang, John Armston, and Ralph Dubayah. 2019. "Improved Forest Height Estimation by Fusion of Simulated GEDI Lidar Data and TanDEM-X InSAR Data." *Remote Sensing of Environment* 221: 621–34 doi: 10.1016/j.rse.2018.11.035.

Rajah, Perushan, John Odindi, and Onisimo Mutanga. 2018. "Feature Level Image Fusion of Optical Imagery and Synthetic Aperture Radar (SAR) for Invasive Alien Plant Species Detection and Mapping." *Remote Sensing Applications: Society and Environment* 10: 198–208 doi: 10.1016/j.rsase.2018.04.007.

Ramachandran, Naveen, Stefano Tebaldini, Mauro M. d'Alessandro, Sassan Saatchi, and Onkar Dikshit. 2019. "Comparative Analysis of Forest Biomass Estimation Using SAR Tomographic Techniques." *In 2019 IEEE MTT-S International Microwave and RF Conference (IMARC)*, Mumbai, India, 1–5: IEEE.

Ramachandran, Naveen, Sassan Saatchi, Stefano Tebaldini, Mauro M. d'Alessandro, and Onkar Dikshit. 2021. "Evaluation of P-Band SAR Tomography for Mapping Tropical Forest Vertical Backscatter and Tree Height." *Remote Sensing* 13 (8): 1485. doi: 10.3390/rs13081485.

Ranson, J. Kenneth, and Guoqing Sun. 1994. "Mapping Biomass of a Northern Forest Using Multifrequency SAR Data." *IEEE Transactions on Geoscience and Remote Sensing* 32 (2): 388–96. doi: 10.1109/36.295053.

Ranson, J. Kenneth, and Guoqing Sun. 2000. "Effects of Environmental Conditions on Boreal Forest Classification and Biomass Estimates with SAR." *IEEE Transactions on Geoscience and Remote Sensing* 38 (3): 1242–52. doi: 10.1109/36.843016.

Ranson, J. Kenneth, Sassan Saatchi, and Guoqing Sun. 1995. "Boreal Forest Ecosystem Characterization with SIR-C/XSAR." *IEEE Transactions on Geoscience and Remote Sensing* 33 (4): 867–76. doi: 10.1109/36.406673.

Rauste, Yrjo, Oleg Antropov, Tuomas Hame, Gernot Ramminger, Sharon Gomez, and Frank M. Seifert. 2013. "Mapping Selective Logging in Tropical Forest with Space-Borne SAR Data." *ESA Living Planet Symposium* 722: 168.

Reiche, Johannes, Carlos M. Souzax, Dirk H. Hoekman, Jan Verbesselt, Haimwant Persaud, and Martin Herold. 2013. "Feature Level Fusion of Multi-Temporal ALOS PALSAR and Landsat Data for Mapping and Monitoring of Tropical Deforestation and Forest Degradation." *IEEE Journal of Selected Topics in Applied Earth Observations and Remote Sensing* 6 (5): 2159–73. doi: 10.1109/JSTARS.2013.2245101.

Reiche, Johannes, Sytze de Bruin, Dirk Hoekman, Jan Verbesselt, and Martin Herold. 2015a. "A Bayesian Approach to Combine Landsat and ALOS PALSAR Time Series for Near Real-Time Deforestation Detection." *Remote Sensing* 7 (5): 4973–96. doi: 10.3390/rs70504973.

Reiche, Johannes, Jan Verbesselt, Dirk Hoekman, and Martin Herold. 2015b. "Fusing Landsat and SAR Time Series to Detect Deforestation in the Tropics." *Remote Sensing of Environment* 156: 276–93 doi: 10.1016/j.rse.2014.10.001.

Reiche, Johannes, Eliakim Hamunyela, Jan Verbesselt, Dirk Hoekman, and Martin Herold. 2018. "Improving Near-Real Time Deforestation Monitoring in Tropical Dry Forests by Combining Dense Sentinel-1 Time Series with Landsat and ALOS-2 PALSAR-2." *Remote Sensing of Environment* 204: 147–61 doi: 10.1016/j.rse.2017.10.034.

Reiche, Johannes, Adugna Mullissa, Bart Slagter, Yaqing Gou, Nandin-Erdene Tsendbazar, Christelle Odongo-Braun, Andreas Vollrath et al. 2021. "Forest Disturbance Alerts for the Congo Basin Using Sentinel-1." *Environmental Research Letters* 16 (2): 24005. doi: 10.1088/1748-9326/abd0a8.

Reigber, Andreas., and Alberto. Moreira. 2000. "First Demonstration of Airborne SAR Tomography Using Multibaseline L-Band Data." *IEEE Transactions on Geoscience and Remote Sensing* 38 (5): 2142–52. doi: 10.1109/36.868873.

Rignot, J. M. Eric, Cynthia. L. Williams, J. Way, and Leslie. A. Viereck. 1994. "Mapping of Forest Types in Alaskan Boreal Forests Using SAR Imagery." *IEEE Transactions on Geoscience and Remote Sensing* 32 (5): 1051–59. doi: 10.1109/36.312893.

Roman Guliaev, Matteo Pardini, and Konstantinos P. Papathanassiou. 2021. A Comparison of Function Bases for Polarization Coherence Tomography in Forest Scenarios. *In 13th European Conference on Synthetic Aperture Radar,* Frankfurt am Main: VDE. https://ieeexplore.ieee.org/servlet/opac?punumber=9472486.

Rüetschi, Marius, Michael Schaepman, and David Small. 2018. "Using Multitemporal Sentinel-1 C-Band Backscatter to Monitor Phenology and Classify Deciduous and Coniferous Forests in Northern Switzerland." *Remote Sensing* 10 (1): 55. doi: 10.3390/rs10010055.

Ruiz, Luis Fernando Chimelo., Laurindo A. Guasselli, João P. D. Simioni, Tássia F. Belloli, and Pâmela C. Barros Fernandes. 2021. "Object-Based Classification of Vegetation Species in a Subtropical Wetland Using Sentinel-1 and Sentinel-2A Images." *Science of Remote Sensing* 3: 100017. doi: 10.1016/j.srs.2021.100017.

Saatchi, S. Sassan., and Kyle. C. McDonald. 1997. "Coherent Effects in Microwave Backscattering Models for Forest Canopies." *IEEE Transactions on Geoscience and Remote Sensing* 35 (4): 1032–44. doi: 10.1109/36.602545.

Saatchi, Sassan, Kerry Halligan, Don G. Despain, and Robert L. Crabtree. 2007. "Estimation of Forest Fuel Load from Radar Remote Sensing." *IEEE Transactions on Geoscience and Remote Sensing* 45 (6): 1726–40. doi: 10.1109/TGRS.2006.887002.

Sandberg, Gustaf., Lars. M. H. Ulander, Johan J. E. S. Fransson, Johan. Holmgren, and Thuy. Le Toan. 2011. "L- and P-Band Backscatter Intensity for Biomass Retrieval in Hemiboreal Forest." *Remote Sensing of Environment* 115 (11): 2874–86. doi: 10.1016/j.rse.2010.03.018.

Santi, Emanuele, Simonetta Paloscia, Simone Pettinato, Giovanni Cuozzo, Antonio Padovano, Claudia Notarnicola, and Clement Albinet. 2020. "Machine-Learning Applications for the Retrieval of Forest Biomass from Airborne P-Band SAR Data." *Remote Sensing* 12 (5): 804. doi: 10.3390/rs12050804.

Sarker, Md. L. R., Janet Nichol, Baharin Ahmad, Ibrahim Busu, and Alias A. Rahman. 2012. "Potential of Texture Measurements of Two-Date Dual Polarization PALSAR Data for the Improvement of Forest Biomass Estimation." *ISPRS Journal of Photogrammetry and Remote Sensing* 69: 146–66 doi: 10.1016/j.isprsjprs.2012.03.002.

Sarker, Md. L. R., Janet Nichol, Huseyin B. Iz, Baharin B. Ahmad, and Alias A. Rahman. 2013. "Forest Biomass Estimation Using Texture Measurements of High-Resolution Dual-Polarization C-Band SAR Data." *IEEE Transactions on Geoscience and Remote Sensing* 51 (6): 3371–84. doi: 10.1109/TGRS.2012.2219872.

Schlund, Michael, Felicitas von Poncet, Steffen Kuntz, Christiane Schmullius, and Dirk H. Hoekman. 2015. "TanDEM-X Data for Aboveground Biomass Retrieval in a Tropical Peat Swamp Forest." *Remote Sensing of Environment* 158: 255–66 doi: 10.1016/j.rse.2014.11.016.

Schlund, Michael, Paul Magdon, Brian Eaton, Craig Aumann, and Stefan Erasmi. 2019. "Canopy Height Estimation with TanDEM-X in Temperate and Boreal Forests." *International Journal of Applied Earth Observation and Geoinformation* 82: 101904. doi: 10.1016/j.jag.2019.101904.

Schutz, B. E., H. J. Zwally, C. A. Shuman, D. Hancock, and J. P. DiMarzio. 2005. "Overview of the ICESat Mission." *Geophysical Research Letters* 32 (21). doi: 10.1029/2005GL024009.

Shimada, Masanobu, Takuya Itoh, Takeshi Motooka, Manabu Watanabe, Tomohiro Shiraishi, Rajesh Thapa, and Richard Lucas. 2014. "New Global Forest/non-Forest Maps from ALOS PALSAR Data (2007–2010)." *Remote Sensing of Environment* 155: 13–31 doi: 10.1016/j.rse.2014.04.014.

Simard, Marc, and Michael Denbina. 2018. "An Assessment of Temporal Decorrelation Compensation Methods for Forest Canopy Height Estimation Using Airborne L-Band Same-Day Repeat-Pass Polarimetric SAR Interferometry." *IEEE Journal of Selected Topics in Applied Earth Observations and Remote Sensing* 11 (1): 95–111. doi: 10.1109/JSTARS.2017.2761338.

Soja, Maciej J., Gustaf Sandberg, and Lars M. H. Ulander. 2013. "Regression-Based Retrieval of Boreal Forest Biomass in Sloping Terrain Using P-Band SAR Backscatter Intensity Data." *IEEE Transactions on Geoscience and Remote Sensing* 51 (5): 2646–65. doi: 10.1109/TGRS.2012.2219538.

Soja, Maciej J., Henrik Persson, and Lars M. H. Ulander. 2015a. "Detection of Forest Change and Robust Estimation of Forest Height from Two-Level Model Inversion of Multi-Temporal, Single-Pass InSAR Data." 2015–2015, 3886–89.

Soja, Maciej J., Henrik Persson, and Lars M. H. Ulander. 2015b. "Estimation of Forest Height and Canopy Density from a Single InSAR Correlation Coefficient." *IEEE Geoscience and Remote Sensing Letters* 12 (3): 646–50. doi: 10.1109/LGRS.2014.2354551.

Soja, Maciej J., Shaun Quegan, Mauro M. d'Alessandro, Francesco Banda, Klaus Scipal, Stefano Tebaldini, and Lars M. Ulander. 2021. "Mapping Above-Ground Biomass in Tropical Forests with Ground-Cancelled P-Band SAR and Limited Reference Data." *Remote Sensing of Environment* 253: 112153. doi: 10.1016/j.rse.2020.112153.

Solberg, Svein, Rasmus Astrup, Terje Gobakken, Erik Næsset, and Dan J. Weydahl. 2010. "Estimating Spruce and Pine Biomass with Interferometric X-Band SAR." *Remote Sensing of Environment* 114 (10): 2353–60. doi: 10.1016/j.rse.2010.05.011.

Solberg, Svein, Rasmus Astrup, and Dan Weydahl. 2013. "Detection of Forest Clear-Cuts with Shuttle Radar Topography Mission (SRTM) And Tandem-X InSAR Data." *Remote Sensing* 5 (11): 5449–62. doi: 10.3390/rs5115449.

Solberg, Svein, Erik Næsset, Terje Gobakken, and Ole-Martin Bollandsås. 2014. "Forest Biomass Change Estimated from Height Change in Interferometric SAR Height Models." *Carbon Balance and Management* 9 (1): 5. doi: 10.1186/s13021-014-0005-2.

Stojanova, Daniela, Panče Panov, Valentin Gjorgjioski, Andrej Kobler, and Sašo Džeroski. 2010. "Estimating Vegetation Height and Canopy Cover from Remotely Sensed Data with Machine Learning." *Ecological Informatics* 5 (4): 256–66. doi: 10.1016/j.ecoinf.2010.03.004.

Sullivan, Martin J. P., Simon L. Lewis, Wannes Hubau, Lan Qie, Timothy R. Baker, Lindsay F. Banin, Jerôme Chave et al. 2018. "Field Methods for Sampling Tree Height for Tropical Forest Biomass Estimation." *Methods in Ecology and Evolution* 9 (5): 1179–89. doi: 10.1111/2041-210X.12962.

Sun, Alexander Y., and Bridget R. Scanlon. 2019. "How Can Big Data and Machine Learning Benefit Environment and Water Management: A Survey of Methods, Applications, and Future Directions." *Environmental Research Letters* 14 (7): 73001. doi: 10.1088/1748-9326/ab1b7d.

Sun, Guoqing, K. J. Ranson, Z. Guo, Z. Zhang, P. Montesano, and D. Kimes. 2011. "Forest Biomass Mapping from Lidar and Radar Synergies." *Remote Sensing of Environment* 115 (11): 2906–16. doi: 10.1016/j.rse.2011.03.021.

Tague, Christina L., Max Moritz, and Erin Hanan. 2019. "The Changing Water Cycle: The Eco-hydrologic Impacts of Forest Density Reduction in Mediterranean (Seasonally Dry) Regions." *WIREs Water* 6 (4). doi: 10.1002/wat2.1350.

Tarazona, Yonatan, Alaitz Zabala, Xavier Pons, Antoni Broquetas, Jakub Nowosad, and Hamdi A. Zurqani. 2021. "Fusing Landsat and SAR Data for Mapping Tropical Deforestation Through Machine Learning Classification and the PVts- B Non-Seasonal Detection Approach." *Canadian Journal of Remote Sensing* 47 (5): 677–96. doi: 10.1080/07038992.2021.1941823.

Tebaldini, Stefano, and Fabio Rocca. 2012. "Multibaseline Polarimetric SAR Tomography of a Boreal Forest at P- and L-Bands." *IEEE Transactions on Geoscience and Remote Sensing* 50 (1): 232–46. doi: 10.1109/TGRS.2011.2159614.

Tebaldini, Stefano, Dinh Ho Tong Minh, Mauro Mariotti d'Alessandro, Ludovic Villard, Thuy Le Toan, and Jerome Chave. 2019. "The Status of Technologies to Measure Forest Biomass and Structural Properties: State of the Art in SAR Tomography of Tropical Forests." *Surveys in Geophysics* 40 (4): 779–801. doi: 10.1007/s10712-019-09539-7.

Thompson, Ian D., Manuel R. Guariguata, Kimiko Okabe, Carlos Bahamondez, Robert Nasi, Victoria Heymell, and Cesar Sabogal. 2013. "An Operational Framework for Defining and Monitoring Forest Degradation." *E&S* 18 (2). doi: 10.5751/ES-05443-180220.

Tiner, Ralph W., Megan W. Lang, and Victor V. Klemas. 2015. *Remote Sensing of Wetlands.* Boca Raton, FL: CRC Press.

Tran, Anh T., Kim A. Nguyen, Yuei an Liou, Minh H. Le, Truong van Vu, and Dinh D. Nguyen. 2021. "Classification and Observed Seasonal Phenology of Broadleaf Deciduous Forests in a Tropical Region by Using Multitemporal Sentinel-1A and Landsat 8 Data." *Forests* 12 (2): 235. doi: 10.3390/f12020235.

Treuhaft, Robert N., and Paul R. Siqueira. 2000. "Vertical Structure of Vegetated Land Surfaces from Interferometric and Polarimetric Radar." *Radio Science* 35 (1): 141–77. doi: 10.1029/1999RS900108.

Treuhaft, Robert N., Søren N. Madsen, Mahta Moghaddam, and Jakob J. van Zyl. 1996. "Vegetation Characteristics and Underlying Topography from Interferometric Radar." *Radio Science* 31 (6): 1449–85. doi: 10.1029/96RS01763.

Treuhaft, Robert, Fabio Gonzalves, Joao R. dos Santos, Michael Keller, Michael Palace, Sren N. Madsen, Franklin Sullivan, and Paulo M. L. A. Graca. 2015. "Tropical-Forest Biomass Estimation at X-Band from the Spaceborne TanDEM-X Interferometer." *IEEE Geoscience and Remote Sensing Letters* 12 (2): 239–43. doi: 10.1109/LGRS.2014.2334140.

Treuhaft, Robert, Yang Lei, Fabio Gonçalves, Michael Keller, João Santos, Maxim Neumann, and André Almeida. 2017. "Tropical-Forest Structure and Biomass Dynamics from TanDEM-X Radar Interferometry." *Forests* 8 (8): 277. doi: 10.3390/f8080277.

Truong-Loi, My-Linh, S. Saatchi, and Sermsak Jaruwatanadilok. 2015. "Soil Moisture Estimation Under Tropical Forests Using UHF Radar Polarimetry." *IEEE Transactions on Geoscience and Remote Sensing* 53 (4): 1718–27. doi: 10.1109/TGRS.2014.2346656.

Tsui, Olivier W., Nicholas C. Coops, Michael A. Wulder, Peter L. Marshall, and Adrian McCardle. 2012. "Using Multi-Frequency Radar and Discrete-Return LiDAR Measurements to Estimate Above-Ground Biomass and Biomass Components in a Coastal Temperate Forest." *ISPRS Journal of Photogrammetry and Remote Sensing* 69: 121–33 doi: 10.1016/j.isprsjprs.2012.02.009.

Udali, Alberto, Emanuele Lingua, and Henrik J. Persson. 2021. "Assessing Forest Type and Tree Species Classification Using Sentinel-1 C-Band SAR Data in Southern Sweden." *Remote Sensing* 13 (16): 3237. doi: 10.3390/rs13163237.

Ulaby, Fawwaz T., Kamal Sarabandi, Kyle McDonald, Michael Whitt, and M. C. Dobson. 1990. "Michigan Microwave Canopy Scattering Model." *International Journal of Remote Sensing* 11 (7): 1223–53. doi: 10.1080/01431169008955090.

UNDP. 2022. "Goal 15: Life on Hand." Accessed March 14, 2022. https://www1.undp.org/content/seoul_policy_center/en/home/sustainable-development-goals/goal-15-life-on-land.html.

Urbazaev, Mikhail, Christian Thiel, Felix Cremer, Ralph Dubayah, Mirco Migliavacca, Markus Reichstein, and Christiane Schmullius. 2018. "Estimation of Forest Aboveground Biomass and Uncertainties by Integration of Field Measurements, Airborne LiDAR, and SAR and Optical Satellite Data in Mexico." *Carbon Balance and Management* 13 (1): 5. doi: 10.1186/s13021-018-0093-5.

Vafaei, Sasan, Javad Soosani, Kamran Adeli, Hadi Fadaei, Hamed Naghavi, Tien Pham, and Dieu Tien Bui. 2018. "Improving Accuracy Estimation of Forest Aboveground Biomass Based on Incorporation of ALOS-2 PALSAR-2 and Sentinel-2A Imagery and Machine Learning: A Case Study of the Hyrcanian Forest Area (Iran)." *Remote Sensing* 10 (2): 172. doi: 10.3390/rs10020172.

Villard, Ludovic, and Thuy Le Toan. 2015. "Relating P-Band SAR Intensity to Biomass for Tropical Dense Forests in Hilly Terrain: γ^0 or t^0?" *IEEE Journal of Selected Topics in Applied Earth Observations and Remote Sensing* 8 (1): 214–23. doi: 10.1109/JSTARS.2014.2359231.

Wang, Haipeng, Kazuo Ouchi, and Ya-Qiu Jin. 2011. "Classification of Typhoon-Destroyed Forests Based on Tree Height Change Detection Using InSAR Technology." *In 2011 IEEE International Geoscience and Remote Sensing Symposium*, Vancouver, Canada, 1247–50: IEEE.

Wollersheim, Michael, Michael J. Collins, and Don Leckie. 2011. "Estimating Boreal Forest Species Type with Airborne Polarimetric Synthetic Aperture Radar." *International Journal of Remote Sensing* 32 (9): 2481–2505. doi: 10.1080/01431161003698377.

Wolter, Peter T., and Philip A. Townsend. 2011. "Multi-Sensor Data Fusion for Estimating Forest Species Composition and Abundance in Northern Minnesota." *Remote Sensing of Environment* 115 (2): 671–91. doi: 10.1016/j.rse.2010.10.010.

Woodhouse, Iain H. 2017. *Introduction to Microwave Remote Sensing.* Boca Raton, FL: CRC Press.

Wulder, Michael A., Christopher W. Bater, Nicholas C. Coops, Thomas Hilker, and Joanne C. White. 2008. "The Role of LiDAR in Sustainable Forest Management." *The Forestry Chronicle* 84 (6): 807–26. doi: 10.5558/tfc84807-6.

Yu, Yifan, and Sassan Saatchi. 2016. "Sensitivity of L-Band SAR Backscatter to Aboveground Biomass of Global Forests." *Remote Sensing* 8 (6): 522. doi: 10.3390/rs8060522.

Yu, Ying, Mingze Li, and Yu Fu. 2018. "Forest Type Identification by Random Forest Classification Combined with SPOT and Multitemporal SAR Data." *Journal of Forestry Research* 29 (5): 1407–14. doi: 10.1007/s11676-017-0530-4.

Zhang, Fengli, Chou Xie, Kun Li, Maosong Xu, Xuejun Wang, and Zhongsheng Xia. 2012a. "Forest and Deforestation Identification Based on Multitemporal Polarimetric RADARSAT-2 Images in Southwestern China." *JARS* 6 (1): 63527. doi: 10.1117/1.JRS.6.063527.

Zhang, Hong, Peifeng Ma, and Chao Wang. 2012b. "A New Function Expansion for Polarization Coherence Tomography." *IEEE Geoscience and Remote Sensing Letters* 9 (5): 891–95. doi: 10.1109/LGRS.2012.2183113.

Zhang, Haibo, Changcheng Wang, Jianjun Zhu, Haiqiang Fu, Qinghua Xie, and Peng Shen. 2018. "Forest Above-Ground Biomass Estimation Using Single-Baseline Polarization Coherence Tomography with P-Band PolInSAR Data." *Forests* 9 (4): 163. doi: 10.3390/f9040163.

Zhang, Qi, Linlin Ge, Scott Hensley, Graciela Isabel Metternicht, Chang Liu, and Ruiheng Zhang. 2022. "PolGAN: A Deep-Learning-Based Unsupervised Forest Height Estimation Based on the Synergy of PolInSAR and LiDAR Data." *ISPRS Journal of Photogrammetry and Remote Sensing* 186: 123–39 doi: 10.1016/j.isprsjprs.2022.02.008.

Zhao, Panpan, Dengsheng Lu, Guangxing Wang, Lijuan Liu, Dengqiu Li, Jinru Zhu, and Shuquan Yu. 2016. "Forest Aboveground Biomass Estimation in Zhejiang Province Using the Integration of Landsat TM and ALOS PALSAR Data." *International Journal of Applied Earth Observation and Geoinformation* 53: 1–15 doi: 10.1016/j.jag.2016.08.007.

Zhao, Feng, Rui Sun, Liheng Zhong, Ran Meng, Chengquan Huang, Xiaoxi Zeng, Mengyu Wang, Yaxin Li, and Ziyang Wang. 2022. "Monthly Mapping of Forest Harvesting Using Dense Time Series Sentinel-1 SAR Imagery and Deep Learning." *Remote Sensing of Environment* 269: 112822. doi: 10.1016/j.rse.2021.112822.

Zhou, Lina, Shimei Pan, Jianwu Wang, and Athanasios V. Vasilakos. 2017. "Machine Learning on Big Data: Opportunities and Challenges." *Neurocomputing* 237: 350–61 doi: 10.1016/j.neucom.2017.01.026.

Zhu, Xiao Xiao, Sina Montazeri, Mohsin Ali, Yuansheng Hua, Yuanyuan Wang, Lichao Mou, Yilei Shi, Feng Xu, and Richard Bamler. 2020a. "Deep Learning Meets SAR." https://arxiv.org/pdf/2006.10027.

Zhu, Xiao Xiao, Cheng Wang, Sheng Nie, Feifei Pan, Xiaohuan Xi, and Zhenyue Hu. 2020b. "Mapping Forest Height Using Photon-Counting LiDAR Data and Landsat 8 OLI Data: A Case Study in Virginia and North Carolina, USA." *Ecological Indicators* 114:106287. doi: 10.1016/j.ecolind.2020.106287.

Index

dynamism 338